Dictionary of the Physical Sciences

DICTIONARY
OF THE
PHYSICAL SCIENCES

Terms · Formulas · Data

CESARE EMILIANI
University of Miami

New York · Oxford
OXFORD UNIVERSITY PRESS

Oxford University Press

Oxford New York Toronto
Delhi Bombay Calcutta Madras Karachi
Petaling Jaya Singapore Hong Kong Tokyo
Nairobi Dar es Salaam Cape Town
Melbourne Auckland

and associated companies in
Beirut Berlin Ibadan Nicosia

Published by Oxford University Press, Inc.,
200 Madison Avenue, New York, New York 10016

Oxford is a registered trademark of Oxford University Press.

Library of Congress Cataloging-in-Publication Data
Emiliani, Cesare.
Dictionary of the physical sciences.
Bibliography: p.
1. Science—Dictionaries. 2. Science—Handbooks, manuals, etc.
3. Engineering—Dictionaries. 4. Engineering—
Handbooks, manuals, etc. I. Title.
Q123.E46 1987   503′.21   85-13594
ISBN 0-19-503651-4
ISBN 0-19-503652-2 (pbk.)

2 4 6 8 9 7 5 3 1

Printed in the United States of America
on acid-free paper

Guidelines

This dictionary consists of text and tables. All areas that are fundamental to understanding our physical world, and life within it, are covered. Included are physics (classical, relativistic, quantum, particle, high-energy); chemistry (inorganic, organic, physical); the geological sciences (geology, geophysics, oceanography, meteorology, paleontology, and related areas of molecular, genetic, and evolutionary biology); and cosmology (which includes astronomy, astrophysics, and the genesis and evolution of the universe).

The text includes not only definitions, but also explanations, formulas, and data. Entries best presented in tabular form are found in the *Tables* section. Any unfamiliar term appearing in the text or the Tables is defined and explained in the text. Acronyms and abbreviations are spelled out, but their definition and additional material, if necessary, will be found under the full term.

All entries in the text as well as all headings of the Tables are in strict alphabetical order, without regard to spaces, punctuations, or numbers appearing as part of an entry or heading. Greek letters are placed at the appropriate locations within the corresponding Latin alphabet. In case of a tie (α, a), precedence is given to the Greek letter.

All quantities in both text and tables are given to the last significant, most recent figure.

An asterisk following a term in the text refers the reader to the pertinent table.

Sources for the illustrations and tables are given in the standard abbreviated form below each pertinent item. The full reference is given in the reference section which follows the tables. The word "From" preceding a given reference indicates that the original material has been modified, augmented, or recalculated. Absence of the word "From" indicates that the material is reproduced as it appeared in the original. If no source is given, the material has been developed, calculated, or otherwised derived by the present author.

Reproduction of materials from published sources is with the written consent of the original authors and publishers.

Acknowledgments and Dedication

The author is grateful to the following persons and publishers:

For suggestions, comments, and criticism:

Dr. Stephen W. Bruenn, Florida Atlantic University; Dr. James S. Clegg, University of California; Dr. Preston Cloud, University of California; Dr. Cecil M. Criss, University of Miami; Dr. Pamela A. Ferguson, University of Miami; Dr. Robert M. Garrels, University of South Florida; Dr. James C. Nearing, University of Miami; Dr. Arnold Perlmutter, University of Miami; Dr. Jeffrey G. Taylor, University of New Mexico; Dr. Kamal Yacoub, University of Miami; Dr. Michael Zeilik, University of New Mexico.

For permission to reproduce original materials:

Academic Press, New York; Dr. Edward Anders, University of Chicago/ Chicago; Edward Arnold Publishers, London; Bell & Hyman, London; Geb. Borntraeger, Berlin; Dr. Geoffrey Burbidge, Kitt Peak National Observatory, Tucson, Arizona; CRC Press, Boca Raton, Florida; Elsevier Science Publishers, Amsterdam; Dr. Gunther Faure, Ohio State University, Columbus, Ohio; W. H. Freeman and Co., New York; Jones and Bartlett Publishers, Boston; Dr. Lynn Margulis, Boston University, Boston; McGraw-Hill Book Co., New York; MIT Press, Cambridge, Massachusetts; Oxford University Press, Oxford, England; Prentice-Hall, Englewood Cliffs, New Jersey; D. Reidel Publishing Co., Dordrecht, Holland; Dr. A. N. Strahler, Santa Barbara, California; Dr. J. K. Tuli, Brookhaven National Laboratory, Upton, New York; Van Nostrand Reinhold Co., New York; John Wiley & Sons, New York; Worth Publishers, New York.

For technical assistance:

Joyce Berry, Joan Bossert, Ellen Leventhal, and Leslie Phillips, Oxford University Press; Jeffrey M. Barrie, University Graphics, Inc., Atlantic Highlands, New Jersey.

This book is dedicated to Rosita Emiliani.

Coral Gables, Florida   C. E.
May 1987

Contents

Illustrations

ILLUSTRATIONS

Tables

TEXT

A

α 1. Activity. 2. Alpha particle. 3. Angular acceleration. 4. Attenuation coefficient. 5. Fine structure constant. 6. Isotopic fractionation factor, equal to R_A/R_B, where R_A = ratio of the less abundant to the more abundant isotope or isotopic species in phase or chemical compound A and R_B = same ratio in phase or chemical compound B. 7. Mineral phase stable at a temperature lower than that of other phases (β, γ, etc.). 8. Referring to the amino acids in which the $-NH_2$ group is attached to the C (the α-carbon) next to the $-COOH$ group. 9. Referring to the helicoidal structure of a polypeptide chain (α helix). 10. Referring to the positions 1,4,5,8 in the naphthalene ring. 11. Referring to the position of attachment of a substitution group when it is the same as that of the substituted group. 12. Right ascension. 13. Semimajor axis of an elliptical orbit.

a 1. Absorbance. 2. Acceleration. 3. Activity. 4. *Annus,* Latin for *year.* 5. Optical depth. 6. Semimajor axis of an elliptical orbit.

a- Prefix meaning *without.*

A 1. Ampere. 2. Atomic mass number. 3. Avogadro. 4. Azimuth.

Å Angstrom (1 Å = 10^{-10} m).

aa *(Hawaiian)* Solidified Hawaiian lava with a rough, jagged surface.

aA Abampere.

a axis The horizontal crystallographic axis oriented front to back. Cf. **b axis, c axis.**

ab- Prefix identifying electrical units in the CGS_{emu} system of units.

Ab Albite.

abampere (aA) The unit of electric current in the CGS_{emu} system, defined as that current that, if flowing through two parallel conductors of negligible cross section and infinite length, placed 1 cm apart in vacuo, would produce on each conductor a force of 1 dyne per centimeter of length. 1 abampere = 1 abcoulomb/s = c statampere (where c = speed of light in cm/s) = 10 ampere.

abcoulomb (aC) The unit of electric charge in the CGS_{emu} system of units, defined as the charge transported by a current of 1 abampere in 1 s. 1 abcoulomb = 1 aA·s = c statcoulomb (where c = speed of light in cm/s) = 10 coulombs.

aberration Imperfect image formation due to geometric imperfections in the optical elements of a system.

abfarad (aF) The CGS_{emu} unit of capacitance, defined as the capacitance of a capacitor that exhibits the potential difference of 1 abvolt between its plates when each is charged with 1 abcoulomb of opposite electricity.

$$aF = aC/aV$$
$$= 10\ C/10^{-8}\ V$$
$$= 10^9\ F$$

abhenry (aH) The CGS_{emu} unit of inductance and permeance, defined as the self or mutual inductance of a closed circuit in which an emf of 1 abvolt is produced when the current changes uniformly at the rate of 1 abampere/second.

$$aH = aV/aA\ s^{-1}$$
$$= 10^{-8}\ V\ s/10\ A$$
$$= 10^{-9}\ H$$

ablation 1. The wasting of glacier ice by any process (calving, melting, evaporation, etc.). 2. The shedding of molten material from the outer surface of a meteorite or tektite during its flight through the atmosphere.

abohm (aΩ) The CGS_{emu} unit of electrical resistance. 1 a Ω = 10^{-9} Ω.

absolute age The age of a natural substance, of a fossil or living organism, or of an artifact, obtained by means of an absolute dating method. See **absolute dating method.**

absolute concentration Concentration of a solute in a solvent expressed in g/g or in mass %. Cf. **formality, molality, molarity, normality.**

absolute dating method Any of the dating methods based on a rate parameter that is invariant with time and insensitive to environmental conditions, such as radioactive decay rates. See **Ab-**

solute dating methods*, argon-40/argon-39 dating method, carbon-14 dating method, fission track dating method, potassium-argon dating method, samarium-neodymium dating method, uranium-lead dating method, uranium-thorium disequilibrium dating method.

absolute density Density in kg/m³ or, more commonly, in g/cm³, both at STP. Cf. **density, relative density.**

absolute gravity See **absolute density.**

absolute magnitude (M) The apparent magnitude of a star reduced to the standard distance of 10 parsecs. See **magnitude.**

absolute temperature Temperature in kelvins (K), starting at the absolute zero.

absolute viscosity See **viscosity.**

absolute zero The temperature at which atomic and molecular translational motion ceases. It is equal to 273.16 K below the triple point of pure water or 273.15 K below the freezing point of pure water at 1 atm.

absorbance (a) The common logarithm of the reciprocal of transmittance:
$$a = \log 1/T$$
where T = transmittance.

absorptance The ratio of radiant flux absorbed to the incident radiant flux.

absorption The intake of matter or energy by a medium.

absorption coefficient See **absorption law.**

absorption law (Bouguer's law) A law giving the flux through a substance in terms of incident flux, coefficient of absorption, and thickness of the substance.
$$I = I_0 e^{-\alpha x}$$
where I_0 = incident flux, I = flux passing through thickness x of the substance, α = absorption coefficient. Cf. **attenuation, attenuation coefficient.**

absorption spectrum Spectrum resulting from the absorption of specific wavelengths when light from a continuous source passes through a given substance.

abvolt (aV) The CGS_{emu} unit of electromotive force or potential difference. 1 aV = 10^{-8} V.

abyssal 1. Defining an igneous intrusion occurring at considerable depth. 2. Defining the oceanic depth zone between 2000 m and 6000 m, i.e. between the bathyal (200–2000 m) and hadal (6000+ m) depth zones.

abyssal hill A common low-relief feature of the deep-sea floor, where sediment cover has not obliterated bedrock topography. Abyssal hills cover 50% of the Atlantic and Indian Ocean floor and 85% of the Pacific Ocean floor.

abyssal plain The flat surface of the ocean floor covered with sediments largely contributed by turbidity currents.

ac Alternating current.

aC Abcoulomb.

A.C. *Ante Christum,* Latin for *before Christ* or the time before January 0d, 0h, 0m, 0s, year A.D. 1 (there is no year 0). See B.C.

acceleration (a) The derivative of velocity with respect to time, or the second derivative of position in space with respect to time.
1. *Rectilinear motion:*
$$a = dv/dt = d^2s/dt^2$$
where a = acceleration, v = velocity, s = position in space, t = time.
2. *Circular motion, uniform:*
$$a_r = v^2/r = \omega^2 r$$
where a_r = radial acceleration, v = velocity, ω = $d\theta/dt$ = angular velocity, r = radius of circle, θ = angular displacement.
3. *Circular motion, uniformly accelerated:*
$$\alpha = d\omega/dt$$
where α = angular acceleration, ω = angular velocity, t = time;
$$a_T = \alpha \cdot r$$
where a_T = tangential acceleration, α = angular acceleration, r = radius of circle;
$$a = (a_r^2 + a_T^2)^{1/2}$$
where a = total acceleration, a_r = $v^2 r$ = $\omega^2 r$ = radial acceleration, a_T = tangential acceleration.

accessory mineral A mineral contributing in a minor way to the bulk of a rock.

accretionary plate boundary The boundary between two plates moving away from each other, where new lithosphere is created.

accuracy A measure of the closeness by which a measurement or a set of measurements approaches the true value.

acetyl The acyl radical $-CH_3CO$.

achondrite A stony meteorite lacking chondrules. Achondrites represent 7.1% of the stony meteorites and are subdivided into Ca-rich (66%, consisting of pyroxene, olivine, and Ca-plagioclase) and

Ca-poor (34%, consisting of pyroxene and olivine). See **Meteorites***.

achromatic Defining an optical system capable of transmitting light without color dispersion.

acid *(Chemistry)* **1. Arrhenius acid** A chemical substance that dissociates in water to give H^+ (or H_3O^+) ions. Cf. **Arrhenius base. 2. Brønsted acid** A chemical substance capable of donating one or more protons. Cf. **Brønsted base. 3. Lewis acid** A chemical substance capable of forming a bond by accepting an electron pair donated by a Lewis base and sharing it with it. Cf. **Lewis base.** *(Petrology)* 1. Defining an igneous rock that contains more than 60% of SiO_2 ("silicic acid"). 2. Referring to any rock primarily composed of light-colored silicate minerals. Syn. **silicic.**

acidic See **acid.**

acme-zone A stratizone characterized by the maximum abundance development of a given taxon.

acoustic basement The deepest seismic reflector below which seismic energy return is poor or absent.

acritarch Any of the single-celled spores or similar structures found in the geological record from Precambrian to Recent.

ACS American Chemical Society.

actinides The 14 elements that follow Ac with identical 5s, 5p, 5d, 6s and 6p subshells but different 5f and/or 6d subshells. See **Elements—electronic configuration*.**

action (S) The integral of the Lagrangian of a physical system with respect to time.
1. *For a rigid body:*

$$S = \int L \, dt$$

where $L = E_k - E_p$, t = time, E_k = kinetic energy, E_p = potential energy.
2. *For a system of particles:*

$$S = \int_{t_1}^{t_2} \sum p_i(t) \dot{q}_i(t) \, dt$$

where the $\dot{q}_i$'s are the time derivative of the generalized coordinates q_i and the p_i's are the conjugate momenta.
3. *For a system of particles in which the coordinates are periodic functions of time:*

$$S = \oint p_q \, dq$$
$$= n_q h$$

where q is one of the coordinates, p_q is its conjugate momentum, n_q is a quantum number taking only integral values, h = Planck's constant.
4. *For an orbital electron:*

$$S = \oint p_x \, dx$$
$$= E/\nu$$
$$= nh$$

where x is the coordinate, p_x is its conjugate momentum (p_x and x form the orthogonal axes of the p-q plane, called *phase space*), E = energy, ν = frequency, n = quantum number taking only integral values, h = Planck's constant.

activation analysis The identification of a stable isotope by rendering it radioactive through bombardment with neutrons, other particles, or radiation.

activation energy The energy above ground state necessary to initiate a process.

active margin A continental margin along a plate margin where subduction, collision, or transform fault motion occurs. Cf. **passive margin.**

active power See **average power.**

activity *(Nuclear Physics)* (A) The average number of radioactive atomic nuclei decaying per unit time t. For a given radionuclide,

$$A = -dN/dt = \lambda N$$

where N = number of atoms, t = time, λ = decay constant. The SI unit of activity is the *becquerel* (Bq) = 1 dps. *(Physical Chemistry)* (a) The ratio of the fugacity of a substance in a solution to the fugacity of the pure substance in the liquid state.

$$a = f_i/f_i^0$$

where f_i = fugacity of substance i in solution, f_i^0 = fugacity of the substance in the pure liquid state.

activity coefficient (γ) The ratio of fugacity to pressure (gases) or activity to mole fraction or to molar or molal concentration (solutions).

acyclic See **aliphatic.**

acyl A radical derived from an organic acid by removal of the hydroxyl group.

A.D. *Anno Domini,* Latin for *year of the Lord,* referring to any year after the birth of Jesus Christ taken as having occurred at the beginning of A.D. 1. The preceding year is year 1 B.C. There is no year 0. Cf. **A.C., B.C.**

adamellite See **quartz monzonite.**

adaptive radiation The diversification of a major taxon into a number of derivative taxa adapted to occupy specialized niches.

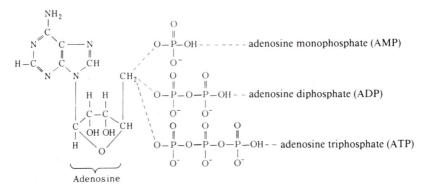

Adenosine phosphates. (King and Stansfield 1985, p. 9)

adenine $C_5H_5N_5$ (mol. mass = 135.128), a nucleic acid base with purine ring structure.

adenosine $C_{10}H_{13}N_5O_4$ (mol. mass = 267.244), a nucleoside consisting of adenine linked to a ribose sugar.

adenosine phosphate See **ADP, AMP, ATP.**

adiabatic Defining a physical or chemical change in a system without transfer of heat.

admittance (Y) The reciprocal of impedance in an ac circuit.

$$Y = G + iB$$

where G = conductance, $i = (-1)^{1/2}$, B = susceptance. It is expressed in siemens.

ADP Adenosine diphosphate ($C_{10}H_{15}N_5O_{10}P_2$, mol. mass = 427.204), consisting of adenosine linked to two phosphate groups. Cf. **AMP, ATP.**

adsorption Adherence of solid, liquid, or gaseous atoms, ions, or molecules to solid or liquid surfaces.

adularia A low-temperature alkali feldspar, $KAlSi_3O_8$.

advection The noncyclical mass motion of air in the atmosphere, water in the ocean, or fluids within the solid Earth. Cf. **convection.**

ae Aeon ($= 10^9$ y).

aeolian See **eolian.**

aeolianite See **eolianite.**

aeon (ae) 1. 10^9 y. 2. One of the two major units of geologic time, Cryptozoic ($4.7–0.590 \cdot 10^9$ y B.P.) and Phanerozoic ($0.590–0 \cdot 10^9$ y B.P.).

aerobic 1. Defining an environment where free oxygen is available. 2. Defining an organism that needs free oxygen for its metabolism.

aerosol A sol in which the dispersant is a gas or a

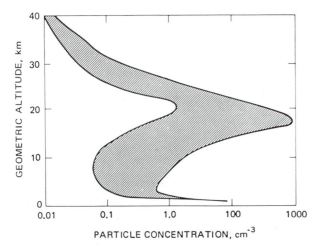

Aerosol. Average stratospheric aerosol particle concentration as a function of altitude. (U.S. Standard Atmosphere 1976, p. 45, Fig. 36)

gas mixture (e.g. the atmosphere) and the dispersed phase consists of solid particles (e.g. smoke) and/or liquid droplets (e.g. water).

aF Abfarad.

aftershock Any of the smaller shocks following the main shock of an earthquake.

agate A variegated form of chalcedony.

age A division of geologic time longer than subage but shorter than epoch, during which the rocks of a stage are formed. Cf. **absolute age.**

agonic line The line connecting the points on the Earth's surface where magnetic declination is zero.

aH Abhenry.

ahermal Defining an organism that does not partake in the construction of a reef. Cf. **hermal.**

ahermatypic See **ahermal.**

A horizon The uppermost soil horizon, consisting of a surface layer (A1) rich in organic matter (the top soil); a layer (A2) in which quartz is the predominant detrital mineral because of leaching of other minerals by humic acids; and a layer (A3) transitional to the B horizon below.

AIP American Institute of Physics.

alabaster A rock consisting of compact, microcrystalline gypsum.

albedo Reflectivity of a nonluminous surface, ranging from 0 (total absorption) to 1.0 (total reflection). It is usually expressed as percent. Representative values: open ocean, smooth, vertical sun, 0.02–0.04; forest, 0.1; grassland, 0.2; desert, 0.2; ice, 0.7; fresh snow, 0.8; clouds, 0.5–0.8; Mercury, 0.06; Venus, 0.72; Earth, 0.39; Moon, 0.068; Mars, 0.16; Jupiter, 0.70; Saturn, 0.75; Uranus, 0.90; Neptune, 0.82; Pluto, ?.

albite The Na end-member of the plagioclase group, $NaAlSi_3O_8$.

alcohols The family of alkyl compounds containing a hydroxyl group, (alkyl)$-$OH. E.g. methyl alcohol, CH_3OH; ethyl alcohol, C_2H_5OH.

aldehydes A family of organic compounds containing the radical $-CHO$. E.g. formaldehyde, $H-CHO$; acetaldehyde, CH_3-CHO.

aldoses A family of carbohydrates containing the aldehyde group. E.g. ribose, $C_5H_{10}O_5$; deoxyribose, $C_5H_{10}O_4$.

Alfvén wave A hydromagnetic shear wave propagating along magnetic field lines. It accelerates charged particles in plasmas and in space.

algal mat An intertidal or supratidal formation along a carbonate coastline or on a carbonate bank consisting of thin detrital laminae alternating with laminae rich in the remains of unicellular or filamentous blue-green algae. A doublet is formed daily (algal lamina during the day, detrital lamina at night), reaching a thickness up to 0.5 mm.

alidade A ruler with a simple or telescopic sight used in plane table mapping. The sight may be mounted on a vertical graduated circle to determine elevation angles.

aliphatic Defining an organic compound in which the carbon atoms are linked in open chains. Cf. **aromatic.**

aliquot 1. An exact fraction of a given quantity. 2. A representative fraction of a given quantity.

alkadienes A family of unsaturated aliphatic hydrocarbons with two $C=C$ bonds.

alkali 1. Defining the hydroxides and carbonates of alkali metals. 2. See **alkali metal.**

alkali feldspar Any of the Na- or K-rich feldspars, including orthoclase, sanidine, microcline, and the plagioclases albite, oligoclase, and andesine.

alkali lake A salt lake containing a high concentration of Na and K halides and carbonates.

alkali-lime index A system of classification for igneous rocks introduced in 1931, based on the increasing mass percent of SiO_2 when the mass percentages of CaO and $Na_2O + K_2O$ are equal. Four classes are recognized: alkaline ($SiO_2 < 51\%$); alkali-calcic ($SiO_2 = 51–56\%$); calc-alkaline ($SiO_2 = 56–61\%$); and calcic ($SiO_2 > 61\%$).

alkali metal The metals in group 1 of the Periodic Table of the Elements, including Li, Na, K, Rb, Cs, and Fr.

alkaline 1. Referring to an alkali. 2. Having a pH > 7.

alkaline cell A primary cell that uses an alkaline electrolyte.

alkaline earth An oxide of an element belonging to group 2 of the Periodic Table of the Elements.

alkanes A family of saturated aliphatic hydrocarbons C_nH_{2n+2}. E.g. methane, CH_4; ethane, C_2H_6. Syn. **methane series.**

alkenes A family of unsaturated aliphatic hydrocarbons C_nH_{2n}, with one $C=C$ bond. E.g. ethylene, $CH_2=CH_2$. Syn. **ethylene series.**

alkyl The radical $-C_nH_{2n+1}$. E.g. methyl (Me), $-CH_3$; ethyl (Et), $-C_2H_5$.

alkyl acids The family of organic acids of the general formula $C_nH_{2n+1}COOH$ (saturated). E.g. palmitic acid, $CH_3(CH_2)_{14}COOH$ (saturated); oleic acid, $CH_3(CH_2)_7CH:CH(CH_2)_7COOH$ (unsaturated). Syn. **fatty acids.**

alkyl amines A family of organic compounds consisting of an alkyl radical and the $-NH_2$ group. E.g. methylamine, CH_3-NH_2.

alkyl halides A family of compounds consisting of an alkyl radical and a halide. E.g. methyl chloride, CH_3Cl.

alkynes A family of unsaturated aliphatic hydrocarbons C_nH_{2n-2}, with one C≡C bond. E.g. acetylene, CH≡CH. Syn. **acetylene series.**

allele Any of the genes belonging to the set that specifies a given physical characteristic of an organism.

allobar An occurrence of an element with isotopic composition, and hence mass, different from the common one.

allochem A carbonate constituent (e.g. an oolite, a shell, etc.) deposited as part of a limestone accumulating away from the site at which the constituent formed.

allochthonous Formed or originated at a place different from that of emplacement.

allogenic Defining a mineral or organic fossil component of a sedimentary rock formed at a site different from that of sedimentation.

allopatric Defining an assemblage of organisms or an evolutionary event occurring at a location different from that under consideration.

allotropic Referring to an element exhibiting allotropy.

allotropy The property of an element to crystallize in two or more different forms depending upon ambient temperature and pressure. Examples: C as graphite and diamond; S as rhombic, monoclinic, and amorphous; Fe as body-centered cubic and face-centered cubic. Cf. **polymorphism.**

alloy A macroscopically homogeneous mixture of (usually) two or more metals. Some nonmetals partake in the formation of alloys (e.g. C in carbon steel). Alloys may be ordered (exhibiting a well-defined multiatomic crystalline unit cell) or disordered (solid solutions).

alluvial fan An apron of detrital sediment deposited by a river issuing from a narrow valley into a broad plain and thus experiencing a sudden reduction in slope.

alluvium The ensemble of loose clastic sediments deposited by stream action in Pleistocene or Holocene time.

almandine A garnet, $Fe_3Al_2Si_3O_{12}$.

alp A high meadow area between snowline and timberline.

alpha decay The emission of an α particle by an atomic nucleus. See **alpha rays.**

alpha helix The spiral structure of a polypeptide chain ($2.3 \cdot 10^{-10}$ m in diameter, positive helicity) forming a fibrous or other protein. Each turn has 3.6 amino acids radiating outwards, is H-bonded to the adjacent turns, and advances the spiral by $5.4 \cdot 10^{-10}$ m.

alpha particle (α) A particle consisting of two protons and two neutrons, identical to the nucleus of ^{4}He. Mass = 4.00150604 u. See **Elementary particles*.**

alpha rays Alpha particles emitted by many unstable radionuclides. Alpha particles energies range from as low as 1.8 MeV (^{144}Nd) to as high as 11.6 MeV (^{212}Po), but are mainly between 4 and 8 MeV. The range of a 5 MeV α particle is about 35 mm in dry air, 35 μm in water, and 25 μm in solids. Upon emerging from the nucleus, the α particle acquires 2 electrons from the electron cloud of the emitting atom, but it tends to lose them during its trajectory. Initial speed is in the order of 10^7 m/s. Kinetic energy is dissipated in ionization and excitation processes and in the acceleration to energies as high as 2000 eV of electrons stripped from atoms along the α particle path (δ rays).

alternating current (ac) An electric current that periodically reverses its direction.

$$I = I_M \sin (\omega t + \theta)$$
$$I_{avg} = 2I_M/\pi = 0.637I_M$$
$$I_{rms} = I_M/2^{1/2} = 0.707I_M$$

where I = current, I_M = maximum current, $\omega = 2\pi f$, f = frequency, θ = phase angle between voltage and current, I_{avg} = average current, I_{rms} = effective current.

alternating voltage Alternating potential difference between two points in a conductor.

$$V = V_M \sin \omega t$$
$$V_{avg} = 2V_M/\pi = 0.637V_M$$
$$V_{rms} = V_M/2^{1/2} = 0.707V_M$$
$$V_M = |Z|I_M$$

where V = voltage, V_M = maximum voltage, $\omega = $

$2\pi f$, f = frequency, V_{avg} = average voltage; V_{rms} = effective voltage = $0.707 V_M$, Z = impedance, I = current.

altitude *(Astronomy)* (**h**) the angular elevation of a celestial body above (+) or below (−) the horizon in the horizon coordinate system. See **coordinate systems.** *(Topography)* The elevation of a topographic or other feature above a given standard datum.

alumel An alloy (95% Ni, 3% Cr, 2% Al, 1% Si) used for thermocouples.

AM Amplitude modulation.

amide An organic compound containing the $-CONH_2$ radical.

amine The radical $-NH_2$.

amines A family of organic compounds derived from NH_3 by replacing with organic groups one H (primary amines; e.g., CH_3-NH_2, methylamine), two H [secondary amines; e.g. $(CH_3)_2-NH$, dimethylamine], or all three H [tertiary amines; e.g. $(CH_3)_3-N$, trimethylamine].

amino acid Any of the acids formed from carboxylic acids, characterized by a common, monovalent $NH_2-CH-COOH$ group and specific side groups attached to the C of the CH component. Molecular mass ranges from 75.067 (glycine) to 240.229 (triptophan). See **Amino acids*.**

ammeter An instrument that measures electric current.

ammonia NH_3 (mol. mass = 17.031).

ammonia clock A clock based on the property of the pyramidal NH_3 molecule to reverse itself with a frequency of $2.387013 \cdot 10^{10}$ Hz. See **atomic clock.**

ammonium The radical $-NH_4$ (mol. mass = 18.039).

Amor A group of asteroids closely approaching the Earth's orbit. Estimated number ~ 1000–2000; diameters ~ 1 to 10 km; collision rate with Earth $\sim 0.5/10^6$ y. See **Apollo, asteroid, Aten.**

amp Ampere.

AMP Adenosine monophosphate ($C_{10}H_{14}N_5O_7P$, mol. mass = 347.224), consisting of adenosine linked to one phosphate group. Cf. **ADP, ATP.**

ampere (A) The SI and MKS unit of electrical current, defined as that current that, if maintained in two parallel conductors of negligible cross section and infinite length, placed 1 m apart in vacuo, would produce on each conductor a force of 10^{-7}

N/m of length (exactly). It is equal to 1 coulomb/second.

ampere/meter (A/m) The SI unit of magnetic field intensity. 1 A/m = $4\pi \cdot 10^{-3}$ oersted. Syn. **ampere-turn/meter.**

Ampere's law The law relating the line integral of the magnetic field **B** around a closed path to the total current i through the circumscribed area.

$$\oint \mathbf{B} \cdot d\mathbf{l} = \mu_0 i$$

where $d\mathbf{l}$ = element of length around the conductor, μ_0 = permeability constant = $4\pi \cdot 10^{-7}$ weber/ampere-meter, and i = current in amperes. It is the integral of Biot-Savart law. See **Biot-Savart law.**

ampere-turn (At) The SI unit of magnetomotive force, equal to the mmf developed when a current of 1 ampere is flowing around a circle. Cf. **gilbert.**

ampere-turn/meter (At/m) Syn. **ampere/meter.**

amphiboles A group of hydroxy Fe–Mg–Ca aluminosilicates, a major component of igneous and metamorphic rocks. Common amphiboles are:

actinolite	$Ca_2(Mg,Fe)_5Si_8O_{22}(OH)_2$
anthophyllite	$(Mg,Fe)_7Si_8O_{22}(OH)_2$
glaucophane	$Na_2(Mg,Fe)_3Al_2Si_8O_{22}(OH)_2$
hornblende	$(Ca,Na)_{2-3}(Mg,Fe,Al)_5(Si,Al)_8 \cdot O_{22}(OH)_2$
tremolite	$Ca_2(Mg,Fe)_5Si_8O_{22}(OH)_2$ (with less Fe than actinolite)

amphibolite A medium grade (450–650°C, 3–10 kb) metamorphic rock consisting mainly of amphibole and plagioclase.

amphidromic point The point of no tide, where the cotidal lines meet and around which the tidal wave rotates.

amphipathic Defining a molecule with a polar head and a nonpolar tail.

amphiprotic Defining a chemical compound that can function both as a Brønsted acid and as a Brønsted base. Cf. **amphoteric.**

amphoteric Defining a chemical compound that can function as an acid or as a base. E.g. amino acids, $NH_2-R-COOH$, characterized by the acid group $-COOH$ and the basic group $-NH_2$. Cf. **amphiprotic.**

amplitude The maximum absolute value attained by a periodic function or phenomenon.

amplitude modulation (AM) The modulation of

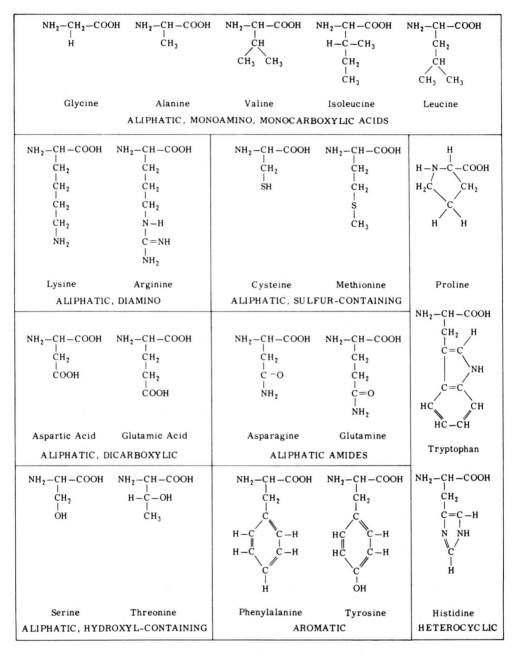

Amino acids. (From King and Stansfield 1985, p. 19)

the amplitude of a carrier wave by a superimposed signal.

amu Atomic mass unit, defined (before 1960) as 1/16th of the mass of the neutral atom of ^{16}O (physical definition) or 1/16th of the average iso-topic mass of oxygen in nature (chemical definition). The chemical definition was 1.000274 larger than the physical definition. Since 1960 the definition of atomic mass unit (symbol **u**) is 1/12th of the mass of the neutral atom of ^{12}C. See **atomic mass unit, u.**

amygdale A gas-formed cavity in an igneous rock that has become filled with secondary minerals.

An Anorthite.

an- Prefix meaning *without*.

ana- Prefix meaning *upwards*.

anabatic Moving upward. Cf. **catabatic**.

anabolism A phase of metabolism that leads to the formation of more complex organic molecules from simpler ones. Cf. **catabolism, metabolism**.

anaerobic 1. Defining an environment that lacks free oxygen. 2. Defining an organism that does not need free oxygen for its metabolism.

anaglacial A climatic phase leading from an interglacial to a glacial age. Cf. **cataglacial**.

analemma A curve showing the difference between Local Time and Mean Time as well as the declination of the Sun. Both parameters change during the year as a result of the combined effects of eccentricity and obliquity. See **equation of time**.

analog A continuously varying quantity proportionally representing another continuously varying quantity.

analog-to-digital conversion The conversion of the values assumed by a continuously varying function into discrete values. Cf. **digital**.

anastigmatic Defining an optical system corrected for astigmatism.

anastomosis The interconnection of vessels, filaments, channels, etc.

anatexis Partial melting of pre-existing rock. Cf. **diatexis**.

anathermal A climatic phase leading from a glacial to an interglacial age. Cf. **catathermal**.

andesine A plagioclase of composition $Ab_{70}An_{30} - Ab_{50}An_{50}$.

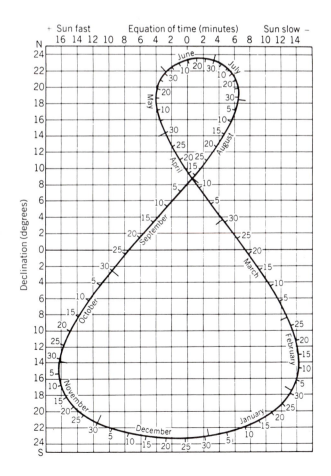

Analemma. (Strahler 1971, p. 43, Fig. 3.11)

andesite A fine-grained volcanic rock, the extrusive equivalent of diorite.

andesite line The petrographic boundary between the basaltic magmas of the Pacific and the andesitic magmas of the circum-Pacific margin.

Andromeda The largest of the Local Group of galaxies to which the Galaxy belongs. Largest diameter = 125,000 l.y.; distance = 2,140,000 l.y.; mass = $3 \cdot 10^{11}$ solar masses.

aneroid Containing or using no water or other liquid.

angiosperm Any plant of the class Angiospermae, characterized by having seeds enclosed in an ovary.

angle of incidence The angle that an electromagnetic or acoustic ray makes with the normal to an interface separating two media.

angle of repose The maximum angle above the horizontal at which loose sediment can support itself without slumping. Cf. **critical slope.**

angstrom (Å) A unit of length equal to 10^{-10} m.

angular acceleration (α) The rate of change of angular velocity.

$$\alpha = d\omega/dt$$

where ω = angular velocity.

angular displacement A measure of the rotation of an object about an axis. If the object rotates in the sense of the head of a normal screw being threaded into a material, the direction of advance of the screw into the material identifies the direction of the vector used to represent angular displacement. This vector is normal to the plane of rotation.

angular frequency (ω) The frequency in hertz of a periodic phenomenon times 2π.

$$\omega = 2\pi f$$

where f = frequency.

angular momentum (L) 1. *Of a particle revolving around an axis:*

$$L = r \times p$$

where r = shortest distance from axis, with vector origin on axis; p = momentum of particle.
2. *Of a body rotating about an axis:*

$$L = I\omega$$

where $I = \int r^2 \, dm$ = moment of inertia, ω = angular velocity, dm = mass increment, r = shortest distance of mass increment dm from axis of rotation. Vector representation as in **angular displacement.**

angular unconformity An unconformity between two sedimentary formations whose bedding planes are at an angle to each other.

angular velocity (ω) The rate of change in angular displacement.

$$\omega = d\theta/dt$$

where θ = angular displacement, t = time. It is measured in degrees or radians per unit of time and is represented vectorially as in **angular displacement.**

anhedral A mineral crystallized without having been able to form euhedral faces.

anhedron A crystal not bound by euhedral faces.

anhydride An oxide that gives an acid or a base when combined with water.

anhydrous Defining a substance or a system that contains no water of hydration or as a chemical combinate.

Animalia The kingdom that includes all multicellular heterotrophs exhibiting extensive tissue differentiation. Cf. **kingdom.** See **Taxonomy*.**

anion A negatively charged ion, i.e. an atom having more electrons than the protons in its nucleus. It moves toward the anode in an electrolytic solution. Cf. **cation.**

anisotropic Defining a system with one or more properties that vary in direction.

anode The positive pole of an electrolytic cell, a primary cell, a battery, or a discharge tube.

anodize To protect a metal surface, functioning as an anode, with a thin film of oxides deposited electrolytically.

anomalistic year The time interval between successive passages of the Earth at perihelion, equal to $365.25964134 + 0.00000304T$ d$_E$ (where T = centuries since 1900.0) = 31,558,433.240 s$_E$. It is 283.247 s$_E$ longer than the sidereal year because the perihelion advances each year in the same direction of the motion of the Earth along its orbit (i.e. counterclockwise as seen from the North), due to planetary perturbations and relativistic effects. Cf. **year.** See **precession of the equinoxes.**

anorthite $CaAl_2Si_2O_8$, the Ca end-member of the plagioclase group.

anorthoclase An alkali feldspar, $(Na,K)AlSi_3O_8$.

anorthosite A plutonic rock consisting mainly of labradorite.

Antarctic Circle See **polar circle.**

antecedent stream A stream existing before local

diastrophism and maintaining its original course during and after diastrophism.

anthracite Coal of the highest metamorphic rank, containing ~90% C and yielding ~7900 cal/g.

antibonding Referring to a molecular orbital characterized by higher energy and lower internuclear electron density, resulting in internuclear repulsion.

anticline An upwardly convex, elongated fold. Cf. **syncline.**

anticlinorium A composite anticlinal structure consisting of a set of parallel anticlines.

anticodon The base triplet carried by tRNA that matches the codon of mRNA. See **codon, mRNA, tRNA.**

anticoincidence The suppression of an event by the occurrence of another.

anticolor The color charge of antiquarks and antigluons, consisting of antired, antigreen, or antiblue.

anticyclone A broadly circular region of high atmospheric pressure with geostrophic winds spiraling out clockwise (northern hemisphere) or counterclockwise (southern hemisphere). Cf. **cyclone.**

antiferromagnetism Property of substances in which the magnetic moments of adjacent atoms align themselves antiparallel to each other, resulting in a net magnetic moment close to zero for the substance. These substances become paramagnetic above the Néel point.

antigluon The antiparticle of the gluon, carrying anticolor.

antimatter Matter made of antiparticles.

antineutrino The antiparticle of the neutrino. It has positive helicity, i.e. it rotates clockwise along its direction of flight.

antineutron The antiparticle of the neutron, having identical mass and spin but opposite magnetic moment.

antinode The point of a standing wave having maximum motion. Cf. **node.**

antinucleon An antiproton or antineutron.

antiparticle The counterpart of a particle, having identical mass, mean life, and spin but opposite, as applicable, charge, magnetic moment, baryon number, color, and strangeness.

antipode The point on a spherical surface diametrically opposite a specified point.

antiproton The antiparticle of the proton, having identical mass and spin but opposite charge and magnetic moment.

antiquark The antiparticle of the quark, having baryon number, electric charge, color, and strangeness of sign opposite that of the quark.

antithetic fault A secondary fault opposite the major fault with which it is associated.

Aouelloul glass A silica-rich glass produced by fusion of country rock resulting from the impact that formed the Aouelloul crater in Mauritania (diameter = 370 m; age = $3.5 \pm 0.5 \cdot 10^6$ y).

A.P. After the Present, meaning the time after 1950. Cf. **B.P.**

apastron The point in a planetary orbit, or in the orbit of a secondary in a binary system of stars, farthest from the star around which the planet or the secondary is revolving. Syn. **aphelion** for the solar planets or other objects in circumsolar orbits. Cf. **periastron.**

apatite A Ca phosphate mineral, $Ca_5(PO_4)_3 \cdot (F,Cl,OH)$.

aperture The diameter of the objective of an optical system.

apex (solar) The point toward which the Sun is moving with respect to the neighboring stars. Coordinates: $\alpha = 271°$, $\delta = +30°$; $l^{II} = 57°$, $b^{II} = +22°$. It is close to the direction of the star Vega. Velocity = 19.7 km/s = 4.15 AU/y.

aphanitic Defining an igneous rock or a groundmass composed of crystals too small to be perceived by the naked eye.

aphelion (Q) The orbital point of a planet, comet, or asteroid farthest from the Sun.

aphotic Defining the depth zone (usually 100–200 m) in the sea below which light is too dim for photosynthesis.

aplanatic Defining an optical system corrected for spherical aberration.

aplite A hypabyssal, granitoid rock with a fine-grained, saccharoidal texture.

aplitic Exhibiting the texture of aplites.

apoapsis The apsis farthest from the center of gravity of an orbiting body. Syn. **aphelion** for the solar planets and other objects in circumsolar orbits. Cf. **apsides, periapsis.**

apochromatic Defining an optical system corrected for chromatic and spherical aberrations.

apogee The point in the elliptical orbit of the Moon or an artificial circumterrestrial satellite farthest from the Earth. Cf. **perigee.**

Apollo Any of a group of asteroids with orbits intersecting the Earth's orbit and with semimajor axis >1 AU. Estimated number with diameter $>$ 0.1 km = 700 $\pm$ 300; mean diameter = 1 to 10 km; collision rate with the Earth = 1.8 $\pm$ 0.8/10^6 y. See **Amor, asteroid, Aten.**

apparent brightness The flux of electromagnetic energy received by a terrestrial observer from a celestial body.

apparent day The time interval between successive local noons.

apparent magnitude See **magnitude.**

apparent power The product of effective current times effective voltage, with no consideration for phase difference between voltage and current.

$$P_a = I_e V_e$$

where P_a = apparent power, I_e = effective current, V_e = effective voltage.

apparent solar time The local time, based on the actual motion of the Sun.

applanation The reduction of topographic relief by erosion of the highs and sediment infilling of the lows.

apsides (sing. **apsis**) The two points at the opposite ends of the major axis of an elliptical orbit.

apsis See **apsides.**

aquiclude A relatively impermeable rock formation incapable of conveying water in amounts sufficient for common usage.

aquifer A rock formation sufficiently permeable to convey water for common usage.

aquifuge A rock formation without interconnecting pores and thus unable to absorb and convey water.

aragonite A high-density (ρ = 2.93) crystalline phase of $CaCO_3$. Cf. **calcite.**

arc 1. A portion of a curve. 2. Expressing an inverse trigonometric function. E.g. arc sin x = angle whose sine is x.

Archean The portion of Precambrian time between 3.8 and 2.7 billion years ago, after the Hadean (4.7–3.8·10^9 y B.P.) and before the Proterozoic (2.7–0.590·10^9 y B.P.).

archetype The ancestral form of a group of related organisms.

Archimedes' principle "In a gravitational field, a body floating or totally immersed in a fluid experiences an upward force equal to the weight of the fluid displaced."

Arctic Circle See **polar circle.**

arenaceous Defining material (rock, exoskeleton, etc.) largely consisting of cemented sand-size particles.

arenite A sedimentary rock formed of more than 50% of sand-size particles. Cf. **lutite, rudite, siltite.**

arête A narrow, jagged rocky ridge above the snowline.

argillaceous Defining a sediment consisting of more than 50% of clay particles.

argillite A weakly metamorphosed claystone.

argon-40/argon-39 dating method An absolute dating method for minerals and rocks based on the decay by K-capture of ^{40}K ($t_{1/2}$ = 1.277·10^9 y) to ^{40}Ar. To avoid problems associated with ambient ^{40}Ar adsorbed during crystallization (mainly along intercrystalline surfaces within the rock), and with ^{40}Ar loss subsequent to crystallization (mainly from the outer portions of the crystals), a rock sample is irradiated with neutrons in a reactor producing the reaction ^{39}K(n,p)^{39}Ar [where (n,p) means "neutron in, proton out"]. ^{39}Ar is radioactive, decaying back to ^{39}K by β^- decay, but its half-life (269 y) is sufficiently long for the isotope to be regarded as stable within the time required for analysis. The irradiated sample is heated stepwise and the ^{40}Ar/^{39}Ar ratio is measured as temperature rises. The ^{40}Ar released at the highest temperature is derived from the inner portions of the crystals and thus represents the true concentration of the radiogenic ^{40}Ar formed since the rock began crystallizing. As the concentration of ^{39}Ar is related to the concentration of ^{39}K which, in turn, is related to the concentration of ^{40}K, the ratio ^{40}Ar/^{39}Ar is sufficient to resolve the age of the sample by providing the concentration of both the isotope ^{40}K and its daughter ^{40}Ar.

argument of perihelion (ω) The angle from ascending node to perihelion in the direction of motion of the orbiting body.

arkose A feldspar-rich ($>25\%$) sandstone.

arkosic Defining a sediment or sedimentary rock rich in feldspar.

aromatic Defining an organic compound in which the carbon atoms are arranged in ring structure. Cf. **aliphatic.**

Arrhenius acid See **acid.**

Arrhenius base See **base.**

Arrhenius equation An equation relating the rate of chemical reactions to absolute temperature:

$$k = Ae^{-Ea/RT}$$

where k = rate constant, A = constant characteristic of specific reaction, E_a = activation energy, R = gas constant, T = absolute temperature.

arroyo The flat-bottomed, steep-sided bed of an episodic torrent in an arid region. Syn. **wadi.**

artesian Defining confined groundwater under hydrostatic pressure.

aryl (Ar) An aromatic ring radical. E.g. phenyl, $-C_6H_5$.

ås Swedish name for **esker.**

åsar Plural of **ås.**

asbestos Commercial name for a group of fibrous silicate minerals, the principal ones of which are chrysotile, $Mg_3Si_2O_5(OH)_4$, a variety of serpentine, and crocidolite (blue asbestos), $Na(Mg,Fe^{2+})_3 \cdot Fe_2^{3+}Si_8O_{22}(OH)_2$, a variety of the amphibole riebeckite.

ascending node See **nodes.**

aseismic ridge A submarine ridge not related to a plate margin, formed by the migration of a plate over a continuously acting hot spot.

ash (volcanic) Unconsolidated pyroclastic material consisting of particles 0.063 to 1.0 mm in size. Cf. **cinder, lapilli, volcanic dust.**

aspect Configuration of the Moon or a planet with respect to the Sun as seen from the Earth.

asphalt A naturally occurring material consisting of a mixture of heavy hydrocarbons and inorganic matter. The organic component softens at about 90°C.

asphaltites Heavy asphalts with softening points between 110° and 300°C.

assise The lithostratigraphic equivalent of substage.

astatic 1. Defining an instrument insensitive to an external, uniform force field. 2. Defining a geophysical instrument with positive feedback reinforcing a deviation from equilibrium induced by the external force being measured.

asteroid Any of the minor planetary bodies orbiting the Sun between the orbits of Mars and Jupiter (included are also the asteroids occupying Jupiter's stable Lagrangian points; see **Trojan**). Asteroids range in diameter from 974 km (Ceres) to less than 1 km. Total number with diameter >1 km $\sim$ 500,000 (7 with diameter >300 km, 200 with diameter >100 km, 2000 with diameter >10 km). The larger asteroids (diameter >200 km) are spherical or spheroidal in shape, while the smaller ones (diameter <20 km) may have very irregular shapes. Asteroids have prograde orbits with median semimajor axis = 2.7 AU, median eccentricity = 0.15, and median inclination = 9.5°. The orbital elements of 3302 asteroids have been determined (1985). Some of the asteroids have orbits of high eccentricity and cross the Earth's orbit (Apollos, Atens) or come close to it (Amors). Asteroidal collision rate with the Earth is about 3.5 collisions per million years. Albedo and spectral measurements indicate that 75% of the asteroids consist of Fe-Mg silicates with some carbon (most common in the central band of the asteroidal belt); 20% consist of Fe-Mg silicates and Fe-Ni metal (most common in the inner portion of the asteroidal belt); and 5% consist of Fe-Mg silicates or Fe-Ni metal. The asteroids are presumed to be the parent bodies of meteorites. See **Amor, Apollo, Asteroids*, Aten, Kirkwood gaps, Trojan.** Cf. **meteorite.**

asteroidal belt A region between the orbits of Mars and Jupiter, ranging from 2 to 4 AU, within which most of the asteroids orbit.

asthenosphere The upper mantle layer from about 65 to about 165 km of depth (suboceanic areas) or from about 120 to about 220 km of depth (subcontinental areas), made softer by a small ($<1\%$?) amount of melting. P and S wave velocities are decreased by about 5% with respect to their velocities in the mantle above. Syn. **low-velocity channel, low-velocity zone.**

astigmatic Defining an optical system with geometrical distortions in the surfaces of its components resulting in a distorted image.

ASTM American Society for Testing and Materials.

astrobleme A scar on the Earth's surface produced by a meteoritic impact. Astroblemes range in diameter from 140 km (Vredefort, South Africa, $1.97 \cdot 10^9$ y old; Sudbury, Canada, $1.84 \cdot 10^9$ y old) to 10 m or less (Haviland, Kansas, $<2 \cdot 10^6$ y old). Barringer (Meteor Crater) in Arizona, 1.2 km in diameter, is about 50,000 y old. See **Astroblemes*.**

astrolabe A graduate vertical circle with a sight to determine the altitude of the Sun at midday or of other celestial bodies.

astronomical unit (AU) The mean distance of the Earth from the Sun (= semimajor axis of the Earth's orbit). It is equal to 149,597,870.7 km = 8.31676 light minutes = 499.004784 light seconds = $1.581284 \cdot 10^{-5}$ light years = $4.848 \cdot 10^{-6}$ parsecs.

asymptote A line approached by a curve at a continuously decreasing rate and reached by it only at infinity.

asymptotic Referring to a curve that approaches a line at a continuously decreasing rate, reaching it only at infinity.

At Ampere-turn.

ataxite A type of iron meteorite consisting of kamacite and taenite with 6–30% Ni, characterized by the absence of macroscopic Widmanstätten figures. Ataxites represent 14% of all iron meteorites. See **meteorite, Meteorites***.

Aten Any of a group of asteroids with orbits intersecting the Earth's orbit with orbital semimajor axis <1 AU and sidereal period <1 year. Estimated number with diameter >0.1 km ~100; mean diameter = 1 to 10 km; rate collision with the Earth ~$0.9/10^6$ y. See **Amor, Apollo, asteroid.**

atm Abbreviation for *atmosphere* as a unit of pressure.

atmosphere The gaseous envelope of a planetary body. The atmosphere of the Earth consists of N_2 (78.084%), O_2 (20.946%), Ar (0.934%), CO_2 (0.032%), other gases (0.004%), and variable amounts (0.004–4%) of water vapor. Mean molecular mass = 28.964 u. It is divided into troposphere (0 to 10 km of altitude at the poles, 0 to 16 km of altitude in the tropics; temperature decreasing with altitude from 260 K to 230 K at the poles, from 300 K to 200 K in the tropics), bound by the tropopause; stratosphere (10–16 km to 50 km, temperature rising with altitude from 200–230 K to 280 K), bound by the stratopause; mesosphere (50 to 85 km, temperature decreasing with altitude from 280 K to 160–190 K), bound by the mesopause; thermosphere (85 to 650 km, temperature rising with altitude from 150–190 K to 500–2000 K), bound by the thermopause; exosphere (650 km to the magnetopause, kinetic temperature rising with altitude to 10^4 K). Ozone layer, 15 to 30 km of altitude. Max. O_3 concentration = $5 \cdot 10^{12}$ molecules/cm^3 at 22 km of altitude; total amount of O_3 = 8–$10 \cdot 10^{18}$ molecules/cm^2 of Earth surface = 3.3-mm-thick layer at STP. See **Atmosphere—**

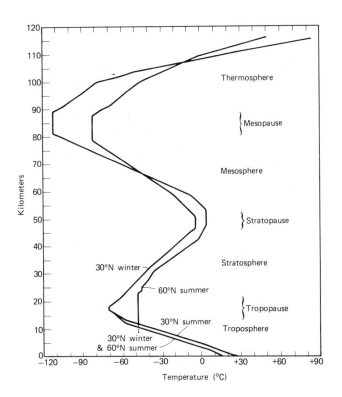

Atmosphere. Thermal structure from 0 to 120 km. Notice the difference between winter and summer in the troposphere at 30°N and the difference between the thermal structure at 30°N versus that at 60°N. Also notice the marked difference in the altitude of the tropopause between 30°N (15 km) and 60°N (10 km). (Data from U.S. Standard Atmosphere Supplements, 1966)

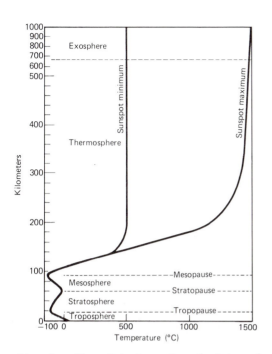

Atmosphere. Thermal structure and zonation between 0 and 1000 km of altitude. (Data from U.S. Standard Atmosphere 1976)

composition*, Atmosphere—physical properties*, Ionosphere*.

atmosphere (atm) A unit of pressure equal to 101,325.0 Pa (exactly) or 760 mmHg (exactly).

atoll A circular, elliptical, or horseshoe-shaped reef on top of an extinct and drowned volcanic cone.

atom The unit of matter, consisting of a nucleus surrounded by one or more electrons. Radius = 0.4–$2.5 \cdot 10^{-10}$ m. Density of atomic species in the solid state = 0.07 (H) to 22.5 (Os) g/cm^3. See **atomic radius, hydrogen atom.**

atomic absorption spectrophotometer An instrument for quantitative element analysis. Light produced by atoms of a given element excited in a discharge lamp is passed through a flame in which the same element from a solution is dispersed. The element's atoms in the flame absorb light from the source and reradiate it in all directions. The power of the source, as seen by the detector, is reduced proportionally to the concentration of the element in the flame and, hence, in the solution.

atomic binding energy See **binding energy.**

atomic clock A clock whose frequency is provided by the natural resonance frequency of a suitable atom or molecule. **1. passive clocks** The system consists of a frequency generator (usually a quartz oscillator), a substance offering an atomic or molecular transition with a specific frequency (NH_3 molecule, $\nu = 2.387013 \cdot 10^{10}$ hertz; ^{133}Cs hyperfine transition, $\nu = 9{,}192{,}631{,}770$ hertz), and a detector. If the frequency generator is exactly tuned to the substance's frequency, power is absorbed by the substance and reradiated in all directions, thus causing the detector to experience a minimum in power received. If the frequency generator drifts, the detector experiences an increase in power received. This signal is fed back to the source, leading to an appropriate correction of the frequency generated. **2. active clocks** The frequency is directly generated by a transition in the atoms of an appropriate substance (most commonly the hyperfine transition of the ground state of Cs, Rb, Tl, or H), followed by power amplification.

atomic core An atom stripped of its valence electrons and thus having only electrons in closed shells.

atomic mass The mass of a neutral atom expressed in atomic mass units. Syn. **atomic weight.** Cf. **amu, u.**

atomic mass number (A) See **mass number**

atomic mass unit (u) The unit of atomic mass, defined as 1/12th of the mass of the neutral atom of ^{12}C. $1\ u = 1\ g/N_A = 1.660540 \cdot 10^{-24}$ g = 931.4943 MeV (where N_A = Avogadro number). Before 1960 the atomic mass unit (symbol **amu**) was defined as 1/16th of the mass of the neutral atom of ^{16}O (physical definition) or 1/16th of the average isotopic mass of oxygen in nature (chemical definition). The chemical amu was 1.000274 larger than the physical amu.

atomic number (Z) The number of protons in an atomic nucleus. It defines the element.

atomic radius One half the interatomic distance (center to center) in diatomic molecules of a given element. Representative radii (10^{-10} m): H = 0.373; O = 0.604; Na = 1.858; S = 0.943; K = 2.272; Ca = 1.973; Fe = 1.241; Rb = 2.475; Sr = 2.151; Ag = 1.445; Cs = 2.654; Os = 1.338; Hg = 1.502; U = 1.385. Atomic radii do not increase very much with atomic number because, as the nuclear charge increases, the increasing electrostatic attraction "pulls in" the electron shells. Cf. **hydrogen atom.**

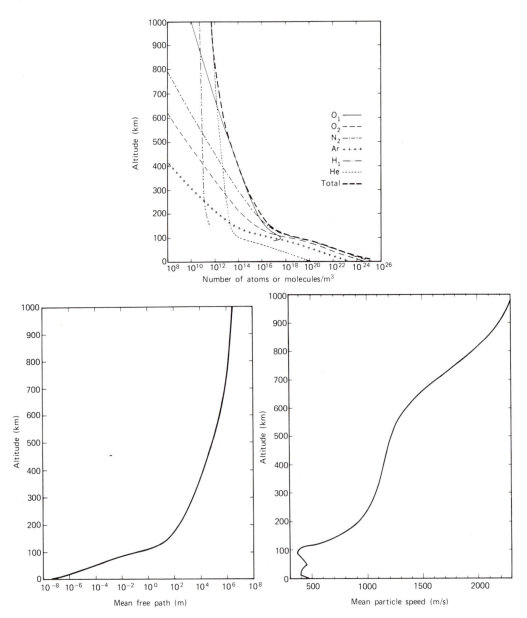

Atmosphere. Concentration of gases, mean free path, and particle speeds as functions of altitude. (From U.S. Standard Atmosphere 1976)

atomic second (s_A) The time interval equal to the duration of 9,192,631,770 periods of the radiation corresponding to the transition between the two hyperfine levels of the ground state of ^{133}Cs.

atomic time (t_A) Time measured in atomic seconds. The epoch of atomic time is taken to be 1958 January 0d, 0h, 0m, 0s, at which time $t_A = t_U = t_E - 32.15$ s.

atomic weight The weight of an atom relative to the weight of 1/12th of the neutral atom of ^{12}C. Syn. **atomic mass.** Cf. **amu, u.**

ATP Adenosine triphosphate, $C_{10}H_{16}N_5O_{13}P_3$

(mol. mass = 507.184), consisting of adenosine linked to three phosphate groups. The "storage battery" of living systems. The photosynthetic or chemical energy stored in the formation of ATP from ADP is 30.6 kJ/mole, or 0.3 eV per bond.

attenuation The decrease in intensity of a quantity away from the source. If the medium through which the quantity is transmitted is homogeneous, the intensity I_x at a distance x along direction x from a source or reference point O, with respect to the intensity I_0 at the source or at the reference point, is given by the equation

$$I_x = I_0 e^{-\alpha x}$$

where α = attenuation coefficient. [Cf. the equation for radioactive decay, where the parameter (number of atoms) decreases with time rather than with distance.] Attenuation is expressed in nepers. The attenuation of voltage V, current I, or power P along direction x on a uniform transmission line, with respect to reference voltage V_0, reference current I_0, or reference power P_0, is given by

$$V = V_0 e^{-\alpha x}$$
$$I = I_0 e^{-\alpha x}$$
$$P = P_0 e^{-2\alpha x}.$$

attenuation coefficient (α) The exponent expressing the decrease in value Q of a quantity, from the original value Q_0, with distance x from a source if the medium through which the quantity is propagating is homogeneous:

$$Q = Q_0 e^{-\alpha x}$$

where α = attenuation coefficient.

atto- Prefix meaning 10^{-18}. Cf. **femto-, micro-, nano-, pico-**.

AU Astronomical Unit.

aubrite An achondrite consisting mainly of enstatite.

audio frequency A frequency within the audio range (15 to 20,000 hertz). See **audio range**.

audio range The audible frequency range between 15 and 20,000 Hz (22 m and 1.66 cm wavelength, respectively, in dry air at 20°C).

Aufbau The process of constructing the electron cloud of an atom by adding electrons at progressively higher energy levels. Cf. **Hund's rule**.

augen structure The structure of some gneisses and metaschists produced by the squeezing together of minerals into ellipsoidal bodies resembling eyes *(augen)* in section.

Auger effect The filling of a vacancy in the electronic structure of an atom by an electron from a higher energy level and the simultaneous emission of a second electron from an energy level such that the ionization energy of the emitted electron equals the energy produced by the filling of the vacancy.

Auger electron An electron emitted from an atom as a result of the Auger effect. See **Auger effect**.

augite A clinopyroxene, $(Ca,Na)(Mg,Fe,Al) \cdot [(Si,Al)O_3]_2$.

aulacogen An extensional trough on a craton, bound by normal faults. Aulacogens are caused by updoming of the craton, radiate from its center, and increase in width outwards.

aureole A zone surrounding an igneous intrusion exhibiting the effect of contact metamorphism on the country rock.

aurora A glow in the upper atmosphere at high northern *(aurora borealis)* or southern *(aurora australis)* magnetic latitudes, most commonly extending from 100 to 200 km of altitude. It results from atmospheric gases being excited by fast protons and electrons of solar and galactic origin channeled toward the Earth's magnetic poles by the Earth's magnetic field.

aurora australis See **aurora**.

aurora borealis See **aurora**.

austausch coefficient A measure of turbulent mixing, given by the eddy diffusion of a fluid normally to its general direction of flow.

austral Referring to the southern latitudes. Cf. **boreal**.

australite An Australian tektite. See **tektite**.

authi- Prefix meaning **there**.

authigenic Defining a mineral formed in place.

auto- Prefix meaning **self**.

autochthonous Defining a formation formed in place.

autoclastic Defining a rock fragmented in place.

autocorrelation The relationship of the value of a variable to the value of the same variable at a different time.

autogeosyncline A parageosyncline without adjacent uplift areas, largely filled with carbonate sediments.

autometamorphism The metamorphism of a rock by reaction of its early-formed minerals with its own fluids as temperature falls and new minerals are formed.

autometasomatism Weak autometamorphism, not involving recrystallization.

autoradiography The imaging of a material on a sensitive plate by its natural or artificial radioactivity.

autotrophic Defining an organism capable of deriving energy by inducing chemical reactions among inorganic substances using electromagnetic interaction as a source of energy.

auxiliary mineral A relatively unimportant, light-colored mineral, part of a mineral assemblage. Cf. **accessory mineral.**

auxiliary plane A plane normal to the fault plane.

avalanche A rapid sequence of events or phenomena of increasing magnitude, related by cause and effect. *(Physics)* A chain reaction by which an electron or other charged particle accelerated in an electrostatic field produces more than one electron or other charged particle from a target, which in turn releases more electrons or other charged particles from additional targets maintained at increasingly greater voltages. *(Geomorphology)* The slumping of a mass of snow.

avdp. Avoirdupois.

average power The average power of an ac circuit.

$$P_{aver} = V_e I_e \cos \phi$$

where V_e = effective voltage, I_e = effective current, ϕ = phase angle between voltage and current.

avogadro (*A*) A unit of number of items (elementary particles, atoms, ions, molecules, objects, organisms, etc.) equal to $6.022136 \cdot 10^{23}$. Syn. **mole.**

Avogadro number (*N_A*) The number of atoms in 12 g of neutral ^{12}C atoms, equal to $6.022136 \cdot 10^{23}$.

avoirdupois (avdp.) (*Old French* for *goods by weight*) Referring to the common English system of units for masses and weights. It includes the pound (= 0.45359237 kg exactly), the ounce (1/16th of a pound), the dram (1/16th of an ounce), the short ton (2000 pounds), and the long ton (2240 pounds).

axial angle See **optical angle.**

axial plane 1. The plane including the optical axes of a biaxial crystal. 2. A plane including two crystallographic axes.

axial surface A surface connecting the hinge lines of a set of superimposed strata that form a fold.

axion A hypothetical, weakly interactic particle with mass $< 10^{-9}$ (?) u, presumed to be formed in weak nuclear interaction (e.g. β^- decay). If real, axions would form abundantly inside stars and would contribute to the mass of the universe, accounting for the arms of spiral galaxies being more tightly bound than apparently allowed by visible mass and even causing the universe to be closed.

axis of inertia Any of the three principal axes of inertia, one about which the moment of inertia is maximum, one about which the moment of inertia is minimum, and the third one perpendicular to the other two.

azeotropic Referring to a solution with vapor pressure higher or lower than that of any of its components. The composition of the vapor at equilibrium remains identical to that of the solution.

azimuth The angle between the North point of an observer's horizon and the vertical projection of a celestial body on that horizon, measured clockwise from the North direction (0° to 360°).

azimuthal projection The projection, in map construction, of the surface of the Earth from the Earth's center on a plane surface tangent to the pole or to any other specified point.

azimuthal quantum number The quantum number representing the orbital angular momentum of an atomic electron. Syn. **orbital angular momentum quantum number.** See **orbital angular momentum.**

B

β 1. Beta particle. 2. Phase of a polymorphic mineral stable at a temperature higher than phase α but lower than phase γ.

β^+ Beta plus particle ($\equiv$ positron).

β^- Beta minus particle ($\equiv$ electron).

b 1. Barn. 2. Semiminor axis of an elliptical orbit.

B 1. Magnetic flux density = magnetic induction. 2. Susceptance.

b^I Galactic latitude in the old IAU system, measured N (+) or S (−) from the galactic equator. It is equal to $b^{II} − 1.40°$.

b^{II} Galactic latitude in the new (1959) IAU system, measured N (+) or S (−) from the galactic equator. It is equal to $b^I + 1.40°$.

back bond A chemical bond between an atom on the surface layer of a solid and an atom in the second layer.

background radiation 1. The radiation relict from the Big Bang. See **microwave background radiation.** 2. The total radiation from outer space. See **cosmic background radiation.** 3. The environmental particle and electromagnetic radiation from natural sources, i.e. from space, from radionuclides in the atmosphere induced by solar and cosmic protons, and from primary and secondary radionuclides in rocks and natural waters.

back reef 1. A reef landward of the fore reef. 2. Referring to the area behind the fore reef.

backscattering The deflection of radiation or particles by angles greater than 90° with respect to the direction of incidence.

backshore The normally dry beach zone shoreward of the high-tide water line.

bacteriochlorophyll $MgN_4O_6C_{52}H_{70}$, a type of chlorophyll present in all photosynthesizing bacteria. No oxygen is evolved in the process of bacterial photosynthesis.

bacteriophage Any of the viruses capable of infecting bacteria. Syn. **phage.**

bacterium Any of the solitary or colonial procaryota forming the kingdom Monera.

backwash The return of the water to the sea following wave uprush on a beach.

badlands Deeply incised and eroded land consisting of bare clays or silts.

ballast resistor A resistor whose resistance is proportional to the current flow.

ball clay A plastic clay rich in organic matter used as a bonding agent in the manufacture of ceramics.

ballistic galvanometer A galvanometer with a long oscillation period, capable of measuring current and voltage pulses.

ball lightning A spherical mass of plasma produced in the atmosphere by electrical discharge.

Balmer series The series of lines in the hydrogen spectrum produced by transitions from $n > 2$ to $n = 2$ energy levels (emission lines) or from $n = 2$ to $n > 2$ (absorption lines), where n is the principal quantum number. Energies range from 1.8892 to 3.4006 eV; corresponding wavelengths range from 0.65628 to 0.36460 μm. See **energy level.**

banded iron formation (BIF) A Precambrian subaqueous formation consisting of thin (micrometers to millimeters) layers of iron oxides alternating with thicker (millimeters to centimeters) layers of chert and carbonate. The iron oxides were formed by the oxidation of ferrous iron in solution by photosynthetic oxygen. The chert and carbonate components are apparently detrital.

band spectrum A spectrum consisting of a continuum of electromagnetic energy across a given energy band. Cf. **continuous spectrum.**

bar *(Physics)* A unit of pressure = 0.9869233 atmospheres = 10^5 Pa or N/m^2 (exactly) = 750.0617 mmHg. *(Geomorphology)* A low, elongated, offshore ridge consisting of loose sediment that may or may not rise above sea level.

baraboo A monadnock buried under glacial sediments and later exhumed.

barchan A crescent-shaped sand dune that is convex windward.

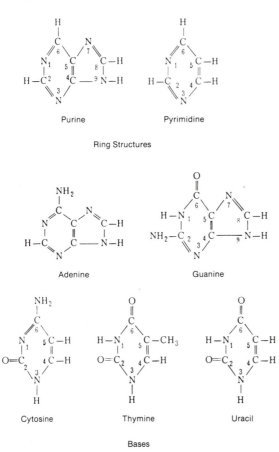

Purine   Pyrimidine

Ring Structures

Adenine   Guanine

Cytosine   Thymine   Uracil

Bases

Bases of nucleic acids. DNA contains adenine, guanine, cytosine, and thymine. RNA contains adenine, guanine, cytosine, and uracil. Their ring structures and the pertinent numerations are shown in the top row.

barn (b) A unit of area used to specify nuclear cross sections. It is equal to 10^{-28} m^2.

barocline A zone in the ocean or the atmosphere through which a change in pressure gradient occurs.

barrier beach An elongated offshore sandy bar rising above high-tide water level, parallel to the shore line and separated from it by a lagoon.

barrier energy Fermi energy plus work function.

barrier island An elongated offshore island parallel to the shoreline and separated from it by a lagoon.

barrier lagoon An elongated lagoon bound by the shoreline and by a barrier beach or barrier island.

barrier reef An elongated reef parallel to the coastline and separated from it by a lagoon.

Barringer crater See **Meteor Crater.**

baryon Any of the nucleons and hyperons, with baryon number = +1, or their antiparticles, with baryon number = −1. See **Elementary particles*.**

baryon number A conserved property of a system of particles, so that no single baryon can be created or destroyed without its antibaryon. Baryon number +1 is assigned to the baryons and baryon number −1 to the antibaryons. Quarks have baryon number +1/3 and antiquarks have baryon number −1/3.

basal conglomerate A coarse beach deposit produced by an ingressive sea and forming the base of a transgressive sedimentary series.

basalt The extrusive equivalent of gabbro. See **Igneous rocks*.**

basanite The effusive equivalent of foid gabbro.

base *(Chemistry)* 1. **Arrhenius base** A chemical substance that dissociates in water to produce

OH$^-$ ions. Cf. **Arrhenius acid.** 2. **Brønsted base** A chemical substance capable of accepting one or more protons. Cf. **Brønsted acid.** 3. **Lewis base** A chemical substance capable of forming a bond by donating an electron pair to a Lewis acid and sharing it with it. Cf. **Lewis acid.** *(Electronics)* The area between emitter and collector in a transistor. *(Biology)* Any of the two purine (adenine, guanine) or three pyrimidine (cytosine, thymine, uracil) ring structures that, together with sugar and phosphate groups, form nucleic acids.

base exchange Syn. **cation exchange.**

base level The lowest level below which erosion is no longer possible.

base line A measured line used as a base in triangulation.

basement The floor of a specified rock sequence.

basement complex The complex of rocks forming the floor of a specified rock sequence.

basement rock The rock forming the floor of a specified rock sequence.

basic *(Chemistry)* Having the chemical characteristics of a base. *(Geology)* Defining an igneous rock with 40% to 50% SiO_2.

basic front In the process of granitization, an advancing front enriched in elements (Ca, Mg, Fe) removed from the parent rock being granitized. Syn. **mafic front.**

basin-and-range A region of elongated, normally faulted blocks forming alternating ranges and basins, caused by regional, elongated updoming.

batholith A large (several kilometers across or more) plutonic mass without apparent base.

bathyal Defining sediments or organisms occurring between 200 and 2000 m of depth in the ocean.

bathypelagic Defining organisms living in sea water between 200 and 2000 m of depth.

bathythermograph An instrument recording the change of water temperature with depth.

bauxite A residual rock rich in aluminum oxides and hydroxides and iron hydroxides, with clay minerals and silica, formed by deep weathering of aluminosilicate rocks under tropical conditions.

***b* axis** The horizontal crystallographic axis oriented left-to-right. Cf. ***a* axis, *c* axis.**

Bayer process A process for obtaining pure Al_2O_3 from bauxite by dissolving it in a hot (140–230°C) NaOH solution.

bayou A water recess in a swamp or a river course along the Gulf Coast.

B.C. Before Christ, referring to the time preceding January 0d 0h 0m 0s of the year A.D. 1. There is no year 0, the end of the year 1 B.C. coinciding with the beginning of the year A.D. 1.

beach The zone of unconsolidated sediment parallel to the coast, ranging from low-tide water level to the line of vegetation or to the line marking the beginning of a different physiography.

beach face The zone of a beach affected by normal wave action.

beach ridge A ridge shoreward of a beach, resulting from storm action.

beachrock A tropical, intertidal beach sand cemented by $CaCO_3$.

bead test A qualitative test for metals, performed by fusing a borax bead on a Pt wire with the powdered mineral. The color of the hot or cold bead after fusing in the oxidizing or reducing portion of the flame is indicative of the metal present. Oxidizing flame: Co, blue (hot and cold); Cr, red (hot), yellowish-green (cold); Cu, green (hot), blue (cold); Fe, red (hot), yellow (cold); Mn, violet (hot), purple (cold); Ni, violet (hot), brown (cold); U, red (hot), yellow (cold). Reducing flame (cold): Co, blue; Cr, bright green; Cu, dull red; Fe, dark green; Mn, pale rose; Ni, gray; U, yellowish-green. Cf. **blowpipe.**

bearing The angle between the geographic North direction and a specified direction, most unequivocally expressed in degrees from 0° (N) to 90° (E), 180° (S), 270° (W), and 360° or 0° (N).

beat A periodic change in amplitude caused by the interference of two waves of similar, but not identical, frequencies.

Beaufort wind scale* A wind scale used at sea and based on the effect of the wind on the sea surface.

becquerel (Bq) The SI unit of activity of a radionuclide. 1 Bq = 1 dps.

bed The smallest unit in the lithostratigraphic hierarchy.

bedding fault A fault parallel to the bedding plane of a rock sequence.

bedding plane A surface separating two sedimentary beds resting on top of each other.

bediasite A tektite from east-central Texas.

bed load The portion of the sediment load carried by a river along or just above its bed, rather than in suspension or in solution.

bedrock The solid rock underlying loose sediment or soil.

bel 1. The common logarithm of the ratio of two physical quantities (current, voltage, power, sound). 1 bel = difference by a factor of 10 between the two quantities. 2. A logarithmic measure of sound level, equal to the logarithm of the ratio of a given sound intensity to the intensity of sound at hearing threshold. The latter is taken as equal to 10^{-16} watt/cm^2 at 1000 Hz. See **decibel, sound level.**

Benioff fault plane See **Benioff zone.**

Benioff zone A seismic zone along the surface of a subducting plate.

benthic Defining a marine or freshwater organism living on the bottom.

benthonic See **benthic.**

bentonite A montmorillonitic clay formed by the alteration of volcanic ash.

benzene ring A planar hexagonal ring of 6 C atoms, numbered 1 to 6 clockwise starting from 12 o'clock. The bonds are alternately single and double, forming a resonance hybrid so that all bonds are in fact equivalent (often represented by a circle inscribed within the planar hexagon). The internal angles between the bonds are 120° and the length of the hybrid bond is $1.39 \cdot 10^{-10}$ m. Cf. **naphthalene ring.**

berg *(German)* Mountain.

Bergmann's rule The "rule" that warm-blooded animals tend to be larger in cold-weather regions than in warmer ones.

Bernouilli's equation An equation relating fluid pressure to fluid velocity and elevation along a pipe in which an incompressible, nonviscous fluid is moving.

$$p/\rho + v^2/2 + gz = k$$

where p = pressure, ρ = fluid density, v = fluid velocity, g = gravitational acceleration, z = change in height, k = constant.

berthollide A nonstoichiometric compound or mixture, such as an alloy.

Bertrand lens A removable lens in the tube of a petrographic microscope used to form an interference figure.

beryl An aluminosilicate of beryllium, $Be_3Al_2 \cdot Si_6O_{18}$.

Bessemer process The production of steel by forcing air through molten cast iron in a converter in order to oxidize impurities.

beta decay Either the beta minus decay or the beta plus decay.

beta minus decay The transformation of a neutron into a proton within an atomic nucleus, accompanied by the ejection of an electron and an antineutrino. One of the d quarks of the neutron changes into a u quark by the emission of a W^- particle (which decays into an e^- and a $\bar{\nu}$), or by the absorption of a W^+ particle derived from the interaction between an e^+ and a ν.

beta minus particle (β^-) The electron.

beta particle (β) The electron or the positron.

beta plus decay The transformation of a proton into a neutron within an atomic nucleus, accompanied by the ejection of a positron and a neutrino. One of the u quarks of the proton changes into a d quark by the emission of a W^+ particle (which decays into an e^+ and a ν), or by the absorption of a W^- particle derived from the interaction between an e^- and a ν.

beta plus particle (β^+) The positron.

betatron A toroidal vacuum chamber used to accelerate electrons by means of a variable magnetic flux normal to the plane of the torus.

BeV Billion (= 10^9) electron volts. Identical to GeV.

beveled Defining a topographic surface truncated by erosion.

B horizon The soil horizon below the A horizon and above the C horizon, characterized by the accumulation of chemical and mineral species (clays, oxides), removed from the A horizon by percolating water.

bias A dc voltage applied to the control electrode of a transistor (base) or a vacuum tube (grid) in order to establish an appropriate electrical reference level. **1. forward bias** A bias applied in the direction of conduction. **2. reverse bias** A bias applied in a direction opposite that of conduction.

biaxial Defining an orthorhombic, monoclinic,

or triclinic crystal, having 2 optical axes and 3 indices of refraction.

BIF Banded iron formation.

Big Bang The primal explosion that occurred about 16.5 billion years ago and that gave origin to the present universe. According to theory, in 3.8 minutes matter was stabilized at 74% H and 26% He. See **element formation, inflation.**

bight A shoreline gently arched and bound by two promontories.

billion 1. 10^9. Syn. **milliard** *(British, French).* 2. 10^{12} *(British).*

billitonite A tektite from the island of Belitung (Billiton), Indonesia.

binary compound A chemical compound containing two elements, each represented by one or more atoms.

binary mixture A mixture of two substances.

binary star A celestial system consisting of two stars orbiting around each other.

binary system A chemical system consisting of two different chemical components.

binding energy The energy holding an atom or a nucleus together. It is an energy which the system has lost in the process of its formation. **1. atomic binding energy** The energy released in the formation of the atom by bringing together its nucleus and electron(s). It ranges from 13.598 eV for ^{1}H to 0.69 MeV for ^{238}U. **2. nuclear binding energy** The energy released in the formation of the atomic nucleus from its individual components (protons and neutrons). It ranges from 2.22 MeV for ^{2}H to 1801.71 MeV for ^{238}U. The nuclear binding energy/nucleon increases from 1.11 MeV for ^{2}H to a maximum in the iron-nickel region (8.79 MeV for ^{56}Fe, ^{58}Fe, and ^{62}Ni) and then it slowly decreases toward the heavier elements (7.57 MeV for ^{238}U, 7.45 MeV for ^{254}Cf).

bio- Prefix meaning *life.*

biocalcarenite A calcarenite consisting of skeletal fragments of organic origin.

biocalcilutite A calcilutite consisting of skeletal elements of organic origin, usually algal needles.

biocalcirudite A calcirudite consisting of skeletal fragments of organic origin.

biocalcisiltite A calcisiltite consisting of skeletal fragments of organic origin.

biochron The time represented by a biozone.

biochronology The chronology based on the succession of organisms in a stratigraphic section.

biochronostratigraphy Chronostratigraphy based on the succession of organisms.

bioclast A fragment of organic origin.

bioclastic Defining a rock or sediment consisting of fragments of organic origin.

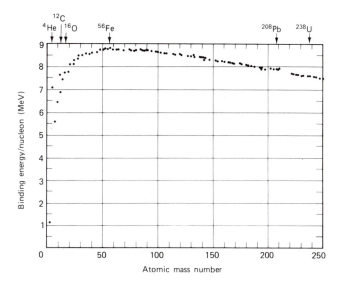

Binding energy curve showing the nuclear binding energy per nucleon (in MeV) as a function of the atomic mass number for the most abundant (or most stable) isotope of each element.

biocoenosis A community of living organisms. Cf. **thanatocoenosis.**

bioerosion Erosion produced by the action of organisms.

biofacies The environmental characteristics of a sedimentary unit as expressed by its fossil content.

biogenic Defining a sediment or deposit formed by biological activity.

biogenous See **biogenic.**

bioglyph The trace left on a sediment surface by the activity of an organism.

bioherm A bioclastic mound.

biohermal Referring to a bioherm.

biohorizon 1. A biostratigraphic surface bounding a biozone. 2. A biostratigraphic unit shorter than a biozone.

biomass The total mass of organic matter in a given environment, region, or worldwide. See **biosphere.**

biomicrite A limestone consisting of variable amounts of organic fragments and of carbonate mud recrystallized into crystals less than 4 μm across (micrite).

biomicrosparite A biomicrite in which the carbonate cement has recrystallized into crystals 5–20 μm across (microspar).

biopelite An organic mud.

bioseries An evolutionary series of fossils belonging to a single species.

biosphere The totality of organisms living on Earth. Total number of living species $\approx 1,330,000$; total number of individuals (mainly unicellular marine algae) $\approx 5 \cdot 10^{22}$; total dry biomass, $\sim 3 \cdot 10^{15}$ kg; net production $\sim 3.6 \cdot 10^{14}$ kg/y (dry).

biostratigraphic unit A sedimentary rock unit characterized by its fossil content.

biostratigraphy Stratigraphy based on fossil content.

biostrome A sediment layer predominantly consisting of the skeletal remains of organisms.

biota A singular name meaning the flora and fauna of a given region.

biotite The black mica, $K(Mg,Fe)_3(AlSi_3O_{10}) \cdot (OH)_2$.

biotope A restricted area characterized by a specific biota.

Biot-Savart law The law giving the magnetic field $d\mathbf{B}$ at a point in space as a function of a neighboring electrical current i through a conductor element $d\mathbf{l}$.

$$dB = \mu_0 i(dl \times r)/4\pi r^3$$

where μ_0 = permeability constant = $4\pi \cdot 10^{-7}$ henry/meter, r = vector distance between the application points of $d\mathbf{B}$ and $d\mathbf{l}$. See **Ampere's law.**

Biot's law A law stating that an optically active substance rotates plane-polarized light by an angle inversely proportional to the square of its wavelength and proportional to the thickness of the substance and, for solutions, also to the concentration of the solute.

bioturbation The reworking of sediment by living organisms.

biozone The smallest stratigraphic unit identifiable worldwide by its fossil content.

bipolar transistor A transistor using both electrons and holes as charge carriers. Cf. **unipolar transistor.**

bird-foot delta A delta consisting of several leveed distributaries radiating seaward.

birefringence 1. The property of crystals, other than those belonging to the isometric system, of transmitting light at different speeds in different directions, with a maximum in one direction and a minimum in another. 2. The difference between the largest and smallest index of refraction of a birefringent crystal.

bitter lake A type of salt lake with a higher concentration of sodium sulfate and a lower concentrate of chlorides than normal salt lakes.

bitumen An asphalt that has been purified of inorganic matter.

bituminous coal A coal with fixed carbon content between 78% and 86%, ranking between subbituminous coal below and anthracite above. Energy content = 6800–8200 cal/g.

bituminous limestone A compact, dark limestone rich in organic matter.

bivalent Defining an element with 2 or 6 valence electrons.

bivalve 1. Defining a shell consisting of two valves. 2. A pelecypod, i.e. a member of the class Bivalvia.

black body A surface capable of absorbing all incident radiation of any wavelength. It is a perfect absorber and also a perfect emitter.

blackbody emission See **blackbody radiation.**

blackbody radiation The radiation emitted by a black body at a given temperature. It has a continuous spectrum with maximum power radiated at $\lambda = 2897.8/T$ μm (cf. **Wien's displacement law**). Blackbody radiancy as a function of wavelength is given by Planck's radiancy law.

black hole A gravitationally collapsed object closed by an event horizon. Black holes are formed in the process of supernova explosion of supermassive stars (>3 solar masses).

black light Ultraviolet light used to produce fluorescence in minerals and other substances.

black shale A shale rich ($>5\%$) in organic matter.

blast furnace A furnace fed with forced air.

-blastic Suffix meaning *bud, germination.*

blasto- Prefix meaning *bud, germination.*

B layer The seismic zone extending from the Mohorovičić discontinuity to a depth of 400 km.

blazer A violently variable quasar or BL Lacertae object.

blende Sphalerite or any of the other metallic sulfides exhibiting resinous luster, such as

antimony blende (kermesite, Sb_2S_2O)
cadmium blende (greenockite, CdS)
hornblende $[Ca,Na)_{2-3}(Mg,Fe,Al)_5(Si,Al)_8O_{22} \cdot (OH)_2]$
pitchblende (massive UO_2).

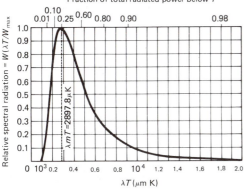

Fraction of total radiated power below T

Blackbody radiation. Relative blackbody radiation as a function of wavelength λ times absolute temperature T. The radiation peak corresponds to $\lambda T = 2897.8$ (Wien's law). (From Eisner 1985, p. 1018, Fig. 2)

BL Lacertae object (BL Lac object) A compact, violently variable (up to a factor of 100 in a few months) quasar-like object lacking emission or absorption spectral lines. Redshift of surrounding gases suggests distances comparable to the nearer quasars.

block faulting Crustal fracturing into elongated blocks bound by normal faults.

blood rain Rain containing desert dust rich in iron oxides.

bloodstone A dark green variety of chalcedony speckled with red jasper.

bloom A sudden flourishing of planktonic algae or other microorganisms resulting from a combination of favorable ecological conditions and an abundant supply of nutrients.

blowhole A vertical hole or fissure extending from the roof of a coastal, partly submerged cave to the ground above. Incoming waves force spray up the duct and off the opening at the top.

blowpipe A tube used to blow air through the flame of a Bunsen burner on a mineral. The color of the flame is indicative of certain elements when present in the specimen (carmine red, Li; purple-red, Sr; orange-red, Ca; yellow, Na; yellowish-green, Ba, Mo; pale green, Te; vivid green, B; emerald green, CuO; bluish-green, phosphates; whitish-blue, As; greenish-blue, Sb; azure-blue, Se, $CuCl_2$; blue, Pb; violet, K). Cf. **bead test.**

blue asbestos Crocidolite, $NaFe_3^{2+}Fe_2^{3+}Si_8O_{22} \cdot (OH)_2.$

blue-green algae Syn. **Cyanobacteria.**

blue hole A submarine sinkhole formed during a lower stand of sea level.

blue ice Air-free, compact glacier ice. Cf. **white ice.**

blue mud A hemipelagic mud rich in organic matter and Fe sulfides.

blue noise Noise in which the amplitudes of the higher frequencies are higher than those of the other frequencies. Cf. **red noise, white noise.**

blueschist A low-temperature (200–350°C), high-pressure (6–10 kbar) metamorphic phase of gabbro-basalt common along converging plate margins. Characteristic minerals are aragonite, epidote, garnet, glaucophane (which imparts the blue color), jadeite, and lawsonite. Cf. **greenschist.**

blueschist facies The set of minerals characteristic of blue schists.

blue shift The apparent decrease in wavelength of light received by an observer when the distance between source and observer decreases. A special case of the Doppler shift. See **Doppler shift.**

bluff A high bank fronting a plain, a river, or a lake.

Bode's law See **Titius-Bode's law.**

body wave A seismic wave (P or S) traveling through a body, as opposed to a seismic wave traveling along a surface or an interface.

bog Waterlogged ground covered with living and decaying organic matter.

boghead coal A sapropelic coal consisting mainly of algal matter.

bog iron ore A limonitic deposit formed in bogs, marshes, or swamps rich in organic matter. It is produced by the oxidation of ferrous oxide in solution by photosynthetic algal oxygen.

Bohr magneton (μ_B) A unit of magnetic moment.
$$\mu_B = eh/4\pi m_e$$
where e = electron charge, h = Planck's constant, m_e = electron mass, c = speed of light. It is equal to $0.9274015 \cdot 10^{-23}$ A m^2 or J/T. See **magneton.**

Bohr radius (α_0) The radius of the H atom in the ground state.
$$\alpha_0 = h^2/4\pi^2 m_e e^2$$
$$= 0.52917725 \cdot 10^{-10} \text{ m}$$
where h = Planck's constant, m_e = electron mass, e = electron charge.

boiling point (bp) The temperature at which a substance changes from the liquid to the gaseous phase at a fixed pressure.

boiling-water reactor A nuclear reactor using water as a coolant. Water pressure is maintained at such a level as to allow water to boil and produce steam.

bolometer An instrument to measure the energy of electromagnetic radiation at specific wavelengths by its heating effect on a thin (2 μm), blackened Pt strip, measured by comparison with a similar strip shielded from the radiation.

bolometric correction Addition, to the visible spectrum, of the energy received in the nonvisible range, in order to obtain the bolometric magnitude.

bolometric magnitude Magnitude of a star including the nonvisible component of the spectrum. See **magnitude.**

Boltzmann constant (k) Two-thirds of the energy required to increase the temperature of a particle in an ideal gas by 1 K.
$$k = R/N_A$$
where R = gas constant, N_A = Avogadro number. It is equal to $1.38066 \cdot 10^{-23}$ J/K.

Boltzmann distribution A distribution law specifying the probable number n_i of particles at energy level E_i in a system of particles in thermal equilibrium at temperature T.
$$n_i = 1/e^{(E_i - \mu)/kT}$$
where μ = chemical potential; k = Boltzmann constant; T = absolute temperature. Cf. **Bose distribution, Fermi distribution.**

bomb (*Physics, Chemistry, Geology*) A container for high-pressure experiments. (*Geology*) A mass of lava ejected into the atmosphere and solidified in flight into a rounded or ellipsoidal shape >64 mm across. Cf. **lapilli.**

bond The force that binds atoms in molecules and crystals. **1. ionic bond** Formed by electron transfer from one atom to another resulting in oppositely charged ions that attract each other. Energies: ~ 3–8 eV. **2. covalent bond** Formed by atoms sharing a pair of electrons of equivalent energy and opposite spin. Energies: ~ 3–8 eV. **3. metallic bond** Formed by valence electrons released by the atoms of a metal and moving freely through the metal lattice. Energies: ~ 1–3 eV. **4. hydrogen bond** Formed by a hydrogen atom, already part of a molecule, with a pair of unshared electrons on an electronegative atom in the same molecule or belonging to an adjacent molecule. Energies: ~ 0.2–0.5 eV. **5. van der Waals bonds** Formed by interaction between instantaneous dipoles when adjacent atoms or molecules distort their charge distribution and induce opposite dipole moments on each other. Energies: ~ 0.02–0.2 eV.

bond energy 1. Single bond energy The energy released in the formation of a chemical bond. **2. Total bond energy** The energy released in the formation of a molecule from its constituent atoms. The heat of formation of the molecule. In diatomic molecules it ranges from 0.05–0.1 eV for noble gases to 11.08 eV for CO.

bonding molecular orbital A molecular orbital with high electron density between the two nuclei.

bond length The distance between the nuclei of two chemically bonded atoms. Bond length averages 1–2.5 Å, with a range from 0.75 Å (H$-$H) to 4.95 Å (Rb$-$Rb).

boomer A strong, intermittent underwater sound source used for seismic work at sea.

borax A hydrated salt of sodium and boron, $Na_2B_4O_5(OH)_4 \cdot H_2O$.

bore A front of rising water resulting from collision between two tidal fronts or from the rushing of the tide upstream within a narrowing estuary or river.

boreal Referring to the northern latitudes. Cf. **austral.**

Born-Haber cycle A cycle to determine the lattice energy of crystals by algebraically adding the enthalpies of sublimation, ionization, dissociation, electron affinity, and crystallization.

Bosanquet's law A law relating magnetic flux Φ, magnetomotive force *mmf,* and reluctance $\mathfrak{R}$.

$$\Phi = mmf/\mathfrak{R}$$

It is the magnetic equivalent of Ohm's law.

Bose distribution A distribution law specifying the probable number n_i of bosons at energy level E_i in a system of bosons in thermal equilibrium at temperature T.

$$n_i = 1/(e^{(E_i - \mu)/kT} - 1)$$

μ = chemical potential; k = Boltzmann constant; T = absolute temperature. Cf. **Boltzmann distribution, Fermi distribution.**

Bose-Einstein statistics The statistics of an assemblage of identical bosons.

boson Any particle with spin angular momentum equal to an integer (or zero) times $h/2\pi$ (where h = Planck's constant). Included are the photon, the mesons, and all nuclei with even number of nucleons.

botryoidal Having a shape or surface resembling that of a bunch of grapes *(botrus).*

bottomset bed Any of the subhorizontal sediment beds upon which a foreset bed advances in deltaic sedimentation.

bottom water The deepest water layer in the ocean.

boudinage The structure of a sedimentary bed that has been squeezed between the overlying and underlying beds so as to be broken into discoidal portions that in cross section (as in exposures) have the appearance of a string of sausages (from *boudin,* French for *sausage).*

Bouguer anomaly The gravity anomaly remaining after corrections for latitude, elevation, and topography.

Bouguer correction The correction of 0.04185 ρh mgal/m at a station at elevation $\pm h$ (in meters) with respect to a reference level in order to compensate for the rock of density ρ (g/cm^3) present above the reference level or absent below it.

boulder A rock fragment more than 256 mm across. See **Wentworth grade scale.**

boulder clay See **till.**

boulder train A line of glacial boulders and pebbles extending downdrift from a rocky outcrop.

boundary current A deep ocean current flowing along a submarine slope bounding an ocean.

Bowen's reaction series A double series of minerals forming in order of decreasing temperature of crystallization from the same subalkaline melt as the melt cools. The two high-temperature end members are olivine and Ca-plagioclase. The common low-temperature member is quartz.

Boyle's law "The product of pressure times volume in a gas at constant temperature is a constant."

$$pV = k$$

where p = pressure, V = volume, k = constant. Because pV has the dimensions of energy ($p = F/l^2$ and $V = l^3$, therefore $pV = F \times l$ = energy, where F = force, l = length), Boyle's law establishes the principle of conservation of energy. Cf. **Charles' law.**

B.P. Before the Present, meaning *before* A.D. *1950,* the datum to which ^{14}C dates are referred.

Brackett series. The series of lines in the infrared region of the hydrogen spectrum, produced by transitions from energy levels $n > 4$ to $n = 4$ (emission lines) or from $n = 4$ to $n > 4$ (absorption lines), where n is the principal quantum number. Energies range from 0.3060 to 0.8501 eV; corresponding wavelengths range from 4.0512 μm to 1.4584 μm.

bradyseism A slow, continued vertical (up and/ or down) movement at a rate of several mm/y.

Bragg angle (ϑ) The angle ϑ in the Bragg equation.

Bragg equation The equation

$$n\lambda = 2d \sin \vartheta$$

where n = any integer, λ = x-ray wavelength, d

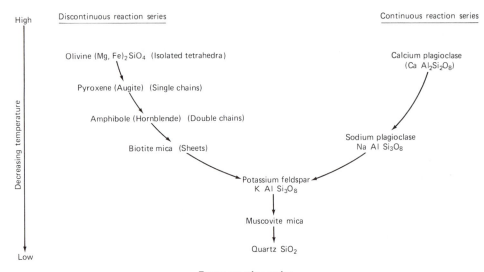

Bowen reaction series.

= crystal plane spacing; ϑ = angle between crystal plane and diffracted x-ray beam.

Bragg's law See **Bragg equation.**

braided stream A stream that subdivides its course into a series of anastomosing channels, common in flood plains.

brain coral Any of the coral species belonging to the genus Diploria.

branching The possibility for a radioactive nucleus to decay by two or more different processes. E.g. ^{40}K by β^- to ^{40}Ca (89.30%), and by K capture to ^{40}Ar (10.70%).

branching fault A fault that splits into two or more faults along its course.

branching fraction The percent fraction of the nuclei of a radioactive isotope decaying by a given branching process.

branching ratio The ratio or ratios to each other of two or more daughter products of a radioactive nucleus undergoing radioactive decay by different processes that have different probabilities.

brass An alloy of Cu and Zn. Yellow brass, 2/3 Cu, 1/3 Zn; red brass, 90% Cu, 10% Zn.

Bravais lattice Any of the 14 possible space lattices.

breccia A rock consisting of coarse, angular fragments in a fine matrix or cement.

breeder reactor A nuclear reactor that produces more fissionable material than it consumes. Free neutrons from the fission of ^{235}U in the reactor (approximately 2.5 neutrons/fission) are captured by ^{238}U to form ^{239}U which decays (β^-, $t_{1/2} = 23.50$ min.) to ^{239}Np which in turn decays (β^-, $t_{1/2} = 2.355$ days) to ^{239}Pu (α, $t_{1/2} = 24,120$ y). ^{239}Pu is fissionable, releasing approximately 2.5 free neutrons/fission.

breeding ratio The ratio of fissionable nuclides produced to fissionable nuclides consumed in a breeder reactor.

bremsstrahlung Radiation emitted by an electron when colliding with an atomic nucleus.

Brewster angle The angle i (measured from the normal to the surface) at which unpolarized light incident on a surface acquires maximum plane polarization.

$$\tan i = n$$

where n = refractive index of the substance forming the surface.

brig See **dex.**

bright field The image formed in transmission electron microscopy by blocking the diffracted beams.

brightness The intensity of light or other radiation. **1. apparent brightness** The intensity of light or other radiation received from a celestial body. **2. intrinsic brightness** The intensity of light or other radiation emitted by a celestial body. Syn. **luminosity.**

Brillouin zone The unit cell of the reciprocal lattice. See **reciprocal lattice.**

brimstone Vernacular name for sulfur.

British system of units A nonmetric system of units based on the foot (= 0.3048 m, exactly), the second, and the pound-mass (= 0.45359237 kg, exactly).

British thermal unit (BTU) 1. A nonmetric thermal unit defined as the amount of heat needed to raise the temperature of 1 lb of air-free water from 60°F to 61°F at the constant pressure of 1 atm. Symbol: $BTU_{60/61}$. It is equal to 1054.68 J = 251.906 cal_{IT}. 2. A nonmetric thermal unit equal to 1/180 of the heat needed to raise the temperature of 1 lb of air-free water from 32°F (= 0°C) to 212°F (= 100°C) at the constant pressure of 1 atm. It is equal to 1055.87 J = 252.190 cal_{IT}. Symbol: BTU_{mean}. 3. A nonmetric thermal unit equal to 1055.05585262 J exactly = 251.99575111 cal_{IT}. Symbol: BTU_{IT}.

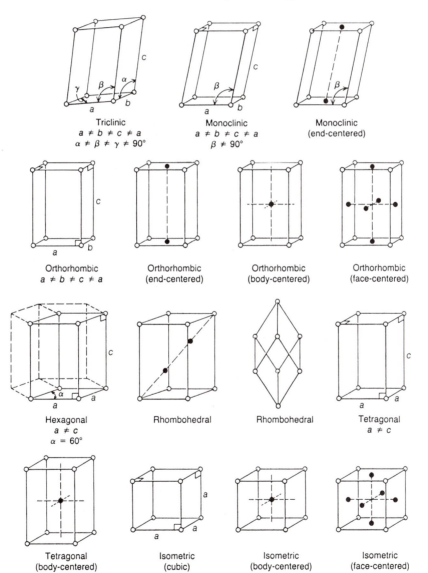

Bravais lattices. The 14 basic types. (Berry, Mason, and Dietrich 1983, p. 18, Fig. 2.7)

Brønsted acid See **acid.**

Brønsted base See **base.**

bronze An alloy of Cu (70–90%) and Sn (10–30%).

bronzite (Mg,Fe)SiO$_3$, an orthopyroxene of composition intermediate between enstatite and hypersthene.

brook A streamlet in a mountain region, smaller than a creek.

brown algae The algae belonging to the phylum Phaeophyta.

Brownian motion Random motion of small particles suspended in a fluid, caused by statistical fluctuations in the net momentum exchange between the molecules of the fluid and the suspended particles.

Brunhes The present epoch of normal magnetic polarity, which began 730,000 y ago.

Brunton compass A compass with sights, a mirror, and a spirit-level clinometer for determining orientations and for measuring horizontal and vertical angles.

Bryophyta A grade of the Kingdom Plantae, characterized by the gametophyte being the major plant body and by the absence of vascular tissue or roots. It includes liverworts (Hepaticae) and mosses (Musci). Cf. **Tracheophyta.**

bubble chamber A chamber containing a superheated fluid in which bubbles are formed by moving charged particles, which thus reveal their trajectories.

bubnoff A velocity unit, equal to 1 mm/1000 y. It is used for tectonic movements and deep-sea sedimentation rates.

bulk modulus See **bulk modulus of elasticity.**

bulk modulus of elasticity The ratio of the change in pressure to the corresponding fractional change in volume of a body.

$$B = -\Delta p/(\Delta V/V)$$

where B = bulk modulus of elasticity, p = pressure, V = volume.

Bullard discontinuity The 450 km-thick seismic velocity transition zone between inner and outer core, extending from 4720 to 5170 km of depth below the Earth's surface. Syn. **F layer.**

bushveld An open, grassy plain with scattered trees in tropical or subtropical areas. Syn. **savanna.**

butte A flat-layered, steep-sided, isolated sedimentary elevation capped with a flat layer of hard rock more resistant to erosion than the underlying layers. The eroded remnant of a mesa.

bytownite A plagioclase of composition Ab$_{30}$An$_{70}$–Ab$_{10}$An$_{90}$.

C

c 1. Curie. 2. Specific heat. 3. Speed of light. 4. Speed of sound.

C 1. Capacitance. 2. Celsius or centigrade. 3. Coulomb. 4. Heat capacity.

C_p Heat capacity at constant pressure.

C_v Heat capacity at constant volume.

C-14 dating See **carbon-14 dating.**

cadmium cell See **Weston cell.**

calc-alkaline Referring to a suite of igneous rocks including the intrusive gabbro-diorite-granodiorite-granite suite and the effusive basalt-andesite-dacite-rhyolite suite, characterized by Ca-rich clinopyroxene, Ca-poor orthopyroxene, hornblende, biotite, feldspars, and quartz. See **alkali-lime index.**

calcarenite An arenite consisting of more than 50% of carbonate particles.

calcareous Defining a rock, sediment, or skeletal part containing a significant amount of $CaCO_3$.

calcic See **alkali-lime index.**

calcilutite A lutite consisting of more than 50% of carbonate particles.

calcimicrite A limestone consisting of carbonate particles smaller than 20 μm with more than 50% micrite.

calcipelite Syn. **calcilutite.**

calcirudite A rudite consisting of more than 50% of carbonate particles.

calcisiltite A siltite consisting of more than 50% of carbonate particles.

calcite The common, low-pressure polymorph of crystalline $CaCO_3$. It crystallizes in the rhombohedral system. Density = 2.71; hardness = 3; refractive index = 1.66. Cf. **aragonite.**

calcium carbonate compensation depth See **compensation depth.**

calcsparite A sparry calcite crystal.

calculus Calculus is that branch of mathematics dealing with the interrelationships between or among continuously changing quantities when the interrelationships are nonlinear and therefore continuously changing. See also **infinitesimal. 1. differential calculus** Differential calculus is concerned with the instantaneous rate of change of a function y with respect to the change of its variable x, called *derivative* of y with respect to x and written dy/dx. It may be a constant or it may be itself a function of the variable and therefore itself variable. For instance, the circumference C of a circle is a function of its radius r ($C = 2\pi r$) and the derivative $dC/dr = 2\pi$ is a constant. On the other hand, the surface A of a circle is also a function of its radius r ($A = \pi r^2$) but the derivative $dA/dr = 2\pi r$ is itself a function of r and is, therefore, variable. Successive differentiation leads to *higher derivatives.* With respect to time t, for instance, velocity v is the first derivative of position S ($v = dS/dt$); acceleration a is the first derivative of v and the second derivative of S ($a = dv/dt = d^2S/dt^2$); and jerk j is the first derivative of a, the second derivative of v, and the third derivative of S ($j = da/dt = d^2v/dv^2 = d^3S/dt^3$). A quantity may be a function of two or more mutually independent variables. If so, there is an independent *partial derivative* (symbol ∂), between function and each one of the variables. The area A of an ellipse, for instance, is given by πxy, where x = semimajor axis, y = semiminor axis. Therefore, $\partial A/\partial x = \pi y$, $\partial A/\partial y = \pi x$. See **derivative. 2. integral calculus** Given the derivative of a function, integral calculus strives to reconstruct the original relationship (called *integral;* symbol $\int$) from previous knowledge obtained via differentiation. Thus, given $dA/dr = 2\pi r$, we have $A = \int 2\pi r\, dr = \pi r^2 + C$ where C, called *constant of integration,* is an indeterminate constant that may have any value. A specific value, however, usually suggests itself as applicable to the specific problem at hand. A quantity may be obtained from n integrations of its nth derivative. In the series S, v, a, j mentioned above, for instance, $S = \int v\, dt = \iint a\, dt^2 = \iiint j\, dt^3$ (neglecting the constants of integration).

caldera A large, circular, bowl-shaped depression produced by a volcanic explosion or by the collapse of a magmatic chamber. Cf. **collapse caldera.**

caliche A soil layer cemented by carbonate incrustations deposited in arid regions by evaporating groundwater solutions.

Callisto See **Jupiter.**

calomel electrode A reference electrode consisting of Hg and Hg_2Cl_2 (calomel) in contact with a solution of KCl.

calorie (cal) A unit of heat energy. **1. food calorie** 1 "calorie" = 1000 g-cal. **2. gram-calorie (g-cal)** The amount of heat required to raise the temperature of 1 g of water from 14.5°C to 15.5°C at a constant pressure of 1 atmosphere (101,325 Pa). It is equal to 4.1855 J. **3. IT calorie (cal_{IT})** The International Table calorie. It is equal to 4.1868 J (exactly). **4. thermochemical calorie** A unit of heat energy equal to 4.184 joules (exactly).

calving The breaking away of ice masses from the front of a glacier or ice shelf.

canada balsam A balsam exuded by incisions in the bark of *Abies balsamica* of Canada and Maine.

cancellate Having a net-like structure.

candela (cd) The SI unit of luminous intensity. It is equal to the luminous intensity of 1/683 watt/steradian emitted by a monochromatic source radiating at the frequency of $540 \cdot 10^{12}$ hertz.

candle Obsolete name for **candela.**

cannel coal A sapropelic coal consisting predominantly of spores.

canonical Defining the simple, standard, most common, or most significant form of a general equation or function.

canonically conjugated variables A generalized position coordinate and its corresponding momentum coordinate. The product of canonically conjugated variables has the dimension of action (energy × time = mass × length squared divided by time) and is invariant under Lorentz transformations. See **generalized coordinates.**

canyon A narrow, steep-sided river valley characteristic of arid regions.

capacitance (C) The ratio of the charge on one of a capacitor's plates to the potential difference between the two plates.

$$C = q/V$$

where q = charge, V = voltage. It is measured in farads.

capacitor An electric circuit element capable of temporarily storing electric charge. It consists of two metal sheets separated by a dielectric.

capacity The ability of running water or wind to transport sediment in terms of mass of sediment per unit volume of the transporting agent. Cf. **competence.**

cape A major projection of land into the sea.

capillary A tube having a sufficiently small internal diameter so as to exhibit capillarity.

capillarity The rising or lowering of a fluid surface inside a capillary when the end of the capillary is dipped into the fluid. The fluid rises inside the capillary with respect to the external free surface, forming a concave meniscus, if the attraction between the molecules of the fluid and the wall of the capillary is greater than the mutual attraction of the fluid's molecules; it is lowered, forming a convex meniscus, in the opposite case.

capillary wave A very small ($\lambda < 1.7$ cm) wave on a liquid surface in which the restoring force is surface tension rather than gravity.

Ca-plagioclase A plagioclase with $>50\%$ anorthite. See **plagioclase.**

capture cross-section (σ) The effective cross section of an atomic nucleus or a particle as regards the capture of neutrons or other particles. It is measured in barns. See **neutron-capture cross-section.**

carat (ct) A unit of weight for precious stones, equal to 200 mg. Cf. **karat.**

carbohydrates Organic compounds of general formula $C_x(H_2O)_y$. E.g. glucose (monosaccharide), $C_6H_{12}O_6$; sucrose (disaccharide), $C_{12}H_{22}O_{11}$; cellulose (a polysaccharide), $(C_6H_{10}O_5)_{2000-4000}$.

carbonaceous chondrite A primitive type of stony meteorite consisting of olivine, pyroxene, and plagioclase chondrules in a matrix of low-temperature phyllosilicates. Composition: 30–60% low-temperature minerals (hydrated silicates, Mg sulfate); up to 3% carbon; and up to 9% water. Only 5.7% of all meteorites are carbonaceous chondrites. See **chondrite, Meteorites*.**

carbonado A cryptocrystalline diamond aggregate.

carbonate compensation depth (CCD) See **compensation depth.**

carbonate ion The CO_3^{2-} ion.

carbonatite A magmatic rock consisting of Ca, Mg, and Na carbonates with secondary feldspar, pyroxene, olivine, and other minerals.

carbon cycle A set of 6 successive nuclear reactions involving C, N, and O and resulting in the synthesis of one ^{4}He nucleus from four ^{1}H nuclei:

$$^{12}\text{C} + {}^1\text{H} \rightarrow {}^{13}\text{N} + \gamma$$
$$^{13}\text{N} \rightarrow {}^{13}\text{C} + e^+ + \nu$$
$$^{13}\text{C} + {}^1\text{H} \rightarrow {}^{14}\text{N} + \gamma$$
$$^{14}\text{N} + {}^1\text{H} \rightarrow {}^{15}\text{O} + \gamma$$
$$^{15}\text{O} \rightarrow {}^{15}\text{N} + e^+ + \nu$$
$$^{15}\text{N} + {}^1\text{H} \rightarrow {}^{12}\text{C} + {}^4\text{He}$$

Energy produced is 25.024 MeV per nucleus of ^{4}He synthesized or $5.98 \cdot 10^{11}$ joules ($= 0.14$ kilotons) per gram of ^{1}H consumed. The carbon cycle occurs in the cores of main-sequence stars and, together with the proton-proton chain, is responsible for the production of thermonuclear energy by these stars. Syn. **carbon-nitrogen cycle, C-N-O cycle.** See **proton-proton chain, Stars—energy production*.**

carbon-14 dating method A radiometric dating method based on the decay of ^{14}C (by β^- to ^{14}N, $t_{1/2} = 5730$ y) in C-containing matter removed from exchange with the mobile carbon reservoir (hydro-atmo-bio-cryo-pedo-sphere). ^{14}C is continuously formed in the tropopause (15–20 km of altitude) by the ^{14}N(n,p)^{14}C reaction. The neutrons required for this reaction are produced by spallation of atomic nuclei of atmospheric gases by energetic (>100 MeV) galactic protons. ^{14}C is rapidly oxidized to CO_2 which becomes part of the mobile carbon reservoir ($5.4 \cdot 10^4$ kg of ^{14}C/$1.5 \cdot 10^{17}$ kg of C in reservoir, equal to a ^{14}C concentration of $3.6 \cdot 10^{-13}$). The activity of ^{14}C in equilibrium with the atmosphere is 13.56 ± 0.07 dpm/g of carbon. The concentration of ^{14}C in a sample is determined by oxidizing the total C in the sample to CO_2 and measuring its radioactivity in a suitably modified and shielded Geiger-Müller counter system, or by

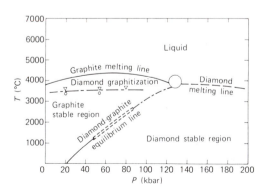

Carbon. Phase diagram. (Jayaraman and Cohen 1970, p. 268, Fig. 9)

converting it to benzene (C_6H_6) and measuring its radioactivity by liquid scintillation counting. Age range of the ^{14}C dating method: 200–40,000 y.

carbonic acid The weak H_2CO_3 acid formed by the solution of CO_2 in water.

carbon-nitrogen cycle See **carbon cycle.**

carbon star Any of the class of low surface temperature (2500–5500 K) red giant stars with a high ratio of C to H and O. The ^{13}C/^{12}C ratio is also high and may range up to 0.25.

carbonyl The group $=C=O$.

Carborundum Trade name for SiC abrasive.

carboxyl The group $-COOH$.

Carnot cycle A hypothetical cycle of four successive reversible processes involving a substance. It consists of an isothermal expansion, requiring addition of heat; an adiabatic expansion; an isothermal compression, releasing heat; and an adiabatic compression, which raises the temperature of the substance to the initial value. The net heat balance would equal the work done in the cycle if no energy were dissipated in friction and turbulence (reversible cycle).

Carnot's theorem "Given two different temperatures, reversible engines operating between them have identical efficiencies while nonreversible engines have lower efficiencies."

carotene Any of a group of highly unsaturated hydrocarbon chains ($C_{40}H_{56}$) with a ring structure at each end. The more unsaturated, the redder the color. Carotenes are widely distributed in plants (in association with chlorophyll and participating in photosynthesis) and in animals.

carpolith A fossil seed or fruit.

carst See **karst.**

Cartesian coordinate system 1. A system of two, usually perpendicular axes on a plane to identify the location of a point on the plane by its shortest distance from the two axes. 2. A system of usually mutually perpendicular axes in space to identify the position of a point in space by its shortest distance from the three planes formed by each pair of axes. Cf. **polar coordinate system.**

casing A heavy, cylindrical metal sleeve lowered into an oil well following drilling to prevent wall cavings, seal out water, and conserve drilling mud.

cassiterite The mineral SnO_2.

cast The reproduction of the surface of a body by

injection of a solidifiable fluid into a mold of that surface.

cast iron A brittle alloy of iron and carbon with 2–4.5% C, 0.5–3% Si, and smaller quantities of S, Mn, and P.

cata- Prefix meaning *downward.*

catabatic Moving downward. Cf. **anabatic.**

catabolism The phase of metabolism involving the breakdown of complex organic molecules into simpler ones. Cf. **anabolism, metabolism.**

cataclastic 1. Defining a clastic rock whose fragments are the product of fragmentation of a preexisting rock. 2. Defining a type of localized metamorphism along a compressional fault or plane.

cataglacial The climatic phase leading from a glacial to an interglacial age. Cf. **anaglacial, anathermal.**

catalysis The action by which specific elements or compounds expedite chemical reactions among other substances by reducing the activation energy.

catalyst A substance capable of performing catalysis. *Homogeneous catalysts* are present in the same phase as the reactants; *heterogeneous catalysts* are present in a different phase. Catalysts are not consumed by catalysis and the quantity needed to expedite a reaction is much smaller than the quantity of the reactants. Catalysts, however, may become "poisoned" with time by impurities and lose their properties. Common catalysts are metals (often pulverized to increase the surface), various inorganic compounds, and a large number of proteins (enzymes).

cataract 1. A series of river rapids. 2. A large waterfall.

catastrophism The theory holding that natural processes proceed by short, intense episodes of activity separated by long periods of stasis.

catathermal The climatic phase leading from an interglacial to a glacial age. Cf. **anaglacial, anathermal.**

cathode 1. The negative electrode of an electrolytic cell. 2. The electron-emitting electrode of an electron tube or cathode-ray tube.

cathode ray A stream of electrons emitted by a cathode.

cathode-ray tube (CRT) An electron tube consisting of a hot, electron-emitting cathode, a set of focusing anodes and plates, and a fluorescent screen.

cation A positively charged ion, i.e. an atom having fewer electrons than the protons in its nucleus. It moves toward the cathode in an electrolytic cell. Cf. **anion.**

cation exchange The replacement of a cation bound to a surface by a different cation in solution.

Ca-Tschermak molecule See **Tschermak molecule.**

Cavendish balance A device to determine the gravitational constant G by measuring the torque exerted by two large masses on two much smaller masses at the ends of a thin bar suspended from a torsion fiber. First used by Lord Cavendish in 1798.

cavitation The sudden formation and collapse of vapor- or gas- filled cavities in a turbulent liquid, due to a sudden reduction in local pressure caused by the turbulence.

cavity radiator A cavity in a metal that radiates energy when the metal is heated. Output radiancy R through opening:

$$R = \sigma T^4$$

where σ = Stefan-Boltzmann constant, T = absolute temperature.

cavity resonator A cavity of appropriate dimensions and constructed of appropriate materials capable of storing electromagnetic or acoustic energy by resonating in response to appropriate frequencies.

c axis The crystallographic axis oriented vertically. Cf. **a axis, b axis.**

cay A small, low island consisting of coral reef and other skeletal remains.

cc Cubic centimeter.

CCD Carbonate compensation depth.

cd Candela.

celestial equator The projection of the terrestrial equator from the center of the Earth onto the celestial sphere. See **coordinate system.**

celestial meridian The great circle on the celestial sphere passing through the celestial poles and the observer's zenith. See **coordinate system.**

celestial poles The projection of the terrestrial poles from the center of the Earth onto the celestial sphere. See **coordinate system.**

celestial sphere Imaginary sphere encompassing

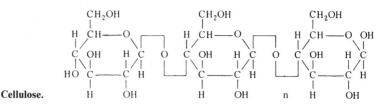

Cellulose.

the entire universe, with center at the center of the Earth. See **coordinate system.**

cellulose A polysaccharide, the main constituent of plant tissue. $(C_6H_{10}O_5)_{2000-4000}$; mol. mass = $(162.142)_{2000-4000}$.

Celsius (C) 1. Defining the temperature scale based on the equation

$$T(°C) = T(K) - 273.15$$

where $T(°C)$ = temperature in degrees Celsius, and $T(K)$ = temperature in kelvins. There are 100 degrees Celsius between the freezing point (273.15 K) and boiling point (373.15 K) of water at the pressure of 1 atmosphere. 2. The degree Celsius, a unit of temperature interval equal to 1 kelvin.

cement A gray powder obtained by grinding the vitrified product of a mixture of limestone (about 70%) and clay (about 30%) heated to about 1500°C. Average composition: 64% CaO, 22% SiO_2, 6% Al_2O_3, 3% MgO, 3% Fe_2O_3, 2% SO_3. About 5% of $CaSO_4$ is added during grinding to control strength, setting time, and other properties.

cenosis See **coenosis.**

cenote A cylindrical opening in carbonate terrain originating as a sinkhole or resulting from the collapse of the roof of a cavern. A cenote has its floor below the local water table.

Cenozoic The geological era following the Mesozoic. It ranges from $65 \cdot 10^6$ y B.P. to the present and is subdivided into the following periods (age of boundaries in million years): 65/Paleocene/ 54.9/Eocene/38/Oligocene/24.6/Miocene/5.1/ Pliocene/1.6/Pleistocene/0.01/Holocene/0. See **Geological time scale*.**

centi- Prefix meaning *1/100th.*

centigrade Syn. **Celsius.**

centimeter (cm) The CGS unit of length, equal to 1/100th of a meter.

centipoise (cP) A unit of dynamic viscosity equal to 0.01 poise.

centistoke (cs) A unit of kinematic viscosity equal to 0.01 stoke.

centrifugal force The force felt by an inertial mass within a rotating frame of reference.

centripetal force The radial force needed to maintain on track a mass moving with curvilinear motion. See **acceleration, circular motion.**

Cepheid Any of the stars of the δ Cephei type, exhibiting rhythmic change in luminosity with a period of 1 to 100 days. The cause of pulsation is ionization of H and He gases in the star's photosphere, leading to expansion, and deionization, leading to contraction. The change in radius is <15%.

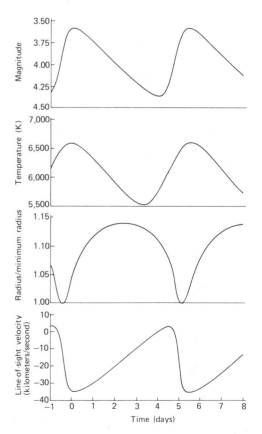

Cepheids. Periodic variations in δ Cephei. (Bowers and Deeming 1984, p. 220, Fig. 11.1)

Cerenkov angle *(Optics)* The angle between the direction of movement of a charged particle traveling through a medium at a speed higher than the speed of light in that medium and the normal to the resulting wavefront in the medium. It is defined by the equation

$$\cos \theta = v/v_s$$
$$= c/nv_s$$
$$= 1/n\beta$$

where v = velocity of light in medium, v_s = velocity of particle in medium, c = velocity of light in vacuo, n = refractive index of medium, $\beta = v_s/c$. *(Sound)* The angle θ between the direction of motion of a source traveling at supersonic speed and the normal to the resulting wavefront in the medium. It is defined by the equation

$$\cos \theta = v/v_s$$
$$= 1/M$$

where v = velocity of sound in medium, v_s = velocity of source, M = Mach number.

Cerenkov radiation Light emitted by a charged particle traveling through a transparent dielectric medium at a speed greater than the speed of light in that medium.

CERN Conseil Européen Recherches Nucléaires.

cerro A craggy, stony hill in the Northamerican desert.

Cf., cf. *Confer,* Latin for *compare with.*

CGS Centimeter-gram-second, a system of units in which the centimeter, the gram, and the second are the fundamental units.

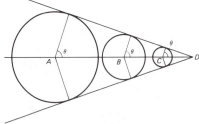

Cerenkov angle. As the particle or body moves from A to D, a wave is generated at each point. If the speed of the particle or body is greater than that of the wave in the medium, the spherical wave fronts generated will form a conic envelope. The Cerenkov angle θ is the angle formed by the normal to the surface of the cone with the axis of the cone. The greater the speed of the particle or body, the smaller the angle will be.

CGS$_{emu}$ See **electromagnetic system of units.**

CGS$_{esu}$ See **electrostatic system of units.**

chain reaction Any self-sustaining nuclear or chemical reaction.

chalcedony Semitranslucent microcrystalline quartz.

chalcophyle Defining an element concentrated in the sulfide phase rather than in the metallic or silicate phase.

chalcopyrite The mineral CuFeS$_2$.

chalcosphere A hypothetical shell at the surface of the Earth's outer core, claimed to be rich in sulfides. Syn. **sulfide layer.**

chalk A fine-grained, poorly cemented calcitic limestone principally consisting of coccoliths and foraminiferal shells.

Chandler wobble The wobble of the Earth's rotational axis around its mean position, with periods of 12 and 14 months. Departure of the North Pole from its mean position ranges up to 8 m.

Chandrasekhar limit The maximum mass that can be supported by electron degenerate pressure. It ranges from 1.2 (Fe-rich stars) to 1.44 (He- and C-rich stars) solar masses. This limit separates the production of white dwarfs (below the limit) from that of neutron stars or black holes (above the limit) during the contraction of stellar cores.

channel bar An elongated sand bar along the course of a river.

channeled scablands Scablands deeply eroded by giant floods.

channel wave An acoustic wave propagating along a specific layer in the solid earth, the ocean, or the atmosphere.

characteristic length The typical length of a given system or body.

charco A small, natural depression into which water accumulates.

charge-coupled device (CCD) An intergrated circuit structurally similar to a MOSFET but with a large number ($\sim$ 1000) of gates between source and drain. A MOS capacitor is formed between each gate and substrate and charge is transferred from capacitor to capacitor by appropriate gate voltages. See **field-effect transistor.**

charged current The exchange of charge in interaction processes mediated by the $W^\pm$ gauge bosons. Cf. **neutral current.**

charge multiplet A pair or a group of similar particles, such as proton and neutron or the three pions, differing in charge. Syn. **isospin multiplet.**

Charles' law "The volume of a gas at constant pressure is proportional to temperature."

$$V = kT$$

where V = volume, k = constant, T = temperature. Cf. **Boyle's law.**

charm One of the six quark flavors. See **quark.** Charm is a quantum number equal to the number of charmed quarks in a particle minus the number of anticharmed quarks.

charmed quark A quark with an electric charge of $+2/3$, 0 strangeness, and charm of $+1$.

charnockite A plutonic rock consisting mainly of quartz, K-feldspar, plagioclase, and orthopyroxene.

chasm A narrow, deep gorge.

chatoyancy The wavy reflection of light by lamellae inside certain minerals. E.g. tourmaline cat's eye.

chelate A heterocyclic ring holding a metal ion between two of its atoms.

chelate compound A chemical compound that includes a heterocyclic ring holding a metal ion.

chelating agent A chemical compound that includes a heterocyclic ring capable of holding a metal ion.

chelation The holding of a metal ion within a single heterocyclic ring.

chemical bond See **bond.**

chemical equilibrium See **equilibrium (chemical).**

chemical potential (μ) The rate of change of the Gibbs free energy of a chemical system with several components, with respect to the change in number of moles n_i of component i, while temperature, pressure, and the number of moles of the other components are kept constant.

chemotaxis Chemically induced taxis.

chemotropism Chemically induced tropism.

chert Opaque compact microcrystalline quartz. Cf. **chalcedony.**

chestnut soil A dark brown soil formed by temperate, subhumid to subarid weathering.

chevron fold A kink fold with limbs of equal length.

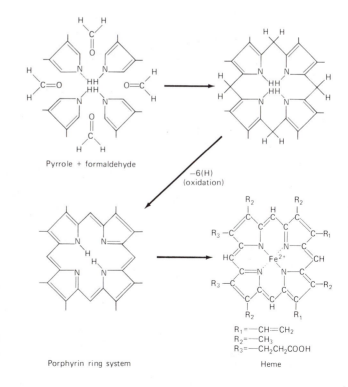

Chelates. Formation from 4 pyrrole rings and 4 formaldehyde molecules. (From Calvin 1969, p. 147, Fig. 7.2)

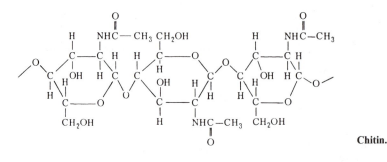

Chitin.

chill zone The edge of an igneous intrusion exhibiting crystals of smaller size than the main body because of more rapid cooling and crystallization.

chinook A catabatic wind blowing down the eastern slope of the Rocky Mountains.

chip An integrated microcircuit mounted on a single substrate.

chirality *(Nuclear Physics)* See **helicity.** *(Chemistry)* The property of left (L) or right (D) handedness in asymmetric molecules which are mirror images of each other.

chitin $(C_{15}H_{26}O_{10}N_2)_n$, a resistant polysaccharide similar to cellulose but with one $-OH$ per ring replaced by a $-NHCOCH_3$ group. Mol. mass = $(394.379)_n$. It is a major constituent of crustacean shells, insect exoskeleta, and the cell walls of some Fungi.

chlorinity The chloride content of seawater in g/kg, including the chloride equivalent of all halides. Standard seawater has a chlorinity of 19.4‰, corresponding to a salinity of 35.5‰. Cf. **chlorosity, salinity.**

chlorite The mineral $(Mg,Fe)_3(Si,Al)_4O_{10}(OH)_2 \cdot (Mg,Fe)_3(OH)_6$, a hydroxy Mg–Fe aluminosilicate mineral common as a low-grade metamorphic alteration of Mg–Fe silicates.

chlorobium chlorophyll A type of chlorophyll that occurs in green sulfur bacteria, together with small amounts of bacteriochlorophyll.

chlorophyll 1. Any of the pigments that mediate photosynthesis, including chlorophyll a, b, c, d, and e, bacteriochlorophyll, and chlorobium chlorophyll. 2. A mixture of chlorophyll a $(MgN_4O_5C_{55}H_{72})$, dark blue, and chlorophyll b $(MgN_4O_6C_{55}H_{70})$, yellow-green, that occur scattered in the cells of blue-green algae and assembled in the chloroplasts of green algae and higher plants. Cf. **bacteriochlorophyll, chlorobium chlorophyll.**

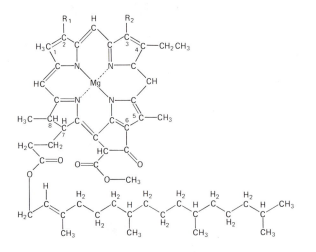

Chlorophylls. Chlorophyll a: $R_1 = -CH=CH_2$, $R_2 = -CH_3$ (in all oxygen-evolving plants); chlorophyll b: $R_1 = -CH=CH_2$, $R_2 = -CHO$ (in higher plants and green algae together with chlorophyll a in the ratio of 70% a to 30% b); chlorophyll c: $R_1 = -CH=CH_2$, $R_2 = -CH_3$, and replace chain attached to C in position 7 with $-CH=CH-COO$ (diatoms and brown algae); chlorophyll d: $R_1 = -CHO$, $R_2 = -CH_3$ (marine red algae).

chloroplast An organelle, about 4–7 μm in size, present in numbers ranging from 1 to several thousand per cell in all photosynthesizing plants except procaryotic ones. Chloroplasts consist of 70% protein, 20% lipids, and up to 7% nucleic acids, and contain chlorophyll in thin membranaceous units (thylacoids). Chloroplast DNA, like bacterial DNA, is bare of proteins. Chloroplasts are self-replicating organelles and may have originated as procaryota symbiotic with early eucaryotes.

chlorosity Chlorinity of seawater expressed in g/liter at 20°C.

choke A high impedance in a circuit blocking the passage of specified frequencies.

chondrite A stony meteorite characterized by olivine-rich chondrules in a matrix of orthopyroxene, olivine, and Fe–Ni microcrystals. Chondrites are subdivided into ordinary chondrites (78.9%, consisting of pyroxene, olivine, and Fe–Ni alloy), carbonaceous chondrites (5.7%, consisting of serpentine, olivine, pyroxenes, sulfates, and organic compounds), and others (14.5%). Chondrites represent 85.7 of all meteorites. See **meteorite, Meteorites***.

chondrule A small (0.5–1 mm across) spherical body consisting of olivine, pyroxene, or plagioclase crystals found in chondritic meteorites. See **chondrite.**

C horizon The horizon below the A and/or B horizon of a soil, consisting of unconsolidated rock material little affected by pedogenesis.

chorography The science describing a region in greater detail than geography but in less detail than by topography.

chott A shallow, brackish, or saline lake in an arid region.

christmas tree A structure consisting of pipes and valves atop the casing of an oil well.

chromatic aberration The failure of an optical system to focus light of different wavelengths on the same point.

chromatin The ensemble of nucleic acids and nucleoproteins in a cell.

chromatography The separation of different chemical constituents of a mixture by passing it through a column containing an absorbing or ion-exchanging medium.

Chromel A trade name for an alloy (90% Ni, 10% Cr) used for thermocouples.

chromite The mineral $FeCr_2O_4$.

chromosome The unit in which the DNA of all living cells is organized. A chromosome consists of a single DNA molecule with molecular mass as high as 10^{10} u or higher. Procaryota have a single chromosome usually forming a loop. Eucaryota have chromosomes with 150–240 base pairs of DNA folded around histones forming bead particles (nucleosomes) about 100 Å across. These are further folded into fibers 25–$30 \cdot 10^{-3}$ μm thick and still further folded to form the chromosome. Chromosomes range in size up to about 3×30 μm. The fruit fly has 8 chromosomes, humans have 46, and radiolaria have more than 800.

chromosphere A layer about 2500 km thick above the solar photosphere, where temperature decreases from 6000 K at the base to 4000 K at 1500 km of altitude, and then increases to 50,000 K at the top. See **corona, photosphere.**

chron The chronological span of a biozone.

chronostratigraphic unit The sedimentary strata formed during a specified time interval. Chronostratigraphic units, with their corresponding time intervals in parenthesis, range from largest to smallest as follows: eonthem (eon), erathem (era), system (period), series (epoch), stage (age), substage (subage).

chronostratigraphy The chronology of stratigraphic units.

chronozone The time span represented by a biozone.

chthonic Defining a sediment derived from pre-existing rocks.

cima The summit of a mountain.

cinder A pyroclastic fragment 1 to 4 mm across, larger than volcanic ash (0.063–1 mm) but smaller than lapilli (4–64 mm).

cinder cone A volcanic cone made predominantly of volcanic cinder and ash.

cipollino A gray-greenish marble containing muscovite.

CIPW Cross-Iddings-Pirsson-Washington, the four scientists who devised (1902) the norm system of expressing the mineral composition of a rock. See **norm.**

circadian Referring to a 24-hour rhythm.

circle See **conic sections.**

circle of longitude Any great circle passing

through the poles of the ecliptic coordinate system. See **coordinate systems.**

circular motion See **acceleration.**

circular polarization See **polarization.**

cirque An amphitheater-shaped recess on a mountain side resulting from erosion by the head of a glacier.

cis- Prefix meaning *on the same side.*

cis-isomer An isomer in which atoms or groups of atoms are attached to the same side of the molecule.

cistron The structural gene, the DNA segment corresponding to a given polypeptide chain plus the start and stop signals. The cistron for common proteins (130–630 amino acids) consists of 390–1890 nucleotides (3 per amino acid). DNA contains the complete cistron system (genotype). mRNA may consist of a single cistron *(monocistronic mRNA)* to encode a single protein, or several cistrons *(polycistronic mRNA)* to encode several proteins which often belong to a specific metabolic pathway.

citric acid cycle See **Krebs cycle.**

clade A lineage derived from a single ancestral form.

cladism The classification of organisms based on their descendance.

cladogenesis The development of a clade.

cladogram A graph showing the genealogical tree of a group of organisms.

clan A group of igneous rocks closely related chemically. Clans are subdivided into families and grouped into tribes.

Clarke-Bumpus sampler A plankton sampler consisting of an opening-and-closing net and a flow meter.

class A taxonomic division below phylum and above order.

classons The photon and the graviton, the massless bosons that are the quanta of electromagnetism and gravitation.

clastic Defining a rock consisting of fragments from a preceding rock or rocks.

clathrate A crystalline or structured liquid substance holding host molecules within its structure as inclusions bonded only by weak van der Waals forces.

clay 1. A sediment consisting of clay mineral particles. See **illite, kaolinite, montmorillonite, vermiculite.** 2. A sediment consisting of particles smaller than 1/256 mm or about 4 μm.

C layer The seismic zone in the Earth's interior extending from 400 to 1000 km of depth.

clay ironstone A sedimentary rock consisting of a mixture of clay minerals and siderite, commonly in nodules, associated with coal deposits.

clay mineral See **illite, kaolinite, montmorillonite, vermiculite.**

claystone A massive pelite.

clean sandstone A sandstone with less than 15% of clay. Cf. **dirty sandstone.**

cleavage 1. The property of a mineral to break along its crystallographic planes, determined by its lattice structure and bonding strength. 2. The property of a rock to break along preferred planes, determined by the orientation of its constituent minerals.

climate The average meteorological conditions characterizing a specified region.

climate optimum The time, about 6000 y B.P., when climate in high northern latitudes was somewhat warmer and more humid than today.

cline 1. A change in the gradient of a physical or chemical property. 2. A gradual change in morphology of an organism resulting from changing environmental conditions or from evolution.

clinoenstatite A clinopyroxene, the monoclinic form of enstatite, $MgSiO_3$.

clinohypersthene A clinopyroxene, the monoclinic form of hypersthene, $(Mg,Fe)SiO_3$.

clinometer An instrument to measure elevations.

clinopyroxene Any of the Ca-rich pyroxenes crystallizing in the monoclinic system. Examples are:

diopside, $CaMg(SiO_3)_2$
hedenbergite, $CaFe(SiO_3)_2$
augite, $(Ca,Na)(Mg,Fe,Al)[(Si,Al)O_3]_2$

Cf. **orthopyroxene.**

clone The totality of descendants from an asexually reproducing organism.

closed shell An atomic or nuclear shell containing the maximum number of electrons or nucleons allowed by the Pauli exclusion principle.

cloud chamber A chamber containing supersaturated gas or vapor in which charged particles leave tracks consisting of droplets. Syn. **Wilson cloud chamber.** Cf. **bubble chamber.**

cm Centimeter.

C–N–O cycle See **carbon cycle.**

coacervate The agglomeration of colloidal particles within a dispersant phase.

coagulation The separation of colloidal particles out of the dispersant phase.

coaltitude See **zenith distance.**

coarse-grained 1. Defining an igneous rock with crystals 5 mm or more in diameter. 2. Defining a sedimentary rock with grains larger than 2 mm.

coarse sand *(Sedimentology)* A sand with grains ranging from 0.5 to 1 mm across. *(Engineering)* A sand with grains ranging from 2 to 4.56 mm across.

coaxial cable A cable consisting of a central metal wire and an outer cylindrical metal sleeve separated by an insulator. Coaxial cables are used mainly for communication transmission.

cobble A rock fragment ranging in size from 64 to 256 mm. See **Wentworth grade scale.**

coccolith Any of the calcitic elements, averaging 2 to 6 μm across, secreted by the Coccolithophoridae.

Coccolithophoridae The Haptophyta, a phylum of unicellular, biflagellate, golden-brown, pelagic protophyta that form coccoliths.

coccosphere The entire coccolithophorid skeleton, consisting of coccoliths.

codon A base triplet of mRNA specific for the corresponding tRNA anticodon which in turn is specific for a given amino acid. The codon AUG, specific for methionine, is also used to start protein chain formation; and three codons ("nonsense triplets"), not specific for given amino acids, are used to stop it. See **DNA, genetic code, mRNA, tRNA.**

coefficient of performance (COP) The ratio of the heat supplied or extracted by a thermodynamic cycle to the work supplied to operate that cycle.

coefficient of variability The standard deviation of a set of data divided by its arithmetic mean.

coelobite An animal living in a cavity, usually within a reef.

coenosis A community of organisms.

coenzyme The nonprotein portion of an enzyme.

coesite A high-pressure phase of silica, SiO_2, stable above 20 kb at room temperature. Density = 2.915 g/cm³.

coherence The existence of a steady relationship between the phases of two or more periodic phenomena or waves.

coherent light Light of the same frequency and with phase that is either the same or exhibiting a constant phase difference.

coherent waves Waves that are either in phase or whose phase differences remain constant in time.

coiling direction The dextral (positive helicity) or sinistral (negative helicity) coiling of trochoidal shells.

coke The solid residue of coal after removal of volatile matter by heating to 1000–1100°C in the absence of air.

colatitude $90° - \phi$, where ϕ = latitude.

cold cathode A cathode from which electrons are drawn at room temperature by a strong (10^7–10^8 V/cm) electrostatic field. See **field emission.**

cold front A sloping zone, about 1 km thick, intersecting the ground and separating advancing colder air below from retreating warmer air above. Cf. **occluded front, stationary front, warm front.**

cold light Light emitted in luminescence.

cold welding The welding of clean metal surfaces by pressure at room temperature in an inert atmosphere or in vacuo.

collapsar A gravitationally collapsed stellar object. See **black hole, neutron star.**

collapse caldera A caldera produced by the collapse of the roof of a magmatic chamber.

collenchyma Supporting tissue in plants, consisting of cells of rectangular, parallelepiped shape with thickenings along their edges.

collenia A domed stromatolitic structure, 10 cm across and less than 3 cm thick, formed in Late Precambrian time by the blue-green algae of the genus *Collenia.*

collider A particle accelerator in which two beams are made to collide with each other. Major colliders in operation (particles involved and maximum collision energies in parentheses): SPS (proton-proton, 30 GeV) and SP$\overline{\text{P}}$S (proton-antiproton, 640 GeV) at CERN, Geneva; PETRA (electron-positron, 46 GeV) at DESY, Hamburg; Tevatron (proton-proton, 40 GeV) at FNAL, Batavia, Illinois; PEP (electron-positron, 30 GeV) at SLAC, Palo Alto, California; VEPP (electron-positron, 7 GeV) at Novosibirsk, USSR. Colliders in construction to be functional in 1986–1990 are: LEP (electron-positron, 100 GeV) at CERN, Geneva; HERA (electron-proton, 350 GeV) at DESY, Hamburg; TeVI (proton-antiproton, $1000 + 1000$ GeV) at FNAL, Batvia, Illinois; SLC (electron-positron, 100 GeV) at SLAC, Palo Alto, California; TRISTAN (electron-positron, 30 GeV) at Kek, Japan; UNK (proton-proton, 600 GeV) at Serpukhov, USSR.

colligative properties Properties depending on the number of molecules present in a system but not on their nature. Colligative properties include vapor pressure lowering, boiling point elevation, freezing point depression, and osmotic pressure.

collision An interaction of short duration between two or more particles or bodies. **1. elastic collision** Total momentum is conserved. **2. inelastic collision** Total momentum is in part or in toto transformed into other forms of energy (especially heat).

collodion A solution of cellulose tri- and tetra-nitrates in ether and alcohol.

colloid A particle consisting of molecular aggregates usually 10^{-1} to 10^{-3} μm across.

collophane Cryptocrystalline apatite.

colluvium Loose waste rock material deposited by rainwash at the foot of a gentle slope.

color 1. The physiological response to the wavelength of visible light. Colors with wavelength boundaries (μm): 0.765/red/0.622/orange/0.597/yellow/0.577/green/0.492/blue/0.455/violet/0.380.2. A property of quarks coupling them to the gluon field. Three colors (red, green and blue) and their anticolors are recognized. The baryons are colorless, being composed of three quarks of different colors. The quark-antiquark pair forming a meson is a color-anticolor pair. Color is a quantum number of quarks, bearing no relationship to the common meaning of the word.

color center A locus in a crystalline lattice where the presence of a foreign atom or ion or of a crystal defect causes one or more bands in the visible spectrum to be absorbed and reradiated with 4π geometry. Transmitted light is thus colored with the nonabsorbed portion of the spectrum. Also called **F center** (F stands for *Farbe,* German for *color*).

color code A system of color bands or dots for the rating of resistors and capacitors. First or second band or dot: black = 0; brown = 1; red = 2; orange = 3; yellow = 4; green = 5; blue = 6; violet = 7; gray = 8; white = 9. The third band or dot indicates the multiplication factor: black = 1; brown = 10; red = 100; orange = 1,000; yellow = 10,000; green = 100,000; blue = 1,000,000; violet = 10,000,000; gray = 100,000,000; white = 1,000,000,000; gold = 0.1; silver = 0.01. Resistance in ohms; capacitance in microfarads.

color force The force holding quarks together to form hadrons. It is carried by gluons. The interaction energy increases with distance, leading to the confinement of quarks within hadrons.

colorimeter An instrument to measure color by determining the intensity of the three primary colors, red, yellow, and blue, in light passing through a colored solution.

color index *(Astronomy)* Apparent magnitude of a star measured at a given wavelength minus its apparent magnitude measured at a longer wavelength. Standard: UBV [ultraviolet (U, 360 nm)—blue (B, 420 nm)—greenish-yellow (V, 540 nm)—red (R, 689 nm)—infrared (I, 825 nm)], expressed as U-B, B-V, V-R, V-I, R-I [U = ultraviolet, B = blue, V (visible) = greenish-yellow, the color band to which the human eye is most sensitive]. *(Petrology)* A number representing the volume percent of dark minerals in a rock. Rocks are classified as leucocratic (color index 0–30), mesocratic (color index 30–60), and melanocratic (color index 60–100).

coma The envelope of ionized gases surrounding the nucleus of a comet. See **comet.**

combination A set of different numbers or objects regardless of ordering. The number of different combinations $C(n, k)$ of k items obtained from a set of n items ($n > k$) is equal to $n!/k!(n - k)! = P(n, k)/k!$ [where $P(n, k)$ = number of permutations of n elements taken k at a time ($n > k$)]. See **permutation.**

combined water Water of hydration or in solid solution in a mineral.

comet Any of the population of small (10^{12} kg ??) bodies probably consisting of a mixture of Fe-Mg silicate and Fe-Ni metal particles and of frozen gases (CN, HCN, CO_2, OH, H_2O, etc.), forming the Oort cloud. Comets captured by the outer planets have their gases volatilized as they cross the asteroidal belt and approach the Sun. They exhibit an expanded (10^4–10^5 km radius) coma, and a long (10^5–10^6 km) heliofuge tail shaped by solar radiation and by the solar wind. See **Comets—chemical composition.**

common lead The stable isotopes of lead, ^{204}Pb (1.42%), ^{206}Pb (24.1%), ^{207}Pb (22.1%), and ^{208}Pb (52.4%) in their normal ratios as seen in young minerals containing no appreciable amounts of U or Th.

common lead dating method An absolute dating method based on the growth of ^{206}Pb, ^{207}Pb, and ^{208}Pb through geologic time because of the decay of, respectively, ^{238}U ($t_{1/2}$ = 4.468·10^9 y), ^{235}U ($t_{1/2}$ = 0.704·10^9 y), and ^{232}Th ($t_{1/2}$ = 14.05·10^9 y).

common logarithm The exponent needed to express a given number as a power of the number 10. Cf. **natural logarithm.**

compensation depth (CD) *(Geophysics)* The depth at which the mass of the overlying rocks is approximately uniform worldwide. It corresponds to the boundary between lithosphere and asthenosphere. *(Marine Geology)* The depth at which the rate of dissolution of a mineral phase balances its rate of deposition. For calcite, the CD is about 5 km deep in the Atlantic and northern Indian Ocean; 4.5 km deep in the Pacific and southern Indian Ocean; and 3.5 km deep in the Southern Ocean. For aragonite, the CD is about 3 km deep in the tropical Atlantic and 2.5 km deep in the tropical Pacific. *(Oceanography and Limnology)* The depth in the ocean or a lake at which the production of oxygen balances its consumption.

competence The ability of a water current or wind to transport particles in terms of their maximum size. Cf. **capacity.**

complementary angle Either of two angles whose sum is 90°. Cf. **supplementary angle.**

complex compound See **coordination compound.**

complexing agent See **chelating agent.**

complex numbers Any of the numbers of the form $a + ib$, where a and b are real numbers and $i = (-1)^{1/2}$. Absolute value *(modulus)* = $(a^2 + b^2)^{1/2}$. See **numbers.**

compressibility factor The ratio pV/RT, equal to 1 for 1 mole of an ideal gas but an often variable function of pressure for real gases.

compressional fault A fault caused by compressional stresses, such as a reverse fault.

compressive strength The maximum compres-

sive stress to which a material can be subjected before failure.

Compton effect The increase in wavelength *(Compton shift)* of x-rays and gamma-rays when scattered by free electrons.

$$\Delta\lambda = (h/m_e c)(1 - \cos\theta)$$

where λ = wavelength, m_e = mass of electron, c = speed of light, θ = scattering angle of photons. For $\theta = 90°$

$$\Delta\lambda = h/m_e c$$

Compton wavelength of the electron = 2.426309· 10^{-12} m. For $\theta = 180°$ (head-on collision),

$$\Delta\lambda = 2h/m_e c$$

Compton electron An electron set in motion by interaction with a photon in Compton scattering.

Compton recoil electron See **Compton electron.**

Compton scattering The scattering of photons by electrons.

Compton wavelength (λ_c). A characteristic of elementary particles, equal to the Compton shift for a scattering angle of 90°.

$$\lambda_c = h/mc$$

where h = Planck's constant, m = mass of particle, and c = speed of light. See **Compton effect.**

conchiolin A fibrous protein ($C_{30}H_{48}N_9O_{11}$) binding the microcrystals in molluscan shells and forming the periostracum.

concordia The curve on a graph plotting the ra-

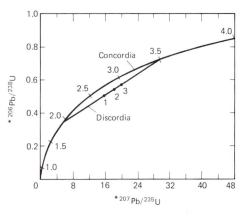

Concordia diagram. Concordia curve with discordia chord for three zircons from a Minnesota granite. The discordia curve intercepts the concordia curve at 3.5·10^9 y (age of the zircons) and at 1.8·10^9 y (termination of Pb loss). Asterisk identifies radiogenic Pb. (From Faure 1986, p. 295, Fig. 18.7)

tios $*^{206}Pb/^{238}U$ versus $*^{207}Pb/^{235}U$, where the asterisks signify the radiogenic portions of the two Pb isotopes. If no Pb or U was lost from, or added to, a U-containing mineral since its crystallization, the two ratios will identify a single point on the concordia curve, representing the true age of the mineral. If, as often is the case, a rock lost Pb during a subsequent episode of metamorphism, different samples of the same mineral within the rock will exhibit different Pb losses. The loss will affect equally the two Pb isotopes and, as a result, the Pb/U ratios will plot along a chord *(discordia)* subtending the concordia curve and intersecting it at two points. The older point gives the age of the original crystallization of the rock (no Pb loss yet), and the younger point gives the age at which Pb loss (and hence metamorphism) occurred (maximum Pb loss).

conductance (G) *(dc circuits)* The reciprocal of resistance. *(ac circuits)* The real part of admittance. It is expressed in siemens.

conduction band An energy band in which electrons can move freely.

conduction electron An electron in the conduction band of a solid.

conductivity Specific conductance. **1. electrical conductivity (σ)** The ability of a substance to conduct electricity.

$$\sigma = -[(dq/dt)/A](dV/dx)$$
$$= J/E$$

where q = charge, t = time, A = area through which current flows, dV = potential difference in the direction x of current flow, J = current density, E = electric field strength. It is measured in siemens $\times$ meter (S·m). Electrical conductivity ranges from 10^8 to 10^6 for metals, from 10^6 to 10^{-6} for semiconductors, and from 10^{-6} to 10^{-14} for insulators. Among the elements in the solid state, Ag has the highest conductivity ($0.6 \cdot 10^8$ S·m at 20°C), yellow sulfur the lowest ($5 \cdot 10^{-14}$ S·m at 20°C). Cf. **resistivity. 2. thermal conductivity (κ)**

$$\kappa = -[(dQ/dt)/A]/(dT/dx)$$

where Q = quantity of heat, t = time, A = area through which heat flows, dT = absolute temperature difference in the direction x of heat flow. Examples of thermal conductivities among elements in the solid state (in W cm^{-1} K^{-1} at 25°C) are: C as diamond = 23.2 (highest); Ag = 4.29; Cu = 4.01; Si = 1.49; Ge = 0.602; S = 0.002 (lowest).

cone A surface represented by either equation (vertex at origin)

$$x^2/a^2 + x^2/a^2 - z^2/c^2 = 0 \quad \text{(circular cone)}$$
$$x^2/a^2 + y^2/b^2 - z^2/c^2 = 0 \quad \text{(elliptical cone)}$$

Sections normal to the c axis (other than through the vertex) are circles in the first case, ellipses in the second. See **conic sections.**

cone-in-cone A structure in calcareous shales consisting of fluted cones with apical angles of 30° to 60° and bases a few centimeters across, resulting from pressure normal to the bedding plane.

conformity A surface separating parallel beds deposited without significant time hiatus.

conglomerate A clastic rock largely composed of transported fragments larger than 4 mm.

conglomerite A completely recrystallized conglomerate.

congruent Coinciding exactly when superimposed.

conic sections Conic sections are generated by a plane intersecting a right circular cone (two nappes with common axis and vertex) at different angles other than through the vertex. For a right circular cone, the conic sections are as follows (A = angle formed by plane with axis; B = generating angle of cone = 1/2 vertex angle):
1. circle $A = 90°$. Equation (center at origin):

$$x^2 + y^2 = 1,$$

2. ellipse $B < A < 90°$. Equation (center at origin, major axis = x axis):

$$x^2/a^2 + y^2/b^2 = 1$$

where a = semimajor axis, b = semiminor axis.
3. parabola $A = B$. Equation (vertex at origin, axis = y axis, coordinate of directrix $p = -y$):

$$x^2 = 4py$$

4. hyperbola $0 < A < B$. Equation (center at origin, transverse axis = y axis):

$$y^2/a^2 - x^2/b^2 = 1$$

where a = semitransverse axis, b = semiconjugate axis.

conjugate Associated with, related to.

conjugate axis The segment passing through the vertex of a hyperbola, parallel to the directrix, bound by the two asymptotes.

conjugated compound A chemical compound with alternating single and double bonds.

conjugated protein A protein containing a nonprotein part.

conjunction The closest, apparent approach of two celestial bodies in the sky.

connate water Water trapped within a sedimentary rock at the time of deposition.

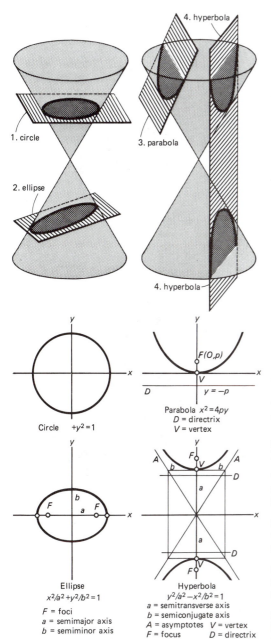

Conic sections: (1) circle; (2) ellipse; (3) parabola; (4) hyperbola.

Conrad discontinuity The boundary between sial and sima within the continental crust, across which P wave velocities increase from about 6 km/s to about 6.5 km/s.

consequent stream A stream whose course is determined by the tectonic structure of the area.

conservative elements Elements in a solution (e.g. seawater) whose concentrations remain constant with time.

conservative force A force that returns to its original value at the end of a process.

conservative property A property of a system that remains constant while the system changes.

constantan An alloy (Cu 55%, Ni 45%) used for thermocouples.

contact metamorphism Metamorphism occurring in a rock close to an igneous rock body.

continent *(Geophysics)* A major crutal block rising above the 900 m isobath. *(Geography)* A major ($>5 \cdot 10^6$ km^2) area above sea level. The six continents are (area and percent of total land surface in parentheses): Asia (44.0 $\cdot 10^6$ km^2 = 29.8%), Africa (30.2 $\cdot 10^6$ km^2 = 20.5%), North America (24.2 $\cdot 10^6$ km^2 = 16.4%), South America (17.8 $\cdot 10^6$ km^2 = 12.1%), Antarctica (13.2 $\cdot 10^6$ km^2 = 8.9%), Europe (10.4 $\cdot 10^6$ km^2 = 7.0%), Australia (7.7 $\cdot 10^6$ km^2 = 5.2%).

continental borderland A folded and faulted continental slope.

continental crust The Earth's crust underlying the continental surface, consisting of sial above and sima below, with or without sediment cover, and ranging in thickness from 20 to 80 km (average 35 km).

continental drift The horizontal (translational and/or rotational) motion of continental plates relative to each other.

continental flexure The hinge line along the continental slope, between continental crust and oceanic crust.

continental margin The zone between shoreline and the oceanic province offshore, including the continental shelf, slope, and rise.

continental platform The area of a craton bordering the continental shield and normally covered with relatively thin, flat-lying marine sediments of Paleozoic or younger age.

continental rise The deeper part of the continental margin, consisting of an apron of sediments derived from the continent or deposited by boundary currents.

continental shelf The area between shoreline (depth = 0 m) and the shelf break (average depth = 130 m).

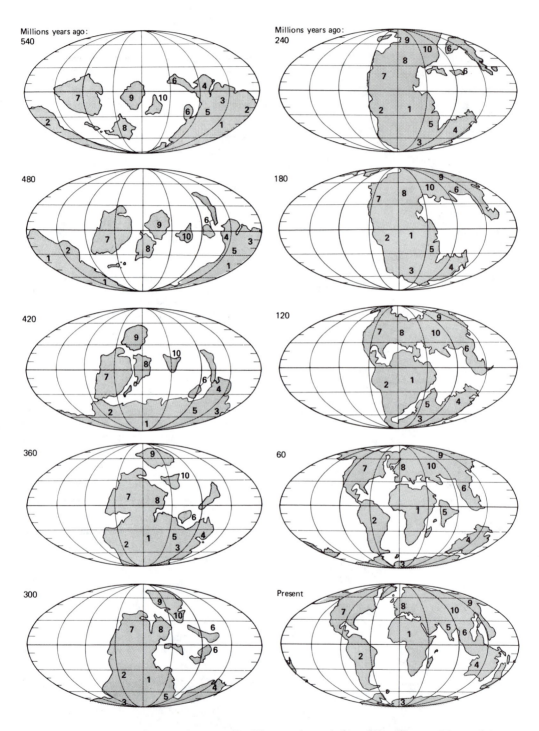

Continents. The motions of the continents at 60-million-year intervals from 540 million years ago to the present. 1 = Africa; 2 = South America; 3 = Antarctica; 4 = Australia; 5 = India; 6 = China; 7 = North America; 8 = Europe; 9 and 10 = Siberia. (From Siever 1983, p. 51, illustration)

continental shield The inner part of a continental craton where Precambrian rocks outcrop.

continental slope The area between the shelf break (average depth = 130 m) and the landward margin of the continental rise or oceanic trench.

continental terrace The area between the shoreline and the foot of the continental slope.

continuity See **equation of continuity.**

continuous spectrum An emission or absorption spectrum of electromagnetic radiation that is continuous over a frequency band. Continuous emission spectra are produced by the capture, by ions, of electrons above the ionization level; by transitions from higher to lower vibrational and rotational states of molecules or hot solids; and by the combination of atoms to form molecules. Continuous absorption spectra are produced by the inverse processes.

contourite A deposit formed by a boundary current flowing along a submarine slope.

control system A dynamic system in which one or more outputs are controlled by modifying the inputs. See **transfer function.**

convection The motion of gaseous, liquid, or solid matter due to density differences produced by differential heating or cooling. The driving force is gravity.

convergence The line where two oceanic water masses of different densities meet, resulting in the sinking of the denser one.

convergent evolution The fortuitous appearance of forms that are physically similar but genetically unrelated.

convergent plate boundary The boundary between two plates moving against each other, resulting in collision (if continent-to-continent or ocean-to-ocean) or subduction (if ocean-to-continent). Cf. **divergent plate boundary.**

conversion factor A coefficient to convert a unit into another one. See **Conversion factors*.**

convolute bedding See **convolute lamination.**

convolute lamination The often intricate folding of thin laminae within an otherwise undisturbed bed resulting from load and internal flow.

coordinate system (celestial) Any of the systems used to locate a celestial body on the celestial sphere. **1. ecliptic coordinate system** Equator = ecliptic; latitude = celestial latitude; longitude = celestial longitude, starting at vernal equinox (Υ) and measured eastward 0° to 360°. **2. equatorial coordinate system** Equator = celestial equator; poles = celestial poles; latitude = declination (δ); longitude = right ascension (RA, α), starting at the vernal equinox (Υ) and measured eastward from 0 to 24 h (1 h = 15°). **3. galactic coordinate system** Equator = galactic equator; poles = intersection of galactic axis with celestial sphere; latitude = galactic latitude; longitude = galactic longitude, starting from the direction of the galactic center ($\alpha = 265°36'$, $\delta = -28°55'$, year 1950) and measured eastward from 0° to 360°. **4. horizon coordinate system** Equator = observer's horizon; pole = zenith; latitude = altitude; longitude = azimuth, starting from the northern direction and measured eastward from 0° to 360°.

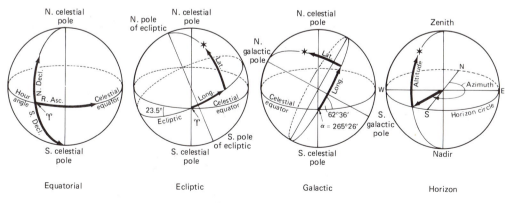

Coordinate systems.

coordinate system (terrestrial) See **geodetic coordinates, geographic coordinates.**

coordination compound A chemical compound consisting of a Lewis acid, usually a central metal atom or cation, to which are bonded one or more Lewis bases, either neutral or anionic species (ligands). See **ligand.**

coordination number The number of ligands bound to a central ion in a coordination compound.

coordination sphere The domain occupied by the anions or anionic molecules in a coordination compound.

COP Coefficient of performance.

copal A group of hard resins from tropical trees.

copalite Fossil copal.

copolymer A polymer formed by the polymerization of two different monomers.

coprolite Fossil excrement of vertebrates, larger than a fecal pellet.

copropel An ooze consisting largely of excrement.

coquina A poorly cemented deposit consisting of shell fragments.

coralgal Defining a carbonate rock consisting of intergrown corals and coralline algae.

coral head A mushroom-shaped living coral colony.

coralline alga Any of the red algae belonging to the family Corallinaceae.

corallite The individual exoskeleton of a coral polyp.

corallum 1. The exoskeleton of a coral colony. 2. The skeleton of a solitary coral.

core *(Astronomy)* The central portion of a star where nuclear reactions take place. *(Physics)* 1. See **atomic core.** 2. A magnetic material placed within an electric coil to intensify the magnetic field produced by a current flowing in the coil. *(Geology)* The central portion of the Earth, bound by the mantle. Radius = 3486 km; mass = $1.90 \cdot 10^{24}$ kg = 31.8% of the mass of the Earth; composition = Fe (90%?), Ni (9%?); mean density = 10.7. **1. outer core** Liquid. Thickness = 1835 km; mass = $1.768 \cdot 10^{24}$ kg = 93.0% of mass of core = 29.6% of mass of Earth; density = 9.9 (outer boundary) to 11.9 (inner boundary); temperature = 3800°C (?) (outer boundary) to 6000°C (?) (inner bound-

ary); pressure = $1.372 \cdot 10^6$ atm (outer boundary) to $3.067 \cdot 10^6$ atm (inner boundary); gravitational acceleration = 10.69 m/s² (outer boundary) to 5.74 m/s² (inner boundary). **2. transition zone** Thickness = 450 km; mass = $0.021 \cdot 10^{24}$ kg = 6.3% of mass of core = 0.2% of mass of Earth; density = 11.9 (outer boundary) to 12.7 (inner boundary); temperature = 6000°C (?) (outer boundary) to 6300°C (?) (inner boundary); pressure = $3.067 \cdot 10^6$ atm (outer boundary) to $3.342 \cdot 10^6$ (inner boundary); gravitational acceleration = 5.74 m/s² (outer boundary) to 4.36 m/s² (inner boundary). **3. inner core** Solid. Radius = 1200 km; mass = $0.129 \cdot 10^{24}$ kg = 38.8% of mass of core = 2.0% of mass of Earth; density = 12.7 (outer boundary), 13.0 (center); temperature = 6300 (?) (outer boundary) to 6600°C (?) (center); pressure = $3.342 \cdot 10^6$ atm (outer boundary) to $3.680 \cdot 10^6$ atm (center); gravitational acceleration = 4.36 m/s² (outer boundary) to 0 (center).

Coriolis acceleration The acceleration needed for an object to maintain its trajectory when moving with respect to a rotating frame of reference. It is equal to $2\omega \times v$, where ω = angular velocity of the rotating frame, v = velocity of the object with respect to the rotating frame.

Coriolis effect The inadequate application of the Coriolis acceleration to an object moving on the surface of a rotating body. If the direction indicated by the body's angular velocity vector is defined as North, the result is a deviation of the object's motion to the right in the northern hemisphere and to the left in the southern hemisphere.

corona The outer layer of the solar atmosphere above the chromosphere, extending from about 2500 km of altitude to several solar radii and beyond. It is separated from the chromosphere below by a transition layer several hundred km thick through which temperature rises from 50,000 K to 500,000 K. The kinetic temperature at a height of 50,000-100,000 km is about $1.5 \cdot 10^6$ K. Mean density of matter in the corona is 10^{-16} to 10^{-19} g/cm³.

corona discharge Rapid flow of electrons into or out of a conductor. Surrounding air or other gases thus become ionized, producing more electrons for additional ionization and continuing discharge.

correspondence principle "Quantum physics reduces to classical physics at very large quantum numbers."

corundum The mineral Al_2O_3 used as an abrasive (hardness = 9 on the Mohs scale).

cos Cosine.

cosec Cosecant.

cosecant See **trigonometric functions.**

cosine See **trigonometric functions.**

cosmic abundances The relative abundances of the elements as determined from spectroscopic analysis of stellar light and from the chemical composition of planetary and terrestrial matter. Relative abundances of the most common elements (Si = 1): H = 27,200; He = 2,180; O = 20.1; C = 12.1; Ne = 3.8; N = 2.5; Mg = 1.1; Si = 1; Fe = 0.9; S = 0.5; Ar = 0.1; Al = 0.08; Ca = 0.06; Na = 0.06; Ni = 0.05; Cr = 0.01; P = 0.01. See **Elements—abundances in the solar system*.**

cosmic background radiation The total radiation from outer space, at all wavelengths, not associated with specific, identifiable sources. Cf. **microwave background radiation.**

cosmic dust 1. Solid particulate matter as microcrystals of small diameter (<1 mm) in interstellar or intergalactic space. 2. Silicate or metal microparticles (radius <1 mm) shed by comets while passing through the inner solar system.

cosmic ray-induced radionuclides* Radionuclides formed by cosmic-ray bombardment of stable nuclides in atmospheric gases, on terrestrial, lunar, or planetary surfaces, or on the surfaces or interiors of meteorites.

cosmic rays A flow of energetic particles from outside the solar system and the products of their collisions with atmospheric gases. **1. primary cosmic rays** Nuclei of the lighter elements (H to Fe) accelerated in galactic fields and reaching the Earth at high speeds. Average flux outside the Earth's atmosphere = 1 particle/cm^2/s; energy from $<10^6$ eV to $>10^{20}$ eV; composition: H = 88.4%, He = 10.6%, C = 0.51%, O = 0.46%, all others = 0.03%. **2. secondary cosmic rays** Particles produced by collisions of primary cosmic rays with atmospheric gases. See **shower.**

cosmic scale factor (R) A measure of the dimension of the universe as a function of cosmological time. It is related to both the Hubble constant H_0 and to the redshift parameter z.

$$(dR/dt)/R = H_0$$
$$R(t_0)/R(t_e) - 1 = z$$

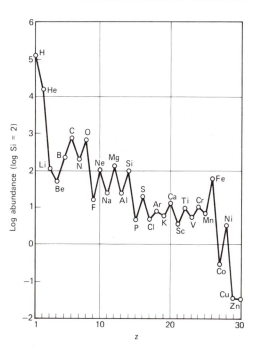

Cosmic rays. Elemental abundances (Si = 100). (Data from Simpson 1983, Table 2 and Fig. 8)

where t_0 = present time; t_e = time when radiation reaching us at time t_0 was emitted.

cosmic spherule A molten silicate or metal droplet ablated from the surface of a meteorite entering the Earth's atmosphere.

cosmological constant (Λ) A term introduced by Einstein in 1916 to stabilize the universe and prevent it from either expanding or contracting.

$$\Lambda = 3H_0^2(\sigma_0 - q_0)$$

where H_0 = Hubble constant, σ_0 = cosmological density parameter, q_0 = deceleration parameter.

cosmological density parameter (σ_0) A dimensionless parameter with the value

$$\sigma_0 = 4\pi G\rho_0/3H_0^2$$

where G = gravitational constant, ρ_0 density of matter and radiation in the Universe, H_0 = Hubble constant.

cosmological distance The distance of a celestial object as derived from the Hubble constant.

cosmological principle "The universe is isotropic and homogeneous when considered across a sufficiently large region of space."

cosmological time The time reckoned since the beginning of the Big Bang. Its major subdivisions are: 0 s/Planckian/5.390·10^{-44} s/Gamowian/11.9· 10^9 y/Hadean/12.8·10^9 y/Archean/13.9·10^9 y/ Proterozoic/16.0·10^9 y/Phanerozoic/16.6·10^9 y. Cf. **Hubble time.**

cosmos The totality of what exists, both visible and invisible, seen as an ordered system. Cf. **metagalaxy, universe.**

cot Cotangent.

cotangent See **trigonometric function.**

cotidal line A line connecting all points where high tide occurs at the same time.

couloir A narrow, vertical cleft in a granitic or granodioritic mountain ridge.

coulomb (C) The SI and MKS unit of electric charge. It is equal to the quantity of electricity transported by a current of 1 ampere in 1 second:

$$C = A \cdot s$$

where A = ampere, s = second. 1 C = 6.24151· 10^{18} electron charges.

Coulomb's constant The constant k in Coulomb's law.

$$k = 1/4\pi\epsilon_0$$

where ϵ_0 = permittivity constant = 8.85418782· 10^{-12} C^2/N m^2 or F/m. It is equal to 8.987552·10^9 N m^2/C^2 or m/F.

Coulomb's law A law relating the force between two electric charges to their magnitude and distance:

$$F = k(q_1 q_2/r^2)$$
$$= (1/4\pi\epsilon_0)(q_1 q_2/r^2)$$

where F = force, k = Coulomb's constant = $1/4\pi\epsilon_0$, q_1 = charge 1, q_2 = charge 2, r = distance between q_1 and q_2, ϵ_0 = permittivity constant = 8.85418782·10^{-12} C^2/N m^2 or F/m.

country rock The local rock intruded by a body of igneous rocks.

covalent bond See **bond.**

covers Coversine.

coversine See **trigonometric functions.**

covolume The volume of a gas actually occupied by its atoms or molecules as physical bodies of finite sizes. It is represented by the constant b in the van der Waals equation.

cP Centipoise.

cps Cycles per second. Syn. **hertz.**

Crab nebula A supernova remnant, a mass of expanding gas in Taurus resulting from the supernova explosion that occurred in A.D. 1054.

craton The portion of a continent that has not been subjected to major deformation since the beginning of the Paleozoic era. It includes continental shield and platform.

creek A water course larger than a brook but smaller than a river.

crescentic beach A beach concave toward the ocean.

creta Latin for *chalk.*

crevasse A vertical fissure in a glacier.

cribrate Defining a structure exhibiting sieve-like perforations.

cristobalite A high-temperature polymorph of quartz, stable between 1470°C and 1713°C (the melting point) at atmospheric pressure and in the absence of impurities. See **silica.**

critical angle The smallest angle of incidence producing total reflection when an electromagnetic or acoustic wave traveling through a medium encounters an interface separating it from another medium having a smaller refractive index.

critical density (ρ_c) The minimum density needed to decelerate the expanding universe and make it collapse into the next singularity.

$$\rho_c = 3H_0^2/8\pi G$$

where H_0 = Hubble constant, G = gravitational constant. For the current value of H_0 = 18 km/s/ 10^6 l.y., ρ_c = 6.5·10^{-30} g/cm^3. See **universe.**

critical distance The distance at which a seismic wave traveling through a surface layer arrives at the same time as a similar wave traveling through a parallel, underlying layer of faster transmission.

$$d_c = 2h \, [(v_2 + v_1)/(v_2 - v_1)]^{1/2}$$

where d_c = critical distance, h = depth of interface between surface layer and deeper layer, v_1 = velocity of seismic wave through surface layer, v_2 = velocity of seismic wave through deeper layer ($v_2 > v_1$). If v_1 and v_2 are known, measurement of d_c will yield the thickness h of the surface layer.

critical level The altitude (650 km) in the atmosphere at which the magnitude of the altitude equals the horizontal mean free path of atmospheric atoms and molecules. It forms the base of the exosphere.

critical mass The mass of fissionable material needed to initiate a nuclear chain reaction. For ^{235}U it ranges from 16 kg as a solid sphere (diameter about 12 cm) to as little as 950 g if properly dispersed and shielded with neutron reflectors.

critical point The temperature and pressure at which the liquid and gaseous phases of a given substance lose their identities and become a single phase.

critical pressure Pressure at the critical point.

critical slope The maximum slope at which loose debris can hold its position without slumping. Cf. **angle of repose.**

critical temperature Temperature at the critical point. No gas above its critical temperature can be liquified by pressure alone.

critical velocity The velocity of a fluid at which the flow changes from laminar to turbulent.

cross-bedding A feature of bedded deposits when air or water currents frequently change direction, obliquely eroding previously deposited beds and depositing new beds at different angles. It is characteristic of dune and tidal deposits.

cross product (×) See **vectorial product.** Cf. **dot product.**

CRT Cathode-ray tube.

crust The outer layer of the solid Earth (including loose sediment cover) above the Mohorovičić discontinuity. **1. continental crust** Thickness = 20–80 km (average = 35 km); composition = sial above and sima below separated by the Conrad discontinuity; density = 2.8 g/cm³. **2. oceanic crust** Thickness = 7 km; composition = gabbro-basalt with 0–5+ km of sediments above; mean density = 2.9 g/cm³.

cryo- Prefix meaning *ice.*

cryogenics 1. The study of phenomena at very low temperature. 2. The technology of producing very low temperatures.

cryosphere The totality of ice on Earth.

cryoturbation A disturbance in soil or sediment caused by frost action.

cryptocrystalline Defining a crystalline rock with crystals smaller than the wavelength of visible light, i.e. smaller than 0.5 μm.

cryptoperthite A perthite with lamellae of thickness below 5 μm and thus clearly detectable only by x-rays. Cf. **microperthite, perthite.**

Cryptozoic The Precambrian time that includes the Archean and the Proterozoic. It ranges from the end of the Hadean ($3.8 \cdot 10^9$ y B.P.) to the beginning of the Paleozoic ($590 \cdot 10^6$ y B.P.). Cf. **Phanerozoic.**

crystal class Any of the 32 possible combinations of symmetry axes intersecting at one point in space.

crystal group See **space group.**

crystal lattice The set of points repeating in space and representing the periodic space distribution of crystal ions or molecules.

crystallite A center of incipient crystallization.

crystalloblastic Defining a metamorphic rock whose crystals have grown entirely as a result of the metamorphic process.

crystallographic axis Any of the axes of orientation in crystal description (*a* axis, front to back; *b* axis, left to right; *c* axis, vertical).

crystal system Any of the 6 major crystal groupings (isometric or cubic, tetragonal, hexagonal, orthorhombic, monoclinic, triclinic) based on the symmetry of crystal axes, planes, and faces.

ct Carat.

cubic See **isometric.**

cuesta The steep scarp of an asymmetric ridge.

cumulate An igneous rock consisting of crystals settled out of a melt under the effect of gravity.

cuprite The mineral Cu_2O.

curie (Ci) A unit of radioactivity equal to that quantity of a radioactive substance that produces $3.7 \cdot 10^{10}$ dps (exactly).

Curie point The temperature above which a ferromagnetic or ferrimagnetic substance becomes paramagnetic. Representative Curie points: Fe, 770°C; Co, 1131°C; Ni, 358°C; magnetite (F_3O_4), 578°C; hematite (Fe_2O_3), 675°C; maghemite (Fe_2O_3), 675°C; ilmenite ($FeTiO_3$), −205°C.

Curie's law "The magnetic susceptibility of paramagnetic substances is inversely proportional to absolute temperature."

Curie temperature See **Curie point.**

curl The curl of a vector **v** is the vectorial product $\nabla \times \mathbf{v}$. Syn. **rotation.** Cf. **del, divergence.**

current *(Mechanics)* The flow of a fluid. *(Electricity)* The flow of charged particles. It is measured in amperes.

current density (*J***)** The electric current per unit cross-section of a conductor:

$$J = i/A$$

where J = current density, i = current, A = cross-section of conductor.

current mark Any mark on a sedimentary layer caused by running water.

cusp Any of the seaward accumulations of beach material separating crescentic beaches.

Cyanobacteria The phylum that includes all photosynthesizing bacteria. These organisms use bacteriochlorophyll and chlorobium chlorophyll rather than the chlorophyll a and b mixture used by Prochlorophyta and all higher plants.

Cyanophyta Syn. **Cyanobacteria.**

cybernetics The mathematical analysis of information flow in physical, biological, and social systems.

cyclic 1. Referring to a chemical compound in which the atoms are arranged in a ring or in a closed chain. 2. Referring to a phenomenon that repeats itself at regular intervals through space and/or through time. 3. Referring to a system that returns to a specific state at regular time intervals.

cycloalkanes Hydrocarbons containing C atoms arranged in a ring structure with only C−C single bonds. General formula, C_nH_{2n}. E.g. cyclopropane, C_3H_6; cyclobutane, C_4H_8.

cycloalkenes Hydrocarbons containing C atoms arranged in a ring structure with one C=C double bond. General formula, C_nH_{2n-2}. E.g. cyclohexene, C_6H_{10}.

cycloalkynes Hydrocarbons containing C atoms arranged in a ring structure with one C≡C triple bond. General formula, C_nH_{2n-4}. E.g. cyclopropyne, C_3H_2. Cycloalkynes are rare because a ring with a triple bond is under strain and therefore unstable.

cycloid The curve generated by a point on a circle as the circle rolls along a straight line.

$$x = r \cos^{-1}[(r - y)/r] - (2ry - y^2)^{1/2}$$

where r = radius of the circle.

cycloidal universe A cosmological model maintaining that the universe had no beginning and will have no end, as it goes through an infinite series of expansions and contractions separated by singularities. One cycle would last about $80 \cdot 10^9$ y.

cyclone A broadly circular system of low atmospheric pressure moving clockwise (northern hemisphere) or counterclockwise (southern hemisphere) at moderate speed (10–25 km/h) but with strong geostrophic winds spiraling counterclockwise (northern hemisphere) or clockwise (southern hemisphere) as they blow from neighboring pressure highs toward the low-pressure center. Cf. **anticyclone.**

cyclothem Any of the rhythmic alternations of continental and shallow marine deposits on a low continental platform, caused by rhythmic marine regressions and ingressions related to glaciation and deglaciation elsewhere. North American cyclothems often include a coal layer.

cyclotron A circular particle accelerator in which the particles, generated at the center, are accelerated spirally outward by an alternating electric field within a constant magnetic field normal to the plane of the spiral path.

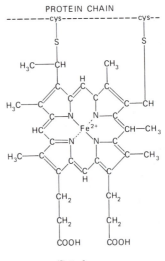

Cytochrome c.

cyst An enclosing and protecting capsule formed by many protozoa and protophyta undergoing a resting stage in order to temporarily survive adverse conditions.

cytidine $C_9H_{13}N_3O_5$ (mol. mass = 243.219), a nucleoside consisting of cytosine linked to a ribose sugar.

cytochromes A group of organic chelates consisting of a heme core with attached protein. The Fe atom in the heme center can be oxidized and reduced. Cytochromes occur mainly in mitochondria and chloroplasts in eucaryotic cells and are important as intermediaries in the electron transfer chain.

cytoplasm The living matter in eucaryotic cells between cell wall and nuclear membrane. Cf. **protoplasm.**

cytosine A nucleic acid base, $C_4H_5N_3O$ (mol. mass = 111.103).

D

δ Declination.

∂ Partial derivative.

d 1. Day. 2. Dextrorotatory. 3. Diameter. 4. Diffuse (see *s, p, d, f*).

d- Deci-

D 1. Deuterium. 2. Dextral chirality. 3. Diffusion coefficient.

d_E Ephemeris day.

da Deca.

dacite The extrusive equivalent of granodiorite.

dalton Syn. **atomic mass unit.**

Dalton's law "The pressure of a gas mixture, in which the components do not react with each other, is equal to the partial pressures of the component gases."

$$P = \Sigma p_i$$

where p_i = partial pressure of component *i*.

dam Decameter.

damped oscillation See **oscillation.**

Daniell cell A voltaic cell with the anode consisting of Cu in a saturated $CuSO_4$ solution and the cathode consisting of Zn in a dilute solution of $ZnSO_4$. The two solutions are separated by a porous partition that allows passage of the Cu^{2+}, Zn^{2+}, and SO_4^{2-} ions. The cell develops a voltage of 1.10 volts at 25°C. Cf. **Leclanché cell, Weston cell.**

darcy A unit of permeability equal to the passage of 1 cm³ of a fluid of 1 cP viscosity flowing in 1 s under a pressure difference of 1 atm through a porous medium having a cross section of 1 cm² and the length of 1 cm. See **permeability** *(Geology).*

dark dwarf A white dwarf star that has cooled and radiates mainly in the infrared.

Darwin glass A highly siliceous (86.34% SiO_2), vesicular glass from Mt. Darwin, western Tasmania. It is associated with an impact crater (Darwin Crater, 1.2 km wide) and has been assigned the age of 720,000 y by fission track. It appears to have been formed at the same time as the Australites, but it is richer in silica and has a higher Fe_2O_3/FeO ratio.

date line See **International Date Line.**

datum 1. A reference level for altitude and depths. See **sea level datum.** 2. A reference time, as identified by worldwide appearances and disappearances of shelled marine plankton and coccoliths.

dB Decibel.

dc Direct current.

de Broglie wave The quantum mechanical wave associated with a material particle. Wavelength:

$$\lambda = h/p$$

where h = Planck's constant, p = momentum of the particle.

debye A unit of electric dipole moment for molecular dipoles. 1 debye = 10^{-18} statC·cm = $3.335641 \cdot 10^{-30}$ C·m.

deca- Prefix meaning *10.*

decay constant (λ) The ratio of a quantity Q or number of items N decaying per unit of time (activity) to the quantity Q or number of items N present.

$$\lambda = -(dQ/dt)/Q$$
$$\lambda = -(dN/dt)/N.$$

decay equation The equation

$$Q = Q_0 e^{-\lambda t}$$

or

$$N = N_0 e^{-\lambda t}$$

expressing the decrease of quantity Q or number of items N from an initial value Q_0 or N_0 at a rate λ through time t, where $\lambda = -(dQ/dt)/Q$ or $-(dN/dt)/N$ = decay constant. See *e*. Cf. **attenuation, growth equation.**

deceleration parameter (q_0) A parameter expressing the change with time in the rate of deceleration of the expanding universe because of self-gravitation.

$$q_0 = 4\pi G\rho_t/3H_t^2$$

where G = universal gravitational constant, ρ_t = mean density of the universe at time t, H_t = value of the Hubble constant at time t. Cf. **critical density**.

deci- Prefix meaning *1/10th*.

decibel (dB) 1/10th of the common logarithm of the ratio of the powers of two signals (usually acoustic or electric) differing by a factor of 10. See **bel**.

declination *(Astronomy)* The angular distance north ($+$) or south ($-$) from the celestial equator = latitude in the equatorial coordinate system (symbol δ). *(Geophysics)* The angle, at any given location, between geographic North and direction of magnetic field measured clockwise 0° to 360°.

décollement The detachment and tectonic displacement of a group of rocks.

dee Either of the two D-shaped chambers of a cyclotron, immersed in a magnetic field and separated by a gap across which a potential difference is reversed at high frequency. See **cyclotron**.

deep-focus earthquake See **earthquake**.

deep-scattering layer A layer of nekton in the ocean that rises from about 800 m of depth during the day to a depth of about 200 m at night, consisting of populations of small fish species. Acoustic waves from echosounders and fathometers are reflected and scattered.

Deep-Sea Drilling Project (DSDP) The largest project of geological exploration of all times, directed at studying the ocean floor by drilling and coring.

deep-sea sediments Sediments, largely transported by wind or water, precipitated from the water column, or resulting from the alteration of volcanic ash, that are deposited on the deep ocean floor. The major types are *foraminiferal ooze* (65% of the deep ocean floor in the Atlantic, 36% in the Pacific, 54% in the Indian Ocean, 47% oceanwide), red clay (26% in the Atlantic, 49% in the Pacific, 25% in the Indian Ocean, 38% oceanwide); *diatom ooze* (7% in the Atlantic, 10% in the Pacific, 20% in the Indian Ocean, 12% oceanwide); *radiolarian ooze* (5% in the Pacific, 0.5% in the Indian Ocean, 3% oceanwide); *pteropod ooze* (2% in the Atlantic, 0.5% in the Pacific, 0.5% oceanwide). Major components are clay minerals derived from land and from the alteration of basaltic volcanic ash, quartz, shells of planktonic foraminifera and radiolaria, diatom frustules, coccoliths. Cosmic spherules and, in specific locations, microtektites are also found in deep-sea sediments.

deep-sea trench See **trench**.

deferent In the Ptolemaic system, the orbit described by the center of the epicycles of a superior planet.

degeneracy A state of a system of fermions in thermal equilibrium at very high density or very low temperature, for which the Fermi distribution rather than the Boltzmann distribution applies.

degenerate Referring to a state of matter in which one or more types of fermions are in a state of degeneracy.

degenerate electron gas An electron gas in a state of degeneracy.

degree of freedom *(Mechanics)* Any of the independent ways in which a particle, a system of particles, or an object can move in space. A single particle free to move in space has 3 degrees of freedom, corresponding to the three cartesian axes. A system of N particles, free to move in space independently of each other, has $3N$ degrees of freedom. *(Chemistry)* Any of the three variables (temperature, pressure, concentration) in a chemical system that must be specified to define the state of the system.

degree of latitude $1°\text{lat} = 111.1334 - 0.5594 \cos 2\phi + 0.0012 \cos 4\phi$ km, where ϕ = degrees of latitude.

deka- See **deca-**.

del (∇) The operator defining the change of the components of a vector along the three axes.
$$\nabla = \mathbf{i}(\partial/\partial x) + \mathbf{j}(\partial/\partial y) + \mathbf{k}(\partial/\partial z).$$

delta An accumulation of sediments at the mouth of a river. Topographic shape ranges from triangular to birdfoot.

delta front The narrow perideltaic belt between sea level and the depth of about 10 m where deltaic deposition is most active.

delta rays Radiation consisting of high energy (100–2000 eV) electrons freed from surrounding atoms by α particles originating from α decay.

demersal Syn. **benthic**.

dendritic Having a branching pattern.

dendrochronology Chronology based on the succession of annual tree rings.

dendroid Having a pattern or shape similar to that of a tree.

density (ρ) Mass/unit volume. Densities in g/cm^3 of common substances at STP are as follows: dry air (20°C) = $1.204 \cdot 10^{-3}$; pure water (20°C) = 0.99984; standard sea water (35‰ salinity, 20°C) = 1.025; crustal rocks = 2.8; mantle rocks = 3.4; iron = 7.87; nickel 8.90; lead = 11.34; gold = 18.88; platinum = 21.45; iridium = 22.42; osmium = 22.48; degenerate matter in stellar cores $\sim 10^7$ to 10^8; nuclear matter = $2.8 \cdot 10^{14}$. See **Density***.

density current A water or air current whose pattern of flow is controlled by its bulk density with respect to the surrounding fluid.

density parameter (Ω) The ratio of the mean density of the universe ($0.9 \cdot 10^{-30}$ g/cm^3) to its critical density ($6.5 \cdot 10^{-30}$ g/cm^3), equal to 0.14 (all uncertain values). See **critical density.**

deoxyribonucleic acid (DNA) See **DNA.**

deoxyribose A pentose sugar, $C_5H_{10}O_4$, identical to ribose in structure but with one less oxygen. See **DNA.**

derivative The ratio $dy/dx = f'(x)$ of the increment Δy of a function y to the increment Δx of the variable x as the value of Δx approaches 0. See **calculus, infinitesimal.**

descending node See **nodes.**

desert varnish A brownish-grayish patina consisting of Mn and Fe oxides deposited on rock surfaces in desert areas by the bacterium Metallogenium.

desiccant A drying substance, such as P_2O_5.

DESY Deutsches Electronen Synchrotron (Hamburg).

deuteric Defining a late reaction between the residual magmatic fluid and the minerals that were crystallized earlier from the magma.

deuterium ^{2}H, a stable isotope of hydrogen with the nucleus consisting of a proton and a neutron. Atomic mass = 2.01410177. Abundance in common terrestrial matter = 0.015% of total H.

deuteron The nucleus of deuterium.

dex Indicating an exponent on base 10. Thus, $10^k = k$ dex. Syn. **brig.**

dextral Clockwise on a plane or having positive helicity in space. Cf. **sinistral.**

dextral coiling See **coiling direction.**

dextral fault See **right-lateral fault.**

dextrorotatory (d) Defining a substance that rotates polarized light clockwise as seen by the eye viewing the substance in transmitted light, or counterclockwise in the direction of light transmission (negative helicity). Cf. **levorotatory.**

dia- Prefix meaning *through.*

diabase An intrusive rock consisting mainly of labradorite and pyroxene.

diachronous 1. Defining a lithologic unit that was deposited at progressively different times along a given direction (e.g. a basal conglomerate). 2. The appearance or disappearance of a species or assemblage at progressively different times in a given direction.

diagenesis The physical and chemical alteration of a sediment after deposition, not exclusive of metamorphism.

diallage A lamellar variety of augite or diospide.

dialysis Ultrafiltration through a semipermeable membrane to separate colloidal particles.

diamagnetic Defining a substance displaying the property of diamagnetism.

diamagnetism Property of substances whose electrons in the ground state are all paired and, therefore, whose atoms or molecules have zero net magnetic moment. The application of an external magnetic field changes the magnetic moments of the individual electrons giving rise to a small, net magnetic moment with a direction opposite that of the applied field. Permeability <1; susceptibility <0. E.g. bismuth: permeability (relative) = 0.9995; susceptibility = $-1.35 \cdot 10^{-6}$ CGS$_{emu}$.

diamictite Lithified diamicton.

diamicton An unsorted terrigenous sediment containing a wide mixture of sizes, resulting from glacial action, slumping, etc.

diamond The high-pressure phase of crystalline carbon. Density = 3.1; hardness = 10 on the Mohs scale; refractive index = 2.4195. Stability field, 20 kb at 0°C to 120 kb at 4000°C. Diamond is the stable form of C inside the Earth at a depth >150 km.

diapir A domal or mushroom-shaped structure resulting from the piercing of overlying rocks by a plastic rock (e.g. salt) forced to flow under rock overburden.

diastem An interruption in sedimentation, shorter than that resulting in a paraconformity, with little or no erosion.

diastrophic Resulting from diastrophism.

diastrophic eustatism Worldwide sealevel change caused by vertical motions of the ocean floor.

diastrophism Deformation of the Earth's crust due to tectonism.

diatexis Total melting of pre-existing rock. Cf. **anatexis.**

diatom A single-celled plant of the class Bacillariophyta, Kingdom Protoctista.

diatomaceous earth See **diatomite.**

diatomite A soft and friable rock consisting mainly of diatom frustules.

diatom ooze An oceanic deposit consisting of at least 30% diatom frustules. See **deep-sea sediments.**

diatreme A volcanic pipe formed by a volcanic gas explosion.

dichroic Exhibiting dichroism.

dichroism The ability of an anisotropic mineral to transmit light of two different colors along two different directions. Cf. **pleochroism.**

dielectric A nonconducting substance, i.e. a substance that "lets through" an electric field.

dielectric constant (κ) 1. Relative permittivity of a substance.

$$\kappa = \epsilon/\epsilon_0$$

where ϵ = permittivity of the substance, ϵ_0 = permittivity constant. 2. The ratio of the capacitance of a capacitor with the space between its plates filled with a substance to the capacitance when the space is empty.

dielectric strength The maximum electric field intensity a dielectric substance can withstand. Examples (kV/mm): vacuum = ∞; mica = 160; amber = 90; Teflon = 60; paper = 14; Pyrex glass = 13; porcelain = 4; air = 0.8.

diesel cycle An internal-combustion engine cycle in which ignition is produced by compression.

differential The infinitesimal change in a function proportional to the corresponding infinitesimal change in the variable, i.e. $dy = f'(x)\,dx$ when $y = f(x)$. See **calculus, derivative, infinitesimal.**

diffraction The effect on wave propagation produced by the edge of an obstacle, or by an aperture with dimension, or a grating with spacings, of the order of magnitude of the wavelength.

diffraction grating A grating consisting of parallel, equidistant lines used to produce spectra by diffraction. Line separation is of the order of magnitude of the wavelength of the incident light.

diffusion coefficient See **diffusivity.**

diffusivity The amount in grams per second of a substance diffusing through an area of 1 cm^2.

digital 1. Defining a discretely varying quantity. 2. Defining the discrete representation of a continuously varying quantity.

dike A sheet-like intrusion of an igneous rock cutting through the bedding or foliation of the country rock. Dikes range from centimeters to meters in thickness. Cf. **sill.**

dike swarm A group of dikes originating from a common magmatic chamber.

dilatancy Microfracturing, pore volume increase, and change from close-packed to open-packed structure, during diastrophism.

dilation Increase in volume without a change in shape.

dimensional formula The expression of a derived quantity in terms of powers of fundamental quantities.

dimensionless number The quotient of two quantities having the same dimension.

dimictic Defining a mid- or high-latitude or altitude lake with two overturns per year (early spring, following surface ice melting; late fall, by surface cooling). Cf. **monomictic, polymictic.**

dimorphic Referring to the property of dimorphism.

dimorphism *(Chemistry)* The property of certain chemical substances to crystallize in two different forms. *(Biology)* The occurrence of two different forms within the same species (e.g. sexual dimorphism).

dinoflagellate Any of the single-celled flagellated aquatic organisms belonging to the phylum Dinoflagellata, kingdom Protoctista.

diode 1. A vacuum tube consisting of a cathode and an anode. 2. A two-terminal semiconductor.

dioecious Having male and female reproductive organs in separate individuals. Cf. **monoecious.**

diopside A clinopyroxene, $CaMg(SiO_3)_2$.

diopter A unit of power for lenses.

$$D = 1/f$$

where D = number of diopters, f = focal length in m.

diorite A plutonic rock consisting mainly of sodic plagioclase, pyroxene, and amphibole.

dip The maximum inclination of a plane or a layer. It is normal to the strike.

diploid Defining a nucleus or cell with two sets of chromosomes. 2. Defining an organisms or generation in which somatic cells have two sets of chromosomes. Cf. **haploid, monoploid.**

dipole An object or a system with oppositely charged or magnetized ends.

dipole antenna An antenna consisting of two symmetric elements connected to the two ends of a transmitting or receiving line.

dipole field *(Physics)* The field produced by an electric or magnetic dipole. *(Geophysics)* The Earth's principal magnetic field, with dipole moment of $7.94 \cdot 10^{22}$ A m^2 or $7.94 \cdot 10^{25}$ gauss cm^3. Its axis coincides with the rotational axis of the Earth if averaged across $\geq$ 10,000 y.

dipole moment The moment of an electric or magnetic dipole. **1. electric dipole moment** The product of one of the charges of a dipole by the distance between the two poles. It is expressed in coulomb$\cdot$meter. **2. magnetic dipole moment** The product of magnetic pole strength times the distance between the two magnetic poles. It is equal to the product NiA in the equivalent electric coil, where N = number of turns, i = current, A = area of turn. It is expressed in ampere$\cdot$m^2

dip pole Either one of the two locations on Earth where magnetic inclination is 90°. North magnetic pole = 77.3°N, 101.8°W; south magnetic pole = 65.8°S, 139.0°E.

dip-slip fault A fault in which the displacement is parallel to the dip.

Dirac monopole A particle proposed by Dirac in 1931 to symmetrize Maxwell's equations. It carries one or more discrete magnetic charges m_0.

$$m_0 = hc/4\pi e$$
$$= e/2\alpha$$
$$= 68.52 \, e$$

where h = Planck's constant, c = speed of light, α = fine structure constant, e = electron charge.

direct See **prograde.**

direct current An electrical current that flows only in one direction.

directrix The line representing the locus of all points whose distance from a point on a conic section is equal to the distance of that point from a focus. See **conic sections.**

dirty sandstone A sandstone with more than 15% of clay.

disaccharides A family of carbohydrates with 12 carbon atoms.

disconformity An unconformity representing a significant time of nondeposition and erosion without diastrophism. Bedding of the older and younger formations thus remains parallel.

discordance A nongenetic term used to indicate lack of parallelism between the bedding planes of two overlying sets of beds.

discordia The line subtending the concordia curve on which minerals that have lost Pb or gained U fall. See **concordia.**

disequilibrium dating method See **uranium disequilibrium dating method.**

dislocation A line defect in a crystal, including edge dislocation (row of atoms from edge of crystal extending only partway in) and screw dislocation (a row of atoms about which a crystallographic plane appears to rotate during crystal growth).

dispersion The scattering of radiation by an obstacle or a grid.

disphotic zone A zone in a natural environment where light is dim, with illumination intermediate between that of the euphotic and aphotic zones.

displacement current The term $\partial D/dt$ in Maxwell's equation describing the magnetic effect of a variable electric field. The name "displacement" derives from an electric field across a dielectric causing a displacement of charges (and also, as Maxwell believed, causing polarization of "ether" in vacuo).

displacement vector A vector describing the change in position of a particle or a body.

distance (cosmic) The distance r of a distant celestial body.

$$r = v_r/H_0$$

where v_r = recessional velocity, H_0 = Hubble constant. See **velocity (recessional).**

distance modulus The difference between apparent and absolute magnitude of a star and, therefore, a measure of its distance.

$$m - M = 5 \log (d/10)$$

where m = apparent magnitude, M = absolute magnitude, d = distance in parsecs.

distribution coefficient The ratio of the concentrations of a substance in a liquid and in a precipitate in equilibrium, or in two immiscible fluids also in equilibrium.

diurnal 1. Occurring once a day. Cf. **semidiurnal.** 2. Occurring during the day.

diurnal tide A tide occurring once in a lunar day.

divergence (div) The divergence of a vector field ϕ is the scalar product $\nabla \cdot \phi$. See **del.**

divergent margin See **divergent plate margin.**

divergent plate margin The boundary between

two plates moving away from each other, along which new lithosphere is formed.

division A major subdivision of the kingdom Plantae, equivalent to phylum in the other kingdoms.

dl A racemic mixture, consisting of equal quantities of dextro- and levorotatory molecules of a given compound.

D layer The seismic zone of the Earth's interior extending from 1000 to 2885 km of depth.

dm Decimeter.

DNA Deoxyribonucleic acid, the carrier of genetic information in all living organisms (except some viruses for which the carrier is RNA). It consists of a chain of 10^4 nucleotides (plasmid) to 10^{11} nucleotides (Homo). Molecular mass and length (stretched molecule) range from $5 \cdot 10^8$ u and 0.7

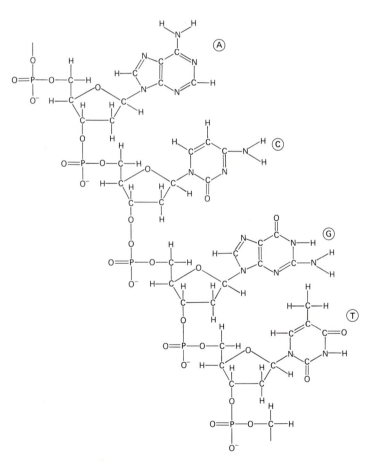

DNA segment. A = adenine; C = cytosine; G = guanine; T = thymine.

μm (plasmid) to $2.7 \cdot 10^{13}$ u and 4 cm (Homo). The width of a DNA molecule is about 20 Å. A single DNA molecule self-duplicates by linking its purines to free nucleotide pyrimidines and its pyrimidines to free nucleotide purines (adenine to thymine, cytosine to guanine) via H bonds and constructing a complementary, antiparallel chain. The double molecule forms a coiled structure exhibiting positive helicity (double helix). Cf. **RNA.**

dolarenite An arenite consisting of dolomitic particles.

dolerite Syn. **diabase.**

dolina A large sinkhole.

Dollo's law "An evolving species does not return to its ancestral form."

dololutite A consolidated dolomitic mud.

dolomicrite A lithified dolomitic mud.

dolomite 1. The mineral $CaMg(CO_3)_2$. 2. A sedimentary rock predominantly consisting of the mineral dolomite.

dolorudite A rudite consisting of dolomitic fragments.

dolosiltite A siltite consisting of dolomitic fragments.

dolosparite Dolomitic sparite.

dolostone Syn. **dolomite** (def. 2).

domain A region within a nonmagnetized ferromagnetic material in which the individual atomic magnetic moments are parallel to each other. Dimensions are 0.1–1 mm across and involve 10^{15}–10^{18} atoms. Thickness of wall between domains = 0.05–0.1 μm.

dome An upward convex, circular or elliptical uplift.

doping The addition of minute ($\approx 10^{-6}$) amounts of an element with 3 or 5 valence electrons to a semiconducting element with 4 valence electrons in its valence shell. See **semiconductor.**

Doppler effect The apparent change in frequency of a sound wave received, with respect to the frequency emitted, when source and receiver are in relative motion.

$$\nu = \nu_0(v + v_m - v_r)/(v + v_m - v_s)$$

where ν = frequency received, ν_0 = frequency emitted, v = speed of sound in medium, v_m = velocity of medium, v_r = velocity of receiver, v_s = velocity of source. In the preceding equation, v is always positive and the signs shown for v_m, v_r, and v_s refer to the case when medium, receiver, and source move in the same direction. The signs are reversed for any opposite direction.

Doppler shift The apparent change in the wavelength of light as seen by an observer when the distance between source and observer along the line of sight is changing. For nonrelativistic speeds:

$$\Delta\lambda/\lambda = (dr/dt)/c$$

For relativistic speeds:

$$\Delta\lambda/\lambda = [(1 + v_r/c)(1 - v^2/c^2)^{-1/2}] - 1$$

In the preceding, λ = wavelength; r = distance between source and observer; c = speed of light; v_r = radial component of velocity v of source with respect to observer. See **blue shift, red shift.**

d* orbital** The orbital of an atomic electron characterized by an orbital angular momentum number of 2. See ***s, p, d, f.

dot product ($\cdot$) See **scalar product.** Cf. **cross product.**

double helix See **DNA.**

doublet 1. Two closely separated spectral lines in alkali spectra resulting from spin and orbital angular momentum interaction of the valence electrons. 2. Two elementary particles with the same baryon number, spin, and parity, but different charge (e.g. proton and neutron). Cf. **triplet, multiplet.**

doughnut The toroidal vacuum chamber of a betatron or synchrotron in which the particles are accelerated.

dpm Disintegrations per minute.

dps Disintegrations per second.

drag mark 1. A groove (wider) or a striation (narrower) made by a rock fragment or other object drawn along a sediment surface. 2. A groove cast.

drewite A fine mud consisting of algal needles.

drift *(Physics)* The movement of electrons and holes in a semiconductor under the influence of an applied voltage (commonly = 0.01–1 m/s/V). *(Geology)* A generic name identifying all materials deposited by a glacier or ice sheet.

drift current *(Physics)* An electric current resulting from the motion of charged particles in response to an applied electric field. Drift current speeds of conducting electrons in metals are a fraction of mm/s compared to thermal speeds around 1500 km/s and wave speeds close to the speed of

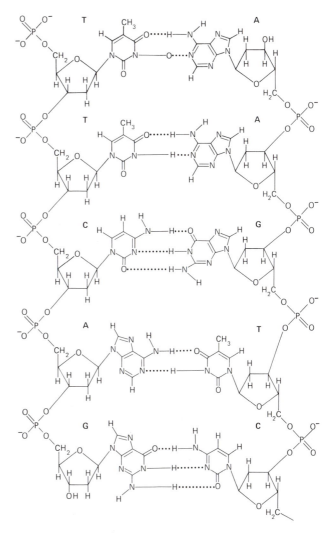

Double helix segment. A = adenine; C = cytosine; G = guanine; T = thymine. Dotted lines represent hydrogen bonds. (From Curtis 1983, p. 288, Fig. 14–11)

light. *(Oceanography)* A broad oceanic current driven by prevailing winds. Normal speeds in open ocean average 0.5–1 knots (0.25–0.50 m/s).

drift speed *(Physics)* The speed of a drift electric current. See **drift current** *(Physics)*. *(Oceanography)* The speed of an oceanic drift current. See **drift current** *(Oceanography)*.

dripstone Any calcitic or aragonitic incrustation deposited by water dripping from the roof of a cave. Included are stalactites, stalagmites, and crusts.

drumlin An ellipsoidal hill consisting of till deposited by an advancing ice margin. It is elongated in the direction of ice flow and the updrift termi-

nus has a steeper slope than the downdrift terminus.

druse A cavity in a vein or rock partly or totally filled with crystals formed from solutions from the host vein or rock.

drusy Having the structure and appearance of a druse.

dry cell A voltaic cell in which the electrolyte is in the form of a jelly or a paste.

DSDP Deep-Sea Drilling Project.

d-**spacing** The distance between successive crystal planes as seen in x-ray analysis. See **Bragg equation.**

ductility The property that enables metals to be drawn into wires.

dune A low ridge consisting of sand grains accumulated by the wind. Maximum height observed (SE Algeria) = 430 m.

dunite An ultramafic rock consisting of olivine with accessory chromite.

duricrust A hard crust on or within a soil in a semiarid region formed by salts deposited by ground water rising by capillary action and evaporating.

dust Particles of solid inorganic matter <62.5 μm in size.

dyn Dyne.

dynamic height The distance, from the geoid, of points on a gravitational equipotential surface.

dynamic metamorphism Metamorphism produced by tectonic movements.

dynamics The branch of mechanics dealing with the motions of bodies under the influence of forces.

dynamic viscosity See **viscosity.**

dynamite An explosive consisting of a mixture of 25–60% nitroglycerin dispersed in diatomite, with small amounts of sawdust and chalk.

dynamo See **generator.**

dyne (dyn) The CGS unit of force, defined on the basis of the Newtonian formula $F = ma$ as that force F that will impart an acceleration $a = 1$ cm/s^2 to a mass $m = 1$ g.

dynode An electrode that emits secondary electrons when struck by an electron.

E

ε Permittivity.

ϵ_0 Permittivity constant = permittivity of vacuum.

η A nonstrange meson. See **Elementary particles***.

e 1. Eccentricity. 2. Electron charge. 3. Symbol for 2.7182818284590 ..., which is the value reached in a unit time by a unit quantity increasing at a rate always equal to its own value. e is used as the base of natural or Napierian logarithms (symbol ln).

E Electric field intensity.

$\mathcal{E}$ Electromotive force.

E_k Kinetic energy.

E_p Potential energy.

Early Referring to the early portion of a chronological or chronostratigraphic unit. Cf. **Late, Lower, Middle, Upper.**

early diagenesis Diagenesis during deposition and immediately after.

Earth The third planet from the Sun. Mean distance from the Sun = 1.000 AU = 149,597,870.7 km = 8.31675 light minutes = 499.004784 light seconds. Sidereal period = 1.0000387 tropical years = 365.25636565 days. Orbital eccentricity = 0.01675104. Sidereal rotational period = 0.99727 days = 23 h 56 m 4.091 s. Angle between rotational axis and the normal to the orbital plane = latitude of tropics = colatitude of polar circles = 23°26′28.0″ (A.D. 1986). Equatorial radius = 6378.139 km; polar radius = 6356.779 km; mean radius = 6371.03 km. Mass = (5.9737 ± 0.0004)·10^{24} kg. Mean density = 5.518 g/cm³. Core: radius = 3486 km; composition (estimated), Fe = 90%, Ni = 9%, Ca + P + S + Co = 1%. Mantle: thickness = 2878 km under oceans, 2850 under continents; composition, Fe–Mg silicates. Crust: oceanic, 7 km thick (mean) consisting of Ca–Fe–Mg silicates; continental, 35 km thick (mean) consisting of Na–K–Ca–Fe–Mg Al-silicates and quartz. Mean oceanic depth = 3729 m; mean continental altitude = 840 m. Land surface = 148.017·10^6 km²; ocean surface = 362.033·10^6 km²; total surface = 510.050·10^6 km². Mean magnetic field = 0.5 gauss. Mean surface temperature = 288 K. Atmospheric pressure = 1.000 atm = 760 mmHg = 1.013250 bar. Gases in atmosphere, N_2 = 78.084%, O_2 = 20.946%, Ar = 0.934%, CO_2 = 0.032%, other gases = 0.004%, and variable amounts (0.004–4%) of water vapor. One satellite, the Moon. See **Atmosphere—composition*, Earth—astronomical and geophysical data*, Earth interior—physical data*.**

earthquake Any of the abrupt motions caused by ruptures in the crust or mantle. Earthquakes are classified in terms of strength (see **Richter scale, Mercalli scale**) and depth (shallow, 0–60 km; intermediate, 60–300 km; deep, 300–720 km). No earthquakes have been registered from depths greater than 720 km. The largest earthquakes, occurring once or twice in a decade, release 10^{18} joules of energy or more. Instrumental earthquakes, numbering about 500,000/y, release a total of about 3.5·10^{14} J/y. The yearly earthquake energy released is about 1.2·10^{20} J, equal to 1/100 of that released as geothermal flux.

ebb and flood The alternating seaward and shoreward ward motion of the tide.

ebb current The outgoing tidal current.

ebb tide The outgoing tide.

eccentricity (e) The ratio PF/PD, where PF = distance of a point from focus of a conic section, PD = distance between point and directrix.

ellipse: $e = (a^2 - b^2)^{1/2}/a = c/a < 1$

where a = semimajor axis, b = semiminor axis, c = semidistance between foci;

hyperbola: $e = (a^2 + b^2)^{1/2}/a > 1$

where a = semitransversal axis, b = semiconjugate axis;

parabola: $e = 1$
circle: $e = 0$

See **conic sections.**

echo ranging The determination of the distance of a target from a source by the two-way travel time of a signal from source to target.

echo sounder An oceanographic instrument used to determine the depth of the sea floor by measuring the time an acoustic pulse takes to reach the bottom and return.

ecliptic The intersection of the plane of the Earth's orbit with the celestial sphere.

ecliptic coordinate system See **coordinate systems.**

eclogite A high-temperature, high-pressure metamorphic phase of gabbro, consisting essentially of omphacite and pyrope. Density = 3.4.

ecocline 1. A genetic gradient of adaptation to an environmental gradient. 2. A gradient in ecologic conditions.

ecome The locus occupied by a species, subspecies, or identifiable population as characterized by its ecologic conditions.

ecophenotype A nongenetic variant of a species produced by specific ecologic conditions.

ecotope The habitat of a particular organism.

ecotype A subspecies adapted to specific ecologic conditions.

écoulement *(French)* The gravitational sliding of a structural unit.

edaphic Pertaining to soil, especially in reference to the organisms living in or on it.

eddy A circular motion within a fluid current, resulting from turbulence. Natural eddies range across from centimeters in rivers to tens of kilometers in oceanic currents.

eddy currents Electric currents induced within a conductor by a change in magnetic flux.

eddy diffusion The diffusion of physical or chemical properties of a fluid normally to its direction of flow, caused by eddies.

edge dislocation See **dislocation.**

Ediacaran The last period of the Proterozoic, ranging from 630 to 590 million years B.P. It is characterized by the appearance of metazoa.

Edicarian See **Ediacaran.**

EDTA The sodium salt of ethylenediaminetetraacetic acid, $(CH_2COO^-Na^+)_2NCH_2 \cdot CH_2N \cdot (CH_2COO^-Na^+)_2$, a strong chelating agent.

effective amplitude The rms (root-mean-square) value of a periodically varying phenomenon. For a sinusoidal wave, the effective or rms amplitude is equal to the maximum amplitude divided by the square root of 2 or multiplied by 0.707107.

effective current The rms (root-mean-square) value of an alternating current, producing the same heating effect as the corresponding value of a direct current. For a sinusoidal alternating current, the effective or rms current is equal to the maximum current divided by the square root of 2 or multiplied by 0.707107.

effective nuclear charge The charge experienced by a valence electron, consisting of the nuclear charge less the shielding effect of the inner electrons.

effective temperature (T_{eff}) Surface temperature of a blackbody having the same radius of a given star or other body and emitting the same total power.

$$T_{eff} = (L/4\pi R^2\sigma)^{1/4}$$

where L = total power (= luminosity), R = radius of star, σ = Stefan-Boltzmann constant = $5.6705 \cdot 10^{-8}$ W m^{-2} K^{-4}.

effective value The rms (root-mean-square) value. For a sinusoidal wave, the effective or rms value is equal to the maximum value divided by the square root of 2 or multiplied by 0.707107.

effective voltage The rms (root-mean-square) value of an alternating voltage. For a sinusoidal alternating current, the effective or rms voltage is equal to the maximum voltage divided by the square root of 2 or multiplied by 0.707107.

e.g. *Exempli gratia,* Latin for *for example.*

eigen *(German)* Characteristic.

eigenfrequency 1. A characteristic frequency at which an oscillator can oscillate. 2. A specific frequency for which the Schroedinger equation has a solution.

eigenfunction A characteristic function of the eigenvalues of the stationary energy states of atomic systems.

eigenvalue A characteristic energy value of the eigenfrequency of a stationary energy state of an atomic system.

Einstein shift The increase in wavelength of light escaping from a massive body.

$$\Delta\lambda/\lambda = Gm/Rc^2$$

where λ = wavelength, G = gravitational constant, m = mass of body, R = radius of body, c = speed of light.

Ekman spiral A spiral describing the vertical change in the velocity vector of an oceanic current. Surface flow is offset by 45° (to the right in the northern hemisphere, to the left in the southern hemisphere) with respect to the wind direction because of the Coriolis effect. The velocity vector is further offset as depth increases until, at a depth called *friction depth,* it reaches a direction opposite that of the wind. As the magnitude of the velocity vector decreases with depth, the net transport (Ekman transport) is at 90° (to the right in the northern hemisphere, to the left in the southern hemisphere) of the wind direction.

Ekman transport   See **Ekman spiral.**

elastic limit   The stress below which deformation of a solid is elastic and above which it is plastic.

E layer   The Earth's outer core.

electret   A dielectric capable of retaining semipermanent electric polarization. Examples are carnauba waxes (from the Brazilian wax palm *Copernicia cerifera*) and some barium titanates.

electric dipole   An atom, molecule, or body in which the center of negative charge distribution does not coincide with the center of positive charge distribution.

electric displacement (D)   The vector

$$\mathbf{D} = \epsilon_0 \mathbf{E} + \mathbf{P}$$

where ϵ_0 = permittivity constant, $\mathbf{E}$ = electric field intensity, $\mathbf{P}$ = electric polarization. It is expressed in C/m^2.

electric field   A region of space in which a point electric charge experiences a force.

electric field intensity (E)   Force per unit charge at a point in space within an electric field. It is expressed in volts/meter.

electric field strength   See **electric field intensity.**

electric polarization (P)   The vector

$$\mathbf{P} = \epsilon_0 (\kappa - 1)\mathbf{E}$$

where ϵ_0 = permittivity constant, κ = dielectric constant, $\mathbf{E}$ = electric field intensity. It is expressed in C/m^2.

electric potential   The work required to bring a unit charge from infinity to a specified position in an electric field.

electroaffinity   The electrode potential for a concentration of 1 gram-ion/liter of ions liberated at atmospheric pressure.

electrochemical series   A series ranking different metals and compounds in terms of their electrode potentials.

electrode potential   The potential between electrode and electrolyte, usually referred to the potential of the hydrogen electrode as a standard. See **standard electrode.**

electroluminescence   The generation of cold light in nonmetallic solids by the application of an electric field. This is best achieved across *p-n* junctions in light-emitting diodes.

electrolysis   The separation of positive from negative ions in a solution or molten material by the application of a voltage, giving rise to an electrical current.

electrolyte   A substance that, to a greater (strong electrolyte) or lesser (weak electrolyte) extent, dissociates into ions in water or other solvents.

electromagnet   A coil wound around an iron core that becomes magnetized when a current flows through the coil. For a given coil and current, the magnetic flux increases in inverse proportion to the reluctance of the core (Bosanquet's law).

electromagnetic field   A field created by changing electric and magnetic fields.

electromagnetic force   See **natural forces.**

electromagnetic induction   The induction of a voltage difference in a conductor by forcing it to move across a magnetic field or by changing the magnetic flux threading it.

electromagnetic interaction   The interaction of elementary particles resulting from coupling of charge with electromagnetic field. See **natural forces.**

electromagnetic radiation   An electromagnetic wave or a system of such waves. See **electromagnetic wave.**

electromagnetic spectrum   The distribution of electromagnetic radiation from a given source in terms of the range and intensities of the wavelengths emitted. See **Electromagnetic spectrum*.**

electromagnetic system of units (emu)   The CGS system of units based on the centimeter, gram, second, and abampere. In this system the permeability constant $\mu_0 = 1$. Cf. **electrostatic system of units.** See **Units*.**

electromagnetic wave   A periodic disturbance in an electromagnetic field created by an accelerated electric charge. It consists of an electric and a magnetic field perpendicular to each other and propagating with the speed of light.

electromotive force (emf)   The electric potential difference causing an electric charge to move within an electric field, defined as the work needed to carry a charge around a closed path within the field divided by that charge.

$$\text{emf} = \int E \cos \theta \, ds$$

where E = intensity of the electric field, θ = angle between the vector representing the direction of the field and that representing the direction of the path. It is expressed in volts.

electron (e) *(Physics)* A stable lepton carrying the natural unit of electric charge, equal to $-1.602177 \cdot 10^{-19}$ coulombs. Classical radius = $2.8179380 \cdot 10^{-15}$; effective radius $<10^{-18}$ m. Rest mass = $0.910939 \cdot 10^{-30}$ kg = 0.0005485799 u = 0.510999 MeV = 1/1836.153 of rest mass of proton. Classical density = $0.97188182 \cdot 10^{11}$ g/cm³. Effective density $>0.9 \times 10^{21}$ g/cm³. Spin angular momentum = 1/2 $(h/2\pi)$ (where h = Planck's constant) = $0.527286 \cdot 10^{-34}$ J s. Spin magnetic moment = Bohr magneton = $0.92740154 \cdot 10^{-23}$ A m². The electron is named after ἤλεκτρον, Greek for "amber," which was defined by Benjamin Franklin as having negative electric charge when rubbed with wool. *(Metallurgy)* An alloy of gold (80%) and silver (20%).

electron acceptor A substance capable of accepting electrons. An oxidizing substance. A Lewis acid. See **acid, base.**

electron affinity The propensity of certain atoms, molecules, and radicals to add an electron and form negative ions. Electron affinity is positive if the negative ion is more stable than the neutral species, in which case ion formation releases energy. Examples (electron affinity in eV): Cl = 3.617; F = 3.399; Br = 3.365; I = 3.059; S = 2.077 OH⁻ = 1.828; O = 1.461; Ag = 1.302 C = 1.263; H = 0.754.

electron charge (e) The elementary charge of negative electricity carried by the electron. $e = -1.602177 \cdot 10^{-19}$ coulombs.

electron Compton wavelength (λ_C) Compton shift of incident x-rays in collision with free electrons, with photon scattering angle of 90°. It is a length characteristic of the electron.

$$\lambda_C = h/m_e c$$

where h = Planck's constant, m_e = rest mass of electron, c = speed of light. It is equal to $2.4263106 \cdot 10^{-10}$ cm. See **Compton effect.**

electron donor A substance capable of donating electrons. A reducing substance. A Lewis base. See **acid, base.**

electronegative 1. Having a negative charge. 2. Referring to an atom or a molecule capable of attracting electrons.

electronic magnetic moment The total magnetic dipole moment produced by the orbital and spin motions of all electrons in an atom.

electron magnetic moment (μ_e) The magnetic moment of the electron resulting from its spin motion. It is equal to $0.928477 \cdot 10^{-23}$ A m² or J T⁻¹. Cf. **Bohr magneton.**

electron microprobe An analytical instrument using a narrow (<1 μm) beam of electrons to excite x-ray emission from a small (<1 μm²) area on a surface. Different elements emit x-rays of different characteristics.

electron microscope A type of microscope that uses a beam of energetic electrons (50–150 kV) passing through a thin or thinly sectioned specimen ($<50 \cdot 10^{-9}$ m thick) to produce an image on a cathode tube screen or on a photographic plate. Resolution is about 0.2–$0.5 \cdot 10^{-9}$ m, compared to 10^{-7} m for high-resolution optical microscopes. Cf. **scanning electron microscope.**

electron multiplier An electron tube containing a sequence of dynodes at successively higher positive potentials. Impinging electrons produce more electrons with each successive collision with a dynode surface, thus creating a cascade. Cf. **photomultiplier tube.**

electron orbit See **orbital.**

electron paramagnetic resonance (EPR) See **electron spin resonance.**

electron rest mass (m_e) The mass of a nonrelativistic electron.

$$m_e = 0.910939 \cdot 10^{-30} \text{ kg}$$
$$= 0.00054857990 \text{ u}$$
$$= 0.51099906 \text{ MeV}$$
$$= 1/1836.153 \, m_p$$

where m_p = rest mass of proton.

electron shell See **shell** *(Physics).*

electron spin resonance (ESR) Resonance between an incident electromagnetic wave of the appropriate frequency and the magnetic moment of unpaired electrons in a paramagnetic substance.

electron transfer chain The transfer of electrons to oxygen by respiratory enzymes (NAD, NADP), with concurrent formation of ATP.

electron transport chain See **electron transfer chain.**

electron volt A unit of energy, equal to the energy acquired by an electron in passing through a potential difference of 1 volt. It is equal to $1.6021773 \cdot 10^{-12}$ erg or $1.6021773 \cdot 10^{-19}$ joules.

electrophoresis The motion of charged particles, especially colloids, through a liquid under the influence of an electric field generated by two electrodes immersed in the liquid.

electrostatic system of units (esu) The CGS system of units based on the centimeter, gram, second, and statcoulomb. In this system the permittivity constant $\epsilon_0 = 1$. Cf. **electromagnetic system of units.** See **Units*.**

electroweak force The force combining the electromagnetic and weak forces at temperatures $> 10^{15}$ K.

electroweak theory A theory relating electromagnetic and weak interactions and their force carriers (the photon, with zero mass and, therefore, infinite range, and the $W^{\pm}$ and Z^0 bosons with masses of, respectively, 86.7 and 99.7 u and, therefore, the limited range of 10^{-18} m). At temperatures $> 10^{15}$ K, the electromagnetic and weak forces combine into a single electroweak force.

electrum See **electron** *(Metallurgy).*

element The set of atoms having the same number of protons in their nuclei, i.e., the same nuclear charge = atomic number (Z). See **Elements*.**

elementary particle Any of the fundamental constituents of matter or energy. See **Elementary particles*.**

element formation (standard model) Time $t = 0$: singularity resulting in Big Bang. Time $t = 10^{-8}$ s after Big Bang, temperature $T = 5.4 \cdot 10^{14}$ K, formation of t quarks; $t = 10^{-6}$ s to 10^{-5} s, $T = 5.4$ to $2.4 \cdot 10^{13}$ K, formation of b and c quarks; $t = 10^{-4}$ to $10^{-3.5}$ s, $T = 5.4$ to $4.3 \cdot 10^{12}$ K, formation of s, u, d quarks and of baryons; $t = 10$ s, $T = 6 \cdot 10^9$ K, formation of electrons; $t = 226$ s, $T = 9 \cdot 10^8$ K, formation of H and He nuclei (74% H, 26% He); $t = 7 \cdot 10^5$ y, $T = 3000$ K, stabilization of neutral atoms (^{1}H, ^{2}H, ^{3}He, ^{4}He). First generation stars produce ^{4}He (via the proton-proton chain) and, at increasingly higher temperatures, ^{8}Be, ^{12}C, ^{16}O, ^{20}Ne, ^{24}Mg, ^{28}Si, ^{32}S, ^{36}Ar, and ^{40}Ca by α-capture as T rises to 10^9 K in their cores, and, at still higher T's, the Fe-group elements. Supernova explosions. Second generation stars: carbon cycle produces ^{4}He from four ^{1}H and, at higher temperatures, free neutrons via the reactions

$$^{12}C(p,\gamma)^{13}N(\beta^+,\nu)^{13}C(\alpha,n)^{16}O$$

and

$$^{14}N(\alpha,\gamma)^{18}F(\beta^+,\nu)^{18}O(\alpha,\gamma)^{22}Ne\ (\alpha,n)^{25}Mg$$

where (x,y) means "x in, y out." The free neutrons are captured by existing nuclei to form all isotopes up to ^{209}Bi by n capture followed by β^- decay. During subsequent supernova explosion, spallation produces abundant free n that are readily captured by ^{209}Bi and other heavy elements to form all remaining isotopes up to ^{254}Cf ($t_{1/2} = 60.5$ days; decay by spontaneous fission) and beyond. The cosmically rare Li, Be, and B are largely spallation products. Continued element formation inside massive stars followed by supernova explosions enriches the interstellar medium in heavy elements as the universe ages. The youngest stars have 100 times more heavy elements that the oldest stars. See **inflation.**

elkhorn coral The fore reef coral *Acropora palmata.*

ellipse For an ellipse centered on the origin and with the major axis on the x axis, the locus of the points represented by the equation

$$x^2/a^2 + y^2/b^2 = 1$$

distance from center to either focus	$= (a^2 - b^2)^{1/2}$
eccentricity	$= (a^2 - b^2)^{1/2}/a = c/a < 1$
ellipicity	$= (a - b)/b$
perimeter	$= 2\pi[1/2(a^2 + b^2)]^{1/2}$ (approximately)
area	$= \pi ab$

In the preceding, a = semimajor axis, b = semiminor axis, c = semidistance between foci. See **conic sections.**

ellipsoid The surface represented by the equation (center at origin, axes on coordinate axes)

$$(x^2/a^2) + (y^2/b^2) + (z^2/c^2) = 1$$

Any section through an ellipsoid is an ellipse. Cf. **ellipsoid of revolution.**

ellipsoid (IUGG 1967) The reference geodetic ellipsoid with semimajor axis = 6378160 m, semiminor axis = 6356775 m, and flattening = $1/298.25 = 0.003353$.

ellipsoid of revolution The surface represented by the equation (center at origin, axes on coordinate axes):

$$(x^2/a^2) + (y^2/a^2) + (z^2/c^2) = 1$$

Sections normal to the z axis within the intercepts are circles, all other ones are ellipses. If $a > c$ (ellipsoid of revolution generated by an ellipse revolving about its minor axis), the ellipsoid of revolution is oblate; if $a < c$ (ellipsoid of revolution generated by an ellipse revolving about its major axis), the ellipsoid of revolution is prolate. Syn. **spheroid.**

elliptical galaxy See **galaxy.**

ellipticity See **ellipse.**

elongation Angular distance between the Sun and a planet or the Moon, as seen from the Earth.

embryophytes The higher plants, characterized by the zygote developing into an embryo within the female sex organ. Cf. **thallophytes.**

emerald Gem beryl, $Be_3Al_2(Si_6O_{18})$. Cf. **oriental emerald.** See **Gems*.**

emery Impure corundum (Al_2O_3), containing some iron oxides.

emf Electromotive force.

emission spectrum Electromagnetic spectrum produced by a transparent gas when its atoms or molecules are excited.

emissivity The emittance of a substance relative to that of a blackbody at the same temperature.

emittance The power per unit area radiated by a substance.

emu Electromagnetic unit. See **electromagnetic system of units.**

emulsion A finely divided mixture of two immiscible liquids.

enantiomorph 1. One of a pair of crystals that is the geometric mirror image of the other and, therefore, nonsuperimposable on it. 2. One of a pair of optically active substances (d or l). 3. One of a pair of molecules with opposite chirality (D or L).

encrinite A limestone consisting of more than 50% crinoidal fragments.

endemic Defining an organism or a group of organisms restricted to a specific location.

endocast The internal cast of a fossil, consisting of lithified mud filling.

endogenous Defining the result of activities or effects from inside the Earth.

endolithic Defining microorganisms (algae, fungi, etc.) living within biogenous carbonates or other rocks.

endomorph A crystal overgrown and surrounded by a different crystal.

endomorphism The result of contact metamorphism on the intruding body by its reaction with country rock.

endorheic Defining a basin with internal drainage.

endoskeleton Internal skeleton.

endothermic Defining a process or chemical reaction that absorbs heat. Cf. **exothermic.**

en échelon *(French)* Imbricated.

energy The quantity of work a system is capable of doing.
1. kinetic energy (E_k) The energy that a body possesses because of its motion. For particles or bodies having mass and moving at nonrelativistic speeds,

$$E_k = \tfrac{1}{2}mv^2 \text{ (rectilinear motion)}$$

or

$$E_k = \tfrac{1}{2}I\omega^2 \text{ (rotational motion)}$$

where m = mass, v = velocity, I (rotational inertia) = $\int r^2 \, dm$ (r = shortest distance between axis of rotation and mass element dm), ω = angular velocity. For particles or bodies traveling at relativistic speeds,

$$E_k = mc^2 \left[(1 - v^2/c^2)^{-1/2} - 1\right]$$

For massless particles,

$$E_k = pc$$

where p = momentum, c = speed of light.
2. potential energy (U) Energy that a system or a body has in storage because of its position within a force field (a mass in a gravitational field, a charge in an electric field, etc.), its configuration (a compressed gas or spring), its temperature, or its chemical or nuclear composition.
3. rest energy (e) The energy equivalent to the rest mass m of a body.

$$e = mc^2$$

where m = mass, c = speed of light.
4. total energy (E)

$$E^2 = m^2c^4 + p^2c^2$$

where E = total energy of particle, m = mass of particle, c = speed of light, p = momentum of particle.

energy level A stationary energy state for a physical system. For the hydrogen atom the energy E_n of the various levels is given by the equation

$$E_n = -e^4 m_e/n^2 h^2 8\epsilon_0^2 \text{ joules}$$

where e = electron charge, m_e = electron mass, n = principal quantum number, h = Planck's constant, ϵ_0 = permittivity constant. For ionization potential = 13.6057 eV and the conversion factor $1.6021773 \cdot 10^{-19}$ joule/eV, the equation reduces to

$$E_n = -13.6057/n^2 \text{ eV}$$

Setting ground level K = 0 eV, the following values for the various excitation energies are calculated from

$$E_n = 13.6057 (1 - 1/n^2) \text{ eV}$$

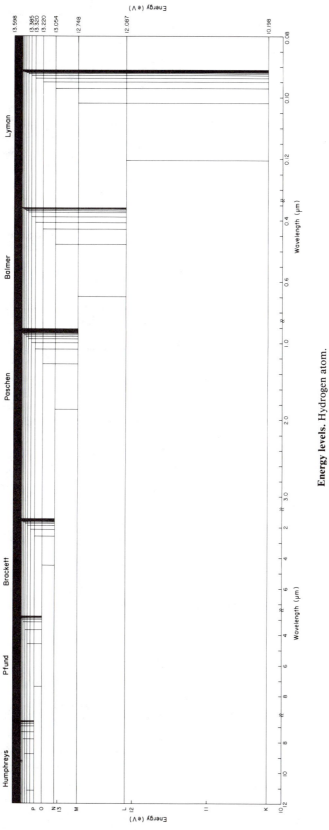

Energy levels. Hydrogen atom.

K ($n = 1$), 0 eV; L ($n = 2$), 10.198 eV; M ($n = 3$), 12.087 eV; N ($n = 4$), 12.748 eV; O ($n = 5$), 13.054 eV; P ($n = 6$), 13.220 eV; Q ($n = 7$), 13.320 eV; R ($n = 8$), 13.385 eV; S ($n = 9$), 13.430 eV; T ($n = 10$), 13.462 eV. . . . Transitions from higher levels to K = Lyman series; to L = Balmer series; to M = Paschen series; to N = Brackett series; to O = Pfund series; to P = Humphreys series. See **Balmer series, Brackett series, Humphreys series, Lyman series, Paschen series, Pfund series.**

ensialic Defining a geosyncline formed on sialic crust.

ensimatic Defining a geosyncline formed on simatic crust.

enstatite An orthopyroxene, $MgSiO_3$.

enthalpy (H) The "total energy" of a system.

$$H = E + pV$$

where E = internal energy of the system, p = pressure, V = volume, pV = external energy of the system.

entropy (S) A function of a reversible heat-transfer process involving a thermodynamic system, equal to the ratio of the heat absorbed by the system to the absolute temperature at which the heat is absorbed.

$$dS = dQ/T$$

where dS = change in entropy in a reversible process, dQ = heat absorbed by the system, T = absolute temperature of the system.

enzyme Any of the catalytic proteins produced by living cells. Enzymes make it possible for biochemical reactions to occur at low (body or ambient) temperature and at high rates (rate increase = 10^4 to 10^{15}). Enzymes are classified into 6 groups: oxidoreductases, which catalyze reactions involving electron or proton transfer; transferases, which catalyze the transfer of chemical groups from one molecule to another; hydrolases, which catalyze hydrolysis; lyases, which catalyze the addition or removal of a chemical group across a double bond; isomerases, which catalyze the conversion of chemical isomers; and ligases, which catalyze the union of molecules and thus synthesize new compounds. Examples of enzymes are catalase (decomposition of H_2O_2 to oxygen and water), pepsin (hydrolysis of proteins to peptides and amino acids), and urease (hydrolysis of urea to ammonia and CO_2).

eolianite A lithified wind deposit.

eon See **aeon.**

eonothem See **eonthem.**

eonthem The chronostratigraphic unit above erathem, consisting of the rocks formed during an aeon.

epeirogeny The uplift of large areas of subcontinental size as contrasted to the narrow, elongated uplift of mountain belts (orogeny).

ephebic The young adult stage of an animal when reproduction becomes possible.

ephemeris 1. Calendar, journal, diary. 2. A compilation of tables showing the positions of celestial bodies at different times during the year.

ephemeris day (d_E) The time interval equal to 86,400 ephemeris seconds.

ephemeris second (s_E) Before 1956: 1 s_E = $1/(24 \cdot 60 \cdot 60)$ = 1/86,400 of mean solar day. From 1956 to 1967: 1 s_E = 1/31,556,925.9747 of tropical year 1900. On October 13, 1967, the ephemeris second was replaced, as the fundamental SI unit of time interval, by the atomic second, which was so defined as to be identical to the ephemeris second then in use. See **atomic second.**

ephemeris time (t_E) Time measured in tropical years from January 0d, 12h, 1900, when the Sun's mean longitude was 279°41′48.04″.

ephemeris year The time interval equal to tropical year 1900, Jan. 0d, 12h, consisting of 365.24219878 d_E or 31,556,925.9747 s_E.

epicenter The point directly above the point (hypocenter or focus) where an earthquake originates.

epicontinental Located on a continent.

epicontinental sea A sea encroaching a continental area, including shelf or interior.

epicycle The apparent loop performed by a superior planet as it moves against the backdrop of the fixed stars. A real loop in the Ptolemaic system.

epidote A low-grade metamorphic mineral, $Ca_2(Al,Fe)Al_2O(SiO_4)(Si_2O_7)(OH)$.

epifauna The marine benthic fauna living on the sediment rather than within it. Cf. **infauna.**

epigenetic Defining an event, such as sulfide ore deposition, taking place after emplacement of the host rock.

epilimnion The upper, wind-mixed layer in a lake, above the thermocline.

epilithic An organism living on, or attached to, a rock surface.

epipelagic Defining the uppermost 200 m of the open ocean.

epiplankton The organisms attached to floating organisms.

epipsammon The organisms living on a sandy surface.

epoch A division of geologic time longer than age but shorter than period, during which the rocks of a series are formed.

equal-area projection A geographic projection in which the relative proportions of the different areas are maintained.

equant *(Astronomy)* The point opposite the Earth with respect to the center of the deferent which, when connected to the center of an epicycle, generates a line that sweeps equal angles in equal times (Ptolemaic system). *(Mineralogy)* Describing a crystal that has similar dimensions in all directions.

equation of continuity An equation expressing the conservation of mass or charge in fluid or electrical motion.

$$\nabla \cdot (\rho \mathbf{v}) = \rho \, (\partial v_x/\partial x + \partial v_y/\partial y + \partial v_z/\partial z)$$
$$= - \, \partial \rho/\partial t$$

where ρ = fluid or charge density, v = velocity, t = time.

equation of state A definition of the physical state of a system based on absolute temperature, pressure, and specific volume or concentration.

equation of time The difference between mean and apparent solar time. Cf. **analemma.**

equator 1. The great circle on a spherical or spheroidal surface equidistant from two opposite points identified as poles. 2. The great circle on the earth's surface equidistant at all points from the two geographic poles. It is 40,075.24 km long. Cf. **parallel.**

equatorial countercurrent A current in the Pacific Ocean flowing eastward along lat. 5°N between the westward North Equatorial Current and South Equatorial Current. See **Ocean currents*.**

equilibrium (chemical) The state of a chemical system in which reactions occur in both directions at the same rate. As a result, the composition of the system remains constant, with the amounts of the various components determined by the equilibrium constants.

equilibrium constant (K) In a reversible chemical reaction at equilibrium, such as

$$mA + nB \rightleftarrows pC + qD$$

the equilibrium constant is

$$K = [C]^p[D]^q/[A]^m[B]^n$$

where A and B are the reactants; C and D are the products; and m, n, p, and q are the coefficients indicating the stoichiometry of the reaction. The brackets indicate that reactants and products are to be expressed as molar concentrations.

equinox 1. Either of the two points where the ecliptic intersects the celestial equator. 2. Either of the two times during the year when the Sun crosses the celestial equator and night and day are approximately of equal duration.

equipotential surface A surface along which the potential is constant. The force is constant and normal to the surface.

equivalence principle See **principle of equivalence.**

equivalent charge The charge carried by an Avogadro number of electrons or protons.

equivalent mass The mass in grams of an element or chemical compound with an equivalent charge available for chemical reaction.

equivalent weight See **equivalent mass.**

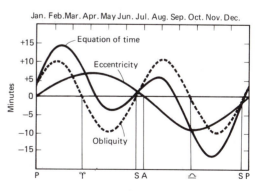

Equation of Time. The Equation of Time shows the difference, in minutes, of Local Time minus Mean Time. This difference results from the combined effects of eccentricity and obliquity. Notice that the effect of eccentricity is zero at perihelion (P) and at aphelion (A), while the effect of obliquity is zero at the vernal (♈) and autumnal (♎) equinoxes and at the solstices (S). The Equation of Time can be obtained from the Analemma by plotting time of the year versus minutes by which Local Time is at variance with Mean Time. (From Payne-Gaposchkin 1954, p. 55, Fig. 3.2)

era A division of geologic time longer than period but shorter than aeon, during which the rocks of an erathem are formed.

erathem The chronostratigraphic unit above system, consisting of the rocks formed during an era.

erg *(Physics)* The CGS unit of energy, equal to 1 dyn·cm = 10^{-7} J. *(Geomorphology)* A vast expanse of sandy desert.

ergosphere The region surrounding a rotating black hole, bound on the outside by the static limit and on the inside by the event horizon. It is theoretically capable of producing energy at the expense of the angular momentum of the black hole.

Eros Minor planet No. 433 (Amor group), 7 × 12 × 36 km in dimensions, with perihelion distance = 1.133 AU, sidereal period = 1.76 y. It does not cross the Earth's orbit, but it came within 0.15 AU from the Earth in 1975.

erratic Referring to a boulder or a block uprooted by glacier ice from an outcrop and deposited at a different site.

ERTS Earth Resource Technology Satellite. See **Landsat.**

ESCA Electron Spectroscopy for Chemical Analysis. See **x-ray photoelectron spectroscopy.**

escape velocity The minimum velocity required for an object to attain a parabolic orbit and thus escape from the vicinity of a celestial body.

$$v_e = (2Gm/r)^{1/2}$$

where v_e = escape velocity, G = gravitational constant, m = mass of celestial body, r = distance from center of gravity of celestial body at which v_e refers. Examples of escape velocities (km/s): Moon = 2.38, Earth = 11.18, Jupiter = 59.5, Sun = 617.7, center of Galaxy = 700. See **Planets—physical data*.**

esker A long, narrow, sinuous, often discontinuous ridge of glacial debris, 5–200 m high and 100 m to 5 km long, deposited by a subglacial stream bound by ice walls on either side.

ESR Electron Spin Resonance.

essence A solution of an essential oil in alcohol.

essential oil A nonfatty volative ethereal oil distilled from a plant.

essexite A plutonic rock consisting mainly of plagioclase, titanaugite, hornblende, and biotite.

esters Organic compounds formed from an alcohol and an organic acid by elimination of H_2O. E.g. methyl acetate, CH_3COOCH_3.

estuary 1. A zone of mixing between fresh river water and seawater. 2. A broadening river mouth opening toward the sea.

esu Electrostatic unit. See **electrostatic system of units.**

Et The ethyl radical $-C_2H_5$.

ET See t_E.

eta (η) A nonstrange meson. See **Elementary particles*.**

étang *(French)* A ponded body of stagnating water.

ether Any of the organic compounds consisting of two hydrocarbon groups linked by an oxygen atom. E.g. ethyl ether, $CH_3CH_2OCH_2CH_3$. 2. Ethyl ether.

ethology The science studying animal behavior.

ethyl (Et) The radical $-C_2H_5$.

ethylene series See **alkenes.**

ethyl ether A solvent and anesthetic substance, $CH_3CH_2OCH_2CH_3$, prepared by reacting ethyl alcohol with sulfuric acid. It is simply called "ether."

Eucaryota One of the two superkingdoms of life on Earth, including all organisms with a well-defined cell nucleus enclosed in a nuclear membrane. Cf. **Procaryota.** See **Taxonomy*.**

eucaryote 1. Referring to a cell with a well-defined nucleus enclosed in a nuclear membrane. 2. Any of the organisms whose cells have a well-defined nucleus enclosed in a nuclear membrane. Cf. **procaryote.**

eucrite 1. A basic gabbro-basalt composed mainly of Ca-plagioclase and clinopyroxene. 2. An achondrite composed mainly of Ca-plagioclase and pigeonite.

eugeosyncline A geosyncline with clastic sedimentation associated with volcanism, located seaward of the miogeosyncline.

euhedral Defining a mineral bound by its crystallographic planes.

eukaryote See **eucaryote.**

eupelagic 1. Referring to a fully oceanic pelagic environment. 2. Defining a sediment consisting wholly of particles settled out of the water column above. Cf. **hemipelagic.**

euphotic zone The zone in the ocean or a lake above the disphotic zone, where photosynthesis thrives (top 100 m on the average).

Europa See **continent, Jupiter, Satellites*.**

eurybathic Defining an aquatic organism tolerant of a wide depth range.

euryhaline Defining an aquatic organism tolerant of a wide salinity range.

eurythermal Defining an organism tolerant of a wide temperature range.

eustasy The phenomenon of worldwide sealevel change resulting from tectonic motions of the sea floor or the waxing and waning of ice on land.

eustatic Defining sealevel changes caused by eustasy.

eustatism See **eustasy.**

eutectic Pertaining to a mixture or alloy consisting of components in such proportions that the melting point is the lowest possible.

eutectic point The melting point of a eutectic mixture.

eutrophic Defining a body of water with a high nutrient content. Cf. **oligotrophic.**

eV Electron volt.

evaporates Minerals and rocks formed by evaporation of solutions. Cf. **evaporite.**

evaporite A marine or lacustrine deposit resulting from the precipitation of chemical compounds from evaporating water.

evapotranspiration Loss of water through evaporation from the ground and transpiration from plants.

evection The periodic inequality in lunar motion caused by a change in the eccentricity of the lunar orbit (from 0.0432 to 0.0666) produced by solar attraction. Longitudinal displacement up to 1°16′ 20.4″; period = 31.807 d.

even-even Defining a nucleus or nuclide with an even number of both protons and neutrons.

even-odd Defining a nucleus or nuclide with an even number of protons and an odd number of neutrons.

event horizon The surface of a black hole, a one-way boundary through which radiation and matter can enter but not escape. Radius = Schwarzschild radius. See **static limit.**

Everest The tallest mountain on Earth, 8848 ± 8 m high, along the boundary between Nepal and Tibet.

everglade A large, flat, marshy, grassy expanse in the tropics.

evolute 1. Defining a planispirally coiled shell in which each whorl is visible from either side. 2. Defining a partially or totally uncoiled shell. Cf. **involute.**

evorsion The carving of potholes on river beds by small eddies and vortices.

exchange coupling The magnetic coupling of adjacent atoms of Fe, Co, Ni, Gd, or Dy at temperatures below their Curie points, creating the property of ferromagnetism. See **ferromagnetism.**

excited state A stationary energy state higher than ground state.

exciton An elementary, localized excited state in a nonmetal that can be propagated through the crystal lattice without transport of charge.

exclusion principle See **Pauli exclusion principle.**

exfoliation The concentric peeling off of surface layers (usually millimeters to centimeters thick) from an igneous or sedimentary rock surface resulting from frost action and thermal weathering.

exoskeleton The external skeleton of an animal.

exosphere The outer layer of the atmosphere, between the altitude of 650 km, where the horizontal mean free path of gas atoms and molecules equals the altitude (critical level), and the magnetopause. H and He atoms receive sufficient energy from the Sun to exceed the escape velocity (10.82 km/s at 650 km of altitude) and leak into outer space.

exothermic Defining a process or chemical reaction that produces heat. Cf. **endothermic.**

exotic *(Biology)* Defining an organism or a population not indigenous to a specified region. *(Geology)* Defining a boulder or rock of foreign origin, emplaced by tectonic processes.

exp Exponential (def. 2).

exponential 1. Pertaining to an exponent. 2. Referring to a given power of *e* (abbr. **exp**). See *e.*

exposure age The length of time a meteorite has been exposed to cosmic rays as determined from

the abundances of cosmic-ray-induced radionuclides and of fission tracks in the body of the meteorite. Constant particle flux is assumed. Exposure ages represent the time since a meteorite was
fragmented to its final size before falling on Earth.
Exposure ages range from $<100 \cdot 10^6$ y to $2.3 \cdot 10^9$
y (commonly a few hundred million years) for iron
meteorites, and from $<1 \cdot 10^6$ y to $100 \cdot 10^6$ y (commonly a few tens of million years) for stony meteorites. The greater exposure ages of the iron meteorites may be related to their greater resistance
to fragmentation.

exsolution The separation into two different
phases of an initially homogenous solid solution.

extant Defining a species or lineage that is still
living.

external cast A reproduction of the external surface of a fossil made by sediment filling an external
mold.

external energy The pV energy of a system

(where p = pressure, V = volume). Cf. **Boyle's
law.**

external mold The mold of the external surface of
a fossil made of sediment hardened or precipitated
around it.

extinctive evolution A theory emphasizing the
role of extinction in the process of evolution, with
extinction resulting, as a rule, from viral or similar
effects and replacement, if any, by species independently evolved in marginal environments.

extinct radionuclides Radionuclides formed at
the time of the origin of the elements in the region
of the solar system but characterized by half-lives
too short to allow their survival to the present
time in measurable amounts, and yet abundant
enough to leave their traces either in the relative
abundances of their daughter products in differentiated planetary materials or in other effects. Included are ^{26}Al, ^{107}Pd, and ^{129}I.

extrinsic semiconductor See **semiconductor.**

F

ϕ 1. Angular displacement. 2. Latitude. 3. Phase angle.

Φ Magnetic flux

f   1. Focal length. 2. Force. 3. Frequency. 4. Fundamental (see **s, p, d, f**).

F   1. Fahrenheit. 2. Farad. 3. Faraday = Faraday constant. 4. Force. 5. Formality.

fabric   1. The spatial arrangement of grains in a sedimentary rock. 2. The spatial arrangement of crystals in a metamorphic rock.

facies   The aspect of a rock unit in terms of the conditions under which it formed.

factor analysis   A multivariate statistical technique aiming at evaluating parameters underlying interrelated sets of data. See **Q-mode factor analysis, R-mode factor analysis.**

factorial (!)   The products of all successive integers from 1 to a specified number. E.g. the factorial of $5 = 5! = 1 \times 2 \times 3 \times 4 \times 5 = 120$.

faecal pellett   See **fecal pellet.**

Fahrenheit (F)   A temperature scale based on an interval of 180° between the freezing point of water set at 32°F and the boiling point of water set at 212°F, both at the pressure of 1 atm. $1°F = 0.555555 \ldots °C$.

$$T(°F) = 1.8T(°C) + 32$$

where T = temperature.

falaise   A high escarpment along a coastline.

fall line   An alignment of waterfalls in a set of parallel rivers flowing from a high plateau to a plain below.

false color   A color different from the natural one, used to emphasize a natural characteristic or feature.

family *(Petrology)* The fundamental grouping of igneous rocks that are closely related chemically. Families are grouped into clans. See **clan.** *(Biology)* A rank in taxonomic classification above genus and below order.

farad (F)   The SI unit of capacitance, defined as the capacitance of a capacitor that exhibits a potential difference of 1 volt between its plates when each is charged with 1 coulomb of opposite electricity.

$$F = CV^{-1}$$

As the farad is too large a unit for most applications, the practical unit of capacitance is the microfarad (μF). $1 \mu F = 10^{-6}$ F.

faraday (F)   A unit of electric charge equal to one Avogadro number of electron charges or of singly charged ions. It is equal to 96,485.3 coulombs. Syn. **Faraday constant.**

Faraday constant   See **faraday.**

Faraday effect   The rotation of plane polarized light passing through specific substances in the direction of a magnetic field. Rotation results from the left and right circular polarized components of the plane polarized light having different indices of refraction, so that one travels faster than the other in the medium.

Faraday's law of electrolysis   A law relating the mass of a substance deposited from an electrolytic solution to the current passing through it.

$$mj/M = It/F$$

where m = mass of substance in gram-atoms or gram-molecules, j = number of gram-equivalents per gram-atom or gram-molecule, M = atomic or molecular mass of substance, I = electric current, t = time, F = faraday.

Faraday's law of induction   A law relating the emf induced in a circuit to the change of magnetic flux threading it.

$$\mathcal{E} = -d\Phi/dt$$

where $\mathcal{E}$ = electromotive force induced in a circuit, Φ = magnetic flux, t = time.

faro   A small, circular or ellipsoidal reef with a central lagoon emplaced on the rim of a larger atoll or on a barrier reef.

Fathometer   A copyrighted name for an echo-sounder model principally used in coastal waters.

fats   Glyceride esters of alkyl acids.

fatty acids   See **alkyl acids.**

fault A break in a rock with displacement of one side with respect to the other. See **normal fault, reversed fault, strike-slip fault, transform fault.**

fault breccia A breccia formed along a fault plane by friction of one side against the other.

fault-plane solution The determination of the direction of a fault plane by analysis, at stations away from the epicenter, of the sense of first motion of P and S waves produced by an earthquake.

fault zone A fractured zone consisting of numerous parallel faults.

fauna 1. The entire animal population living in an area. 2. The entire fossil animal population in a given sedimentary bed or set of beds.

faunizone An animal biozone. Cf. **biozone.**

faunule The fossil fauna obtained from a single outcrop. Cf. **florule.**

fayalite The mineral Fe_2SiO_4, the Fe end-member of olivine $(Mg,Fe)_2SiO_4$. Cf. **forsterite.**

F-corona The outer layer of the solar corona.

fecal pellet A microscopic (50–250 μm), ovoidal aggregate of invertebate fecal matter covered with a pellicle.

feldspar A group of K or Na-Ca Al-silicates. The most common are orthoclase, microcline, sanidine, and the plagioclase series. See **plagioclase, Minerals*.**

feldspathic Defining a rock or sediment containing more than 10% feldspar.

feldspathoid Any of a group of low-silica Na, K, Ca, Al-silicates, the most common of which are leucite, nepheline, cancrinite, and sodalite. See **Minerals*.**

felsic Indicating the presence of feldspar, feldspathoids, and silica.

felsic index (FI) An expression of the concentration of Na and K with respect to Ca in minerals. FI = $100 (Na_2O + K_2O)/(Na_2O + K_2O + CaO)$, ranging from 25 (basalt) to 100 (rhyolite).

femic Defining an igneous rock rich in ferromagnesian minerals.

femto- Prefix meaning 10^{-15}. Cf. **atto-, micro-, nano-, pico-.**

Fermat's principle "An electromagnetic wave traveling between two points will follow the path of minimum travel time."

fermentation The incomplete oxidation of organic molecules by anaerobic respiration of bacteria and yeasts.

fermi A unit of length equal to 10^{-15} m.

Fermi-Dirac statistics The statistics of a system of fermions.

Fermi distribution A distribution law specifying the probable number of particles n_i at energy level E_i in a system of fermions in thermal equilibrium at temperature T.

$$n_i = 1/(e^{(E_i - \mu)/kT} + 1)$$

where μ = chemical potential, k = Boltzmann constant, T = absolute temperature. $n_i = 1/2$ for $E_i = \mu$. Cf. **Boltzmann distribution, Bose distribution.**

Fermi energy (E_f) The energy of the Fermi level. In a system of ideal, free fermions at $T = 0$ K, the Fermi energy is equal to the chemical potential μ if there are no gaps in the energy spectrum.

$$E_f = (h^2/8m) (6\rho/\pi g)^{2/3}$$

where h = Planck's constant, m = mass of fermion, ρ = number of fermions per unit volume (density), g = spin degeneracy. The average energy of a system of ideal, free fermions at 0 K is 3/5 of the Fermi energy (nonrelativistic fermions) or 3/4 of the energy of the Fermi energy (relativistic fermions). At room temperature, the Fermi level for Cu (with 1 conduction electron/atom and about $4 \cdot 10^{22}$ filled energy levels/cm³) is 7.04 eV and the work function is 4.46 eV, giving a barrier energy of 11.50 eV. See **barrier energy.**

Fermi gas A system of fermions, e.g. free electrons in metals.

Fermi level (E_0) The energy level separating completely filled quantum states below from empty quantum states above in a system of ideal, free fermions at 0 K. It is the highest energy level in such a system. If there are no gaps in the energy spectrum, $E_0 = E_f = \mu$, where E_f = Fermi energy, μ = chemical potential.

fermion Any particle with spin angular momentum equal to a half integer times $h/2\pi$. Included are the neutrino, the electron, the muon, the proton, the neutron, the hyperons, other baryons, and all nuclei with odd number of nucleons.

Fermi surface The surface of the Fermi level. It is the surface in momentum space (the space defined by the three orthogonal momentum components, p_x, p_y, and p_z) separating energy states in metals or semiconductors that are filled by free electrons from those that are empty. It is the surface of constant energy in wave vector space (the space defined by the components k_x, k_y, and k_z of the wave vector $\boldsymbol{k}$; $\boldsymbol{k} = \boldsymbol{p}/(h/2\pi)$.

Fermi temperature Fermi energy divided by the Boltzmann constant.

ferralite A soil rich in iron oxides and hydroxides derived from the weathering of basic rocks under humid, tropical conditions.

Ferrel cell Either one of the two mid-latitude atmospheric circulation cell systems. Air rises to the upper troposphere at 60° of latitude, it moves toward the lower latitudes on a 225° (northern hemisphere) or 315° (southern hemisphere) course, it sinks at 30° of latitude, and returns to 60° of latitude on a 45° (northern hemisphere) or 135° (southern hemisphere) course. See **Hadley cell, polar cell.**

ferretto A mid-latitude interglacial fossil soil formed under warmer and more humid conditions than today.

ferrian Containing ferric iron (Fe^{3+}).

ferric Characterized by the presence of iron in the trivalent state (Fe^{3+}).

ferricrete A conglomerate cemented with iron oxides.

ferricrust A duricrust cemented with iron oxides.

ferrimagnetism A form of magnetism weaker than ferromagnetism, resulting from antiparallel alignment of the magnetic moments of adjacent ions in a crystal, with the ions having the stronger moments determining the net resultant field. Cf. **ferromagnetism.**

ferrite *(Physics)* A member of the class of magnetic metallic oxides consisting of $Fe_2O_4{}^{2+}$ with a suitable bivalent metallic ion (e.g. Mg, Mn, Fe, Co, Ni, Cu, Zn, Cd) attached and having a spinel, garnet, or perovskite structure. Ferrites combine high resistivity (10^{14} Ω m) with high permeability. *(Geology)* 1. The component, rich in iron oxides, of the groundmass of a porphyritic rock. 2. A sediment cemented with iron oxides.

ferroan Containing iron in the bivalent state (Fe^{2+}).

ferroelectricity The property of some materials to display electric polarization in the absence of an externally applied electric field. Electric polarization results from the spontaneous displacement of atoms within the crystalline unit cell. An example is $BaTiO_3$, which has a dielectric constant that is very high (2000 at room temperature, rising to a maximum of 11,000 at 121°C, the Curie point),

compared to a dielectric constant of 2 to 10 for most nonconducting substances.

ferromagnesian Defining a mineral containing Fe and Mg.

ferromagnetism Property of Fe, Co, Ni, Gd, and Dy, and alloys of these and other elements, resulting in the coupling of the magnetic moments of adjacent atoms and the grouping of these atoms into domains (about 10^{15}–10^{18} atoms each) with a strong net magnetic moment. The application of an external magnetic field orients the magnetic moments of the domains parallel to each other producing permanent magnetization that can be destroyed only by raising the temperature above the Curie point or by introducing an equivalent amount of mechanical energy. Cf. **ferrimagnetism.**

ferromanganoan Defining a mineral or a sediment rich in Fe and Mn oxides.

ferrous Pertaining to or containing iron in the bivalent state (Fe^{2+}).

ferruginous Defining a sediment or rock containing iron oxide in disperse form, as a cement or cement component in a sedimentary rock or as a surface coating of igneous rocks exposed to the atmosphere.

FET Field-effect transistor.

fetch An area of the ocean where wind conditions are constant, creating a uniform wave system.

FI Felsic index.

fiber optics The transmission of optical images through long, thin fibers made of transparent materials (glass, plastics).

field A region of space in which a specific property, invariable or variable through time, exists at each point. See **force field.**

field-effect transistor (FET) A three-electrode transistor in which current flow between two electrodes (source and drain) is controlled by the voltage applied to the third terminal (the gate). Cf. **junction field-effect transistor, metal-oxide-semiconductor field-effect transistor.**

field emission The emission of electrons by a metal surface under the influence of a strong (10^7–10^8 V/cm) electrostatic field. The field emission current I is given by the Fowler-Nordheim equation:

$$I = 1.55 \cdot 10^{-6}\,(E^2/W)\,e^{-(6.85 \cdot 10^7 W^{3/2}/E)}$$

where I = field emission current in A/cm^2, E = field strength in V/cm, W = work function in eV.

field of force See **force field.**

field quantum The quantum of field energy. The four field quanta of the four force fields are the gluon (strong force), gauge boson (weak force), photon (electromagnetic force), and graviton (gravitational force). See **force field.**

filament *(Physics)* A metallic wire or ribbon that is heated by the passage of an electrical current to produce emission of light or electrons. *(Astronomy)* A string of galactic clusters. Filaments are the largest known structures in the universe. The largest is over $1 \cdot 10^9$ l.y. long.

field theory Any of the theories describing force fields and their interactions with matter.

filter feeder An animal that catches food by filtering seawater through body filaments.

fine sand Sand with particle sizes between 0.125 and 0.250 mm.

fine structure The close splitting of a spectral line caused by the coupling of the spin and orbital angular momenta of an electron.

fine-structure constant (α) The fundamental, dimensionless constant of atomic physics and QED.

$$\alpha = 2\pi e^2/hc \; (\text{CGS}_{esu})$$
$$= \mu_0 c e^2/2h \; (\text{SI})$$
$$= 0.007297353$$
$$= 1/137.036$$

where e = electron charge, h = Planck's constant, c = speed of light, μ_0 = permeability constant = $4\pi \cdot 10^{-7}$ henry/meter (exactly).

firn Snow recrystallized in granules, a transitional phase between snow and glacier ice.

firn ice Partly recrystallized firn, a transitional phase between firn and glacier ice.

firn line The highest level to which snow on a glacier surface retreats during the summer.

first arrival The first arrival of a seismic wave (P, S, surface, etc.) at a receiving station. First arriving waves have traveled through paths providing the highest velocities of transmission for a given distance between source and receiver.

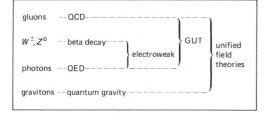

Field theories.

first law of thermodynamics "No energy can be gained or lost by an isolated system."

$$dQ - dW = dU$$

where dQ = heat added to the system, dW = work done by the system, dU = change in the internal energy of the system. Cf. **Charles' law.**

first-order reaction A chemical reaction in which the reaction rate of a component is proportional to its concentration.

$$[A] = [A_0] \, e^{-kt}$$

where $[A]$ = concentration of reactant; $[A_0]$ = initial concentration; k = reaction rate constant; t = time. Cf. **radioactive decay.** See **order of reaction.**

First Point of Aries (Υ) The vernal equinox which, at the time of Hipparchos (late 2nd century B.C.), who first used the term, lay in the constellation of Aries but which now lies in Pisces because of the precession of the equinoxes.

firth A long, narrow, usually sinuous estuary.

fish *(Biology)* Any of the organisms belonging to the classes Agnatha, Acanthodii, Placodermi, Chondrichthyes, or Osteichthyes. *(Oceanography)* An oceanographic instrumentation package towed behind a ship.

fissile Referring to a solid breakable along closely spaced planes.

fissility The property of a solid to break along closely spaced planes.

fission The breakup of heavy atomic nuclei into two major fragments and several minor ones including free neutrons. The two major fission fragments are around masses 90–100 and 132–142 for ^{235}U, and around masses 100 and 132 for ^{239}Pu. Cf. **spallation, spontaneous fission.**

fission track The track left in a crystal lattice by the fragments produced by the natural fission of ^{238}U. ^{238}U decays by α emission (99.275%; $t_{1/2}$ = $4.468 \cdot 10^9$ y) as well as by spontaneous fission (0.725%; $t_{1/2}$ = $10.1 \cdot 10^{15}$y). The number of ^{238}U atoms decaying by spontaneous fission, given by the ratio of the two half lives, is 1/2,260,000. Fission tracks are about 10 μm long but only 10^{-4} μm wide. They are made visible under high-power optical microscope by etching with suitable solutions (HF, HCl, HNO$_3$, NaOH) because the damaged region is more soluble than the undamaged one.

fission-track dating method An absolute dating method based on the number of fission tracks produced by the spontaneous fission of ^{238}U in a crys-

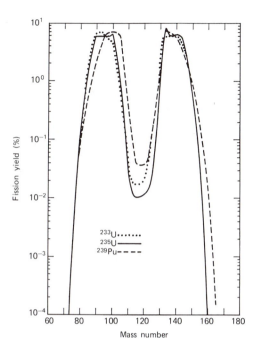

Fission. Mass distribution of fission products for the thermal-neutron-induced fission of ^{233}U, ^{235}U, and ^{239}Pu. (From Loveland 1972, p. 8.257, Fig. 8g.5)

tal relative to the amount of ^{238}U present in that crystal:

$$N = N_0 e^{-\lambda t}$$

where N = number of ^{238}U atoms present in the crystal, N_0 = number of ^{238}U atoms originally present = N + number of fission tracks, λ = decay constant, t = time.

fissure A large crack in a rock.

fissure eruption A volcanic eruption from a fissure rather than from a cylindrical vent.

fissure volcano Any of a series of volcanoes aligned along a major crustal fissure.

FitzGerald-Lorentz contraction The one-dimensional contraction of a body, as seen from a frame of reference moving at relativistic speed v, in the direction of motion of the frame.

$$L_v = L_0 (1 - v^2/c^2)^{1/2}$$

where L_v = dimension in the direction of motion as seen from the moving frame; L_0 = same dimension as seen from the body's frame; c = speed of light.

fjord A drowned glacial valley.

flag See **flagstone.**

flagstone 1. A fine-grained sandstone that easily breaks into tabular sections. 2. A tabular section broken off from a fine-grained sandstone.

flame photometer An instrument for quantitative element analysis. The element from a solution is excited in a flame. The intensities of the characteristic spectral lines are proportional to the concentration of the element in the flame and, therefore, in the solution. The output is calibrated with artificially prepared standard solutions.

flame photometry Elemental analysis using a flame photometer. See **flame photometer.**

flare A short-lived (minutes to hours) brightening within the solar chromosphere, 5000–50,000 km across, emitting intense radiation and ejecting matter at speeds of 500–1000 km/s.

flavor A quark property. See **quark.**

F layer The 450-km-thick seismic velocity transition zone between inner and outer core, extending from 4720 to 5170 km of depth below the Earth's surface. Syn. **Bullard discontinuity.**

flint A dark variety of chalcedony or chert.

flocculation The coagulation of colloidal particles in electrolytic solutions.

floe A floating, tabular section of pack ice.

flood basalt A large outpouring of basalt from a fissure or fissures, flooding a large area and solidifying in approximately horizontal layers.

flood plain An alluvial surface formed by a river during flooding.

florizone A plant biozone. Cf. **faunizone.**

florule The fossil flora obtained from a single outcrop. Cf. **faunule.**

flos ferri Arborescent aragonite encrusting hematite.

fluid inclusion A cavity in a crystal, 1–100 μm across, containing liquid or gas trapped during crystallization.

fluidity The reciprocal of viscosity.

fluorapatite The mineral $Ca_5(PO_4)_3F$.

fluorescence The emission of light from atoms excited by ultraviolet or x-ray radiation.

fluorescence x-ray Characteristic x-rays emitted by atoms excited by x-rays of shorter wavelength.

fluorine dating A dating method based on the increasing absorption of fluorine by fossil bones with time. The amounts range from 0.3% (Recent) to 2% (early Pleistocene).

fluorometer An instrument to measure fluorescence.

flute A synsedimentary, spoon-shaped depression formed by current scour of soft mud sediment. It is 5-15 cm long, with greatest depth at the upstream end and major axis parallel to the current.

flute cast The infilling of a flute as seen on the underside of the bed deposited on it.

fluvioglacial See **glaciofluvial.**

flux *(Physics)* A flow of matter or energy. *(Metallurgy)* In smelting, a substance to facilitate the flowing of another substance, as when $CaCO_3$ is added to ores to facilitate the separation of metal from gangue. In soldering or brazing, a substance coated on a surface to be soldered or brazed so as to prevent the formation of oxides.

flux-gate magnetometer A doubly relative instrument (calibration to zero point and calibration of the rate of change are needed) for magnetic measurements on land, at sea, and airborne. It consists of two identical, rectangular strips of high-susceptibility metal (e.g. Mumetal) placed adjacent and parallel to each other. These are wound with primary coils in series, but in opposite directions, through which an ac current flows. The strips follow the same hysteresis curve but in opposite directions, reaching saturation at the same time. A secondary is wound around both primaries. In the absence of an external field, no voltage is induced through the secondary. If there is an external field with a component in the direction of the two strips, a voltage is developed along the secondary as this component would reinforce the exciting field in one strip and oppose it in the other. Three assemblies at 90° from each other may be used to determine the three components of the magnetic field. Precision is $\sim 1\ \gamma$.

flysch An assemblage of sedimentary rocks consisting of largely unfossiliferous, interbedded graywackes, pelagic limestones, and clays. Rapid sedimentation into a deep basin with an abundant supply of sediment from a neighboring area is indicated.

FM Frequency modulation.

FNAL Fermi National Accelerator Laboratory (Fermilab) at Batavia, Illinois. It houses the world's most powerful accelerator, a proton synchrotron with diameter of 2 km.

f number (*f*) In optical systems:
$$f = L/a$$
where L = focal length, a = aperture.

focal depth The distance between hypocenter (focus) and epicenter.

focal distance See **focal length.**

focal length The distance from the center of a lens or a concave mirror to its principal focus. See **diopter.**
1. *Concave mirror:*
$$F = r/2$$
where F = focal length, r = radius of curvature.
2. *Thin symmetrical lens:*
$$1/F = 1/O + 1/I$$
where F = focal length, O = object distance from lens center, I = image distance from lens center.

focal mechanism See **fault-plane solution.**

focal plane A plane passing through the principal focus of an optical system normally to the optical axis.

focal point See **principal focus.**

focus *(Geometry)* The point that, together with a directrix, defines a conic section. *(Geology)* The point of origin of an earthquake. *(Optics)* The point to which rays converge or from which they appear to diverge.

foid Feldspathoid.

foot A nonmetric unit of length equal to 30.48 cm (exactly).

foot-pound-second (fps) The British system of units (q.v.) based on the foot, the pound-mass, and the second.

foram Colloquial name for *foraminifer.*

foraminifer Any of the protozoa belonging to the phylum Foraminifera.

Foraminifera A phylum of brackish and marine protozoa. Of the living species, approximately 32 are planktonic and more than 2000 benthic.

foraminiferal number The number of foraminiferal shells >100 μm in size per gram of dry sediment.

foraminiferal ooze A calcitic (>30% $CaCO_3$), oceanic mud consisting of shells of planktonic foraminifera, clay, and coccoliths.

forbidden band A band of unallowed energy levels for electrons in solids. Syn. **energy gap.**

forbidden energy band See **forbidden band.**

forbidden line A spectral line representing a transition of low probability.

forbidden transition A low-probability transition between two quantum states.

f orbital The orbital of an atomic electron characterized by an orbital angular momentum quantum number of 3. See s, p, d, f.

force A vector quantity equal to the derivative of momentum with respect to time:

$$\mathbf{F} = d\mathbf{p}/dt$$

where $\mathbf{F}$ = force, $\mathbf{p}$ = momentum, t = time. For constant mass:

$$\mathbf{F} = m\mathbf{a}$$

where $\mathbf{F}$ = force, m = mass, $\mathbf{a}$ = acceleration.

force constant The constant k in the equation

$$F = -kx$$

where F = force and x = displacement, describing the dynamics of a spring or analogous device or system.

force field A region of space throughout which a specific force exists.

forced oscillation See **oscillation**.

forces of nature See **natural forces**.

foredeep An elongated trough oceanward of a coastal orogenic belt or an island arc.

foreland A tectonically stable continental area (usually a cratonic margin) against which folding of a marginal mobile belt occurs.

fore reef The seaward edge and front of a reef.

foreset bed Any of the inclined beds on the front of an advancing delta or dune, progressively covering the horizontal bottomset bed.

foreshock A smaller shock or any of such shocks preceding the main shock of a earthquake.

forma The smallest identifiable variant in a monospecific population.

formal charge The charge assigned to an atom in a covalent compound by dividing equally the bonding electrons among the participating atoms.

formality (F) Formula weight/liter.

formation An assemblage of sedimentary strata exhibiting a clearly identifiable environmental and genetic identity. Formation is the fundamental unit of lithostratigraphic classification.

formula weight The sum of the atomic masses of the elements forming a compound, each multiplied by the number of times the element appears in the compound.

forsterite The mineral Mg_2SiO_4, the Mg endmember of olivine $(Mg,Fe)_2SiO_4$. See **fayalite**.

forward bias See **bias**.

four-pi Pertaining to all directions in isotropic space. It refers to 4π being the solid angle subtended by the center of a sphere. Cf. **two-pi**. See **rationalized unit**.

fourth-power law See **Stefan-Boltzmann law**.

fps Foot-pound-second.

fractional crystallization The successive crystallization, out of a cooling magma, of minerals with decreasing melting points.

Frasch process The process of mining subsurface sulfur deposits by means of three concentric pipes within a well casing. Hot water is injected down the outer pipe and air pressure applied within the inner pipe, forcing sulfur up in the space between the intermediate and the inner pipe.

Fraunhofer lines The more than 25,000 dark lines in the solar spectrum resulting from absorption, by cooler gases in the solar chromosphere, of the continuous spectrum emitted by the solar photosphere. The stronger lines are due to H, Na, Mg, Ca, and Fe. They appear dark to an observer because the absorbed radiation is reradiated in all directions so that only a minute fraction reaches the observer's instrument.

free-air anomaly The gravity anomaly remaining after the free-air correction has been applied to a gravity measurement.

free air correction A correction for the altitude of a station with respect to the surface of the geoid, assuming that the space between station and geoid is empty space ("free air"). It amounts to 0.3086 mgal/m.

free energy (G) A measure of a system's capacity of doing work at constant temperature and pressure.

$$G = H - TS$$

where H = enthalpy, T = absolute temperature, S = entropy.

free-fall velocity The velocity of a body freely falling in a gravitational field in vacuo.

$$v_z = v_{oz} - gt$$

where v_z = velocity along vertical (z) axis at time t, v_{oz} = initial velocity along vertical (z) axis, g = gravitational acceleration, t = time.

free oscillation 1. An oscillation of a system not constrained by external conditions. 2. The low-frequency vibration of the entire Earth following a major earthquake. See **oscillation.**

free radical An atom or molecule with an unpaired electron.

freezing point The temperature at which a pure substance changes from the liquid to the solid phase at a fixed pressure. Syn. **melting point.**

frequency (ν) Number of cycles per unit time.

$$\nu = 1/T$$
$$= \omega/2\pi$$

where T = period, ω = angular velocity.

frequency band A range of frequencies.

frequency modulation (FM) The modulation of the frequency of a carrier wave by an input signal.

freshet A sudden rise in a small river caused by a cloudburst or sudden snowmelt in the headwater or upstream region.

friction The tangential force needed to slide a body in contact with another under normal load. Friction is a surface phenomenon depending essentially upon the physical condition of the surfaces.

friction breccia A breccia produced by friction, such as along a fault plane.

friction depth See **Eckman spiral.**

fringing reef A reef attached to the shore.

front *(Meteorology)* A sloping zone, usually 1 km thick, separating a colder air mass below from a warmer air mass above. See **cold front, occluded front, stationary front, warm front.** *(Petrology)* A metamorphic zone surrounding an igneous intrusion.

frontal moraine A moraine in front of a glacier.

frost The sublimation of atmospheric moisture on a surface that is below the freezing point.

frost action The breakup of rocks under successive freezing and thawing of water in the rock's pores and cracks.

frosting The pitted surface developed on quartz and other mineral grains or surfaces by wind action in desertic environments.

frustule The siliceous exoskeleton of a diatom, consisting of two halves.

ft Foot, feet.

fuel cell A cell that converts fuel directly into electricity, without first converting it into heat and mechanical energy. Common reactions include the oxidation of H_2 and CO. Potential developed = 0.9–1.1 V. Current density = 500–1000 A/m^2. Power developed by fuel cell systems = 100–2000 W.

fugacity A property of real gases and liquids similar to vapor pressure in ideal gases or osmotic pressure in ideal dilute solutions.

fulgurite An irregular, glassy, tubular structure produced by lightning fusing together sand grains in a beach, dune, or desertic environment.

fuller's earth A very fine clay consisting mainly of montmorillonite and palygorskite, originally used in England to full (i.e. to shorten and thicken) wool fibers.

fully developed sea A sea condition in which all possible wavelengths consistent with the prevailing wind speed have developed to maximum energy.

fumarole A fissure in volcanic territory from which volcanic gases escape.

fumarolic stage The late stage of volcanic activity.

function A quantity y that varies in accord with another quantity x (the variable).

fundamental frequency The lowest eigenfrequency of a system.

fundamental quantity A quantity, such as mass, length, and time, that cannot be derived from other quantities. See **Units*.**

fundamental strength The maximum stress a body can withstand for an indefinite amount of time without creep or flow.

fundamental unit The unit of a fundamental quantity.

Fungi One of the 5 kingdoms. See **kingdom.**

fusiform Shaped like a spindle.

fusion The combination of lighter nuclei into heavier ones. Mass is lost and energy produced. The fusion of four ^{1}H into one ^{4}He causes a mass loss of 0.0276 u (= 0.685%), and an energy release of 25.71 MeV.

fusion crust A glassy crust, 1 mm or less in thickness, formed by the fusion of the surface material of a meteorite while in flight through the atmosphere.

G

γ 1. Activity coefficient. 2. Gamma ($= 10^{-5}$ oersted). 3. Gyromagnetic ratio. 4. Mineral phase stable at temperatures higher than the α and β phases. 5. Photon. 6. Surface tension.

Γ 1. Gauss. 2. Width of resonant state ($\Gamma = h/2\pi\tau$, where τ = mean life).

(γ,n) *Gamma in, neutron out*, a symbolism describing a photonuclear reaction in which a photon of appropriate energy (>5 MeV) is absorbed by a nucleus which becomes excited and emits a neutron. E.g. $^9Be(\gamma,n)^8Be$, $^{16}O(\gamma,n)^{15}O$.

g 1. Gram. 2. Gravitational acceleration of the Earth (which ranges from 9.780 m/s^2 at the equator to 9.832 m/s^2 at the poles).

G 1. Conductance. 2. Gauss. 3. Gibbs free energy. 4. Gravitational constant.

g_0 Standard *g*. $g_0 = 980.665$ gal (exactly).

gabbro A plutonic rock consisting mainly of Ca-plagioclase and clinopyroxene, the intrusive equivalent of basalt. Density about 3.0 g/cm^3. $V_p = 7.5$ km/s.

gal The CGS unit of acceleration, equal to 1 cm/s^2.

galactic axis The axis passing through the galactic poles.

galactic center The center of the Galaxy. Coordinates: $l'' = 0.0°$, $b'' = 0.0°$; $l' = 327.68°$, $b' = -1.40°$; $\alpha = 264.83°$, $\delta = -28.90°$ (1900); $\alpha = 265.60°$, $\delta = -28.92$ (1950).

galactic coordinate system See **coordinate systems**.

galactic equator The intersection of the galactic plane with the celestial sphere. The galactic equator forms an angle of 62.60° with the celestial equator. Position of the ascending node: $l'' = 33.00°$; $\alpha = 282.25°$ (1950). See **coordinate system (celestial)**.

galactic halo A spherical region of space centered on the center of the Galaxy, with radius similar to that of the galactic disc (50,000 l.y.) and a population of globular clusters.

galactic latitude See b^l, b^{ll}.

galactic longitude See l^l, l^{ll}.

galactic plane The major plane of the Galaxy, normal to the galactic axis and passing through the galacitic center. Inclination on the celestial equator = 62.60°. See **cordinate system (celestial)**.

galactic pole (North) The northern intersection of the galactic axis with the celestial sphere. Coordinates: in the new (1959) IAU system: $\alpha = 191.65°$, $\delta = +27.67°$ (1900) or $\alpha = 192.25°$, $\delta = +27.40°$ (1950); in the old IAU system: $\alpha = 190.0°$, $\delta = +28.0°$ (1900) (Ohlson Pole). See **coordinate system (celestial)**.

galaxy Any of the large, gravitationally bound systems of stars that form the universe. Galaxies are grouped in clusters (10^2–10^3 members each) and in superclusters (10^4–10^5 members each). Total number of galaxies in the visible universe $\approx 10^{11}$. Galaxies exhibit elliptical, spiral, or irregular shapes, and range in largest diameter from 5000 to 150,000 l.y. **1. elliptical galaxies** Elliptical galaxies are gas-poor and their stars are metal-poor, indicating that star formation has ceased. Mass ranges from 10^6 solar masses (dwarf ellipticals) to 10^{12} solar masses (giant ellipticals). Elliptical galaxies represent 20% of the brighter galaxies, 60% of the total. **2. spiral galaxies** Spiral galaxies are gas-rich and exhibit continuing star formation. Their youngest stars contain ~ 100 times more metals than the oldest stars. Mass ranges from 10^{10} to 10^{11} solar masses. Spiral galaxies represent 75% of the brighter galaxies, 30% of the total. **3. irregular galaxies** Irregular galaxies have no clearly defined geometric organization, are gas-rich, and exhibit continuing star formation. Irregular galaxies represent 5% of the brighter galaxies, 10% of the total.

Galaxy The galaxy of which the Sun is a member. It is a spiral galaxy 100,000 l.y. across and 1000 l.y. thick in the center, embedded in a sphere of globular clusters (halo) 50,000 l.y. in radius. The center and spiral arms are rich in gas and consist mainly of Population I stars. The globular clusters of the halo consist mainly of Population II stars. Mean density of matter = 10^{-24} g/cm^3. The Galaxy contains $\sim 1.8 \cdot 10^{11}$ stars. It was formed about $10-15 \cdot 10^9$ y ago.

galena The mineral PbS.

galilean Referring to any of the four larger satellites of Jupiter, discovered by Galileo in 1610 and named by him, in order of increasing distance from Jupiter, Io (radius = 1816 km), Europa (radius = 1563 km), Ganymede (radius = 2638 km), and Callisto (radius = 2410 km). See **Jupiter, Satellites***.

galvanic electricity Electricity represented by flow of electrical current, as opposed to static electricity.

galvanometer An instrument to measure small electric currents.

gamete A haploid cell capable of fusing with another such cell to give rise to a zygote.

gametophyte The haploid phase in the life cycle of a plant, arising from a spore and producing gametes. The gametophyte is the dominant phase only in Bryophyta and primitive algae. Cf. **sporophyte.**

gamma (γ) A unit of magnetic field strength equal to 10^{-5} oersteds.

gamma counter An instrument to detect gamma radiation using the electrons liberated by the radiation.

gamma radiation Electromagnetic radiation with $\nu > 10^{21}$ s^{-1} and $\lambda < 3 \cdot 10^{-13}$ m.

gamma rays Energetic (1 keV to 100 MeV) photons produced by nuclei while dropping from higher to lower energy levels or by particle-antiparticle annihilation. γ rays interact with surrounding matter by means of the photoelectric effect (ejection of an electron from the K shell), Compton scattering (interaction with atomic electrons and consequent loss of energy resulting in an increase in wavelength), and, at energies >1.02 MeV, pair production (production of an electron and a positron). The γ ray is completely absorbed in the photoelectric effect and in pair production. Greatest penetration of γ rays in Pb is 2 cm for 3 MeV γ rays (penetration is less for both lower and higher-energy rays). See **Electromagnetic spectrum***.

gamma-ray spectrometry The study of the energy spectrum of gamma radiation from nuclei.

Gamowian The division of time from the end of the Planckian (cosmological time $t = 5.390 \cdot 10^{-44}$ s) to the beginning of the Hadean

(4.7·10⁹ y B.P.). Named after George Gamow (1904–1968).

gangue The rock matrix of ores.

Ganymede See **Jupiter, Satellites***.

gara A mushroom-shaped rock in an arid region resulting from wind erosion of less resistant strata underlying more resistant strata.

garnet Any of a group of high-grade metamorphic Ca, Al, Mg, Fe, Mn, Cr, and V silicates of the general formula $X(SiO_4)_3$, where X = Fe$_3$Al$_2$ (almandine), Ca$_3$Fe$_2$ (andradite), Ca$_3$V$_2$ (goldmanite), Ca$_3$Al$_2$ (grossularite), Mg$_3$Al$_2$ (pyrope), Mn$_3$Al$_2$ (spessartite), and Ca$_3$Cr$_2$ (uvarovite).

garnetite A dense igneous rock (zero pressure ρ = 3.76 g/cm^3) consisting of 90 vol. % garnet and 10 vol. % stishovite, believed to be the major constituent of the Earth's mantle between 450 and 600 km of depth. Cf. **perovskitite.**

gas constant (R) Two-thirds of the energy needed to raise the temperature of 1 Avogadro number of noninteracting particles (ideal gas) by 1 K.

$$\begin{aligned} R &= pV/T \\ &= 1.9859 \text{ cal K}^{-1} \\ &= 8.3145 \text{ J K}^{-1} \\ &= 82.058 \text{ atm cm}^3 \text{ K}^{-1} \end{aligned}$$

Cf. **Boltzmann constant.**

gas electrode A finely divided or spongy metallic electrode capable of holding a gas liberated in an electrolytic solution.

gas law (ideal) "The external energy pV of a gaseous system is proportional to the absolute temperature T and the number n of moles present."

$$pV = nRT$$

where pV = external energy (p = pressure, V = volume), n = number of moles, R = gas constant, T = absolute temperature.

gastrolith A polished pebble from the stomach of some vertebrates, possibly used as a food mill.

gate 1. The control electrode in a CCD, FET, MOS, MOSFET, or SCR. It is analogous to the base in a transistor or the grid in a vacuum tube. 2. A multiple-input, single output electronic device in which the output is activated by specific combinations of input signals.

gauge The absolute value of a reference point or level.

gauge boson Any of the quanta associated with a gauge field, such as the quanta of weak interaction,

$W^\pm$ (mass = 86.7 u, charge = $\pm e$) and Z^0 (mass = 99.7 u, charge = 0). See **Elementary particles***.

gauge invariant Defining a quantity (e.g. gravitational or electrostatic potential difference) that is independent of gauge transformations.

gauge theory Any of the field theories whose predictions are unchanged by gauge transformations.

gauge transformation A transformation of the internal (non space–time) component of a set of quantum fields.

gauss (G, Γ) The CGS_{emu} unit of magnetic flux density = magnetic induction, equal to 1 maxwell/cm^2 = 10^{-4} tesla = 1 CGS_{esu} unit of magnetic flux density/c (where c = speed of light in cm/s). A magnetic field of 1 oersted produces a magnetic induction of 1 gauss in a medium of magnetic permeability = 1.

Gauss' law "The closed integral of the electrical vector over a surface is equal to the total net charge within the surface divided by the permittivity constant."

$$\oint \cdot E\, d\mathbf{S} = Q/\epsilon_0$$

where $\mathbf{E}$ = electrical vector, $\mathbf{S}$ = surface vector taken normal to the surface element and pointing outward, Q = total net charge, ϵ_0 = permittivity constant.

Gay-Lussac's law "The volumes of reacting gases and of the gas produced are in simple proportions represented by small integers."

g-cal Gram-calorie.

geanticline An anticline developed within a geosyncline by lateral compression.

Geiger counter See **Geiger-Müller counter.**

Geiger-Müller counter A particle counter consisting of a metallic tube with a wire along its axis. The tube is filled with gas and a voltage almost sufficient to cause a spark is maintained between tube (cathode) and wire (anode). A particle crossing the gas produces ionization along its track causing a gas discharge, which thus reveals the particle.

gel A colloidal system in which the dispersed phase has combined with the dispersant to form a jelly.

gelifluction Soil creep on a permanently frozen surface.

gelifraction The fracturing of rocks by freezing of percolating water.

geliturbation Reworking of soil or unconsolidated sediment by water freezing and melting.

gendarme *(French)* A sharp pinnacle on an arête.

gene A hereditary unit occupying a fixed site on a chromosome, responsible for a specific physical or chemical characteristic of the phenotype, and subject to converting to a different allele by mutation. A gene is a specific section of a DNA molecule (cistron), corresponding to a specific protein. Cf. **DNA.**

generalized coordinates Any set of independent coordinates used to describe the configuration of a system. For a rigid body with three translational and three rotational degrees of freedom in space, the generalized coordinates are the three coordinates q_1, q_2, and q_3 specifying position, the three angular coordinates q_4, q_5, and q_6, specifying orientation, and the six coordinates q'_1, q'_2, q'_3, q'_4, q'_5, and q'_6, which are the derivatives with respect to time of the previous six coordinates, or the six coordinates p_1, p_2, p_3, p_4, p_5, and p_6, representing momenta (3 linear and 3 angular). Either system of 12 coordinates (for a three-dimensional space plus time) completely describes the state of the body at a given instant in time. A position coordinate and its corresponding momentum coordinate are said to be canonically conjugated.

generalized force The ratio of change in potential energy E_p to change in the generalized coordinate q_i.

generalized velocity The first derivative with respect to time of a generalized position coordinate.

general relativity See **relativity.**

generator A conductor or set of conductors made to rotate in a magnetic field to produce electric energy.

genetic code The code relating each of the 64 codons of DNA to specific amino acids and to start and stop signals for protein chain formation. See **codon.**

genetic drift The result of random fluctuations in gene frequencies occurring through time in small populations.

genocline A hybridization gradient between subspecies occupying neighboring regions.

genome 1. The total DNA content of a haploid or monoploid organism. 2. The DNA that contains one complete set of genes in a diploid organism. 3. The genetic endowment of a species.

genotype The genetic makeup of a species. Cf. **phenotype.**

2nd →	U	C	A	G	3rd ↓
U	PHE PHE LEU LEU	SER SER SER SER	TYR TYR STOP STOP	CYS CYS STOP TRP	U C A G
C	LEU LEU LEU LEU	PRO PRO PRO PRO	HIS HIS GLN GLN	ARG ARG ARG ARG	U C A G
A	ILE ILE ILE MET	THR THR THR THR	ASN ASN LYS LYS	SER SER ARG ARG	U C A G
G	VAL VAL VAL VAL	ALA ALA ALA ALA	ASP ASP GLU GLU	GLY GLY GLY GLY	U C A G

Genetic Code. One or more triplets of DNA bases (U = uracil, C = cytosine, A = adenine, G = guanine) code for a specific amino acid or for a start (AUG, which also codes for methionine) or stop signal.

genus A rank in taxonomic classification above species and below family.

geo- Prefix meaning *earth*.

geocentric Referring to the motion of the Moon or an artificial satellite that has the Earth as center of mass.

geode A partly hollow spherical or ovoidal concretion consisting of an outer layer of chalcedony serving as base for inward-growing and projecting megacrystals, usually of quartz or calcite. Geodes are normally found in limestones.

geodesic The shortest line between two points along any mathematically defined surface.

geodesy The science studying the shape of the Earth and that of its gravitational field.

geodetic coordinates Latitude = angular distance N or S from equator; longitude = angular distance E or W from meridian passing through Greenwich, England; elevation = vertical distance from mean sea level.

geographic coordinates A subset of the geodetic coordinates, including only latitude and longitude.

geographic pole Either one of the two points where the rotational axis of the Earth intersects the Earth's surface.

geoid The theoretical, equipotential sealevel surface enveloping the Earth everywhere, normally to the direction of the gravitational field ("reference geoid").

$$r_0 = a[1 + (2f - f^2)\sin^2\phi/(1 - f)^2]^{-1/2}$$
$$\simeq a(1 - f\sin^2\phi)$$

where r_0 = distance between center of Earth and a given latitude on geoid; a = equatorial radius; f = flattening = $(a - c)/a$ = 0.00335282; ϕ = latitude; c = polar radius.

geological time scale An absolute time scale dating the boundaries between the successive time units of the Earth's history. It ranges from the origin of the solar system (4.7 billion years ago) to the present. The major subdivisions (with boundary ages in years) are: $4.7 \cdot 10^9$/Hadean/$3.8 \cdot 10^9$/Archean/$2.7 \cdot 10^9$/Proterozoic/$590 \cdot 10^6$/Paleozoic/$248 \cdot 10^6$/Mesozoic/$65 \cdot 10^6$/Cenozoic/0. See also **Cenozoic, cosmological time, Geological time scale*, Mesozoic, Paleozoic, Quaternary.**

geologic time A commonly used expression to indicate the time since the formation of the Earth ($4.7 \cdot 10^9$ y. B.P.).

geomagnetic axis The axis of the dipole field most closely approximating the magnetic field of the Earth. It is inclined 11° with respect to the rotational axis but it coincides with it if averaged across ~10,000 y.

geomagnetic equator The line of zero geomagnetic latitude. Cf. **magnetic equator.**

geomagnetic field The geocentric magnetic dipole field most closely approximating the Earth's magnetic field. Geomagnetic dipole moment = $7.90 \cdot 10^{22}$ A m^2 = $7.90 \cdot 10^{25}$ G cm^3 (1985). The geomagnetic field reverses itself at intervals ranging from <10,000 y to >$25 \cdot 10^6$ y. See **dipole field, geomagnetic pole, magnetic field, polarity epoch, polarity event.**

geomagnetic latitude The latitude of any given point on the Earth's surface referred to the geomagnetic poles. Cf. **magnetic latitude.**

geomagnetic longitude The longitude of any given point on the Earth's surface referred to the geomagnetic field, with base meridian extending south from the north geomagnetic pole. Cf. **magnetic longitude.**

geomagnetic pole Either one of the two points where the geomagnetic axis intersects the Earth's surface. North geomagnetic pole: 78°30′N, 68°50′W; south geomagnetic pole: 78°30′S, 111°10′ E. The geomagnetic poles are antipodal and come to coincide with the geographic poles if their position is averaged across ~10,000 y. See **geomagnetic field, magnetic field.**

geopetal Referring to any rock feature that identifies the original orientation of the rock in terms of top and bottom.

geophone A coil rigidly attached to a frame within the field of a magnet, which is attached to the frame by springs, capable of measuring vertical ground motion by the voltage developed in it.

geopotential The gravitational potential energy at any given point within the gravitational field of the Earth.

geopotential height The vertical distance between geoid and a given geopotential surface.

geopotential surface A surface within the Earth's gravitational field on which all points have the same gravitational potential energy.

geostrophic Defining a wind or ocean current in which the direction of motion is governed by a balance between driving force (e.g. a pressure difference) and the Coriolis effect.

geosuture A plane along which two plates collided and became welded to each other.

geosynclinal facies A sedimentary facies characterized by a predominance of largely unfossiliferous clastic sediments including graded bedded graywackes and shales, and by a great sediment thickness. The geosynclinal facies is characteristic of rapid sedimentation in a geosyncline.

geosyncline A subsiding trough along a continental margin.

geotaxis Taxis resulting from the effect of the Earth's gravitational field.

geothermal Referring to the internal heat of the Earth.

geothermal energy Energy produced by the internal heat of the Earth. See **Geothermal energy*.**

geothermal flux The flow of heat from the interior of the Earth through the surface. Continental average = 0.0565 W/m^2 = 1.35 μcal/cm^2/s; oceanic average = 0.0782 W/m^2 = 1.87 μcal/cm^2/s; worldwide average = 0.0699 W/m^2 = 1.67 μcal/cm^2/s. Higher values are measured along the midocean ridges (average 0.0837 W/m^2 = 2.00 μcal/cm^2/s, but ranging up to 0.3351 W/m^2 = 8.00 μcal/cm^2/s). Cf. with 160 W/m^2 = 3.8 mcal/cm^2/s for the average solar energy flux at the Earth's surface.

geothermal gradient The increase in temperature from the surface of the Earth downward. It averages 0.8°C/km through the mantle, but it ranges as high as 10–100°C/km in the crust. See **Earth interior—physical data*.**

geotropism Tropism guided by the gravitational field of the Earth, as the downward growth of aerial roots of tropical plants.

getter A substance that binds gases to its surface, thus increasing the vacuum in a vacuum system.

GeV Gigaelectronvolt (= 10^9 electron volts). It is identical to BeV.

geyser An intermittent emission of water and steam from an underground reservoir above a mass of hot rock. The long (several tens of meters) neck connecting the reservoir to the surface acts as a cork as the water below becomes heated to the boiling point at the pertinent pressure. As the water in the neck expands, it begins to overflow. This pressure release converts the entire water column into steam with consequent blowout. Recharge is from percolating waters.

geyserite Hydrated silica deposited by a geyser.

g factor The negative ratio of magnetic moment (in Bohr magnetons) to angular momentum (in units of $h/2\pi$) for electrons, other elementary particles, atomic nuclei, or atoms.
1. **electron orbital g factor:**
$$g_l = -\mu_l h/2\pi L \mu_B$$
$$= 1.001159652193$$
where μ_l = orbital magnetic dipole moment vector, h = Planck's constant, L = orbital angular momentum vector, μ_B = Bohr magneton.
2. **electron spin g factor:**
$$g_s = -\mu_s h/2\pi S \mu_B$$
$$= 2.002319304386$$
where μ_s = spin magnetic dipole moment vector, h = Planck's constant, S = spin vector, μ_B = Bohr magneton.
3. **nuclear spin g factor:**
$$g_i = \pm \mu_i h/2\pi I \mu_N$$
where μ_i = spin magnetic moment dipole vector, h = Planck's constant, I = nuclear spin angular momentum vector, μ_N = nuclear magneton. Cf. **gyromagnetic ratio.**

Ghyben-Herzberg ratio The ratio 1/40 between the static head of fresh water above sea level in coastal regions and the depth of the boundary with the underlying seawater. The value of the ratio (1/40) derives from fresh water being about 1/40th less dense than seawater (ρ = 1 vs. ρ = 1.026).

Thus, for a hydrostatic head of 1 m above sea level, the boundary is at 40 m below sea level.

giant planets The planets Jupiter, Saturn, Uranus, and Neptune. Syn. **Jovian planets.**

Gibbs free energy (G) See **free energy.**

Gibbs phase rule See **phase rule.**

giga- Prefix meaning 10^9.

gigayear (Gy) A unit of time equal to 10^9 y. Syn. **aeon.**

gilbert The CGS_{emu} unit of magnetomotive force. 1 gilbert = $1/4\pi$ abampere-turn = $10/4\pi$ ampere-turn.

glacial age An interval of glaciation between two interglacial ages.

glacial epoch A period of glaciation consisting of alternating glacial and interglacial ages.

glacial groove A groove incised on bedrock under a glacier by a rock fragment embedded in the ice. It is larger than a glacial stria.

glacial lake A lake dammed by glacial deposits.

glacial-marine Defining marine sediments containing an appreciable amount of ice-rafted material.

glacial milk Milky water issuing from under a glacier, carrying a load of clay-size mineral particles abraded from bedrock.

glacial period A geological period characterized by repeated glaciations.

glacial stria A striation incised on bedrock under a glacier by a rock fragment embedded in the ice. It is smaller than a glacial groove.

glacial striation See **glacial stria.**

glaciation A major expansion of ice cover over land and over polar oceans during an ice age. There appear to have been $\backsim$ 30 glaciations during the past 3 million years. The most recent one ended 10,000 y ago. Additional glaciations are expected in the future.

glaciofluvial Defining deposits formed by rivers issuing from glaciers.

glass A material exhibiting total or near-total crystallographic disorder while sufficiently rigid to maintain mechanical shape for an appreciable length of time. Because the amorphous state has higher energy than the crystalline state, glass tends to crystallize.

glassy metal Any of a class of metallic alloys with disordered structure achieved by ultrarapid cooling (10^6–10^8 °C/s).

glauconite An authigenic, sedimentary K, Na, Fe and Mg hydroxy Al-silicate, $(K,Na)(Al,Fe^{3+}, Mg)_2(Al,Si)_4O_{10}(OH)_2$.

glaucophane A metamorphic hydroxy mineral, $Na_2Mg_3Al_2Si_8O_{22}(OH)_2$.

glaucophane schist See **blue schist.**

glaucophane schist facies See **blueschist facies.**

G layer The layer inside the Earth, from 5170 km of depth to the center of the Earth, occupied by the inner core.

glide plane A plane of symmetry in a crystal consisting of a reflection and a translation.

glitch A sudden (minutes), small ($2 \cdot 10^{-6}$) decrease in the rotational period of a pulsar, believed to result from internal readjustment (starquake).

globigerina ooze A calcareous (>30% $CaCO_3$) deep-sea sediment largely consisting of planktonic foraminiferal tests, coccoliths, and red clay. See **deep-sea sediments.**

globular cluster Any of the spherical clusters, 60 to 600 l.y. across, of Population II stars distributed throughout the galactic halo. More than 125 have been catalogued. Number of stars/cluster = 10^5 to 10^7; mean age of stars = 10^{10} y; density of stars in cluster centers = $30/(l.y.)^3$.

globule Any of the dense (10^3–10^4 particles/cm^3), dark, rotating ($\approx 5 \cdot 10^{-14}$ rad s^{-1}) globular concentrations of interstellar matter, with radius = 0.3–30 l.y. and mass = 1–500 solar masses. See **star formation.**

gluon The carrier of the color force and field quantum of strong interaction between quarks. Mass = 0; charge = 0; color = red, green, or blue; spin = 1.

glyceride ester An ester derived from glycerin.

glycerin See **glycerol.**

glycerol $CH_2OH \cdot CH(OH) \cdot CH_2OH$. Mol. mass = 92.095.

glyptolith A stone shaped by wind in a desertic area. Syn. **ventifact.**

GMAT Greenwich Mean Astronomical Time.

GMT Greenwich Mean Time.

gneiss A metamorphic rock consisting of feldspar, quartz, and biotite with the crystals variously flattened in a common direction. Gneisses may result from ultrametamorphism of graywackes and/or recrystallization of granitic intrusives under unidirectional compression.

golden-brown algae Largely unicellular or colonial planktonic algae characterized by carotenoid pigments in addition to chlorophyll, belonging to the phylum Chrysophyta. Included are the classes Xanthophyceae, Chrysophyceae, and Bacillariophyceae (diatoms).

Gondwana The Late Paleozoic southern supercontinent consisting of South America, Africa, Madagascar, peninsular India, Australia, and Antarctica. Cf. **Laurasia, Pangea.**

Gondwanaland Syn. **Gondwana.**

gossan A weathered layer consisting of iron oxides and hydroxides capping a sulfide deposit.

graben A structural trough bound by normal faults, resulting from elongated updoming. Cf. **aulacogen.**

grad Gradient.

graded bed A clastic bed with coarse elements below grading into finer elements above. It results from turbidity current deposition. Cf. **turbidite.**

gradient (grad) The gradient ϕ of a scalar field ϕ (x,y,z) that changes continuously in space.

$$\text{grad } \phi = \nabla\phi$$
$$= \mathbf{i}\phi_x + \mathbf{j}\phi_y + \mathbf{k}\phi_z$$

where $\mathbf{i}$, $\mathbf{j}$, and $\mathbf{k}$ are unit vectors along the x, y, and z axes. It represents the direction along which the scalar changes most rapidly.

gradualism A theory holding that evolution proceeds gradually.

Graham's diffusion law "The rates of diffusion of gases vary inversely to the square of their densities."

$$D_1/D_2 = (\rho_2/\rho_1)^{1/2}$$

where D = rate of diffusion, ρ = density.

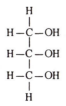

Glycerol.

gram (g) The CGS unit of mass, equal to 1/1000th of the mass of the kilogram.

gram-atom (mol) An Avogadro number of atoms.

gram-atomic weight The weight of a gram atom.

gram-calorie See **calorie.**

gram-equivalent The amount of a substance having an Avogadro number of valences.

gram-ion An Avogadro number of ions.

gram-molecular weight The weight of a gram-molecule.

gram molecule (mol) An Avogadro number of molecules.

Grand Unified Theory (GUT) Any of the theories attempting to find a common derivation for quarks and leptons and for the color and electroweak forces.

granite A plutonic rock consisting mainly of quartz, orthoclase, Na-plagioclase, and biotite.

granitization The transformation of a sedimentary rock (graywacke or shale) into granite.

granoblastic Defining a metamorphic rock in which the crystals are equidimensional.

granodiorite A plutonic rock consisting mainly of quartz, Na-plagioclase, orthoclase, and biotite.

granule *(Astronomy)* The bright top of a convection cell in the solar photosphere, about 1000 km across, separated from adjacent tops by darker furrows representing descending limbs. Ascending velocity is about 500 m/s and spreading velocity from the granule center is about 250 m/s. *(Geology)* A mineral or rock fragment ranging in size from 2 to 4 mm.

granulite A high-temperature (650–900°C), low-to-high pressure (1–10 kbar) metamorphic rock consisting mainly of feldspars, quartz, and biotite in crystals of equal dimensions.

granuloblastic Defining a metamorphic rock in which the crystals are equidimensional but smaller than in granoblastic rocks.

grapestone A small clump of calcareous grains with incipient cementation occurring in modern carbonate environments (e.g. the Bahama Banks).

graphite The low temperature, low pressure phase of crystalline carbon. Density = 2.27 g/cm³. See **diamond.**

grating 1. A grid of parallel lines for diffraction

studies. The distance between the lines is about 3 times greater than the wavelength of the radiation to be studied. 2. A grid of crossed wires to reflect and focus microwaves.

gravel A loose accumulation of rock fragments larger than 2 mm.

gravimeter An instrument used to measure the gravitational field of the Earth at any given point. It basically consists of an inertial mass suspended from a spring, on which the force $F = mg$ is acting (m = mass, g = gravitational acceleration). The position of the mass with respect to a vertical reference scale is a function of the ambient gravitational acceleration.

gravitation The attraction between masses. It is one of the four natural forces; it has infinite range; and it obeys the inverse square law.

gravitational constant (G) The constant needed to express Newton's law of gravitation, $F = Gm_1m_2/r^2$ (where F = force between masses m_1 and m_2 separated by distance r), in units based on his second law.

$G = Fr^2/m_1m_2$

$\quad = 6.67206 \cdot 10^{-11}$ N m^2 kg^{-2} or m^3 s^{-2} kg^{-1}

$\quad = 6.67206 \cdot 10^{-8}$ dyn cm^2 g^{-2} or cm^3 s^{-2} g^{-1}.

gravitational field The field created by a mass. The gravitational field g of the Earth varies principally with latitude because of the flattening of the Earth. International Gravity Formula (before 1967)

$g = 9.780490 \, (1 + 0.0052884 \sin^2 \phi$
$\qquad\qquad\qquad - 0.0000059 \sin^2 2\phi)$

Geodetic Reference System (GRS 67) Formula (since 1967)

$g = 9.78031846 \, (1 + 0.005278895 \sin^2 \phi$
$\qquad\qquad\qquad + 0.000023462 \sin^4 \phi)$

In the preceding, ϕ = degrees of latitude. The Earth's gravitational field ranges from 9.7803185 m/s^2 at the equator to 9.8321776 m/s^2 at the poles (standard $g = g_0 = 9.80665$ m/s^2). See also g_0, **gravitational force, graviton, gravity standard.**

gravitational force One of the four natural forces, caused by the presence of mass. See **graviton, natural forces.**

gravitational mass The mass m of a body obtained from the equation $F = mg$:

$$m = F/g$$

where F = force and g gravitational acceleration. According to the principle of equivalence, it is identical to the inertial mass obtained from the

equation $F = ma$, where F = force, a = acceleration. See **principle of equivalence.**

gravitational shift See **Einstein shift.**

graviton The postulated field quantum of gravitational interaction, with mass = 0, charge = 0, spin = 2, coupling constant K (from $Km^2/\hbar c$, where m = mass, $\hbar = h/2\pi$ and h = Planck's constant, c = speed of light) = $0.53 \cdot 10^{-38}$. See **Elementary particles*.**

gravity The force of gravitation, one of the four natural forces.

gravity anomaly The anomaly remaining after a gravity measurement at a given site has been corrected for any one of several parameters (vertical distance above or below the geoid, attraction of intervening rocks if any, the existence and geometry of mountain roots, etc.).

gravity standard (Potsdam gravity) The value of the gravitational acceleration at the Pendelsaal of the Geodetic Institute in Potsdam, East Germany. The value accepted until 1967 was 9.81274 m/s^2 or 981.274 gal. The revised value is 9.81260 m/s^2 or 981.260 gal.

gravity tectonics Tectonic motions resulting from uplift followed by gravitational sliding.

graywacke A dirty sandstone, consisting of quartz, feldspar, and femic minerals in a clay matrix, characteristic of geosynclinal sedimentation.

great circle Any circle on a spherical surface with its center at the center of the sphere. A great circle passing through two points on the surface represents the shortest distance between the two points (geodesic).

Great Red Spot A large (15,000 $\times$ 30,000 km), oval, counterclockwise vortex in Jupiter's atmosphere, 22° south of Jupiter's equator, consisting of a mass of warm gases rising 8 km above the surrounding cloud floor. The red color may result from organic molecules, including amino acids, produced by electrical discharges. It has been stable for more than 200 years.

green algae A chlorophyll-rich group of algae, from unicellular to large seaweeds, belonging to the phylum Chlorophyta.

greenalite 1. Greenish, hydrated iron silicate. 2. A fine-grained, dark-green sedimentary rock containing greenalite in a matrix of carbonate and chert.

greenhouse effect The heating of a planet's surface by the trapping of infrared backradiation from the planet's surface by atmospheric gases (especially H_2O and CO_2).

green mud A hemipelagic mud rich in organic matter and sulfides of iron and manganese.

greensand A marine sand consisting in part of glauconite.

greenschist A low-temperature (300–400°C), low-pressure (1–5 kbar) metamorphic phase of gabbro-basalt common along converging plate margins. Characteristic minerals are chlorite, epidote, and actinolite. Cf. **blueschist**.

greenschist facies The set of minerals characteristic of greenschists. Cf. **blueschist facies**.

greenstone A low-grade metamorphosed basic igneous rock, containing chlorite, epidote, and actinolite.

greenstone belt A belt in Precambrian terrain consisting of greenstone festoons, each with an upward mafic-to-felsic gradation.

Greenwich Mean Astronomical Time (GMAT) Greenwich Mean Time counted from noon at the Greenwich meridian.

Greenwich Mean Time (GMT) Universal Time (UT) = Zulu (Z) time. Mean solar time counted from midnight at the Greenwich meridian.

Greenwich meridian The meridian for reckoning longitude, passing through the original location of the Royal Astronomical Observatory at Greenwich, near London, England.

Greenwich Sidereal Time (GST) Sidereal time at Greenwich, England, based on the longitudinal coordinate of the vernal equinox in the equatorial system of coordinates.

Gregorian Calendar A modification of the Julian calendar instituted by Pope Gregory XIII in 1582 in which century years are leap years only when divisible by 400.

greisen A pneumatolytically altered granite, consisting of quartz and muscovite with accessory tourmaline, fluorite, and topaz.

Grenzhorizont A horizon of nondeposition in a northern bog, often marked by a layer of clay. Syn. **recurrence horizon.**

greywacke See **graywacke.**

grid An electrode between anode and cathode that controls the flow of electrons in electron tubes.

grit A sandstone consisting of angular fragments of similar sizes in fine-grained cement.

groove cast An imprint, on the underside of the overlying bed, of a groove on the silty or clayey surface of the underlying bed that was filled with coarser sediment during the deposition of the overlying bed.

grossular Syn. **grossularite.**

grossularite See **garnet.**

grotto A small cave.

ground A low-resistance conductor directly or indirectly connected with the earth.

groundmass The fine material between larger crystals in a porphyritic rock.

ground moraine A sheet of till left by the melting of a glacier, consisting of fragments previously embedded in the ice or carried on the ice's surface.

ground state The lowest stationary energy state of a particle or system of particles.

groundwater Subsurface water that is part of the mobile surface reservoir. Syn. **phreatic water.**

group *(Mathematics)* A set S of elements a, b, c . . . in which a binary operation (i.e. an operation involving two members of the set) produces a third element which is also a member of the set. The set is then said to be closed with respect to that operation. In the set of positive integers, for instance, the operation 2×3 yields 6 which is also part of the set but the operation 2/3 does not. The group must also satisfy the properties of associativity [e.g. $(2 \times 3)4 = 2(3 \times 4)$]; identity ($ai = ia = a$, where i = identity; for instance, in the operation of multiplication, $i = 1$ because $a \times 1 = 1 \times a = a$; in the operation of addition, $i = 0$ because $a + 0 = 0 + a = a$); and inversion (for each element a there is an inverse element a^{-1} so that $a \times a^{-1} = i$). Groups satisfying the commutative law for addition ($a + b = b + a$) and multiplication ($a \times b = b \times a$) are called "abelian groups" (after the Norwegian mathematician Abel, 1802–1829). *(Chemistry)* Any of the columns in the Periodic Table of the Elements. *(Geology)* The rank above formation in the lithostratigraphic hierarchy.

group theory The branch of mathematics that studies the properties of groups.

group velocity 1. The velocity of propagation of a group of waves forming a wave packet. See **wave packet.** 2. The common velocity of a group of acoustic waves of different frequencies, when velocity is not a function of the wavelength. See **phase velocity.**

growth curve 1. A curve representing a growth function. 2. The curve representing the growth of the daughter product of a radioactive nuclide, following segregation of the parent nuclide in a different geochemical system. See **growth equation.**

growth equation The equation

$$Q = Q_0 e^{kt}$$

expressing the growth of the quantity Q, from an initial value Q_0, at the rate k through time t, where $k = (dQ/dt)/Q$ = growth constant. See e (def. 3). Cf. **attenuation, decay equation.**

growth line Any of the thin ridges on the outer surface of a molluscan shell, solitary coral calyx, or other exoskeleton, representing stasis during growth.

growth ring The layer of wood produced by a tree in one year.

growth ruga A thick growth line, representing a longer period of stasis than a growth line.

grus A granular mass produced by the in-situ disintegration of granite or gneiss.

GSA Geological Society of America.

GST Greenwich Sidereal Time.

guanine A nucleic acid base with purine ring structure, $C_5H_5N_5O$ (mol. mass = 151.128).

guano An indurated and leached bird excrement found in arid coastal regions or islands.

guided wave An acoustical or optical wave confined within a particular transparent medium.

guide fossil A preferably abundant fossil confined within a clearly defined, restricted segment of geologic time.

gulch A ravine with steep sides, larger than a gully.

gully 1. A small depression carved by water running down a slope. 2. A small depression on a submarine or sublacustrine delta front produced by sediment slumping.

gumbotil A dark gray, leached, deoxidized B horizon in a mature soil with poor drainage, largely consisting of illite.

GUT See **Grand Unified Theory.**

Gutenberg discontinuity The seismic discontinuity at the boundary between mantle and core (2885 km of depth).

guyot A submarine volcanic cone whose top was truncated by weathering and wave action and which subsequently subsided.

G wave A long period (1–4 min) Love wave in the upper mantle, observed under both continents and oceans.

gymnosperm Any of the plants characterized by having their seeds not enclosed in an ovary. Cf. **angiosperm.**

gypsum Hydrated Ca sulfate, $CaSO_4 \cdot 2H_2O$.

gypsum plate A plate of selenite in a polarizing microscope giving first-order red interference color.

gyre A broad, circular motion of surface seawater in each of the major oceanic basins, produced by geostrophic winds radiating from the subtropical high-pressure regions.

gyrocompass A direction reference device consisting of a rotor in a pendular case capable of swinging azimuthally. If the rotor's axis is aligned with the Earth's axis, it will remain so as the Earth rotates since it will not experience any torque. Any offset introduces torques that bring about realignment.

gyromagnetic ratio (γ) The ratio of the magnetic dipole moment of a particle to its angular momentum. Cf. g **factor.**

gyttja A freshwater mud rich in organic matter deposited in lakes or marshes.

H

h 1. Celestial altitude. 2. Hecto-. 3. Hour. 4. Planck's constant ($= 6.626175 \cdot 10^{-34}$ J s $= 4.135669 \cdot 10^{-15}$ eV s).

$\hbar$ *h* bar $= h/2\pi = 1.0545726 \cdot 10^{-34}$ J s $= 6.5820727 \cdot 10^{-16}$ eV s.

H 1. Enthalpy. 2. Hamiltonian. 3. Henry. 4. Magnetic field intensity.

H_0 Hubble constant.

HI Neutral hydrogen.

HII Ionized hydrogen.

Haber process A nitrogen-fixation process to produce NH_3 by reacting N_2 with H_2 at high pressure (100–1000 atm) and temperature (400–550°C) in the presence of a suitable catalyst (commonly Fe with small amounts of K and Al).

habit The characteristic form of a crystal.

hadal Defining the deepest (>6000 m) oceanic environment.

Hadean A division of geologic time from the origin of the solar system ($4.7 \cdot 10^9$ y B.P.) to the beginning of the Archean ($3.8 \cdot 10^9$ y B.P.). No rocks of Hadean age have yet been discovered on Earth. See **Geological Time Scale***.

Hadley cell Either one of the two low-latitude atmospheric circulation cell systems on either side of the equator. Warm air rises to the upper troposphere off the equator, it moves poleward on a 45° course (northern hemisphere) or 135° course (southern hemisphere), it sinks to the lower troposphere at 30° of latitude, and returns to the equator on a 225° course (northern hemisphere) or 315° course (southern hemisphere). See **Ferrel cell, polar cell.**

hadron Any of the strongly interacting particles, consisting of either a quark-antiquark pair (mesons) or 3 quarks (baryons). See **Hadrons—quark structure*.**

haem See **heme.**

haem- See **hemo-.**

hail Accretionary, concentrically layered, spheroidal ice pellet, millimeters to centimeters across, forming in cumulonimbus storm clouds with vigorous updraft.

half-life ($t_{1/2}$) The time needed for a given amount of a radioactive substance to be reduced to one-half by radioactive decay.

$$t_{1/2} = \ln 2/\lambda$$
$$= 0.6931472/\lambda$$

where λ = decay constant.

half-wave rectifier A rectifier that admits alternating current flow only during alternating half-cycles.

halite The common salt, NaCl.

Hall effect The development of an electric field normal to both a magnetic field threading a conductor and the direction of current flow in the conductor.

Halley's comet A comet with a period of 76 years, first recorded in 239 B.C. Recent appearances (year of closest approach to Earth): A.D. 1066, 1145, 1222, 1301, 1378, 1456, 1531, 1607, 1682, 1759, 1835, 1910, 1986.

Hall process An electrolytic process to recover Al from Al_2O_3 dissolved in molten cryolite (Na_3AlF_6).

halmeic Defining an authigenic deep-sea precipitate.

halmyrolysis The reaction between seawater and sediment on the ocean floor before burial.

halo *(Astronomy)* A large (radius 50,000 l.y.) region of space centered on the center of the Galaxy and populated by globular clusters. *(Meteorology)* A circular band of colored light around the Sun or the Moon, caused by solar or lunar light being reflected and refracted by atmospheric ice crystals. *(Geology)* See **pleochroic halo.**

halocline A change in the salinity gradient in the sea or a salt lake.

halogen Any of the nonmetallic elements of group 17 of the Periodic Table of the Elements, including F, Cl, Br, I, and At.

halophyte A plant adapted to high salinity soils.

Hamiltonian (H) A function of the generalized

coordinates and momenta of a dynamic system of particles.

$$H(q,p,t) = \Sigma \, \dot{q}_i p_i - L(q,\dot{q},t)$$

where q = generalized coordinate, $\dot{q}$ = generalized velocity, p = generalized momentum, t = time, L = Lagrangian.

hammada A rock desert on a bare plateau.

hammock A karstic depression in a tropical area that has accumulated soil and is supporting lush vegetation including hardwoods.

hanging tributary A tributary occupying a hanging valley.

hanging valley 1. A glacial valley with floor at its mouth above that of the valley of which it is a tributary. It arises from the tributary glacier having been unable to erode its floor as deeply as the major glacier. 2. A river valley with its lower course and mouth higher than the river valley of which it is a tributary. It arises when rocks of the tributary riverbed are harder to erode than those of the major riverbed.

haploid Having half as many chromosomes as the full complement, a characteristic of spores, gametes, and gametophytes. Cf. **monoploid.**

hard *(Physics)* Defining radiation consisting of particles or photons of sufficient energy to penetrate at least 10 cm of lead or 1.3 m of water. *(Chemistry)* Defining water containing >60 mg/liter of dissolved carbonates expressed as $CaCO_3$.

hardground A submarine carbonate surface free from sedimentation and modified into a hard phosphatic or ferromanganoan crust.

hardness *(Mineralogy)* The resistance of a mineral to scratching, ranging from 1 (talc) to 10 (diamond) on the Mohs scale. See **Mohs scale.** *(Materials)* The resistance of solid materials to indentation, measured by various standardized techniques. *(Chemistry)* The concentration of Ca and Mg ions in water, expressed as ppt or ppm of $CaCO_3$.

hardpan A hard soil layer resulting from cementation of particles by carbonates, iron oxides, and hydrated silica left behind by evaporating solutions.

hard-rock geology The geology and petrology of igneous and metamorphic rocks. Cf. **soft rock geology.**

hard water See **hard** *(Chemistry).*

Hardy-Weinberg law "The proportion of alleles at any given locus remains constant in a large, genetically isolated population mating at random." This law assumes no mutation and no selection.

harmonic frequency A frequency that is an integral multiple of a fundamental frequency.

harmonic motion A periodic motion in which the restoring force is proportional to the displacement.
1. simple harmonic motion:.

$$-kx = m(d^2x/dt^2)$$

where $-kx$ = restoring force, m = mass.

period	T	$= 2\pi/\omega$
frequency	ν	$= 1/T$
		$= \omega/2\pi$
displacement	x	$= A \cos(\omega t + \phi)$
velocity	v	$= dx/dt$
		$= -\omega A \sin(\omega t + \phi)$
maximum velocity	v_{max}	$= \omega A$
acceleration	a	$= dv/dt$
		$= -\omega^2 A \cos(\omega t + \phi)$
maximum acceleration		$= \omega^2 A$
potential energy	U_x	$= \frac{1}{2}m\omega^2 A^2 \cos^2(\omega t + \phi)$
kinetic energy	K_x	$= \frac{1}{2}mv^2$
		$= \frac{1}{2}m\omega^2 A^2 \sin^2(\omega t + \phi)$
total energy	E_x	$= U_x + K_x$
		$= \frac{1}{2}m\omega^2 A^2$
		$= \frac{1}{2}mv_{max}^2$

In the preceding, $\omega = 2\pi\nu = 2\pi/T$ = angular frequency; t = time; $(\omega t + \phi)$ = phase; ϕ = phase angle at time $t = 0$; A = amplitude, m = mass.
2. damped harmonic motion:

$$-kx - b(dx/dt) = m(d^2x/dt^2)$$

where $-kx$ = restoring force, $-b(dx/dt)$ = damping force, m = mass, d^2x/dt^2 = acceleration.

harzburgite A plutonic rock consisting mainly of olivine and pyroxene.

haversine One half of versine.

Hawaiian-type eruption The nonexplosive eruption of basaltic lavas of low viscosity.

haze A suspension of aerosols in the atmosphere.

head The nucleus and coma of a comet.

headland A high and prominent promontory.

heat capacity The quantity of heat needed to raise the temperature of a specific amount of a given substance or system by 1°C.

heat conduction The flux of thermal energy through a conductor.

$$dQ/dt = \kappa A(dT/dx)$$

where Q = heat, t = time, T = temperature, κ =

thermal conductivity, A = cross section of heat conductor, dT/dx = temperature gradient.

heat conductivity See **conductivity**.

heat flow See **geothermal flux**.

heat-flow unit (HFU) The unit of geothermal flux, equal to 10^{-6} cal/cm²/s = 0.041868 W/m².

heat of combustion The amount of heat produced by the oxidation of a specific quantity of a substance at constant pressure or volume.

heat of condensation The amount of heat released by a specific amount of a substance during the process of condensation at constant pressure.

heat of crystallization The amount of heat released by a specific amount of a substance during the process of crystallization at constant pressure.

heat of evaporation The amount of heat absorbed by a specific amount of a substance during the process of evaporation at constant pressure.

heat of formation The amount of heat released or absorbed when a specific quantity of a substance is formed from its elements.

heat of fusion The amount of heat absorbed by a specific amount of a substance during the process of melting at constant pressure.

heat of hydration The amount of heat released or absorbed by a specific amount of a substance during the process of hydration at constant temperature and pressure.

heat of solution The amount of heat released or absorbed by a specific amount of a substance during the process of solution at constant pressure.

heat of sublimation The amount of heat absorbed by a specific quantity of a substance during the process of sublimation at constant pressure.

heat transfer coefficient (h) Heat transfer across a unit area of a medium or system per second per °C of temperature difference between the boundaries of the medium or system.

heave The horizontal displacement of a fault.

heavy mineral 1. A heavy ($\rho > 3$) rock-forming mafic mineral. 2. A heavy ($\rho > 2.85$) accessory mineral (magnetite, ilmenite, rutile, zircon, tourmaline, biotite, etc.) in a sediment or sedimentary rock.

heavy water Water in which one or both of the hydrogen atoms in the molecule are deuterium.

hecto- (h) Prefix meaning *100*.

hedenbergite A Ca-rich clinopyroxene, $CaFe \cdot (SiO_3)_2$.

Heisenberg uncertainty principle See **uncertainty principle**.

helicity 1. The combination of rotational and translational motion. Helicity is termed *dextral* or *positive* if the rotational angular velocity vector and the translational velocity vector are parallel. It is termed *sinistral* or *negative* if they are antiparallel. 2. The component of the spin of a particle along the direction of its momentum vector. It is considered positive or right-handed if the spin angular momentum vector is parallel to the linear momentum vector; negative or left-handed if the spin angular momentum vector is antiparallel to the linear momentum vector.

heliocentric Referring to the motion of a planet, asteroid, or comet around the Sun.

helion The nucleus of ⁴He.

heliopause See **heliosphere**.

heliosphere The region of expanding solar wind. Radius ≈ 100 AU = 14 light hours, bound by the heliopause where particle density equals that of interstellar space.

hem- See **hemo-**.

hema- See **hemo-**.

hematite The mineral Fe_2O_3.

heme The prosthetic group of hemoglobin, myoglobin, and cytochromes, consisting of a porphin ring with an iron at its center. $Fe^{2+}N_4O_4C_{34}H_{32}$. Mol. mass = 616.499.

hemera The time interval corresponding to an acme-zone.

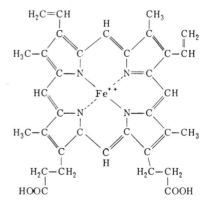

Heme. (From King and Stansfield 1985, p. 172)

hemipelagic Defining an oceanic sediment with more than 25% of redeposited neritic sediments.

hemo- Prefix meaning *blood.*

hemocyanin The blue, oxygen-carrying pigment in the blood of many mollusks and arthropods. It consists of a porphin ring structure with two Cu atoms at its center, surrounded by a protein and repeated n times. Mol. mass $= (\sim 70,000)_n$ u, for a total of $2\text{-}6 \cdot 10^6$ u.

hemoglobin The red, oxygen-carrying pigment in the blood of most animals, consisting of a heme surrounded by a protein (globin). Typical composition: $(FeS_2N_{203}O_{208}C_{738}H_{1166})_4$, corresponding to a molecular mas of 65,322.361 u. Endocellular hemoglobin may range as low as 32,000 u in molecular mass (Nastomastus, an annelid), while extracellular hemoglobin may range as high as $3 \cdot 10^6$ u (Arenicola, another annelid).

henry (H) The SI unit of inductance and permeance, defined as the self- or mutual inductance of a closed circuit in which an emf of 1 volt is produced when the current varies uniformly at the rate of 1 ampere/s.

Henry's law "The solubility of a nonreactive gas in a dilute solution is proportional to its partial pressure above the solution."

Henry's law constant See **Henry's law.**

herm A mound or reef of either clastic or bioclastic material. (Originally a stone pillar supporting the semblance of Hermes). Cf. **bioherm.**

hermal 1. Defining a reef-like structure. 2. Defining a reef-building organism. Syn. **hermatypic.**

hermatypic See **hermal.**

hertz (Hz) The SI and CGS unit of frequency, equal to 1 cycle/s.

Hertzprung-Russell diagram A graph showing the relationship between luminosity and spectral type of stars (and, therefore, surface temperatures).

heterochronous Referring to a fauna, flora, or lithofacies appearing at increasingly later times with increasing distance from the region of origin or of earliest appearance.

heterocyclic Referring to a ring structure containing different types of atoms.

heterodyne To beat. The mixing of a locally produced RF signal with an incoming signal to create a beat frequency suitable for detection and amplification.

heterogeneous catalyst A catalyst present in a phase different from that of the reactants. Cf. **homogeneous catalyst.**

heterogeneous equilibrium Equilibrium between substances in more than one phase. Cf. **homogeneous equilibrium.**

heteropolar Referring to a covalent bond in which the two atoms share the electrons unequally (dipole moment $\neq 0$). Cf. **homopolar.**

heterosis Genetic vigor exhibited by hybrids.

heterozygote A zygote having two different alleles at the same genetic locus in a diploid cell. Cf. **homozygote.**

heterozygous Having two different alleles at the same genetic locus in a diploid cell. Cf. **homozygous.**

hexagonal One of the six crystal systems, consisting of a vertical axis with threefold or sixfold symmetry and, perpendicular to it and of different length, three identical axes intersecting at 120° angles.

hexahedrite A type of iron meteorite consisting of kamacite with 5–6.5% Ni. Hexahedrites represent 10.5% of all iron meteorites or 0.6% of all meteorites. See **Meteorites*.**

HFU Heat-flow unit.

hiatus A break in the continuity of the stratigraphic record.

Higgs particle The yet unobserved quantum of the postulated Higgs field. This particle has minimum energy when the field strength is above zero, which leads to isospin symmetry breaking and the consequent existence of massive $W^{\pm}$ and Z^0 particles and neutral current interaction. The Higgs particle has no charge, no color, no spin. Suggested frequency: $\nu = 2.0 \cdot 10^{24}$ Hz $= 8.9$ u. See **technicolor.**

high-calcium limestone A limestone with less than 2.3% $MgCO_3$.

high-energy *(Physics)* Referring to the branch of physics that studies particle interactions at energies of hundreds of MeV's or higher. *(Geology)* Referring to an environment subject to strong mechanical effects by winds, waves, or currents.

highlands 1. Elevated regions on the surface of any of the inner planets, such as the terrestrial continents. 2. The elevated regions of the Moon, also called Terrae. They consist largely of anorthosites, norites, and troctolites, with an average of

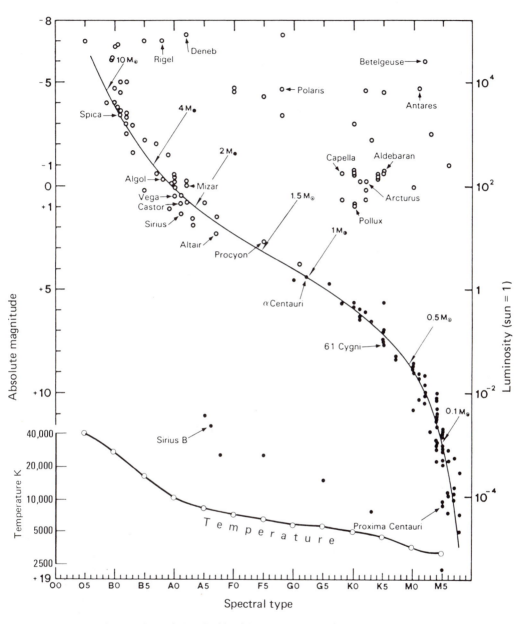

Hertzsprung-Russell diagram. Open circles: the 90 brightest stars; closed circles: the 82 stars located within 20 light years from the Sun; central line: Main Sequence; upper left region: blue giants; upper right region: red giants; bottom center: white dwarfs. (From Rigutti 1984, p. 212, Fig. 84)

70% Ca-plagioclase. Rock ages range from 3.7 to $4.6 \cdot 10^9$ y (most commonly 3.9 to $4.2 \cdot 10^9$ y), which is considerably greater than mare basalt (3.1 to $4.0 \cdot 10^9$ y).

high-magnesium calcite Calcite containing more than 4% $MgCO_3$. It is unstable in diagenesis and

converts to low-magnesium calcite or to dolomite. See **magnesian calcite.**

high quartz A high-temperature polymorph of quartz stable between 573° and 867° at atmospheric pressure and in the absence of impurities. See **silica.**

hill A topographic elevation no higher than 300 m above surrounding land.

histone A basic protein. It forms the core of nucleosomes.

holarctic Distributed throughout the Arctic regions.

hole A vacant electron energy state in the valence band of a solid. It behaves as a positively charged particle.

holo- Prefix meaning *whole.*

Holocene The most recent age of the Quaternary, ranging from 10,000 y B.P. to the present, Together with the Pleistocene ($1.6 \cdot 10^6$–10,000 y B.P.) it constitutes the Quaternary sub-era. See **Geological time scale***.

hologram An interference pattern on a photographic plate produced by the superposition of a split beam of coherent light, one part of which (the reference wave) reaches the plate directly from the source (e.g. a laser), while the other part (the object wave) reaches the plate after reflection by an object. The image is reconstructed by passing coherent light through the hologram.

holography The technique of producing holograms. It is applicable to electromagnetic waves ($\lambda = 10^{-10}$ to 10^{-1} m) and to acoustic waves.

holoplankton Collective name for the organisms that remain planktonic throughout their life cycle. Cf. **meroplankton.**

holostratotype The original stratotype chosen by the original author in establishing a stratigraphic name.

holotype The original specimen chosen by the original author in establishing a taxonomic name.

homo- Prefix meaning *one and the same.*

homeoblastic Defining a crystalloblastic structure in which the individual crystals have similar sizes.

homeomorphic Defining crystals with similar crystalline structure but different chemical composition.

homeostasis The maintenance, in higher animals, of constant internal conditions (e.g. temperature) independently of the environment.

homoclinal Defining a set of beds having the same inclination.

homocyclic Referring to a ring structure containing the same type of atom all around. Cf. **heterocyclic.**

homogeneous catalyst A catalyst present in the

same phase as the reactants. Cf. **heterogeneous catalyst.**

homogeneous equilibrium Equilibrium among substances in the same phase. Cf. **heterogeneous equilibrium.**

homogranular Defining a sedimentary rock in which the constituent grains have similar sizes.

homologous Defining structural elements evolved from the same ancestral element but performing different functions.

homologous chromosome Either of the two chromosomes that pair during meiosis.

homolographic projection An equal-area projection.

homonym A pre-emptied taxonomic name newly applied to a different taxon, usually by mistake.

homopolar Referring to a covalent bond in which the two participating atoms share the electrons equally (dipole moment = 0).

homozygote A zygote with two similar alleles at the same genetic locus in a diploid cell. Cf. **heterozygote.**

homozygous Having two similar alleles at the same genetic locus in a diploid cell. Cf. **heterozygous.**

Hooke's law "The strain of a solid is proportional to the applied stress." The proportionality factor is Young's modulus.

horizon 1. A stratigraphic surface defined by a specific lithological or paleontological characteristic. 2. A stratigraphic unit shorter than a zone. See **biohorizon, lithohorizon**.

horizon A A cherty horizon of Middle Eocene age located between 300 and 500 m below the floor of the western North Atlantic.

horizon B A cherty horizon of Cretaceous age in the western North Atlantic beneath horizon A.

horizon distance The distance reached by light at any given time since cosmological time $t = 0$.

horizontal throw See **heave.**

hornblende The most common amphibole, $(Ca,Na)_{2-3}(Mg,Fe,Al)_5(Si,Al)_8O_{22}(OH)_2$.

hornblendite A plutonic rock consisting almost entirely of hornblende.

hornfels A fine-grained, homeoblastic rock resulting from contact metamorphism.

hornfels facies A low-pressure (<1.5 kbar), medium-temperature (300–600°C) facies of contact metamorphism. Characteristic minerals are andalusite, biotite, cordierite, sillimanite, or wollastonite.

horse latitudes The zones of high pressure and oceanic calms at 30°N and 30°S, caused by the converging and descending limbs of the Hadley and Ferrel cells.

horsepower (hp) A nonmetric unit of power, equal to 745.700 watts.

horst An elongated, uplifted crustal block bound by normal faults.

hot spot A center of heat in the Earth's mantle, underlying a major volcanic district. Hot spots remain stationary with respect to the overlying plates. Episodic activity produces chains of volcanoes which become extinct as they move off the area directly above the hot spot. An example is the Hawaiian-Emperor Chain, which ranges in age from $<2 \cdot 10^6$ y (Hawaii) to $72 \cdot 10^6$ y at its NW end where subduction under the Aleutians is in progress.

hot spring A spring issuing water warmed by geothermal flux.

hour (h) A unit of time. 1 h = 60 m = 3600 s.

hour angle (t) Angle (in hours, minutes, and seconds) measured westward along the celestial equator from the observer's meridian (0 hours) to the hour circle of a celestial body.

hour circle Any of the great circles on the celestial sphere passing through the celestial poles.

hr Hour. Syn. **h** (def. 3).

H-R diagram See **Hertzsprung-Russell diagram.**

Hubble constant (H_0) A constant giving the rate of expansion of the universe per unit distance. A commonly accepted value is 18 ± 5 km/s/10^6 l.y. The value of 18 km/s/10^6 l.y., corresponding to $1.90275 \cdot 10^{-18}$ s^{-1} or $6.0 \cdot 10^{-11}$ y^{-1}, is adopted for this dictionary.

Hubble distance The horizon distance r at the present time, equal to the speed of light multiplied by the Hubble time.

$$r = c(1/H_0)$$
$$= 16.6 \cdot 10^9 \text{ l.y.}$$

Hubble's law "The recessional velocity of a distant extragalactic object is proportional to its distance." The proportionality factor is the Hubble constant.

$$v = H_0 r$$

where v = recessional velocity, H_0 = Hubble constant, r = distance.

Hubble time The age of the universe based on the

Hubble constant. Hubble time = $1/H_0$ = $5.255 \cdot 10^{17}$ s = $16.6 \cdot 10^9$ y. See **Hubble constant.**

humic acid The portion of humus extracted with KOH solution and precipitated with HCl.

humidity The water-vapor content of the atmosphere. **1. absolute humidity** The water content of air in g/g or g/cm^3. **2. relative humidity** The ratio of the water content of air to the saturation value at the same temperature and pressure, usually expressed as percent.

hummock A small, rounded rise on a level surface.

Humphreys series A series of lines in the far infrared region of the hydrogen spectrum, representing transitions between $n > 6$ and $n = 6$ energy levels, where n is the principal quantum number. Energies range from 0.100 to 0.378 eV; corresponding wavelengths range from 12.3680 to 3.2814 μm.

humus The nonliving, finely divided organic matter in soil.

Hund's rule A rule stating that electrons distribute themselves among orbitals in a subshell so that the number of unpaired electrons with parallel spins is maximum (and hence the total spin angular momentum is also maximum). This state has the lowest energy and, therefore, the highest probability.

hurricane A tropical cyclone with wind speed of at least 75 miles per hour = 121 km per hour = 33.5 m/s. Cf. **typhoon.**

Huygens principle "All points on a light wavefront may be considered to originate secondary wavelets whose envelope represents the new wavefront."

hyaline Glassy.

hyalite A mass of colorless opal.

hyalo- Prefix meaning *glassy.*

hyaloclastite A deposit formed by the shattering of hot submarine basalt in contact with cold seawater.

hyalocrystalline Defining an igneous rock in which the crystals equal in volume the glassy groundmass.

hybrid *(Physics)* Referring to an orbital that results from the combination of two or more orbitals of equivalent energy. See **hybridized orbital.** *(Biology)* An individual resulting from crossbreeding.

hybrid computer A computer designed to handle both analog and digital data.

hybridized orbital A molecular orbital resulting from the combination of two or more orbitals of equivalent energy.

hydrate A substance containing water as part of its chemical composition.

hydration A physicochemical reaction incorporating water in the structure of a substance.

hydraulic action The mechanical effect of flowing water.

hydraulic force The force exerted by flowing water.

hydraulic head The height of the free water surface above subsurface water or above the free water surface at a lower point in a stream.

hydro- Prefix meaning *water.*

hydrocarbon A chemical compound consisting exclusively of C and H.

hydrogen atom The simplest atom, consisting of a proton and a single orbiting electron.

Bohr radius: $r = h^2n^2/4\pi^2me^2$
$$= 0.52917725\cdot10^{-10} \text{ m}$$
$$(n = 1)$$
$$= 25.92967\cdot10^{-10} \text{ m}$$
$$(n = 7)$$

orbital angular velocity
of the electron: $\omega = 8\pi^3me^4/h^3n^3$
energy of the orbiting
electron: $E = 2\pi^2me^4/h^2n^2$

where h = Planck's constant, n = principal quantum number, m = electron mass, e = electron charge.

hydrogen bond A weak ($\sim$0.2–0.5 eV) bond formed by a hydrogen atom, already bound to a molecule, with a pair of shared electrons on an electronegative atom in the same molecule or belonging to an adjacent molecule.

hydrogen electrode A hydrogen gas electrode (on metallic Pt) in a 1 M solution of H^+ ions at 25°C and 1 atm producing the reaction

$$H_2 = 2H^- + 2e^-$$

The potential, taken as zero, is used as reference for measuring electrode potentials.

hydrogen-ion concentration The amount of H^+ ions in moles/liter in an aqueous solution. It is expressed as pH = $\log_{10}$ of the hydrogen ion concentration. Cf. **pH.**

hydrogen line The 21.11 cm wavelength (= $1.43\cdot10^9$ Hz = $5.873\cdot10^{-6}$ eV = $9.410\cdot10^{-25}$ J) radiation emitted by the H atom when its electron switches from the higher-energy parallel to the lower-energy antiparallel orientation of its spin vector with respect to the orientation of the proton spin vector. The transition can occur at a temperature as low as 100 K.

hydrogeology The science dealing with ground water.

hydrographic chart A navigational chart showing coastlines and/or water depths.

hydrology The science that deals with the mass movements of continental waters (liquid, solid, and vapor).

hydrolysis A decomposition reaction produced by water.

hydrolyzates Compounds or minerals readily decomposed by hydrolysis.

hydrometamorphism Low-temperature, low-pressure metamorphism produced by water solutions.

hydronium The ion H_3O^+, formed by the ionization of water:

$$2H_2O \rightleftharpoons H_3O^+ + OH^-$$

Dissociation constant (K):
$$K = [H_3O^+][OH]^-/[H_2O]$$
$$= 1.0\cdot10^{-14} \text{ (25°C)}.$$

hydrophone A detector sensitive to pressure change, used to measure underwater sound.

hydrophyte A sessile or freely floating metaphyte adapted to a wetland habitat.

hydrophytic Referring to a plant or a flora adapted to a wetland environment.

hydrosphere The totality of free water on Earth = $1.72\cdot10^{21}$ kg. Included are seawater ($1.37\cdot10^{21}$ kg = 80% of total), pore water in sediments ($330\cdot10^{18}$ kg = 18.8%), water as ice ($20\cdot10^{18}$ kg = 1.2%), fresh water in lakes and rivers ($30\cdot10^{15}$ kg = 0.002%), and atmospheric water ($13\cdot10^{15}$ kg = 0.0008%).

hydrothermal Referring to the action or effect of hot water.

hydrothermal stage The last stage of petrogenesis, during which the magma cools. The residual fluids are enriched in water and other volatiles and still interact significantly with the cooling magma. The hydrothermal stage follows the pneumatolytic stage.

hydroxy Containing the hydroxyl radical OH^-.

hydroxyl The radical OH⁻.

-hyet- Prefix or suffix meaning *rain*.

hyetal Pertaining to rain.

hygroscopic Defining a substance capable of absorbing water.

hypabyssal Defining an intrusive rock at a depth intermediate between abyssal or plutonic and the surface.

hyper- Prefix meaning *over, above*.

hyperbola See **conic sections**.

hyperborean Referring to the far north.

hypercharge A quantum number equal to the sum of baryon number and strangeness. It is conserved in strong and electromagnetic interactions.

hypercomplex number Syn. **quaternion**. See **numbers**.

hyperfine structure The splitting of spectral lines due either to the interaction of the nuclear magnetic moment with the magnetic moment of the electron cloud, or to the presence of different isotopes of the same element.

hyperfine transition The transition between the higher-energy parallel and the lower-energy antiparallel orientation of the nuclear magnetic moment with respect to the net magnetic moment of the electron cloud.

hyperon An unstable baryon with mass greater than that of a nucleon.

hypersaline Having a salinity greater than that of normal seawater.

hypersthene A common orthopyroxene, $(Mg,Fe) \cdot SiO_3$.

hypidioblast A late-formed metamorphic mineral only partly bound by its characteristic faces.

hypidioblastic Referring to a hypidioblast.

hypidiotopic Referring to a sedimentary rock formed by precipitation or recrystallization in which the crystals are subhedral. Hypidiotopic is intermediate between idiotopic and xenotopic.

hypo- Prefix meaning *below, beneath*.

hypocenter The site at depth of an earthquake. Syn. **focus**. Cf. **epicenter**.

hypocrystalline Defining the texture of an igneous rock consisting mainly (60–80%) of crystals in a glassy matrix.

hypogenetic Defining a process occurring, or a rock formed, at depth below the surface.

hypolimnion The bottom-water layer in a lake, below the metalimnion.

hypostratotype A stratotype representative of the holostratotype in a different region.

hypotype A specimen supplementing the description of the holotype.

hypsithermal Any of the short ($<10,000$ y), high-temperature intervals separating the Late Cenozoic ice ages from each other.

Hypsithermal A name formally defining the climate optimum.

hypso- Prefix meaning *high*.

hypsographic Describing a line or a curve representing equal altitudes.

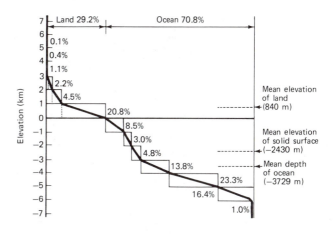

Hypsographic curve. Percent of Earth's surface at increments of 1 kilometer in elevation.

hysteresis 1. The lagging of magnetization change in a substance as the applied magnetic field changes. 2. The lagging in deformation of a body as the applied stress changes.

hystrichosphaerids An informal group of resis- tant single-cell wall structures, including cysts, ranging in age from the Precambrian to the Holo- cene. The group includes acritarchs and dinoflagellates.

Hz Hertz.

i 1. $(-1)^{1/2}$. Syn. ***j***. 2. Electronic current. 3. Inclination (astronomical).

I 1. Electric current. 2. Ionic strength. 3. Luminous intensity. 4. Moment of inertia.

IAT International Atomic Time.

IAU International Astronomical Union.

IC Integrated circuit.

Icarus Minor planet No. 1566 (Apollo group), 1.9 km across, with perihelion distance 0.187 AU, aphelion distance 1.718 AU, inclination = 22.945°, and sidereal period = 1.12 y. It crosses the Earth's orbit and it came within 0.04 AU of the Earth in 1968.

ice The solid phase of water, consisting of 11 polymorphs (ice I to ice XI). Density of pure ice (cubic) at 0°C and 1 atm = 0.91647 g/cm³. Crystallographic system: hexagonal. Cell dimensions (0°C): a = 4.5239 Å, c = 7.3690 Å. Ice polymorphs X and XI are vitreous phases stable below 123 K.

ice age Any of the episodes of intense glaciation within a glacial period or epoch.

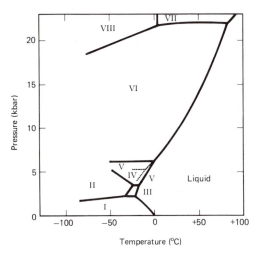

Ice: phase diagram (Ice IV is metastable). (Eisenberg and Kauzman 1969, p. 93, Fig. 3.11)

iceberg A thick (tens to hundreds of meters), tabular mass of ice broken off from an ice shelf and freely floating in the ocean.

ice cap An ice dome up to 50 km in radius spreading out in all directions from a center, as opposed to a glacier that flows down a topographic gradient in one direction. An ice cap grades into an ice sheet.

icecrete A frozen mixture of ice and sediment.

ice epoch See **glacial epoch.**

ice pack The polar sea ice. See **pack ice.**

ice period See **glacial period.**

ice point The freezing point of air-saturated pure water at 1 atmosphere of pressure. It is equal to 0.00°C or 273.15 K.

ice sheet A large (radius > 50 km) ice cap or a series of welded ice caps. The Pleistocene ice sheets covered 30.0% of the land surface in contrast to the present 10.1%.

ice shelf The seaward extension of an ice cap or ice sheet, grounded on the continental shelf or slope and freely floating further offshore.

ice wedge A wedge of ice tapering downward in frozen ground, resulting from water filling cracks in frozen soil and loose sediment.

ichnofossil A trace fossil, i.e. the trace left on a sediment surface or within the sediment by the action of an organism. Syn. **lebensspur.**

ideal gas A gas consisting of point masses having no other effect on each other than perfectly elastic collisions.

ideal gas constant See **gas constant.**

ideal gas law See **gas law.**

idio- Prefix meaning *individual, personal.*

idioblastic Defining the texture of a metamorphic rock in which the crystals are bound by their own crystal faces.

idiomorphic Defining an igneous rock texture in which the component crystals are euhedral.

idiotopic Referring to a sedimentary rock formed by precipitation or recrystallization, in which the crystals are euhedral.

i.e. *Id est,* Latin for *that is.*

igneous Defining a rock or mineral solidified from a magma in the interior of the Earth.

igneous rock A rock solidified from a magma. See **Igneous rocks—fundamental types.***

ignimbrite A sedimentary rock formed by the consolidation of volcanoclastic deposits.

ignitron A gas tube rectifier with a mercury pool as cathode. A current pulse from a thyratron through an ignition point dipping into the mercury pool initiates a conducting gas discharge until anode potential reduces to zero. Conduction starts again at the next cycle.

illite A name applied to a group of clay minerals common in shales and having approximately the formula

$$(K,H_3O)(Al,Mg,Fe)_2(Si,Al)_4O_{10}[(OH)_2,H_2O]$$

Illite has a layered structure similar to that of mus-

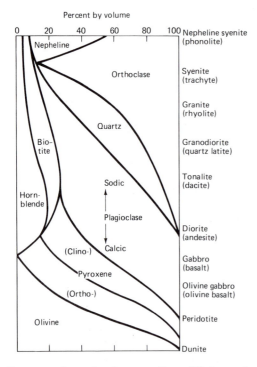

Igneous rocks: mineral composition. Effusive rock names are shown in parentheses. (From Mason and Moore 1982, p. 96, Fig. 5.3)

covite. It consists of two layers of SiO_4 tetrahedra with vertices pointing toward each other and forming Si_4O_{10} sheets; these are joined by Al and OH ions which, together with O from the SiO_4 tetrahedral vertices, form an intervening $Al_4O_4(OH)_8$ sheet. About 15% of the Si in the tetahedra is replaced by Al; the resulting charge deficiency is balanced by K ions bonding the facing bases of adjacent tetrahedral sheets. Total thickness of a packet = 10.0 Å. Illite is an intermediate weathering product between the original K–Mg–Fe–Al silicate minerals and kaolinite, and is the dominant clay mineral at middle latitudes. Cf. **kaolinite, montmorillonite.**

illuminance The density of luminous flux on a surface.

illumination See **illuminance.**

illuvial Referring to the process of illuviation.

illuviation The transport of soluble substances by percolating water from the surface soil layer to a deeper soil layer.

ilmenite The mineral $FeTiO_3$. Ilmenite is a dark, opaque, common accessory mineral in basic igneous rocks and the derived sands. It is the principal ore of titanium.

image The reproduction of the appearance of an object by focusing electromagnetic or acoustic waves emitted or reflected by the object.

imaginary number See **numbers.**

imino The $-NH$ group attached to 1 or 2 C atoms, $=C-NH$ or $-C\cdot NH\cdot C$.

immature Defining a landscape that has not yet been deeply eroded.

impact breccia A breccia formed during an impact event.

impact crater A crater formed on the solid surface of the Earth or other planet or satellite by the impact of a meteorite, asteroid, or comet. Cf. **astrobleme.** See **Astroblemes*.**

impactite The fused or crystallographically altered target rock and rock fragments formed by the impact of a meteorite, asteroid, or comet on the rocky surface of a planet or satellite. See **impact metamorphism.**

impact melt Rock melted by an impact event. The molten rock, normally covered with impact breccia, solidifies slowly and acquires macrocrystalline structure.

impact metamorphism Shock metamorphism

produced during impact on the solid surface of a planet or satellite by the impact of a meteorite, asteroid, or comet. Pressure is raised to 500–5000 kbars and temperature to 2000°C or more for a few microseconds to a few minutes. Fused mineral phases (lechatelierite, fused feldspars) and high density phases (coesite, stishovite) are formed.

impact structure A general term for *astrobleme* or *impact crater.*

impedance (Z) The combined opposition (resistance and reactance) of a circuit to alternating current. It is expressed in ohms.

$$Z_R = R$$
$$|Z_C| = 1/\omega C$$
$$|Z_L| = \omega L$$
$$|Z_{RL}| = [R^2 + (\omega L)^2]^{1/2}$$
$$|Z_{RLC}| = [R^2 + (\omega L - 1/\omega C)^2]^{1/2}$$

where Z_R = resistive impedance; R = resistance; Z_C = capacitive impedance; $\omega = 2\pi f$ = angular frequency in radians/second; C = capacitance; Z_L = inductive impedance; L = impedance; Z_{RL} = resistive-inductive impedance; Z_{RLC} = resistive-inductive-capacitive impedance.

impulse (J) The integral of a force over a time interval $t - t_0$, equal to the change in momentum $m(\boldsymbol{v} - \boldsymbol{v}_0)$ of a mass m to which the force is applied.

$$J = \int_{t_0}^{t} \boldsymbol{F}\, dt$$

incertae sedis Latin for *of uncertain place,* referring to a fossil or living organism whose taxonomic classification is uncertain.

inch A nonmetric unit of length, equal to 2.54 cm (exactly).

inclination (i) *(Astronomy)* The angle between the orbit of a planetary body and the plane of the ecliptic, or between the orbit of a satellite and that of its primary. *(Geophysics)* The angle between the horizontal and the direction of the magnetic field lines at any point on Earth.

inclined fold A fold with its axial plane inclined to the vertical.

inclinometer 1. Any instrument to determine the angle between the horizontal and any other given direction. 2. An instrument to determine the inclination of the magnetic field of the Earth.

index fossil A widely distributed, common fossil identifying a specific time interval.

index of refraction The ratio of the phase velocity

of light in vacuo to that in a medium. Characteristic indices of refraction (for Na light, $\lambda = 0.5893$ μm): vacuum, 1; air (dry, 0°C, 1 atm) = 1.0002926; pure water (20°C) = 1.33335; sea water (35‰ salinity, 20°C) = 1.339; fused quartz = 1.4584; diamond = 2.42; rutile = 2.61; iodine = 3.34.

indicatrix A surface whose distance from the center represents at any point the index of refraction of a given substance. This surface is spherical for an isotropic crystal; an ellipsoid of revolution for a uniaxial crystal; and an ellipsoid for a biaxial crystal.

induced radioactivity The transformation of a stable nuclide into an unstable one by means of natural or artificial nuclear reactions. E.g. $^{14}N(n,p)^{14}C$.

induced radionuclide Any of the radionuclides formed by means of naturally or artificially induced nuclear reactions. Among the radionuclides formed by natural processes (cosmic-ray bombardment) are ^3H, ^{10}Be, ^{14}C, ^{26}Al, ^{32}Si, and Cl^{36}. Cf. **primary radionuclide, secondary radionuclide. See Cosmic-ray-induced radionuclides*.**

inductance (L) The negative ratio of the induced emf to the rate of change in current:

$$L = -E/(dI/dt)$$

where L = inductance, E = emf, I = current, t = time. The unit of inductance is the henry = 1 volt·second/ampere.

induction The production of emf by the motion of a conductor through a magnetic field or by a change in the magnetic flux threading the conductor.

inductive coupling The coupling of two circuits by means of magnetic flux.

inertia The resistance of mass to a change in momentum.

inertial force Any force felt by an inertial mass in a noninertial frame. Syn. **pseudoforce.**

inertial frame 1. A frame of reference within which a body, not acted upon by a force, remains at rest or moves with uniform, rectilinear motion. 2. A frame of reference in uniform, rectilinear motion with respect to any other such frame.

inertial mass The mass m of a body

$$m = F/a$$

obtained from the Newtonian equation $F = ma$, where F = force, m = mass, a = acceleration. Cf. **gravitational mass, principle of equivalence.**

infauna The fauna living within, rather than on, the sediment in the marine or freshwater environment. Cf. **epifauna.**

inferior planet Either the planet Mercury or Venus, lying between the Sun and the Earth. Cf. **superior planets.**

infinitesimal An arbitrarily small, extensive quantity (space, time, mass or any combinations thereof). Infinitesimals are a fundamental concept of calculus. Calculus deals with change through space (including geometry), or through time, or both. It deals, therefore, with the physical world. Although, theoretically, a line segment can be divided into an infinite number of infinitesimally small segments, a unit of time can be divided into an infinite number of infinitesimally small time intervals, and a unit of mass can be divided into an infinite number of infinitesimally small masses, in practice the shortest length is the Planck length $(= 1.616 \cdot 10^{-35}$ m), the shortest time is Planck time $(= 5.390 \cdot 10^{-44}$ s), and the smallest stable mass is that of the electron $(= 0.91 \cdot 10^{-30}$ kg). These quantities (from which all other quantities can be derived) are so small that in practice space, time, and mass may be considered continuous. As a result, when a derivative is itself a function of the variable, it can be evaluated only across an infinitesimal of space, time, or mass. Notice that in calculus, which deals with physical quantities, one cannot add quantities having different dimensions. While in algebra $x^3 + x^2 = 150$ for $x = 5$, in geometry, which deals with space, one cannot add a square (x^2) to a cube (x^3). In geometry $x^3 + x^2 \equiv x^3$ or generally $x^n + x^{n-1} + x^{n-2} + \cdots + x^{n-n+1} \equiv x^n$. Similarly, infinitesimals of higher order are not commensurate with infinitesimals of lower order. Thus $dx + dx^2 \equiv dx$ or, generally $dx^n + dx^{n+1} + dx^{n+2} + \cdots + dx^{n+n} \equiv dx^n$. See **calculus.**

infinities There is an infinite number of infinities. The two most common are: **1. countable infinities ($\aleph_0$)** An infinite set that can be put into 1-to-1 correspondence with the set of positive integers (e.g. the set of rational numbers). **2. uncountable infinities (c)** An infinite set that cannot be put into 1-to-1 correspondence with the set of rational numbers (e.g. the set of real numbers).

inflation A cosmological theory according to which the universe started from a state of zero radius, infinite energy density and infinite temperature; it slowly expanded from cosmological time $t = 0$ to $t = 10^{-34}$ s reaching a radius $r = 10^{-35}$ m while cooling to 10^{22} K; it underwent a phase transition from $t = 10^{-33}$ s to $t = 10^{-32}$ s, which raised

temperature to 10^{27} K and caused a very rapid expansion to $r = 10$ cm; and from then on it followed the evolution of the standard Big Bang model. In the inflation model the horizon distance (the distance reached by light at any given time since $t = 0$) attained a value some 10^{25} times greater than the radius of the observable universe at time $t = 10^{-32}$ s and has remained so since. See **Big Bang, element formation (standard model).**

infra- Prefix meaning *below.*

infrasonic Defining a sound frequency below the audio range, i.e. below 15 Hz.

ingression The inland advance of the sea. Cf. **regression, transgression.**

inlet A small water passage connecting two water bodies.

inner core The central part of the Earth's core, from 5170 km of depth to the center of the Earth at 6371 km. It is solid because of the high pressure at the given temperatures and it has a density increasing from 12.7 g/cm^3 at its outer boundary to 13 g/cm^3 at the center. See **core** *(Geology),* **Earth interior*.**

inner planets The planets Mercury, Venus, Earth, and Mars, with orbits within the asteroidal belt. Cf. **outer planets, inferior planets, superior planets.**

inner quantum number (J) A quantum number for the total angular momentum of an atom, less the nuclear spin. It is called *inner* because it was believed to result from the motions of the atomic core.

inosilicates See **silicates.**

inselberg A hill consisting of harder rock, resulting from greater erosion of the surrounding terrain. .

insolation Irradiation from the Sun. See **solar constant.**

instantaneous dipole The transient magnetic dipole created in a molecule by the motions of its electrons.

instar 1. A growth stage in the life of an arthropod, shedding a molt. 2. The molt shedded by an arthropod during growth.

insulator 1. An electrical component having high resistivity ($>10^6$ ohm·m). 2. A substance with the energy band full and separated from the conduction band by an energy gap of several eV. See **conductivity.**

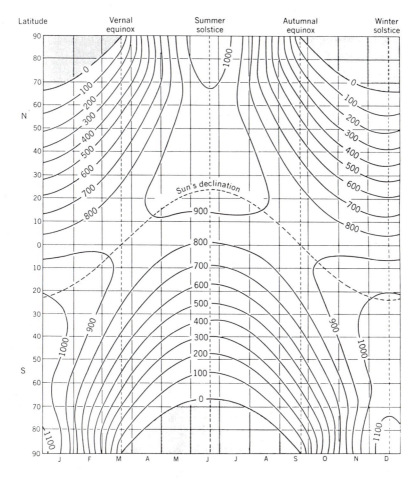

Insolation. The amount of solar radiation received by a horizontal surface outside the terrestrial atmosphere (langleys/day). (Strahler 1971, p. 201, Fig. 13.4, from List 1968, p. 419.)

integral See **calculus.**

integrated circuit (IC) A set of electronic circuits on a common substrate.

inter- Prefix meaning *between.* Cf. **intra-.**

interaction See **natural forces.**

interference The interaction of two or more waves of similar frequencies at a given point in space or time.

interferometer A device to measure interference.

interglacial The time interval between two successive ice ages.

intermediate 1. Referring to an igneous rock with 50% to 60% SiO_2, intermediate in composition between acid and basic. 2. Referring to an earthquake with focal depth between 60 and 300 km.

intermediate vector boson See **gauge boson.**

intermediate water The oceanic water layer above deep water, formed at the subpolar convergence zones.

intermolecular forces See **van der Waals force.**

internal cast Lithified filling of the internal cavity of a shell or other organic structure. Syn. **steinkern.**

internal energy (U) The energy of a system resulting from the motions of its constituent atoms and molecules. Cf. **external energy.** See **enthalpy.**

internal mold The mold of the internal surface of a fossil. Cf. **external mold.**

internal seiche A seiche occurring along an interface separating two water layers of different densities in a lake or bay.

internal wave A wave along an interface separating two solid, liquid, or gaseous layers having different densities. Because the density difference is usually small, amplitude, wavelength and period are usually large.

International Atomic Time (IAT, TAI) Ephemeris time t_E − 32.15 s (January 0d, 0h, 0m, 0s, 1958), adopted from January 1, 1972.

International Date Line The line, largely following the 180° meridian, to the east of which the calendar date is one day earlier than to the west.

International System of Units (SI) A system of units based on the meter (m), kilogram (kg), second (s), ampere (A), kelvin (K), candela (cd), and mole (mol).

interpluvial The interval of time between successive pluvial ages in the African Pleistocene.

interstadial A time interval of higher temperature within a glacial age.

interstellar cloud A region within a galaxy (especially irregular and spiral galaxies) characterized by a higher concentration of matter than surrounding interstellar space. Radius = 1–500 l.y. Mass = 10–10^6 solar masses. Mass gaseous matter/mass of solid matter = 50 ± 20. **1. Diffuse clouds** Mean particle density ≈ 10–100 cm^{-3}; composition: H, H$_2$, He; mean temperature = 80 K. **2. Dense clouds** Mean particle density = 10^3–10^6 cm^{-3}; composition: primarily H, H$_2$, He, and secondarily (10^{-6}–10^{-10} fractional abundance) CO, HCN, OH, NH$_3$, SO$_2$ etc. (see **Molecules—interstellar***, **Molecules—interstellar, relative abundances***). Mean temperature = 10 K. Magnetic field in the denser regions ≈ 0.5 milligauss. Angular velocity ≈ 1–$6 \cdot 10^{-14}$ rad s^{-1}.

interstellar medium (ISM) Atomic, molecular, and ionic gaseous matter (see **Molecules—interstellar***, **Molecules—interstellar, relative abundances***) and μm-size grains (C, Fe, Fe–Mg silicates) between stars within galaxies, either diffused in interstellar space or gathered into clouds, especially common (≈ 10% of total mass) in irregular galaxies or within arms of spiral galaxies. Mean particle density (cm^{-3}): between clouds = 0.2; within diffuse clouds = 10–100; within dense

clouds = 10^3–10^6. Mean magnetic field = 1–4 μgauss. See **interstellar cloud.**

interstellar molecules Inorganic and organic molecules that occur in interstellar space. More than 50 different types have been detected, plus a number of ionic and isotopic species. **Molecules—interstellar***, **Molecules—interstellar, relative abundances***

intertidal 1. Defining the zone between high- and low-tide water levels on a coastline. 2. Defining the organisms living in the zone between the high- and low-tide water levels on a coastline.

intra- Prefix meaning *within.* Cf. **inter-.**

intraclast A limestone component consisting of a lump of material penecontemporaneously torn from the surface sediment layer and redeposited.

intradeep A trough within a geosynclinal belt resulting from incipient folding.

intraformational Referring to a deposit formed penecontemporaneously with the enclosing sedimentary strata.

intrinsic semiconductor See **semiconductor.**

intrusion Emplacement of magma within a preexisting rock.

intrusive Describing rocks or processes related to intrusion.

Invar Trade name for an alloy of steel (64%) and nickel (36%) characterized by low heat expansion coefficient.

invariable plane The plane normal to the vector representing the total angular momentum of the solar system, including rotational and revolutional angular momenta. Longitude of ascending node Ω = 106°44' + 59'T. Inclination of the ecliptic to the invariable plane = 1°39' − 0.3'T. (T = centuries from 1900.0.)

invariant Referring to a property that does not change if the frame of reference is changed. For instance, the distance between two points in three-dimensional space does not depend upon the reference system of coordinates. Cf. **gauge invariant.**

inverse square law A law relating the intensity of a phenomenon to the inverse of the square of its distance from the source. The radial flux of a quantity expanding isotropically outward from a center remains constant through each centered spherical surface of increasing radius. The flux per unit surface of such spheres is inversely proportional to the square of the distance because the surface of a sphere is $4\pi r^2$ (the constant 4π cancels out because it applies equally to all spherical surfaces).

inverted See **overturned.**

inverted relief Topographic relief opposite the geological structure (i.e. valleys emplaced on anticlines, mountain ridges emplaced on synclines).

involute Describing a coiled shell in which the last whorl extends to the umbilicus. Cf. **evolute.**

Io See **Jupiter, Satellites*.**

ion An atom that has an excess (anion) or deficiency (cation) of electrons.

ion exchange column A column filled with an ion-exchanging substance (resin, zeolite), used to separate chemical mixtures into their components.

ionic bond See **bond.**

ionic crystal A crystal held together by ionic bonds between its constituent atoms.

ionic equilibrium The condition in which the rate of molecules dissociating into ions equals the inverse rate.

ionic radius The effective radius of an ion resulting from its electrostatic charge. Examples (values in Å): $H^- = 1.54$, $C^{4-} = 2.60$, $C^{4+} = 0.16$, $N^{3-} = 1.71$; $N^{3+} = 0.16$, $N^{5+} = 0.13$, $O^{2-} = 1.32$, $Na^+ = 0.97$, $Mg^{2+} = 0.66$, $Si^{4-} = 2.71$, $Si^{4+} = 0.42$, $P^{3-} = 2.12$, $P^{3+} = 0.44$, $S^{2-} = 1.84$, $S^{4+} = 0.37$, $Cl^- = 1.81$, $K^+ = 1.33$, $Ca^{2+} = 0.99$, $Fe^{2+} = 0.74$, $Fe^{3+} = 0.64$, $Cu^+ = 0.96$, $Cu^{2+} = 0.72$, $Rb^+ = 1.47$, $Sr^{2+} = 1.12$, $Cs^+ = 1.67$, $Ba^{2+} = 1.34$, $Pb^{2+} = 1.20$, $U^{4+} = 0.97$, $U^{6+} = 0.80$.

ionic strength (μ,I) A measure of the level of electrical forces within an electrolytic solution.

$$\mu = \Sigma \tfrac{1}{2} m_i z_i^2$$

where μ = ionic strength, m = molality of component i, z = ionic charge of component i.

ionium Early name for ^{230}Th ($t_{1/2} = 75,380$ y).

ionium dating methods Two dating methods based on the activity of ^{230}Th (ionium). **1. disequilibrium method** The ionium disequilibrium dating method is based on the ^{230}Th (ionium) excess deposited on the sea floor, upon formation from ^{234}U in seawater, because of its insolubility. Excess ^{230}Th, not being in equilibrium with the ^{234}U present in the sediment, disappears with the half-life

of ^{230}Th (75,380 y). The ratio of ^{230}Th to ^{232}Th (ionium-thorium dating method) or to any other suitable element (e.g. Ti) will be a characteristic of the age of the sediment if the rate of sedimentation has remained constant. **2. growth method** A dating method based on the growth of ^{230}Th to equilibrium with ^{234}U in aragonitic material, which rejects Th but accepts U during crystallization.

ionization constant (K) The ratio of the products of the concentrations of the ions to the concentration of the molecule.

$$K = [C^+][A^-]/[CA]$$

where $[C^+]$ = concentration of the cation, $[A^-]$ = concentration of the anion, $[CA]$ = concentration of the molecule.

ionization energy The energy needed to remove a given electron from an atom or an ion. Characteristic values (in eV) of first ionization potential (removal of an electron from a neutral atom) are: $H = 13.6057$, $He = 24.587$ (highest of all elements), $C = 11.260$, $N = 14.534$, $O = 13.618$, $Na = 5.139$, $Cl = 12.967$, $K = 4.341$, $Ca = 6.113$, $Mn = 7.435$, $Fe = 7.870$, $Co = 7.860$, $Ni = 7.635$, $Cs = 3.894$ (lowest of all elements). Removal of the 20th electron from the Ca atom requires 5469.738 eV. Cf. **ionization potential.** See **Ionization energy—first electron*.**

ionization potential The energy needed to remove a given electron from an atom or ion, usually expressed in eV.

ion microprobe An instrument for element analysis in which high-energy (5000–20,000 eV) ions bombard a target and the secondary ions emitted by the target are analyzed by mass spectrometry.

ionosphere The layer of the thermosphere where ionization is important.

ion product The product of the concentrations of ions in an aqueous solution. Cf. **solubility product.**

IR 1. Infrared. 2. Insoluble residue.

iron meteorite Any of the meteorites consisting of Fe–Ni alloys. Three groups are recognized: octahedrites (75.4%), consisting of kamacite and taenite with 6.5–16% Ni; hexahedrites (10.6%), consisting of kamacite with 5–6% Ni; and ataxites (14.0%), consisting of taenite and kamacite with 6–30% Ni. Exposure ages range from $<100 \cdot 10^6$ y to $2.3 \cdot 10^9$ y, most commonly several hundred million years. These ages are greater than the exposure ages of stony meteorites, indicating, as ex-

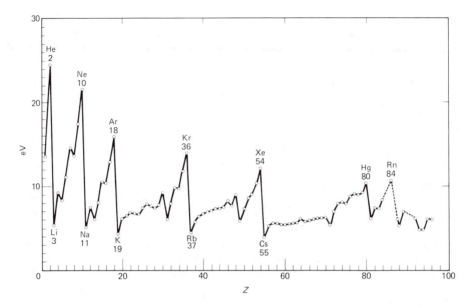

Ionization energy of the elements (first electron only).

pected, that iron meteorites are more resistant to collision fragmentation than stony meteorites. Iron meteorites comprise 5.7% of all meteorites. See **Meteorites***.

iron oxides Fe_2O_3 (hematite, maghemite, Fe/O = 0.67); Fe_3O_4 (magnetite, Fe/O = 0.75); FeO (wüstite, Fe/O = 1); $2Fe_2O_3 \cdot 3H_2O$ (rust).

ironstone An iron-rich sedimentary rock.

irradiance See **radiant flux density.**

irradiation The exposure of materials to ionizing radiation (gamma radiation, x-rays, UV radiation).

irrational number See **numbers.**

irregular galaxy See **galaxy.**

isenthalpic With no change in enthalpy.

isentropic With no change in entropy.

island arc A chain of volcanic islands, usually convex toward the open ocean.

islet A small island.

ISM Interstellar medium.

iso- Prefix meaning *equal, similar.*

isobar *(Physics)* Any of the set of nuclides having the same number of nucleons and, therefore, the same mass number, regardless of charge. *(Meteorology)* A line or surface connecting points of equal pressure.

isobaric 1. Referring to an isobar. 2. Having constant pressure.

isobath A line or surface connecting points of equal water depth.

isochore A line or surface connecting points of equal volume.

isochoric Without change in volume.

isochron 1. A line or surface connecting points of equal age as derived by a given dating method. 2. A straight line on a graph plotting the ratio of a parent radioactive isotope to a pertinent nonradiogenic isotope (e.g. $^{87}Rb/^{86}Sr$) on the abscissa versus the ratio of the daughter isotope to the same nonradiogenic isotope (e.g. $^{87}Sr/^{86}Sr$) on the ordinate. It is used when minerals from the same rock have initial, different concentrations of the parent isotope as well as a uniform, initial ratio of the daughter isotope to the nonradiogenic isotope. The inclination of the line gives the age of the rock and the intercept on the ordinate gives the initial isotopic ratio of the daughter isotope to the nonradiogenic isotope.

isochronous Having the same age.

isoclinal Describing a fold with parallel limbs.

isocline 1. A line connecting points of equal magnetic inclination. 2. A fold with parallel limbs.

isodynamic Having equal force.

isogam A line connecting points of equal magnetic field intensity.

isogon A line connecting points of equal magnetic declination. Cf. **agonic line.**

isogonic line Syn. **isogon.**

isohaline 1. Having the same salinity. 2. A line or surface connecting points of equal salinity.

isohyet A line connecting points of equal precipitation.

isoline Syn. **isopleth.**

isomer *(Physics)* Any of the set of atomic nuclei

of identical composition but different intrinsic energy. *(Chemistry)* Any of the set of molecules of identical composition but different structure.

isometric 1. One of the six crystal systems characterized by four axes of threefold symmetry and three equal axes of either fourfold or twofold symmetry at right angles to each other. Syn. **cubic.** 2. Having the same dimensions in all directions.

isomorph Any of a set of crystals of similar form but different composition.

isomorphic Having similar form.

isomorphism *(Mineralogy)* The similarity in crystal form of substances with different chemical composition. *(Biology)* The similarity in form between two organisms of different lineage.

isopach A line connecting points on a map where a given geological formation has equal thickness.

isophot A line or a surface of equal light intensity.

isopleth A line connecting points having the same physical property.

isopycnic A line or surface connecting points of equal density.

isospin (I) A quantum number having values of 0, 1/2, 1, 3/2, 2 . . . , and grouping hadrons in multiplets regardless of charge. Isospin is conserved in strong interactions. Syn. **isotopic spin.**

isospin multiplet See **charge multiplet.**

isospin orientation quantum number (I_3) The orientation of the isospin vector **I** in isospin space.

$$I_3 =$$
$$-\mathbf{I}, -\mathbf{I} + 1, -\mathbf{I} + 2, \ldots, 0, \mathbf{I} + 1, \mathbf{I} + 2, \ldots.$$

Together with the hypercharge quantum number, it identifies the components of a charge multiplet in the isospin-hypercharge plane.

isostasy The condition of buoyant equilibrium arising from Archimedes' principle.

isostatic anomaly The gravitational anomaly remaining after application of the isostatic correction.

isostatic compensation surface The surface along which the pressure exerted by the overlaying matter (rocks, water, ice, atmosphere) is constant.

isostatic correction A correction to account for the lighter rocks beneath mountains needed for isostatic support.

isostatic equilibrium The state by which matter

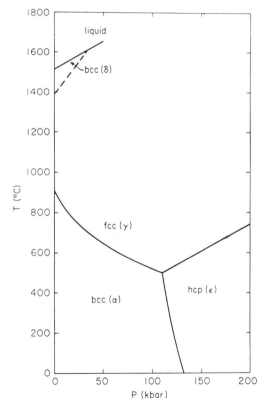

Iron: phase diagram. bcc = body-centered cubic; fcc = face-centered cubic; hcp = hexagonal close packing. Greek letters identify different solid phases. (Jayaraman and Cohen 1970, p. 261, Fig. 4)

in a region is neutrally buoyant on the isostatic compensation surface.

isotherm A line or surface connecting points of equal temperature.

isothermal Describing a process that takes place at constant temperature.

isotone Any of the set of atomic nuclei having the same number of neutrons but different number of protons.

isotope Any of the set of atomic nuclei having the same number of protons but different number of neutrons. See **Isotope chart***.

isotope dilution An analytical method based on the dilution of an unknown amount of an element in a system by a radioactive isotope of the same element. The decrease in radioactivity of an aliquot from the mixture yields the concentration of the original element in the system.

isotope fractionation The fractionation of iso-topes of the same element in physical or chemical processes due to their slight mass-related differences in properties.

isotopic spin See **isospin**.

isotropic Having identical properties in all directions.

isthmus A narrow land bridge between two land masses.

IUGG International Union of Geology and Geophysics.

IUGS International Union of Geological Sciences.

IUGS classification The international classification of igneous rocks, based on the modal proportions of Q (quartz), A (alkali feldspars), P (plagioclase with more than 5% anorthite), F (feldspathoids), and M (mafic minerals).

IUPAC International Union of Pure and Applied Chemistry.

J

j 1. $(-1)^{1/2}$. Syn. *i*. 2. Electric current density. 3. Inner quantum number.

J 1. Electric current density. 2. Inner quantum number. 3. Joule. 4. Total angular momentum quantum number.

jade A compact gemstone consisting of either jadeite or nephrite.

jansky (Jy) A unit of flux density, equal to 10^{-26} W m^{-2} Hz^{-1}.

jasper A chert containing iron oxides.

jaspilite A banded chert containing layers enriched in iron oxide.

javaite A tektite from Java.

JD Julian Date.

Jeans mass The minimum mass of interstellar gas capable of gravitational self collapse as a function of its temperature and density.

$$M_J = \rho(\pi kT/\mu\rho G)^{3/2}$$

where M_J = Jeans mass, k = Boltzmann constant, T = absolute temperature, μ = molecular mass, ρ = density, G = gravitational constant. As $\mu\rho$ = volume, $kT/\mu\rho$ = pressure within the cloud.

jerk The rate of change of acceleration, the third derivative of position with respect to time.

jet A dense, black coal that can be polished and used for jewelry.

jet stream A relatively narrow, high-velocity wind current flowing eastward in the upper troposphere at midlatitudes in both hemispheres. Average latitude: summer, 42°; winter, 25°. Average altitude: summer, 12 km; winter, 13 km. Average speed: summer, 60 km/hr; winter, 150 km/hr. Highest speed: 400 km/hr. The jet stream results from conservation of angular momentum of air masses that drop in altitude along the boundary between the higher tropical tropopause and the lower polar tropopause. As the contrast is greater during the winter, the jet stream is better developed during that season.

JFET See **junction field-effect transistor.**

jiffy A unit of time equal to Planck time (= $5.390 \cdot 10^{-44}$ s).

joint One of (usually) a set of parallel or intersecting fractures in a rock.

josephinite A mineral consisting of a natural alloy of Fe and Ni from Josephine County, Oregon, once claimed to be derived from the Earth's core.

Josephson effect The tunneling of electron pairs through a Josephson junction at a temperature sufficiently low to allow superconductivity.

Josephson junction A thin (<20 Å) insulating oxide layer within a conductor or joining two conductors that are capable of exhibiting superconductivity.

joule (J) The unit of energy or work in the SI and MKS systems of units, equal to 1 newton·meter.

Joule heating The heating of a substance when electrical current flows through it. Cf. **Joule's law.**

Joule's law A law relating electric power to current and resistance:

$$P = I^2R$$

where P = power, I = electric current, R = resistance.

Jovian planets The planets Jupiter, Saturn, Uranus, and Neptune. Syn. **giant planets.**

Julian calendar The calendar instituted by Julius Caesar in 46 B.C., in which the year is divided into 12 months and 365 days, with a leap year of 366 days every fourth year. Cf. **Gregorian calendar.**

Julian Date (JD) Julian Day, the number of days elapsed since noon GMT on January 1, 4713 B.C., numbered consecutively. The fundamental time 1900 January 0d, 12h t_E, is Julian Day 2,415,020.0, while January 1, A.D. 2000, 0h, will be Julian Day 2,451,543.5. This system was devised by the French scholar Joseph Scaliger in 1582.

Julian Day (JD) See **Julian Date.**

junction field-effect transistor (JFET) A transistor with the gate diffused into the conduction channel. Voltage applied to the gate controls the current flow between source and drain. Cf. **field-effect transistor, metal-oxide-semiconductor field-effect transistor.**

junction theorem See **Kirchhoff's first law.**

Jupiter The largest planet and the fifth from the Sun. Mean distance from the Sun = 5.202561 AU Sidereal period = 11.8623 y; sidereal rotational period at equator = 9.841 h. Equatorial radius = 71,492 km; polar radius = 66,854 km. Mass = $1899.728 \cdot 10^{24}$ kg; mean density = 1.33 g/cm^3. Internal structure (estimated): Fe–Ni metal and silicate core with radius = 10,000 km and a 40,000 km-thick mantle of metallic H. Magnetic field = 3–14 gauss. Atmospheric temperature at 1 bar = 165 K; thickness of atmosphere = 17,000 km; atmospheric pressure $\gg$100 bar; gases in atmosphere, H$_2$ = 90%, He = 10%. Sixteen satellites, the four largest of which (Galilean satellites) are, in order of increasing distance from the planet, Io (radius = 1816 km, mass = $8.9169 \cdot 10^{22}$ kg, density = 3.55 g/cm^3), Europa (radius = 1563 km, mass = $4.873 \cdot 10^{22}$ kg, density = 3.04 g/cm^3), Ganymede (radius = 2638 km, mass = $1.490 \cdot 10^{23}$ kg, density = 1.93 g/cm^3), Callisto (radius = 2410 km, mass = $1.064 \cdot 10^{23}$ kg, density = 1.81 g/cm^3), Two flat (<30 km thick), tenuous rings consisting of micron-size (?) particles extending from close to the planet's equatorial surface to 359,000 km away. See **Great Red Spot, Planets—atmospheres*, Planets—physical data*, Planets—ring systems*, Satellites*.**

juvenile Referring to water and gases that are fresh from the interior of the Earth, not recycled from the surface.

J wave A shear (S) wave traveling through the solid inner core of the Earth.

Jy Jansky.

K

κ Electric conductivity.

k 1. Boltzmann constant. 2. Kilo-.

K 1. Equilibrium constant. 2. Kelvin. 3. Kinetic energy

kali- Prefix to an igneous rock name indicating <5% plagioclase in composition.

kamacite α-Fe-Ni (body-centered cubic) alloy (5–7% Ni) in iron meteorites. Cf. **taenite.**

kame A stratified, low mound of glaciofluvial sand, silt, and gravel.

kame terrace A terrace of stratified glaciofluvial sand, silt, and gravel built between a stagnating or melting glacier and the confining valley side.

kaolin A sediment consisting of kaolinite.

kaolinite A clay mineral, $Al_4Si_4O_{10}(OH)_8$, formed by the weathering of feldspars and other Al-silicates. Kaolinite has a layered structure consisting of a layer of SiO_4 tetahedra bound to a layer of Al and OH ions. Thickness of a two-layer unit = 7.37 Å. Cf. **illite, montmorillonite.**

kaon ($K^\pm$, K^0) A strange mesons. See **Elementary Particles*.**

karat (kt) The proportion of pure gold in an alloy, based on 24 karat = 100% gold. Not to be confused with carat, which is a unit of mass = 0.2 g. See **carat.**

karren Solution grooves in karstic terrain averaging centimeters in width and depth.

karrenfeld A limestone surface incised by karren.

karroo An elevated tableland in South Africa.

karst A topography on a limestone surface characterized by dissolution features (sinkholes, karren, dolinas, caves, etc.).

kata- See **cata-.**

katabatic See **catabatic.**

kb Kilobar.

K-cal Kilocalorie (= 1000 g-cal).

K capture The capture by an atomic nucleus of an electron from the K shell, resulting in the transformation of a proton into a neutron and the emission of a neutrino. One of the u quarks of the proton is transformed into a d quark by the emission of a W^+ particle (which interacts with an e^- in the K shell producing a ν), or by the absorption of a W^- particle derived from the interaction between an e^- from the K shell and a $\bar{\nu}$. Cf. **beta plus decay.**

K electron An electron in or from the K shell of an atom.

kelvin (K) The fundamental SI and MKS unit of temperature interval, equal to 1/273.16 of the absolute temperature of the triple point of pure water. 1 K = 1°C in magnitude.

Kelvin temperature The temperature in kelvins (K), starting at the absolute zero. 0 K = −273.15°C.

Kelvin temperature scale The temperature scale starting at the absolute zero. $T(K) = T(°C) + 273.15$, where T = temperature.

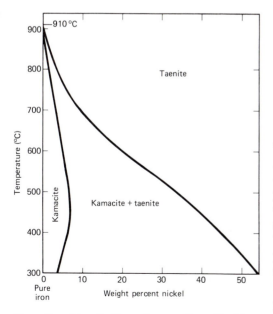

Kamacite and taenite. Phase diagram. (From Hutchison 1983, p. 109, Fig. 6.1)

Kepler's laws The three laws of planetary motion developed by Kepler. **1. First law** "The orbit of a planet is an ellipse with the Sun at one of the two foci." It establishes the shape of the planetary orbits. **2. Second law** "A planet revolves around the Sun with the connecting line sweeping equal areas in equal times." It establishes that planets move faster at perihelion than at aphelion and it establishes by how much. **3. Third law** "The square of the sidereal period of a planet is proportional to the cube of the semimajor axis of its orbit." It establishes that the planetary year increases with planetary distance from the Sun and it states by how much.

Kepler's star The galactic supernova of 1604 that was studied by Kepler.

kernel An atom stripped of any incomplete outer shell.

kerogen Insoluble organic residue in sediments, averaging 80% C, 12% O, 8% H, and some N.

Kerr effects *(Electro-optics)* The inducing of birefringence in a substance (especially liquid or gaseous) by the application of an electric field. *(Magneto-optics)* The slight rotation of polarized light incident on the polished surface of a ferromagnetic material.

ketones A family of organic compounds consisting of two radicals (aliphatic or aromatic) joined by a C=O group.

kettle 1. A bowl-shaped depression on the surface of a glaciofluvial deposit, formed by the melting of a block of ice. 2. A bowl-shaped depression on the rocky surface of a stream bed caused by the abrading action of sand and gravel.

keV Kiloelectronvolt (= 1000 eV).

K-feldspars The minerals orthoclase, microcline, and sanidine, all $KAlSi_3O_8$.

kg Kilogram.

kieselguhr See **diatomite**.

kilo- Prefix meaning *1000*.

kilocalorie (K-cal) 1000 g-cal, the caloric unit of foodstuff.

kilogram (kg) The SI and MKS unit of mass, equal to the mass of the Pt-Ir International Prototype Kilogram kept at Sèvres, S.-et-O., France.

kilometer (km) A metric unit of length equal to 1000 m.

kiloton A unit of energy equal to the energy released by 10^3 tonnes of an explosive rated at 10^3

cal/g. It is equal to 10^{12} cal $= 4.1868 \cdot 10^{12}$ J (exactly). Cf. **megaton.**

kilowatt (kw) 1000 watts.

kilowatt-hour (kWh) A unit of energy equal to a $3.6 \cdot 10^6$ joules (exactly).

kimberlite A serpentinized porphyritic peridotite.

kinematics The study of motion without reference to mass or force.

kinematic viscosity The viscosity of a fluid divided by its density.

kinetic energy (E_k) The energy that a body possesses because of its motion. See **energy.**

kingdom The highest taxonomic category. Five kingdoms are recognized: Monera (solitary or colonial unicellular procaryota); Protoctista (single-celled microorganisms and their immediate multicellular descendants); Plantae (multicellular autotrophs with chlorophylls and exhibiting tissue differentiation); Fungi (fungi and molds); Animalia (multicellular heterotrophs exhibiting extensive tissue differentiation). See **Taxonomy*.**

kink fold A V-shaped fold, with flat sides and a sharp hinge.

Kirchhoff's laws Two laws referring to electric currents at a junction (Kirchhoff's current law) and voltages along a closed loop (Kirchhoff's voltage law). **1. Kirchhoff's current law** "The algebraic sum of the currents flowing at any one instant into (+) or out (−) of a junction is zero."

$$\Sigma I_i = 0$$

where I = current. **2. Kirchhoff's voltage law** "The algebraic sum of the voltages around a loop is at any one instant equal to zero."

$$\Sigma V_i = 0$$

where ΔV = voltage change.

Kirchhoff's rules See **Kirchhoff's laws.**

Kirkwood gaps Gaps in the radial distribution of the asteroids within the asteroidal belt, corresponding to the absence of orbits with periods represented by simple fractions (1/4, 1/3, 1/2, 2/5, etc.) of the orbital period of Jupiter. Asteroids with such orbits would come close to Jupiter (and therefore subject to perturbations) each 4th, 3rd, 2nd, etc. passage at aphelion.

klint See **falaise.**

klippe The erosional remnant of the forward portion of a nappe.

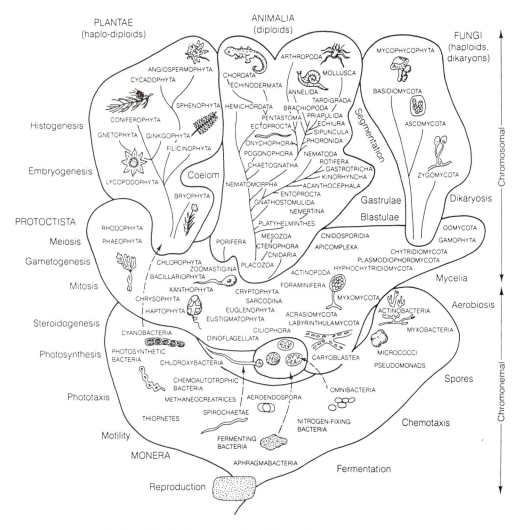

Kingdoms. The five kingdoms of the living world. (Margulis 1981, p. 354)

klystron An electronic tube producing microwave radiation with wavelength between 1 mm and 1 m, i.e. between infrared and short radio waves.

km Kilometer.

knickpoint A break in the slope of a stream or a river.

knoll A small, rounded elevation, either subaerial or submarine.

knot A unit of velocity used in navigation, equal to 1 nautical mile/hour.

Kohlrausch's law "The conductivity of a completely ionized electrolyte is the sum of the conductivities of its component ions."

$$\Lambda_0 = \lambda_0{}^+ + \lambda_0{}^-$$

where Λ_0 = equivalent conductivity at infinite dilution, $\lambda_0{}^\pm$ = equivalent ionic conductivity at infinite dilution.

komatiite An effusive rock consisting of ultramafic lavas exhibiting spinifex structure.

koppje A knoll-size erosional remnant in the South African veld.

kraton See **craton.**

Krebs cycle A cycle within the process of respiration (oxidation of organic compounds to CO_2 + H_2O), in which the acetyl group CH_3CO- is combined with oxaloacetic acid ($COOHCH_2 \cdot COCOOH$) to form citric acid [$COOHCH_2COH \cdot (COOH)CH_2COOH$], which is oxidized in steps to CO_2 and H_2O yielding again oxaloacetic acid. The energy released by the oxidation process is used to form ATP molecules (2 per cycle).

KREEP A nonmare lunar basalt rich in potassium (K), rare-earth elements (REE), and phosphorus (P).

krotovina An animal burrow in soil filled with material from a different soil horizon.

kryoturbation See **cryoturbation.**

Kullenberg corer The original type of piston corer.

kummerform An organism, skeleton, or skeletal part that is stunted because of environmental stress.

kurtosis (K) A measure of the peakedness of a set of data normally distributed about a mean.

$$K = (\Sigma\, x^4 f(x)/N)\sigma^4$$

where $x = x - \bar{x}$ = distance of class interval x from mean $\bar{x}$, $f(x)$ = frequency of items in class interval x, σ = standard deviation.

kW Kilowatt ($= 1000$ watts).

K wave A pressure (P) wave in the Earth's outer core.

kWh Kilowatt-hour.

L

λ 1. Decay constant. 2. Wavelength.

$λ_C$ Electron Compton wavelength.

$λ_{C,n}$ Neutron Compton wavelength.

$λ_{C,p}$ Proton Compton wavelength.

l 1. Length. 2. Levorotatory. 3. Liter. 4. Orbital angular momentum quantum number (syn. **azimuthal quantum number, subsidiary quantum number;** see **quantum number**).

L 1. Angular momentum. 2. Inductance. 3. Lagrangian. 4. Lambert. 5. Left-handed chirality. 6. Luminance. 7. Luminosity. 8. Self inductance.

l^I Galactic longitude in the old IAU system, measured eastward on the celestial equator from the ascending node of the galactic equator. $l^I = l^{II} - 32.32°$.

l^{II} Galactic longitude in the new (1959) IAU system, measured eastward from a point representing the projection of the galactic center on the celestial sphere from the point of view of the Earth. $l^{II} = l^I + 32.32°$.

labradorite 1. A dark gray, calcic plagioclase ($Ab_{50}An_{50}$ to $Ab_{30}An_{70}$). 2. An anorthosite largely composed of the mineral labradorite.

laccolith An igneous intrusion between sedimentary beds, approximately circular in shape, flat-floored and domed, a few hundred meters to a few kilometers across, a few meters to more than a hundred meters thick, fed from below through a duct.

laevo- See **levo-**.

lag deposit A residual deposit consisting of coarser fragments left behind by erosion and transport.

lagoon 1. A coastal body of seawater separated from the open ocean by a bar. 2. The body of water enclosed by an atoll.

Lagrange's equations The set of equations
$$d(∂L/∂\dot{q}_i)/dt - ∂L/∂q_i = 0$$
derived from the Lagrangian L and relating the generalized coordinates q_i, the time-derivatives of the generalized coordinates $\dot{q}_i = dq_i/dt$, and the generalized forces.

Lagrangian (L) The difference L between the total kinetic energy E_k and the potential energy E_p of a dynamic system of particles.
$$L = E_k - E_p$$
expressed as a function of the generalized coordinates and their time-derivatives. Also called *Lagrangian function*.

Lagrangian function See **Lagrangian**.

Lagrangian points Five points on the plane of two bodies orbiting around each other where stable orbits of minor bodies are possible. These are: the inner Lagrangian point L_1 where the opposite gravitational attractions are in balance; the two outer Lagrangian points L_2 and L_3 on the opposite sides of each body along their common axis; and the two orbital Lagrangian points preceding (L_4) and following (L_5) each of the orbiting bodies at a distance of 60°, which thus form equilateral triangles with the two orbiting bodies. L_1, L_2, and L_3 are positions of unstable equilibrium. L_4 and L_5 are positions of stable equilibrium if the mass ratio of the larger to the smaller orbiting body is $≥25$. For Jupiter in the Sun-Jupiter system (mass ratio = 1047), L_4 and L_5 are occupied by the two Trojan groups of asteroids.

laguna Lagoon.

lahar A volcanoclastic mudflow down the slope of a volcano.

lake An inland body of water, larger than a pond. Lakes with surface >15.000 km² are (surface and average depth in parenthesis): Caspian Sea (370,800 km², 1025 m), Superior (82,103 km², 406 m), Victoria (69,484 km², 82 m), Aral Sea (64,500 km², 67 m), Huron (59,570 km², 229 m), Michigan (57,757 km², 281 m), Tanganyika (32,900 km², 1470 m), Baykal (31,500 km², 1620 m), Great Bear (31,328 km², 446 m), Malawi (28,880 km², 695 m), Great Slave (28,570 km², 614 m), Erie (25,667 km², 64 m), Winnipeg (24,390 km², 18 m), Ontario (19,554 km², 244 m), Balkhash (18,430 km², 26 m), Ladoga (17,702 km², 225 m), Chad (16,300 km², 7 m).

lamarckism The theory claiming that function can direct the evolution of the genome.

lambert (L) A unit of luminance, defined as the luminance of a body radiating or reflecting light at the rate of 1 lumen/cm^2.

lamella 1. A thin lamina. 2. A thin (<2 mm) sedimentary layer recognizable by some specific physical characteristic.

lambellibranch Syn. **bivalve** (def. 2), **pelecypod.**

lamina 1. A thin plate. 2. A thin (2–10 mm) sedimentary layer recognizable by some specific physical characteristic.

laminar flow Fluid flow without turbulence.

lamprophyre A hypabyssal porphyritic rock rich in mafic minerals as both phenocrysts and groundmass.

Landsat An artificial satellite transmitting Earth images to receiving stations, formerly called ERTS.

landslide A massive, downslope movement of soil, loose sediment, and rocks.

langley (ly) A unit of illuminance, equal to 1 cal/cm^2.

Langmuir circulation Vertical circulation of the surface ocean water in parallel, helicoidal cells 20–30 m wide and a few meters thick under the influence and in the direction of constant wind. The cells are bound by parallel strips of alternating upwelling and downwelling.

lanthanides The 14 elements that follow La (nos. 58, Ce, to 71, Lu) with identical 5 (except Gd and Lu) and 6 shells but different 4f subshell.

lapilli Small size (4–64 mm) pyroclastic fragments derived from (a) fragmentation of volcanic rocks, (b) solidification of lava fragments while in flight, or (c) accretion from volcanic ash while in suspension within a volcanic cloud. Cf. **bomb, cinder** (Geology)

lapillite An indurated deposit consisting of abundant lapilli in a matrix of volcanic ash.

lapilli tuff Syn. **lapillite.**

lapis lazuli A crystalline rock composed mainly of lazurite and calcite.

Laplace operator See **Laplacian.**

Laplacian (∇^{2}) The operator
$$\nabla^2 = \partial^2/\partial x^2 + \partial^2/\partial y^2 + \partial^2/\partial z^2$$
(for a function of three variables).

lapse rate The decrease of temperature with altitude.

Large Magellanic Cloud See **Magellanic Clouds.**

larger Foraminifera Collective name for a group of foraminiferal families that exhibit richly multicameral shells.

laser Light amplification by stimulated emission of radiation. A device consisting of a solid, liquid, or gaseous system whose atoms or molecules are excited to a higher energy level and then stimulated to radiate in phase at the same frequency as that of the stimulating wave. Picosecond pulses with power output of 10^{12} W have been obtained. The luminance of an average laser beam is $2 \cdot 10^8$ Cd/cm^2, or about 1250 times that of the Sun at meridian (as viewed from the Earth). Cf. **maser.**

Late Referring to the middle portion of a chronological or chronostratigraphic unit. Cf. **Early, Lower, Middle, Upper.**

lateral fault A fault separating two blocks that have slid by each other. Cf. **left-lateral fault, right-lateral fault.**

lateral moraine A moraine deposited along the side of a glacier, bound by the confining valley side.

laterite A deeply weathered tropical soil rich in Al- and Fe-oxides.

lath A narrow, thin strip of mineral or other matter.

latite The extrusive equivalent of monzonite.

latitude (ϕ) The angular distance of a given point from the equator, either north or south.

lattice An open framework, as that formed by ions or molecules in a crystal.

lattice energy The energy required to disperse the ions of an ionic crystal to infinite distance. Examples of lattice energies (kJ/mol or $1.03643 \cdot 10^{-2}$ eV/ion): Al_2O_3 = 15,916; Fe_2O_3 = 14,774; SiO_2 = 13,125; MnO_2 = 12,970; TiO_2 = 12,150; ZrO_2 = 11,188; $Ca_3(PO_4)_2$ = 10,479; $MgCO_3$ = 3122; $CaCO_3$ = 2810; $SrCO_3$ = 2688; $CaSO_4$ = 2480; NaCl = 786; KCL = 715.

Laurasia The Late Paleozoic, northern supercontinent, including North America and Eurasia, that separated from Gondwana in Triassic time. The northern half of Pangea. See **Gondwana, Pangea.**

lava A molten or solidified extrusive rock mass.

lava lake A lake of molten lava in a volcanic crater.

law of Malus See **polarization (optical)**.

law of superposition "Barring an overturn, an overlying sedimentary or extrusive volcanic layer is younger than the underlying one."

lawsonite A low-temperature, high pressure metamorphic mineral, $CaAl_2(Si_2O_7)(OH)_2 \cdot H_2O$.

lazurite A deep blue Na-Ca aluminosilicate mineral, $(Na,Ca)_8(AlSiO_4)_6(SO_4,S,Cl)_2$.

lb Pound.

LC circuit A circuit having impedance and capacitance in series.

LCR circuit A circuit having impedance, capacitance, and resistance in series.

leaching The process of removing soluble substances from soils, rocks, and other substances by percolating water and water solutions.

lead-lead dating method An absolute dating method based on the $^{207}Pb/^{206}Pb$ ratio. This ratio changes with time because the parent isotopes (^{235}U and ^{238}U, respectively) have different half lives ($704 \cdot 10^6$ y and $4.468 \cdot 10^9$ y, respectively), resulting in ^{207}Pb being added more rapidly than ^{208}Pb.

lead-thorium dating method An absolute dating method based on the $^{*208}Pb/^{232}Th$ ratio, where the asterisk identifies the radiogenic component of ^{208}Pb.

lead-uranium dating method An absolute dating method based on the $^{*206}Pb/^{238}U$ and/or the $^{*207}Pb/^{235}U$ ratio, where the asterisk identifies the radiogenic component of ^{206}Pb or ^{207}Pb.

lead-210 dating method An absolute dating method based on the content of ^{210}Pb ($t_{1/2} = 22.3$ y) in various substances. ^{210}Pb is part of the decay series of ^{238}U. It is formed from ^{222}Rn ($t_{1/2} = 3.8235$ d) via a series of short-lived decay products (^{218}Po, $t_{1/2} = 3.11$ m; ^{214}Pb, $t_{1/2} = 26.8$ m; ^{214}Bi, $t_{1/2} = 19.9$ m; ^{214}Po, $t_{1/2} = 164.3$ μs). Rn is a gas produced by Ra and released to the hydroatmosphere. ^{210}Pb is removed upon formation and included in precipitation, skeletal materials, etc., where it decays to ^{206}Pb (stable) via ^{210}Bi ($t_{1/2} = 5.013$ d) and ^{210}Po ($t_{1/2} = 138.376$ d).

leap year A year containing an extra day (February 29). The Julian calendar had a leap year every fourth year. The Gregorian calendar established that century years are leap years only when divisible by 400.

least square method A method of fitting a curve to a set of points by minimizing the squares of their distances from the curve.

lebensspur The trace on soft sediment of the activity of an organism living on it. Syn. **ichnofossil**.

lechatelierite Fused silica in fulgurites or impactites.

Le Châtelier principle "A system in equilibrium reacts to an external force so as to minimize its effect."

Leclanché cell The common dry cell, a primary cell with zinc as cathode, carbon as anode, NH_4Cl as electrolyte, and MnO_2 as depolarizer; emf = 1.5 volt. Cf. **Daniell cell, Weston cell**.

lecto- Prefix meaning *selected*.

lectostratotype A stratotype selected after the original definition. Cf. **stratotype**.

lectotype A taxonomic type selected after the original description, replacing the holotype.

LED Light-emitting diode.

leeward Downwind.

left-hand rules A set of unnecessary "rules" to describe the vectorial product

$$\mathbf{F} = I\mathbf{l} \times \mathbf{B}$$

where $\mathbf{F}$ = force, I = current (either consisting of positive ions or taken as having a sense opposite that of electron flow), $\mathbf{l}$ = length of straight current-carrying conductor, $\mathbf{B}$ = external magnetic field. 1. (Direction of the magnetic field created by a *negative* charged particle moving on a straight line or a *negative* current flowing through a straight conductor.) If the fingers of the left hand are wrapped around the path of the particle or around the conductor and if the thumb is extended in the direction of motion of the particle or of the current flow, the fingers will indicate the direction and circular shape of the magnetic field created. 2. (Force on a *positive* charged particle or a conductor carrying a *positive* current in an external magnetic field.) If the thumb, first, and second finger of the left hand are extended 90° to each other, and if the second finger indicates the direction of motion of the charged particle or the current flowing through the conductor and the first finger indicates the direction of the external magnetic field, the thumb indicates the direction of the force experienced by the particle or the conductor. See **right-hand rules**.

left-lateral fault A fault in which a block appears to have moved to the left when viewed from the opposite block. Cf. **right-lateral fault**.

lenad A collective name for leucite and nepheline.

Lenz's law "An induced emf has always a sign such as to oppose the action that produces it."

lepido- Prefix meaning *scale, flake.*

lepidoblastic A homeoblastic structure with parallel orientation of the crystals.

lepto- Prefix meaning *thin, fine, small.*

leptokurtic Defining a distribution more peaked than normal. Cf. **platykurtic**.

lepton Originally a fermion with mass smaller than that of the proton, i.e. the muon, the electron, and the neutrino. Included is now the tauon, with mass greater than the proton. Leptons are weakly interacting particles. See **Elementary Particles***.

leucite A feldspathoid mineral, $KAlSi_2O_6$.

leucitophyre An extrusive, porphyritic rock consisting mainly of leucite, nepheline, and clinopyroxene.

leuco- Prefix meaning *pale, white-colored.*

leucocratic Defining a light-colored igneous rock.

levee A sedimentary embankment on either side of the lower course of a river or submarine valley.

levo- Prefix meaning *sinistral.*

levorotatory (l) Defining a substance that rotates polarized light counterclockwise as seen by the eye viewing the substance in transmitted light, or clockwise in the direction of light transmission. Cf. **dextrorotatory**.

Lewis acid See **acid**.

Lewis base See **base**.

Lewis structure The representation of a covalent bond (involving two electrons) by a dash, and of an unshared electron by a dot.

lherzolite A plutonic rock consisting mainly of olivine, orthopyroxene, and clinopyroxene.

libration Any of the apparent oscillations in the motion of the Moon caused by its rotation and the eccentricity (0.0549) of its orbit around the Earth; by the $5°8'43''$ inclination of the lunar orbit to the ecliptic, which oscillates $\pm 9'$ with a period of 173 days; by the $6°40'44''$ inclination of the lunar equa-

tor to the lunar orbit; by the parallactic effect to which a terrestrial observer is subjected because of the rotation of the Earth; and by the small physical libration of the Moon caused by the attraction of the Earth on the 1.09 km-high lunar bulge that points toward the Earth. The result is that 41% of the surface of the Moon is always visible, 41% never visible, and 18% alternating visible and invisible. See **physical libration**.

Libyan glass Fused silica glass from the Libyan desert, probably an impactite. Fission track age = $28.5 \cdot 10^6$ y.

lichenometry A geochronometric method based on the growth rate of lichens.

ligand An anion or a molecule with a pair of unshared electrons forming a bond with a central metallic cation in a coordination compound by means of these electrons. The atom in the molecule actually forming the bond with the metal cation is called *donor*.

light 1. Visible electromagnetic radiation ($\lambda = 0.40$ to 0.72 μm). 2. Electromagnetic radiation of any wavelength (see **Electromagnetic spectrum***). Light travels in vacuo at the invariable speed $c = 299,792,458$ m/s (exactly), but at the lower speed c/n through media with refractive index n. See **electromagnetic radiation**.

light cone See **world line**.

light-emitting diode (LED) A *p-n* junction emitting light when biased forward.

lightning A large, natural electrical discharge through the atmosphere. Length = 0.5–5 km; width of channel = 10 cm; descending stroke (negative): duration = 0.5–2 ms in steps of 50 m (duration = 1 μs each, separated by pauses of 50 μs); average speed = 1000 km/s; average speed of positive return stroke (in a single step through the same channel) = 10,000 km/s; number of secondary strokes = 30–40; total duration of lightning event = 0.5 s; voltage = 10^8–10^9 V; amperage: average = 10^4 A, maximum = $5 \cdot 10^4$ A; electric charge transported = 25–250 C (average 37 C); maximum temperature = 30,000°C; electron density within channel = 10^{17} to 10^{18}/cm^3; pressure within channel = 10 atm.

light year (l.y.) A unit of distance equal to the distance traveled by light in vacuo in 1 tropical year. It is equal to 9,460,528,404,879,358.8126 m = $9.4605284 \cdot 10^{12}$ km.

lignin The main noncarbohydrate component of wood, providing support for the cellulose fibers.

lignite A brownish mass of dead plant matter intermediate in coalification between peat and coal.

LIL Large ion lithophyle elements, such as K, Rb, and Cs.

lime CaO.

limestone A sedimentary rock consisting largely of $CaCO_3$. See **Sedimentary rocks***.

limnal Referring to a lake.

limnetic Referring to the open water portion of a lake.

limnite Bog iron ore.

limnology The science that studies lakes.

limonite The mineral $FeO(OH) \cdot nH_2O$.

lineage The line of descent of an organism.

lineation A linear structure in a rock due to depositional dynamics or to postdepositional (sedimentary rocks) or postgenetic (igneous and metamorphic rocks) deformation.

line defect A crystal defect occurring along a line.

line of apsides The straight line connecting the two apsides.

line of force Any of a system of conventional lines whose number per unit area is set to be proportional to the intensity of a given field and whose tangent at any point indicates the direction of the field.

line of nodes The straight line connecting the two nodes.

lipids A large group of substances, including fats and oils, that can be dissolved from organic matter by means of nonpolar solvents.

liquid crystal A liquid consisting of rod-like organic molecules that tend to orient parallel to each other. Liquid crystals are more structured than liquids but less structured than organic crystals.

liquidus The curve (in a binary system), surface (in a ternary system), or volume (in a quaternary system) separating the solid from the liquid phase in a temperature vs. composition diagram. Cf. **solidus.**

Lissajous figure Any of the cyclical paths on a plane followed by a point made to vibrate by two transversal waves perpendicular to each other. The shape of the planar cyclic path of the point depends upon the amplitude, frequency, and phase angle of the two waves.

listric surface A spoon-shaped surface beginning horizontal and terminating with a steep slope.

liter (l) A metric unit of volume or capacity, equal to 1 dm^3 = 1000 cm^3 = 10^{-3} m^3.

lith- See **litho-**.

-lith Suffix meaning *stone.*

litharenite An arenite with >25% of rock fragments.

lithic 1. Stony. 2. Referring to a sedimentary or pyroclastic rock containing >25% of fragments of derived rocks.

lithification The transformation of loose sediment into indurated sedimentary rock.

litho- Prefix meaning *stone.*

lithofacies The ensemble of lithic characteristics of a rock unit.

lithogenesis The formation of sedimentary rocks from loose sediments. Cf. **petrogenesis.**

lithographic limestone A micritic limestone.

lithoherm A lithified herm.

lithohorizon A stratigraphic horizon characterized by a specific lithofacies.

lithology The field description of rocks.

lithophile Defining any of the elements that tend to concentrate in the silicate phase. Cf. **chalcophile, siderophile.**

lithosome A body of rocks of uniform lithologic characteristics.

lithosphere 1. The more rigid outer layer of the Earth overlying the asthenosphere. 2. The stony portion of the Earth, extending from the base of the mantle to the surface of the crust (including loose surface sediment and soil). Together with the siderosphere, hydrosphere, and atmosphere, it is one of the four major "spheres" that form the Earth.

lithostatic pressure The pressure exerted by the rock burden. The gradient is equal to about 0.27 atm/m in the crust and 0.34 atm/m in the mantle.

lithostratigraphic unit A rock unit identifiable by its lithologic characteristics.

lithothamnion ridge An algal ridge built by Lithothamnion and other red calcareous algae and occurring on the ocean side of the fore reef.

Little Ice Age A period of unusually cold weather

in northern Europe from the XVIth to the XVIIIth century.

littoral Referring to the coastal zone between high- and low-tide sea levels. Syn. **intertidal.**

llano A semiarid tropical plain (South America).

lm lumen.

ln Natural or Napierian logarithm, which uses the number e as a base. See e, **logarithm.**

load cast A syndepositional imprint (sole mark) made by the protrusion of a lump of coarser material from the bottom of the overlying bed into the finer-grained surface of the underlying bed.

loadstone See **lodestone.**

loam A smooth, richly organic soil consisting of sand, silt, and clay in similar amounts.

Local Group The cluster of galaxies to which the Galaxy belongs. It consists of 25+ galaxies within a distance of $2.4 \cdot 10^6$ l.y. and includes Andromeda and the Magellanic Clouds. The mean density of matter in the group is 10^{-27} g/cm^3.

Local Sidereal Time (LST) GST + 1 hr/15° of longitude if location is east of Greenwich, or GST − 1 hr/15° of longitude if location is west of Greenwich.

Local Time (LT) Local mean time, equal to the mean solar time at a given location. It is equal to GMT + 1 hr/15° of longitude if location is east of Greenwich, or GMT − 1 hr/15° of longitude if location is west of Greenwich.

lodestone A magnetized piece of magnetite (Fe_3O_4).

loess A homogeneous, nonstratified, semiconsolidated, light tan-colored periglacial eolian deposit consisting of silt-size particles of glacial origin.

log Logarithm.

logarithm The exponent needed for a given number to represent another number. Common logarithms use the number 10 as a base. Natural or Napierian logarithms use the number $e = 2.7182818284590...$.

loma A gently-rising hill on a plain (southwest United States).

lomita A small loma.

London forces Attractive interatomic or intermolecular forces caused by interaction among instantaneous atomic or molecular dipoles.

longitude The angular distance east or west of the Greenwich meridian.

longitudinal wave A wave in which the parameter involved changes in the same direction as that of wave propagation. See **P wave, S wave, transversal wave.**

long-period variables Variable stars, mainly Mira stars, with periods ranging from 90 to 600 days and magnitude changes averaging 6 for variables with periods <400 days or 9 for variables with periods >400 days. See **Mira Ceti.**

loran Long-range navigation, a system based on the determination from a receiver of the phase difference between synchronized pulses emitted by two shore-based transmitters.

Lorentz contraction See **FitzGerald-Lorentz contraction.**

Lorentz-FitzGerald contraction See **FitzGerald-Lorentz contraction.**

Lorentz force The force experienced by a charged particle moving through a region in which both an electric and a magnetic field are present.

$$\mathbf{F} = q\mathbf{E} + q\mathbf{v} \times \mathbf{B}$$

where $\mathbf{F}$ = force, q = charge, $\mathbf{E}$ = electric field intensity, $\mathbf{v}$ = velocity, $\mathbf{B}$ magnetic field intensity. If only the magnetic field is present, the equation reduces to

$$\mathbf{F} = q\mathbf{v} \times \mathbf{B}$$
$$= I\mathbf{l} \times \mathbf{B}$$

for a length $\mathbf{l}$ of a straight conductor carrying the current I. Cf. **left-hand rules** (def. 2), **magnetic induction.**

Lorentz frame Any of the inertial frames with three space coordinates and one time coordinate used to describe the motions of nonaccelerated relativistic objects. Lorentz frames are in uniform motion with respect to each other.

Lorentz relation An equation describing the force experienced by a charged particle moving through a region in which both an electric and a magnetic field exist. See **Lorentz force.**

Lorentz transformation A set of equations used to transform the coordinates $xyzt$ of an event in one Lorentz frame into the coordinates $x'y'z't'$ in another Lorentz frame. For two frames with common origin at time $t = t' = 0$, parallel axes, and in relative, uniform motion in the x and x' direction, the Lorentz transformation equations are:

$$x' = (x - vt)/(1 - v^2/c^2)^{1/2}$$
$$y' = y$$

$$z' = z$$
$$t' = (t - vx/c^2)/(1 - v^2/c^2)^{1/2}$$

where v = relative velocity of the two frames, t = time, c = speed of light.

Love wave A shear interface wave.

low-energy *(Physics)* Referring to the branch of physics that studies particle interactions at energies lower than a few MeV. Cf. **high-energy** *(Physics)*. *(Geology)* Referring to a coast or coastal environment in which wave and current action is limited. Cf. **high-energy** *(Geology)*.

Lower Referring to the stratigraphic position of a chronostratigraphic unit deposited during the Early portion of the corresponding chronological unit. E.g. the Lower Dwyka Series, deposited during the Early Permian. Cf. **Early, Late, Middle, Upper.**

lower mantle The portion of the mantle below 670 km from the Earth's surface.

low-grade metamorphism Metamorphism at low temperature and pressure.

low-level Referring to weak radioactivity.

low-level counting The counting of particles emitted by a radioactive element of low activity.

low-magnesium calcite Calcite containing <4% $MgCO_3$. See **magnesian calcite.**

low quartz The polymorph of quartz stable below 573°C at atmospheric pressure. See **silica.**

low-velocity channel See **asthenosphere.**

low-velocity zone See **asthenosphere.**

loxodrome See **rhumb line.**

LR circuit A circuit having impedance and resistance in series.

LRC circuit A circuit having impedance, resistance, and capacitance in series.

LST Local Sidereal Time

LT Local Time.

lumachella 1. A variety of marble consisting of abundant, recrystallized shell fragments. 2. A poorly cemented, stratified accumulation of shells and shell fragments.

lumen (lm) The SI unit of luminous flux, equal to the luminous flux emitted within 1 steradian from a point source having the intensity of 1 candela.

luminance (L) Luminous intensity per unit area normal to the emitting surface.

$$L = dI/dA \cos \theta$$

where I = luminous intensity, A = emitting area, θ = angle between normal to the emitting area A and direction from which A is viewed. Common luminances are (in stilb = Cd/cm^2): laser beam, $2 \cdot 10^8$; solar disc at meridian as viewed from the Earth, 160,000; tungsten filament in 100 W lamp, 1200; white paper illuminated perpendicularly by Sun at meridian, 2.7; fluorescent light, 0.6–1.5; clear sky at noon, 1.1; lunar disc at meridian, 0.25; star-lit sky, $5 \cdot 10^{-8}$.

luminescence Light emission resulting from physical or chemical processes at room temperature. It includes fluorescence and phosphorescence.

luminosity (L) 1. The total power of a light source. 2. The total power of a celestial body, equal to the total energy emitted per second.

$$L = 4\pi R^2 \sigma T_{eff}^4$$

where R = radius of the body, σ = Stefan-Boltzmann constant, T_{eff} = effective temperature.

luminous flux (Φ) The rate of light flow.

$$\Phi = dQ/dt$$

where Q = luminous energy (quantity of light). It is measured in lumens. A point source radiating in all directions with a light intensity of 1 candela emits 4π lumens.

luminous intensity (I) Luminous flux per steradian.

$$I = d\Phi/d\omega$$

where Φ = luminous flux, ω = solid angle through which flux from point source is radiated. The unit is the candela = lumen/steradian.

lunar day 1. The rotational period of the Moon, equal to 27.32167 days. 2. The time between two successive transits of the Moon across the local meridian, equal to 24.8411h or 24h 50m 28.2s. Cf. **tidal day.**

lunar month The interval of 29.5305882 + 0.00000016T d_E between succesive new Moons, where T = time in centuries from 1900.0. Syn. **synodic month.**

lunar year A year of 12 lunar (synodic) months, equal to 354.3670596 days.

lutaceous Referring to a sediment or substance formed by silt-and/or clay-size particles.

lutite A sedimentary rock consisting of silt- and/ or clay-size particles. Cf. **arenite, rudite, siltite.**

lux (lx) The SI unit of illumination, equal to 1 lumen/m^2.

L wave See **surface wave.**

lx Lux.

ly Langley.

l.y. Light year.

Lyman series A series of lines in the ultraviolet region of the spectrum of hydrogen, representing transitions between $n > 1$ and $n = 1$ energy levels, where n is the principal quantum number. Energies range from 10.2045 to 13.6057 eV; corresponding wavelengths range from 0.1215 to 0.0911 μm. See **energy level.**

lysosome A globose organelle in eucaryota, averaging 0.5 μm in diameter and containing hydrolytic enzymes. It performs the function of digestion, including the breakdown of foreign bodies (e.g. bacteria) entering the cell.

M

μ 1. Chemical potential. 2. Dynamic viscosity. 3. Ionic strength. 4. Magnetic moment. 5. Magneton. 6. Micron. 7. Permeability. 8. Proper motion.

μ_μ Muon magnetic moment.

μ_0 Permeability constant = permeability of vacuum = $4\pi \cdot 10^7$ henry/meter.

μ_r Relative permeability ($= \mu/\mu_0$).

μF Microfarad.

μm Micrometer.

m 1. Apparent magnitude. 2. Electromagnetic moment. 3. Mass. 4. Meter. 5. Minute. 6. Molal concentration.

m- Meta- *(Chemistry)*.

M 1. Absolute magnitude. 2. Magnetization. 3. Mass (of celestial body). 4. Mega-. 5. Messier number. 6. Molar concentration or molarity. 7. Mutual induction.

m_μ Muon rest mass.

m_π Pion rest mass.

m_{bol} Apparent bolometric magnitude (see **magnitude**).

m_e Rest mass of electron.

m_l Magnetic orbital momentum quantum number.

m_n Neutron rest mass.

m_0 Rest mass.

m_p Proton rest mass.

m_s magnetic spin angular momentum quantum number.

m_v Apparent visual magnitude. See **magnitude.**

M_v Absolute visual magnitude. See **magnitude.**

m_{vis} Apparent visual magnitude. See **magnitude.**

M_{vis} Absolute visual magnitude. See **magnitude.**

Ma Mega-annus = 10^6 y.

maar A wide, low volcanic crater.

Mach number The ratio of the speed of an object in a medium to the speed of sound in that medium. See **Cerenkov angle** *(Acoustics)*.

Mach principle The conjecture that the inertia of a particle or a body is not an intrinsic property but results from (instantaneous?) interaction with all other particles or bodies in the universe.

macigno A sequence of graded sandstone and shale beds (northern Apennines).

macro- Prefix meaning *long* (space or time), *tall, far, distant,* but commonly understood to mean *large, big,* and therefore to be synonymous with **mega-.**

macrocrystalline Referring to a rock consisting of crystals visible to the naked eye.

macroevolution The evolutionary development of higher taxa by wide evolutionary steps (macromutations).

macrofauna *(Invertebrate)* An assemblage of living or fossil animals or animal remains that are > 5 mm in size and thus visible and identifiable with the naked eye. Cf. **microfauna** *(Invertebrate)*. 2. *(Vertebrate)* An assemblage of living or fossil animals larger than 5 cm. Cf. **microfauna** *(Vertebrate).*

macroflora The flora visible to the naked eye.

macrofossil A fossil visible and identifiable with the naked eye.

macromolecule A molecule with a high molecular mass, such as a natural or artifical polymer.

macromutation A hypothetical, wide evolutionary step producing a progeny sufficiently different to be assigned to a different higher taxon.

macroplankton Plankton ranging in size from 1 mm to 1 cm.

macula A spot on a planet or satellite.

maelstrom A stormy sea created by interaction between tidal currents and wind-generated waves (Lofoten Islands, Norway).

mafic 1. Defining a ferromagnesian mineral. 2. Referring to a rock containing ferromagnesian minerals. Cf. **basic.**

mafic front See **basic front.**

mafic index (*MI*) An expression of the concentration of Fe with respect to Mg in mafic minerals.

$$MI = 100[(FeO + Fe_2O_3)]/(MgO + FeO + Fe_2O_3)$$

Cf. **felsic index.**

Magellanic Clouds Two small, irregular galaxies satellite to the Galaxy. The Large Magellanic Cloud is about 40,000 l.y. in diameter and is 160,000 l.y. away in Dorado. The Small Magellanic Cloud is about 30,000 l.y. in diameter and is 180,000 l.y. away in Toucan.

maghemite A mineral consisting of γ-Fe_2O_3. It is dimorphic with hematite.

magic numbers The numbers 2, 8, 20, 28, 50, 82, and 126 which, when representing the number of protons or neutrons in atomic nuclei, indicate a nucleus that is particulary stable and cosmically more abundant than its neighbors.

magma A rock melt formed inside the crust or mantle of the Earth or other planetary body.

magma chamber A cavity within the crust or mantle of the Earth or other planetary body containing magma.

magnesian calcite Calcite containing up to 4% $MgCO_3$ (low-magnesium calcite) or 4–19% $MgCO_3$ (high-magnesium calcite).

magnesian dolomite A dolomite with 50–66.6% $MgCO_3$.

magnesian limestone A limestone with up to 5% $MgCO_3$.

magnesite The mineral $MgCO_3$.

magnetic azimuth The azimuth meaured eastward from the magnetic north through 360°.

magnetic dipole A magnetic or magnetized object or an electrical circuit producing a magnetic field similar to that of a bar magnet.

magnetic dipole moment (*μ*) See **magnetic moment.**

magnetic domain See **domain.**

magnetic elements The 7 elements of the Earth's magnetic field, three of which are sufficient to define the magnetic vector. They are: declination (*D*); inclination (*I*); intensity (*F*); and the horizontal (*H*), vertical (*Z*), north (*X*), and east (*Y*) intensity components.

magnetic epoch See **polarity epoch.**

magnetic equator The line of zero magnetic inclination.

magnetic event See **polarity event.**

magnetic field (*H*) *(Physics)* The field produced by a magnetic or magnetized substance, by an electric current, by the flow of charged particles, or by a time-varying electric field. *(Geology)* The magnetic field of the Earth, consisting of the dipole (see **geomagnetic field**) and nondipole components. The latter apparently results from convective motions in the outer core. The nondipole component drifts westward at an average rate of 0.2°/y.

magnetic field intensity (H) The intensity of the magnetic field at a given point *P*.

$$\mathbf{H}(P) = \mathbf{B}/\mu_0 - \mathbf{M}$$

where $\mathbf{B}$ = induction vector, μ_0 = permeability constant, $\mathbf{M}$ = magnetization (magnetic moment density). It is expressed in ampere/meter (SI) or oersted (CGS_{emu}). 1 A/m = $4\pi \cdot 10^{-3}$ oersted.

magnetic flux (Φ) The integral over a given surface of the magnetic induction vector component normal to the surface.

$$\Phi = \int \mathbf{B} \cdot d\mathbf{S}$$

where $\mathbf{B}$ = magnetic induction, $\mathbf{S}$ = surface. It is expressed in weber (SI) or maxwell (CGS_{emu}).

magnetic flux density Magnetic flux per unit surface. It is expressed in tesla (SI; 1 tesla = 1 Wb/m^2) or gauss (CGS_{emu}; 1 gauss = 1 Mx/cm^2). 1 tesla = 10^4 gauss.

magnetic flux quantum (Φ₀) The quantum of magnetic flux.

$$\Phi_0 = h/2e$$

where h = Planck's constant, e = electron charge. It is equal to $2.067834 \cdot 10^{-15}$ Wb.

magnetic hysteresis See **hysteresis.**

magnetic induction (B) A vector quantity expressing the strength and direction of a magnetic field.

$$\mathbf{F} = q_0 \mathbf{v} \times \mathbf{B}$$

where $\mathbf{F}$ = force experienced by positive test charge q_0 moving with velocity $\mathbf{v}$ in magnetic field of magnetic induction $\mathbf{B}$. The magnitude of $\mathbf{B}$ is expressed in tesla (SI) or gauss (CGS_{emu}). 1 tesla = 10^4 gauss.

magnetic induction line Any of the imaginary lines everywhere tangent to the magnetic induction vector $\mathbf{B}$. In the magnetic field created by a magnet, the direction of $\mathbf{B}$ is, by convention, from north (defined as "positive") to south (defined as "negative") outside the magnet, and from south to north inside it. Also by convention, one line of magnetic induction represents a unit of magnetic flux density (1 Wb/m^2 = 1 tesla in the SI system; 1 Mx/cm^2 = 1 gauss in the CGS_{emu} system; 1 tesla = 10^4 gauss).

magnetic latitude The latitude of any given point on the Earth's surface referred to the magnetic poles. Cf. **geomagnetic latitude.**

magnetic longitude The longitude of any point on the Earth's surface referred to the magnetic field, with base meridian extending south from the north magnetic pole. Cf. **geomagnetic longitude.**

magnetic moment (μ) A measure of the strength of a magnetic dipole, as evidenced by the torque τ when the dipole is placed in a homogeneous external magnetic field **B**.

$$\tau = \mu \times \mathbf{B}$$

where μ = magnetic dipole moment (or simply magnetic moment), **B** = intensity of external magnetic field. It is expressed in J/T or A m^2.
1. *Of a revolving electron:*

$$\mu_l = g_l \mu_B [l(l + 1)]^{1/2}$$

where g_l = orbital g factor, m_B = Bohr magneton, l = orbital angular momentum quantum number.
2. *Of a spinning electron:*

$$\mu_s = 2\pi g_s \mu_B \mathbf{S}/h$$

where g_s = spin g factor, μ_B = Bohr magneton, **S** = spin vector, h = Planck's constant.
3. *Of an electric current loop:*

$$\mu = IA$$

where I = current, A = area described by loop.

magnetic monopole See **Dirac monopole.**

magnetic orbital angular momentum quantum number (m_l) The quantum number that specifies the orientation of an atomic electron orbit. m_l = $2l + 1$, where l = orbital angular momentum quantum number.

magnetic permeability See **permeability.**

magnetic polarity The direction S to N of a magnetic field, where S = south-seeking end of magnetic needle, N = north-seeking end. As a result of this definition, the Earth's north magnetic pole is a S pole, and the Earth's south magnetic pole is a N pole. See **magnetic pole.**

magnetic pole 1. Either one of the two ends of a bar magnet. 2. Either one of the two sites on the Earth's surface where inclination is 90°. North magnetic pole: 77.3°N, 101.8°W; south magnetic pole: 65.8°S, 139.0°E. Cf. **dip pole, geomagnetic pole.**

magnetic quantum number See **magnetic orbital angular momentum quantum number.**

magnetic resonance The absorbance of energy at specific resonant frequencies by an atom when subjected to a magnetic field alternating at frequencies synchronous with the natural frequencies of the spin system of the atom.

magnetic spin angular momentum quantum number (m_s) The quantum number specifying the orientation of the magnetic spin vector of an atomic electron. m_s = $\pm 1/2$ (in units of $h/2\pi$), parallel (+) or antiparallel (−) relative to the direction of the magnetic field of the atom.

magnetic stratigraphy See **magnetostratigraphy.**

magnetic susceptibility See **susceptibility.**

magnetic viscosity The resistance of ferromagnetic substances to adjusting to a change in the ambient magnetic field.

magnetite The strongly magnetic mineral Fe_3O_4. Susceptibility ~ 1 CGS_{emu}; Curie point = 578°C.

magnetization (M) Magnetic dipole moment μ per unit volume = magnetic dipole moment density.

$$\mathbf{M} = \mu/V$$
$$= \chi_m \cdot \mathbf{H}$$

where V = volume, χ_m = magnetic susceptibility, **H** = magnetic field intensity. It is expressed in ampere/meter.

magnetization curve A curve showing the relationship between applied magnetic field and the induction produced in a magnetic material.

magnetohydrodynamics The study of conducting fluids in magnetic fields.

magnetomotive force (mmf, F_m) The line integral of the magnetic field strength around a closed path divided by the permeability constant.

$$\text{mmf} = \mu_0^{-1} \oint H \cos \theta \, ds$$

where μ_0 = permeability constant, H = intensity of magnetic field, θ = angle between magnetic lines and direction of path. It is expressed in ampere-turns (SI) or gilberts (CGS_{emu}). 1 ampere-turn = $4\pi/10$ gilbert.

magneton (μ) 1. The Bohr magneton (μ_B), a unit of magnetic moment for subatomic particles.

$$\mu_B = eh/4\pi m$$

where e = charge of the particle, h = Planck's

constant, m = mass of the particle. Included are the Bohr magneton (μ_B, referring to the electron) and the nuclear magneton (μ_N, referring to baryons and nuclei). 2. The Weiss magneton (μ_W), a unit of magnetic moment for molecules.

$$\mu_W = 1.853 \cdot 10^{-21} \text{ erg/oersted.}$$

magnetopause The outer edge of the magnetosphere.

magnetosphere The pear-shaped region of space surrounding a planet, where the paths of charged particles from the Sun and elsewhere are affected by the magnetic field of the planet. The magnetosphere of the Earth extends to 5–10 terrestrial radii toward the Sun and to several hundred terrestrial radii in the opposite direction. Density of matter in the magnetosphere $\sim 10^{-21}$ g/cm^3.

magnetostratigraphy Stratigraphy based on the reversals of the magnetic field of the Earth.

magnetostriction The deformation of ferromagnetic materials subjected to a magnetic field.

magnetron A vacuum tube used to generate high-power microwaves.

magnitude (astronomical) The brightness of a celestial body. **1. absolute magnitude (M).** The apparent magnitude of a celestial body reduced to the standard distance of 10 parsecs. The absolute magnitude of the Sun is +4.8. **2. apparent magnitude (m).** The brightness of a celestial body as seen from the Earth. It ranges from −1.45 for Sirius (the brightest star in the sky) to +25 for the faintest objects detected with the 200-in. Mt. Palomar reflector. The difference of 1 magnitude corresponds to a ratio of 2.512 in light intensity received (Pogson ratio). The apparent magnitude of the Sun is −26.74. **3. bolometric magnitude (m_{bol}).** Apparent magnitude based on all wavelengths, including the nonvisible portion of the spectrum. **4. photoelectric magnitude.** Apparent magnitude based on the light intensity received by a photoelectric cell with filters for different colors. See **color index. 5. photographic magnitude (m_{pg}).** Apparent magnitude based on the optical intensity on ordinary photographic film, which is most sensitive to blue light. **6. photovisual magnitude (m_{pv}).** Apparent magnitude based on the optical intensity on film most sensitive to the greenish-yellow light, the optical band to which the human eye is most sensitive. **7. visual magnitude.** The apparent (m_V) or absolute (M_V) magnitude in the wavelength range to which the human eye is most sensitive.

magnitude (seismic) See **earthquake, Mercalli scale, Richter scale.**

magnon The quantum of spin wave propagation in magnetically ordered crystals. Cf. **phonon.**

Main Sequence The principal sequence of stars in the Hertzprung-Russell diagram, running from the upper left region of high temperature (40,000 K) and high luminosity (−7 M_V) to the lower right region of low temperature (2400 K) and low luminosity (+17 M_V), and containing about 90% of all visible stars. Stars in the Main Sequence are undergoing their first evolutionary phase, fusing H to He in their cores. See **spectral classification.**

major axis The axis of an ellipse passing through the two foci.

majority carrier The dominant carrier in a semiconductor (electrons in n-type semiconductors, holes in p-type semiconductors).

major planets The 9 planets of the solar system. Cf. **minor planets.**

malachite The mineral $Cu_2CO_3(OH)_2$.

malacology The science that studies mollusks.

malleability The property that enables metals to be hammered or pressed into plates, sheets, or foils.

Malus, law of See **polarization.**

manometer An instrument to measure pressure.

mantle *(Geology)* The zone of the Earth's interior extending from the base of the crust (20–80 km deep under the continents, 7 km deep under the ocean floor) to the top of the outer core (2885 km below the Earth's surface). From top to bottom of the mantle temperature increases from about 500°C to about 2900°C, density increases from 3.3 to 5.5 g/cm^3, and pressure increases from 9800 to 1,372,000 atm. A well defined transition zone between 650 and 700 km of depth separates the more heterogeneous upper mantle, in which various mineral rearrangements and phase changes take place as pressure increases and the denser phases become stable (pyroxenes → olivine + stishovite and olivine → spinel at 400 km of depth; spinel → periclase + wüstite + stishovite at 670 km of depth), from the more homogeneous lower mantle believed to consist mainly of pyrolite. *(Biology)* The outer body wall lining the shell of many invertebrates.

marcasite Orthorhombic iron sulfide, FeS_2. Cf. **pyrite.**

mare Any of the several broad basaltic plains on the lunar surface, formed by partial melting of ultramafic rocks 150–450 km below the surface of the Moon.

mare basalt The basalt that forms the floor of the lunar maria. Mare basalts range in age from 3.1 to $3.8 \cdot 10^9$ y.

marginal crevasse A crevasse extending upstream at a 45° angle, from the margin of a glacier toward its middle.

maria Plural for *mare.*

Markov process A stochastic process in which the state $n(t)$ at a time t of a system resulting from a number of random events depends only on the state $n(t-1)$ of the system at the time $t-1$, as the outcome of each contributing event depends upon the outcome of its predecessor.

marl An unstratified sediment consisting of 50% clay and 50% carbonate mud.

marlstone An indurated marl.

marmorization The metamorphism of limestone into marble.

Mars The fourth planet from the Sun. Mean distance from the Sun = 1.523688 AU. Sidereal period = 686.980 d; sidereal rotational period = 24.6229 h. Equatorial radius R_{eq} = 3398 km; mass = $0.6421 \cdot 10^{24}$ kg; mean density = 3.970 g/cm³. Internal structure (estimated): Fe-Ni core (radius = $0.32R_{eq}$), silicate mantle. Magnetic field = $4 \cdot 10^{-5}$ gauss. Surface temperature = 250 K (day), 218 K (average), 170 K (night). Atmospheric pressure = 0.007 bar; gases in atmosphere, CO_2 = 95.7%, N_2 = 2.7%, Ar = 1.6%. Two exceedingly small satellites, Phobos (size = $19 \times 21 \times 27$ km; mass = $9.6 \cdot 10^{15}$ kg; density = 1.8 g/cm³; sidereal period = 0.31891 days) and Deimos (size = $11 \times 12 \times 15$ km; mass = $2.0 \cdot 10^{15}$ kg; density = 2.0 g/cm³; sidereal period = 1.26244 days). See **Olympus Mons, Planetary atmospheres*, Planets—physical data*, Satellites*.**

Marsden square A square 10° lat. $\times$ 10° long., each divided into 100 1° subsquares. It is used to locate oceanographic and meteorological data.

marsh A wetland with abundant hydrophytic vegetation. See **wetland.**

marsh gas Methane produced by the decay of vegetable matter under water.

mascon Mass concentration, describing the concentration of mass beneath the lunar maria responsible for the observed 100–200 mgal positive gravitational anomalies.

maser Microwave amplification by stimulated emission of radiation. A device consisting of a gaseous or solid system whose atoms or molecules are excited to a higher energy level and then stimulated to radiate in phase at the same frequency as that of the stimulating wave. Maser amplification ranges from optical to audio frequencies. Cf. **laser.**

mass (*m*) A measure of inertia, i.e. the resistance of a particle or a body to a change in momentum.

mass action law "The rate of reaction in a chemical system at constant temperature is proportional to the concentration of the reactants."

mass defect 1. The difference between the mass of an atomic nucleus and its mass number. 2. The difference between the sum of the masses of the individual nucleons in the free state and the corresponding mass of the assembled nucleus. It is equivalent, in this sense, to nuclear binding energy.

mass-luminosity relation (*M-L* relation) An approximate mass-to-luminosity relation in the Main Sequence of the Hertzprung-Russell diagram.

$$\log (L/L_{Sun}) = 3.5 \log (M/M_{Sun})$$

where L = luminosity, M = mass.

mass number (*A*) The number of nucleons in an atomic nucleus.

$$A = Z + N$$

where Z = number of protons, N = number of neutrons.

matrix *(Igneous rocks)* See **groundmass.** *(Sedimentary rocks)* The finer particles surrounding and cementing the coarser ones.

mature 1. Referring to a clastic sedimentary rock consisting only of the more resistant mineral particles. 2. Describing a landscape with subdued topography resulting from a long period of weathering and erosion.

maxwell (Mx) The CGS_{emu} unit of magnetic flux. 1 maxwell = $1 \cdot 10^{-8}$ weber (exactly).

Maxwell-Boltzmann distribution A distribution law specifying the number of particles having a given speed interval in a gaseous system in thermal equilibrium.

$$dN/N = 4\pi v^2 (m/2\pi kT)^{3/2} e^{-mv^2/2kT} \, dv$$

where dN = number of particles having speeds between v and $v + dv$, N = total number of particles, m = mass of particle, k = Boltzmann's constant, T = absolute temperature.

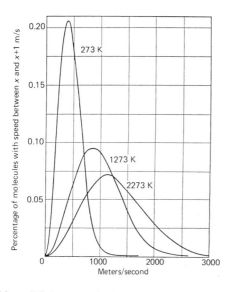

Maxwell-Boltzmann distribution of molecular speeds (m/s) for nitrogen gas at different temperatures. (From Moore 1950, p. 185, Fig. 7.12)

Maxwell's equations Four equations that describe the laws of electromagnetism. These are, in vectorial integral and differential form:
1. Magnetic effect of a current (generalized Ampere's law):
$$\oint_P H \cdot dl = \int_S (J + \partial D/\partial t) \cdot n \, dS$$
$$\nabla \times H = J + \partial D/\partial t$$
2. Electrical effect of a changing magnetic field (Faraday's emf law):
$$\int_P E \cdot dl = -\int_S \partial B/\partial t \cdot n \, dS$$
$$\nabla \times E = -\partial B/\partial t$$
3. The magnetic field (Gauss' law for magnetism):
$$\oint B \cdot n \, dS = 0$$
$$\nabla \cdot B = 0$$
4. Charge and electric field (Gauss' law for electricity):
$$\oint D \cdot n \, dS = \int_v \rho \, dV$$
$$\nabla \cdot D = \rho$$
(Explanation of the symbols: P = perimeter of closed curve spanned by arbitrary surface S, both stationary in the observer's frame of reference; H = magnetic field intensity vector in A/m; l = length; S = arbitrary surface; J = current density vector in A/m^2; $D = \epsilon E$ = electric displacement vector in C/m^2; n = unit vector normal to S; t = time; E = electric field intensity vector in V/m; B

= magnetic-induction vector in Wb/m^2; ρ = volume density of charge in C/m^3; ϵ = permittivity of medium.)

maxwellian gas A gas whose particles follow the Maxwell-Boltzmann distribution of velocities.

mb Millibar.

md Millidarcy.

M discontinuity Mohorovičić discontinuity.

mean A single number representative of a set of numbers. **1. arithmetic mean** The average of a set of values. If the distribution of values is normal, the mean coincides with the mode (the most common value) and the median (the value that divides the distribution into two halves). If the distribution is skewed, mean, mode, and median do not coincide. **2. geometric mean** The nth root of the product of n numbers.

meander A freely developed loop in a stream course with low proclivity in both of the two plains bordering the river on either side and in their intersection forming the river bed axis. Meandering is initiated by any minor topographic irregularity and develops laterally until the downstream component of the gravitational force overcomes the momentum of the moving water.

mean equinox The mean position of the vernal equinox, obtained by correcting the position of the true equinox for the effect of nutation.

mean free path The average distance traveled by a particle between successive collisions.
$$L = 1/\pi 2^{1/2} n d^2$$
where L = mean free path, n = number of particles per unit volume, d = diameter of particle. The mean free path of atmospheric molecules (mean molecular mass = 28.964 u) at a pressure of 1 atm and a temperature of 15°C is $6.6332 \cdot 10^{-6}$ cm or about 200 molecular diameters, with $6.9189 \cdot 10^9$ collisions/s. At the altitude of the aurora (av. 150 km), the mean free path is 33 m, with 2.3 collisions/s (pressure = $4.5 \cdot 19^{-9}$ atm = $3.4 \cdot 10^{-6}$ mmHg). See **critical level***.

mean high water (MHW) The average of the high water levels at a given location over a period of 19 years.

mean life (τ) The average time a radioactive nuclide is expected to live as such. It is the reciprocal of the decay constant:
$$\tau = 1/\lambda$$

where λ = decay constant. Mean life τ and half life $t_{1/2}$ are related:

$$t_{1/2} = (\ln 2)\tau$$
$$\tau = 1.443\, t_{1/2}$$

See **decay constant, half life.**

mean low water (MLW) The average of the low water levels at a given location over a period of 19 years.

mean sea level (MSL) The average position of sea level across all tidal periods during a 19-year interval. It is the datum to which altitudes above sea level and depths below it are referred.

mean sidereal time Time based on the mean equinox.

mean solar day The time interval between successive meridian passages of the mean sun = average length of the apparent solar day = apparent solar day corrected for the equation of time = 24h 3m 56.555s of mean sidereal time (in 1900).

mean solar time Mean time, equal to local hour angle of the mean sun + 12h.

mean sun The hypothetical position of the Sun if it were moving at a constant rate and completed a circular path between successive vernal equinoxes in 1 tropical year. It is the base for mean solar time.

mean tide level (MTL) The mean level between mean high water and mean low water at a coastal marine location.

mean time See **mean solar time.**

mechanical weathering The breaking down of rocks by the mechanical action of alternating high and low temperatures, freezing of interstitial water, wind action, etc.

medial moraine A moraine carried along the middle of a glacier and formed by coalescence of the proximal lateral moraines of tributary glaciers.

median The value that divides a frequency distribution in two halves, each one containing an equal number of values. Cf. **mean.**

medium-grained 1. Referring to an igneous rock with crystal sizes in the range of 1 to 4 mm. 2. Referring to a sedimentary rock with particle sizes in the range of 1/16 to 2 mm.

meerschaum Massive sepiolite.

mega- 1. Prefix meaning 10^6 (symbol M). 2. Prefix meaning *large, big, great.* Cf. **macro-.**

megaparsec (Mpc) 10^6 parsecs = $3.261633 \cdot 10^6$ light years.

megaton A unit of energy equal to the energy released by 10^6 tonnes of an explosive rated at 10^3 cal/g. It is equal to 10^{15} calories = $4.1868 \cdot 10^{15}$ joules (exactly). Cf. **kiloton.**

megawatt (MW) 10^6 watts.

meiofauna The fauna ranging in size from 1 to 5 mm and thus intermediate between microfauna and macrofauna (invertebrate).

meiosis The process of cell division in diploid or polyploid organisms leading to the formation of haploid gametes or spores. It consists of a first division, during which the homologous maternal and paternal chromosomes line up, exchange DNA material, and duplicate, and a second division during which the hybridized chromosomes segregate in separate cells without further division. Thus, one diploid cell forms four haploid cells (gametes or spores).

mel A unit of pitch, equal to 1/1000 of pitch of a 1000 Hz frequency 40 dB above hearing threshold. Pitch > frequency at frequencies <1000 Hz; pitch < frequency at frequencies >1000 Hz.

mela- Prefix meaning *black, dark.* Cf. **leuco-.**

mélange A mixture of different rocks and rock types brought together by tectonic or sedimentary processes.

melanocratic Dark colored.

melanophyre A dark-colored porphyritic rock.

melting point The temperature at which a pure substance changes from the solid to the liquid phase at a fixed pressure. Syn. **freezing point.**

member A lithostratigraphic unit, part of a formation.

membrane A monomolecular, bimolecular, or plurimolecular layer or set of such layers of inorganic or organic compounds capable of selectively admitting specific chemical compounds or particles. Most biological membranes consist of proteins and lipids. The basic bilayer membrane consists of phospholipid molecules arranged nonpolar hydrocarbon tail to nonpolar hydrocarbon tail, with the opposite polar ends facing ambient water on either side. Phospholipids dispersed in water tend to aggregate and form micelles.

Mendeleev's Table See **Periodic Table of the Elements.**

Mendel's laws Two laws establishing the basic principles of genetics. **1. Law of Segregation** "There are two alleles for each characteristic, each segregating in a different gamete." **2. Law of Independent Assortment** "Alleles segregate in different gametes independently of each other."

Mercalli scale A scale of earthquake intensity, ranging from I (instrumental detection only) to XII (total destruction). Cf. **earthquake, Richter scale.**

mercaptans See **thiols.** (The name *mercaptan* comes from *mercurium captans,* Latin for *capturing mercury,* referring to the ability of thiols to react with mercuric ions and ions of other heavy metals to form precipitates.)

Mercator projection A map constructed by projecting the Earth's surface from the center of the Earth on a cylindrical surface tangent to the equator. Loxodromes (lines that intersect successive meridians at the same angle) are straight lines, making the Mercator projection invaluable for navigation. See **rhumb line.**

Mercury The innermost planet of the solar system. Mean distance from the Sun = 0.387099 AU. Sidereal period = 0.24085 tropical years = 87.969 d; sidereal rotational period = 58.65 d. Equatorial radius R_{eq} = 2439 km. Mass = $0.3302 \cdot 10^{24}$ kg; mean density = 5.48 g/cm^3. Internal structure (estimated): Fe-Ni core (radius = 0.47 R_{eq}), silicate mantle. Magnetic field = $1-2 \cdot 10^{-3}$ gauss. Surface temperature = 775 K (day, at perihelion), 430 K (average), 90 K (night). Atmospheric pressure = $2 \cdot 10^{-15}$ bar. Gases in atmosphere, He = 98%; H = 2%. No satellites. See **Planets—atmospheres*, Planets—physical data*.**

mercury cell A primary cell consisting of a zinc anode, an HgO cathode, and an electrolytic solution absorbed on a paper strip.

mercury lamp A lamp emitting bluish light, rich in ultraviolet radiation, from mercury vapor ions energized by a discharge between two electrodes.

meridian Half of any great circle passing through the poles. Terrestrial meridians are reckoned east or west of the meridian passing through Greenwich, England.

mero- Prefix meaning *part, portion.*

meromictic lake A lake in which the bottom water does not mix with the surface layer.

meroplankton Plankton exhibiting planktonic habit only during a portion of the life cycle.

mesa An elevated, flat landmass rising above the surrounding country.

meseta A small mesa.

meso- Prefix meaning *medial.*

meson Any of the strongly interacting elementary particles with baryon number 0. Mesons consist of a quark-antiquark pair and range in mass from 0.144888 u (π^0 meson) to 5.6620 u (B^0 meson). See **Elementary particles*.**

mesopause The boundary between mesosphere and thermosphere at 85 km of altitude. See **atmosphere.**

mesopelagic Defining the pelagic environment between 100 and 500 m of depth.

mesosiderite Any of a group of stony-iron meteorites consisting of orthopyroxene, plagioclase, kamacite, taenite, and troilite. Mesosiderites are 60% of all stony-iron meteorites, or 0.9% of all meteorites. Cf. **Meteorites*.**

mesosphere *(Meteorology)* The layer between stratosphere and thermosphere, extending from 50 to 85 km of altitude in the standard atmosphere. See **atmosphere.** *(Geology)* The lower mantle, between 670 km of depth and the surface of the outer core. Syn. **lower mantle.** See **mantle.**

Mesozoa A metazoan phylum of small, parasitic, worm-like organisms.

Mesozoic The geological era following the Paleozoic and preceding the Cenozoic. It ranges from 248 to 65 million years B.P. It is subdivided as follows, with ages of boundaries in 10^6 y: 248/Triassic/213/Jurassic/144/Cretaceous/65. See **Geological time scale*.**

messenger RNA (mRNA) The strand of DNA-templated RNA that carries genetic information from DNA in the nucleus to the ribosomes where protein synthesis takes place. mRNA may consist of a single cistron *(monocistronic mRNA)* to encode a single protein, or several cistrons *(polycistronic mRNA)* to encode several proteins which often belong to a specific metabolic pathway. See **cistron.**

Messier Catalogue A catalogue of 103 celestial nebulae (both galactic and extragalactic) published by Charles Messier in 1784. Cf. **NGC.**

meta- Prefix meaning *between, among, after.*

(Chemistry) Prefix indicating the 1,3 positions in the benzene ring. Cf. **ortho-, para-** *(Chemistry)*. *(Petrology)* Prefix indicating metamorphism.

metabolism The set of chemical transformations that occur in cells, consisting of catabolism and anabolism.

metacenter The intersection of the vertical lines passing through the center of buoyancy of an object when the object is floating undisturbed and when it is tilted. The metacenter must be above the center of gravity for stability.

metacentric height The distance between metacenter and center of gravity of a floating object.

metagalaxy A rarely used name for the total physical universe, both visible and invisible. Cf. **cosmos, universe.**

metal Any of the electropositive elements. Metals form the majority (76%) of the elements. Metallic properties (ductility, malleability, and electrical and thermal conductivity) generally decrease from left to right across the Periodic Table of the Elements. Cf. **nonmetal.**

metalimnion The layer in a lake below the epilimnion (surface layer) and above the hypolimnion (deep layer), characterized by the presence of the thermocline.

metallic bond See **bond.**

metalloid Any of the elements exhibiting both metallic and nonmetallic properties. E.g. As, Ge, Sb, Si, Te.

metal-oxide semiconductor (MOS) A field-effect transistor with the gate insulated from the conduction channel by a thin oxide film.

metal-oxide-semiconductor field-effect transistor (MOSFET) A unipolar transistor consisting of source and drain regions on either side of a p or n conduction region, and a gate insulated from the conduction channel by a thin layer of silicon oxide. Voltage applied to the gate controls current flow, which occurs along or close to the surface of the substrate. Source and drain are interchangeable and, in addition, less power is dissipated than in bipolar transistors where conduction occurs through the bulk of the semiconducting material.

metamict Defining a mineral whose crystal structure has suffered damage because of radiation from radionuclides present in trace amounts in the crystal lattice.

metamorphic 1. Referring to the process of metamorphism. 2. Referring to a mineral or a rock that has undergone metamorphism.

metamorphism The physicochemical, mineralogical, and structural change undergone by a rock when submitted to a different physicochemical environment, including higher temperature and pressure, at depth within the crust. Metamorphism is classified, in order of increasing change, as low-grade metamorphism, high-grade metamorphism, and ultrametamorphism.

metaphyte A multicellular plant.

metasomatism Replacement of minerals by dif-

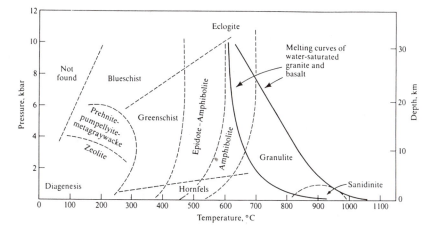

Metamorphic facies as functions of temperature and pressure. (Ehlers and Blatt 1982, p. 613, Fig. 20-1)

ferent ones within a rock by the slow action of percolating fluids.

metathesis reaction A reaction of the type A^+B^- + $C^+D^- \rightleftarrows A^+D^- + C^+B^-$ in which cations and anions exchange partners.

Metazoa The set of multicellular animals functioning as organisms.

metazoan 1. Referring to Metazoa. 2. Any of the members of the set Metazoa. Syn. **metazoon.**

metazoon Any of the members of the subkingdom Metazoa.

meteor A bright streak of ionized gas 7 to 20 km long and 1 m across generated between 115 and 70 km of altitude by a silicate, metal, or ice meteoroid, <1 to 10 mm in size, that vaporizes completely upon entering the atmosphere. Entry velocities range from 11.18 km/s (the Earth's escape velocity) to 72 km/s (the highest velocity observed).

Meteor Crater A crater in central Arizona, 1200 m across and 180 m deep, formed 50,000 y ago by the impact of a large (30 m across, 110,000 tons) iron meteorite (an octahedrite called *Canyon Diablo*).

meteorite Any of the silicate or metal pieces falling on the Earth's surface from outer space. Meteorites are classified into stony (92.8%), stony-iron (1.5%), and iron (5.7%). Stony meteorites are subdivided into chondrites [85.7%, further subdivided into ordinary chondrites (67.6%), carbonaceous chondrites (5.7%), and others (12.5%)], and achondrites [7.1%, further subdivided into Ca-rich (4.7%) and Ca-poor (2.4%)]. The stony-iron meteorites are subdivided into mesosiderites (0.9%), pallasites (0.5%), and others (0.1%). The iron meteorites are subdivided into octahedrites (4.3%), hexahedrites (0.6%), and ataxites (0.8%). See **Meteorites*.**

meteoroid A small (1–10 mm across) meteoritic body orbiting the Sun. These bodies are generally prograde, with orbits of low inclinations and with density distribution varying inversely with the square of the distance from the Sun. See **meteor.**

meteor shower The marked increase in the rate of appearance of meteors when the Earth crosses a meteor stream. Major showers in order of decreasing display are: Quadrantids (max. January 3; α = 231°, δ = +49°); Perseids (max. August 12; α = 46°, δ = +58°); Geminids (max. December 13; α = 112°, δ = +32°; δ Aquarids (max. July 30; α = 339°, δ = −10°); Orionids (max. October 21; α = 95°, δ = +15°); η Aquarids (max. May 4; α = 336°, δ = 0°).

meteor stream The trail of particles shed by a comet and left along its orbit. Diameter = 10–50· 10^6 km.

meter The SI and MKS unit of length, equal to the distance traveled by light in vacuo in 1/299,792,458 s (exactly). See **Units*.**

methane CH_4, a gas commonly formed in marshes.

methane series See **alkanes.**

methyl The radical $-CH_3$.

metric system A decimal system of measures based on the meter as a unit of length and the kilogram as a unit of mass. It was first proposed in a report by the French Academy to the National Assembly in 1791. The meter was defined as 10^{-7} of the Earth's meridional quadrant (present definition: 1 m = 1/299792458 of distance covered by light in vacuo in 1 second). The kilogram was defined as the mass of 1 dm^3 of pure water at 4°C (present definition: 1 kg = mass of Pt-Ir International Prototype Kilogram kept at Sèvres, S.-et-O., France). The metric system was officially adopted in France in 1801.

MeV Million electron volts.

mgal Milligal.

mho Reciprocal of ohm. See **siemens.**

MHW Mean high water.

mi Mile.

miarolitic Referring to an igneous rock containing small cavities in which crystals of the rock-forming minerals protrude euhedrally.

mica Any of the igneous and metamorphic rock-forming phyllosilicates. See **biotite, muscovite.**

micelle A spherule of amphipathic molecules in water, with the polar heads forming the surface and the nonpolar tails pointing toward the interior.

micrite Crystalline limestone matrix with crystal sizes <4 μm. Cf. **sparite.**

micro- 1. Prefix meaning 10^{-6}. Cf. **atto-, femto-, nano-, pico-.** 2. Prefix meaning *small.*

microclimate The local, groundlevel climate as conditioned by the local environment, including artificial structures if any.

microcline A triclinic dimorph of orthoclase, $KAlSi_3O_8$.

microcrystalline Describing the texture of a rock whose crystals are too small to be seen with the naked eye.

microenvironment The ensemble of environmental conditions over a very restricted area (a few meters across or less).

microfacies The facies of a rock resulting from microenvironmental factors.

microfauna *(Invertebrate)* An assemblage of living or fossil animals or animal remains <1 mm in size and thus too small to be visible or identifiable with the naked eye. Cf. **macrofauna** *(Invertebrate)*, **meiofauna**. *(Vertebrate)* An assemblage of living or fossil animals smaller than 5 cm. Cf. **macrofauna** *(Vertebrate)*.

microflora An assemblage of plant organisms and/or plant parts (e.g. spores) <1 mm in size. Cf. **macroflora**.

microforaminifera Foraminifera that failed to grow beyond the neanic stage.

microfossil A fossil too small to be seen or studied without the aid of a microscope. Cf. **macrofossil**.

micrometeorite A small (<1 mm in size) silicate, metal, or ice body derived from the ablation of a meteorite or shed by a comet. Cf. **meteoroid**.

micrometer (μm) A unit of length equal to 10^{-6} m.

micron (μ) See **micrometer**.

microperthite A perthite with lamellae (5–100 μm wide) visible only with the aid of a microscope. Cf. **cryptoperthite, perthite**.

microphyric Describing a porphyritic rock with phenocrysts of microscopic size (<2 mm across).

microplankton Plankton ranging in size from 50 μm to 1 mm, larger than nanoplankton and ultraplankton.

microprobe See **electron microprobe**.

microseism Any of the long-period (>1 s), small-amplitude motions of the solid Earth induced by the motions of the hydroatmosphere, not by earthquakes.

microspar Fine calcitic matrix in limestones, with crystal sizes between 5 and 25 μm.

microsparite A limestone with fine matrix recrystallized to microspar.

microtektite A small (<1 mm), spherical to subspherical tektite found in deep-sea sediment layers deposited at the time of a major tektite fall. See **tektite**.

microwave An electromagnetic wave within the 0.3 to 30 cm wavelength band. See **Electromagnetic spectrum***.

microwave background radiation The radiation relic from the Big Bang. It has a blackbody spectrum that ranges from 0.02 to >10^3 cm wavelength, with a power peak at the 0.2 cm wavelength that corresponds to a temperature of 2.75 K. Energy density = $4 \cdot 10^{-14}$ J/m^3.

midden A mound of food and artifact refuse revealing an ancient human settlement. Cf. **tell**.

Middle Referring to the middle portion of a chronological or chronostratigraphic unit. Cf. **Early, Late, Lower, Upper**.

mid-oceanic ridge Any of the approximately median, volcanic, almost entirely submarine ridges extending through the Arctic, Atlantic, Indian, and South Pacific oceans.

migma A mixture of solid rock and magma.

migmatite A rock consisting of a mixture of igneous and metamorphic rocks.

migmatization The formation of a migmatite by

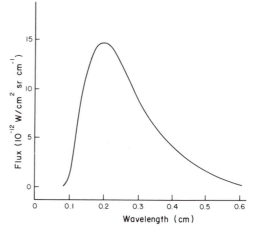

Microwave background radiation. The spectrum of the 2.75 K microwave background radiation (based on data points, with interpolations and extrapolations).

injection of magma into solid rock or by partial melting of the rock.

Milankovitch theory A theory of glaciation according to which the ice ages are related to variations in the precession and obliquity of the Earth's axis and in the eccentricity of the Earth's orbit.

mile See **nautical mile, statute mile.**

Milky Way The band of stars across the sky representing the accumulation of stars along the galactic plane.

Miller indices A set of letters and integers used to define the orientation of a crystal face or internal crystal plane.

milli- Prefix meaning *1/1000th.*

milliard *(British, French)* 10^9. Syn. **billion.**

millidarcy (md) A unit of permeability equal to 1/1000 darcy. See **darcy, permeability** *(Geology).*

milliliter (ml) A metric unit of volume equal to 10^{-3} liters = 1 cm^3.

millimeter (mm) A metric unit of length equal to 10^{-3} m.

mineral Any of the naturally occurring, chemically definable substances, either solid crystalline (e.g. quartz) or noncrystalline (e.g. opal, coal), liquid (petroleum, water), or gaseous (methane, air). See **Minerals*.**

mineral water A natural water enriched in mineral salts or gases. -

minette 1. A lamprophyre consisting principally of biotite in a matrix of biotite and alkali feldspar. 2. A deposit of ferruginous oolites.

minor planet An asteroid.

minute (m) 1. A time interval equal to 60 s. 2. An angle equal to 1/60 of a degree or 1/21,600 of a circle.

miogeosyncline The nonvolcanic geosyncline landward of the eugeosyncline in an orthogeosynclinal belt.

Mira Ceti A variable red giant, 130 l.y. away in Cetus, with a period averaging 330 days. Apparent magnitude ranges from a minimum of about 9 with surface temperature of 1900 K to a maximum of about 3 with surface temperature of 2600 K. Diameter is 420 times that of the Sun and it changes by 20% during a cycle. Mean density is about 10^{-7} g/cm^3.

Mira star Any of the class of long-period variable stars of which Mira Ceti is an example.

mitochondrion An organelle, 1–2 μm long and 0.5 μm wide, present in the cytoplasm of all aerobic plants and animals in numbers ranging from hundreds to thousands. Mitochondria contain enzymes for the oxidation of foods and the production of ATP (respiration). Mitochondria contain their own DNA which, like bacterial DNA, is bare of nucleoproteins. Mitochondria are self-reproducing and may have originated as procaryota symbiotic with early eucaryota.

mitosis Cell division, including chromosome duplication and segregation of the products into the daughter cells which thus receive the same number of chromosomes as the parent cell. Cf. **meiosis.**

MKS Meter-kilogram-second, the system of measurement based on these units.

MKSA Meter-kilogram-second-ampere, the system of measurement based on these units.

MKSΩ Meter-kilogram-second-ohm, the system of measurement based on these units.

ml Milliliter.

MLW Mean low water.

mm Millimeter.

mmf Magnetomotive force.

mmHg Millimeters of mercury, a measure of pressure. 760 mmHg = 1 atmosphere.

mobile belt A long, narrow crustal region characterized by present or past tectonic acivity.

modal analysis The determination of the mineral composition of a rock in terms of the relative volumes or masses of the component minerals.

mode *(Statistics)* The most common value of a set of data. Cf. **mean, median.** *(Petrology).* The composition of an igneous rock in terms of the volumes or masses of the component minerals.

modulus *(Mathematics)* The absolute value of a complex number. See **complex number.** *(Physics)* A coefficient expressing the degree to which a substance or a system possesses a given property.

modulus of elasticity See **bulk modulus of elasticity.**

Moho Abbreviation for *Mohorovičić discontinuity.*

Mohorovičić discontinuity The seismic, petrologic, and probably geochemical discontinuity separating the Earth's crust from the mantle.

Mohs scale A scale of mineral hardness ranging from 1 to 10. 1, talc; 2, gypsum; 3, calcite; 4, fluorite; 5, apatite; 6, orthoclase; 7, quartz; 8, topaz; 9, corundum; 10, diamond.

moiety A portion of a chemical compound having a specific structural or chemical identity.

moiré effect The appearance of a new set of curves when two sets of curves of similar pattern cross each other at an angle $<45°$ (French from *mohair*, a shiny fabric).

mol Mole.

molal (*m*) Referring to a solution containing 1 mole of solute in 1 kg of solvent.

molality (*m*) The concentration of a solute in a solution in moles/kg of solvent.

molar (*M*) Referring to a solution containing 1 mole of solute in 1 liter of solution.

molarity (*M*) The concentration of a solute in a solution in moles/liter of solution.

molar volume The volume of 1 mole of a solid, liquid, or gaseous substance at STP. It is equal to 22.4141 liters for an ideal gas.

molasse A paralic sedimentary deposit consisting of thick, ungraded, coarse sandstones, shales, and conglomerates. It is characteristic of the terminal, postorogenic filling of marginal basins.

mold The cavity produced by the solution of an original fossil embedded in a sedimentary rock. Cf. **cast.**

mole An SI, MKS, and CGS unit of quantity, equal to one Avogadro number ($6.022136 \cdot 10^{23}$) of items (particles, atoms, molecules, objects, organisms, etc). Syn. **avogadro.**

molecular and atomic velocities For gaseous molecules or atoms:
$$v_{mean} = (8kT/\pi m)^{1/2}$$
$$v_{rms} = (3kT/m)^{1/2}$$
where v_{mean} = mean velocity, v_{rms} = root mean square velocity, k = Boltzmann constant, T = absolute temperature, m = atomic or molecular mass. Examples of mean velocities (m/s at 0°C): ^1H = 2395; ^4He = 1202; N_2 = 454; O_2 = 425; Ar = 380; CO_2 = 362; ^{222}Rn = 161.

molecular crystal A crystal formed by molecules bound to each other by van der Waals forces.

molecular mass The mass of a molecule in atomic mass units (u).

molecular volume The volume of 1 mole of a solid, liquid, or gaseous substance at STP. It is equal to 22.4141 liters for an ideal gas. Syn. *molar volume.*

molecular weight 1. The weight of 1 mole of a substance relative to 1/12 of the weight of 1 mole of ^{12}C. 2. The weight of a molecule relative to 1/12 of the weight of the neutral ^{12}C atom. Cf. **molecular mass.**

mole fraction The ratio M_x/M_i, where M_x = number of moles of component x in a system, M_i number of moles of all components in the system.

moment (*M*) The product of a quantity times its distance from a reference point.

moment of inertia (*I*) A measure of the resistance of a body to angular acceleration.
1. *Of a particle:* The product of the mass m of the particle times the square of the shortest distance r from an axis of rotation.
$$I = mr^2$$
2. *Of an object:* The sum of the products of the particles forming the object times their distances from the axis of rotation.
$$I = \Sigma \, m_i r_i^2$$
$$= \int r^2 \, dm$$
3. *Of a rotating sphere of uniform density:*
$$I = \tfrac{2}{3}MR^2$$
where M = mass of sphere, R = radius of the sphere.

moment of momentum Syn. **angular momentum.**

momentum (p) A fundamental quantity related to the motions of a physical system, conserved in the absence of external forces.
1. *Nonrelativistic particles or objects:*
$$\mathbf{p} = m\mathbf{v}$$
where m = mass, $\mathbf{v}$ = velocity.
2. *Relativistic particles:*
$$p = m_0v(1 - v^2/c^2)^{-1/2}$$
where m_0 = rest mass, v = particle velocity, c = speed of light.
3. *Zero-mass particles:*
$$p = hv/c$$
where h = Planck's constant, v = frequency, c = speed of light.

monadnock Syn. **inselberg.**

monazite A phosphate of rare earth elements, $(REE)PO_4$.

Monera One of the 5 kingdoms. It includes all procaryota. See **kingdom.**

mono- Prefix meaning *one, single.*

monoclinic One of the 6 crystal systems, characterized by a single twofold axis of symmetry and a plane of symmetry normal to it.

monoecious Having male and female reproductive organs in separate structures in the same individual. Cf. **dioecious.**

monomer A molecular unit that can partake in the formation of a polymer.

monomictic *(Limnology)* Describing a lake with only one overturn per year. Polar lakes overturn during the summer; tropical lakes during the winter. *(Geology)* Defining a sedimentary rock or a breccia in which the fragments have the same composition.

monomineralic Defining a rock consisting exclusively or almost exclusively of a single mineral.

monophyletic Derived from a single ancestor.

monoploid Referring to an organism with a single chromosome set (haploid) normally associated with a diploid population (e.g. males in ants and bees that arise from unfertilized eggs). Cf. **haploid.**

monopole See **Dirac monopole.**

monoprotic Refering to an acid that can only donate one proton per molecule. E.g. HCl.

monosaccharides A family of carbohydrates with 5 or 6 carbons. E.g. ribose, $C_5H_{10}O_5$; deoxyribose, $C_5H_{10}O_4$; glucose, $C_6H_{12}O_6$.

monotypic 1. Defining an assemblage consisting of only one species. 2. Defining a species represented by a single type. 3. Defining a genus that includes only one species.

monsoon A wind system that reverses its direction in opposite seasons. The largest monsoon system is over India and the Arabian Sea, with winds that blow from the northeast during the winter and from the southwest during the summer.

montmorillonite A clay mineral, $(Na, Ca) (Al, Mg)_6(Si_4O_{10})_3 (OH)_6 \cdot nH_2O$. It consists of two layers of SiO_4 tetrahedra pointing toward each other and jointed by Al and OH ions which, together with O from the SiO_4 tetrahedral vertices, form an intervening octahedral layer. Total thickness of a packet is 9.6 Å. A layer of water molecules and exchangeable cations is, howver, commonly present between the facing bases of the SiO_4 tetrahedra, expanding the thickness of a packet to 21.4 Å. Mont-

morillonite is commonly derived from the alteration of ferromagnesian minerals including basaltic ash. Syn. **smetcite.** Cf. **illite, kaolinite.**

monzonite A plutonic rock consisting of approximately equal amounts of K-feldspar and plagioclase, augite, and little quartz. It is intermediate in composition between syenite and diorite.

Moon The single, natural satellite of the Earth. Mean distance from the Earth = 384,401 km = 1.2822237 light seconds. Sidereal period = 27.32166 d. Mean radius = 1738.2 km. Mass = $73.49 \cdot 10^{21}$ kg; mean density = 3.343 g/cm³. Internal structure (estimated): Fe-Ni liquid metallic core (radius = 200–300 km), ultramafic silicate mantle (1400 km thick, with a seismic zone in its middle), and a 60-km-thick crust of anorthosite and gabbro-basalt. Ages of lunar rocks: mare basalts = $3.1–3.8 \cdot 10^9$ y, terra anorthosites = $3.7–4.6 \cdot 10^9$ y. Surface gravity (mean) = 1.62 m/s². Escape velocity = 2.38 km/s. Surface temperature = 400 K (day), 115 K (night). Albedo = 0.068. Atmospheric pressure = $2 \cdot 10^{-14}$ bar. Gases in atmosphere: Ne = 40%, Ar = 40%, He = 20%. See **Moon—crater ages and rhegolith thickness*, Moon—major rock types*, Moon—major rock types—chemical composition*, Moon—physical data*, moonquake, Satellites*, Tycho.**

moonquake Any of the weak seismic disturbances occurring either at shallow depth within the lunar crust or, more commonly, at depths of 900–1000 km in the lunar interior. The deeper quakes are triggered by tidal stresses at perigee.

moraine A subglacial (ground moraine), supraglacial (lateral or medial moraine), or frontal (end moraine) accumulation of rock debris formed and transported by a glacier.

MORB Mid-oceanic ridge basalt.

morass ore Syn. **bog iron ore.**

morphotype A type representative of a population exhibiting a characteristic morphology.

mortar A mixture of cement or lime with sand and water.

MOS Metal-oxide semiconductor.

mosaic evolution The evolution at different rates of different morphological characteristics of organisms within a lineage.

MOSFET Metal-oxide-semiconductor field-effect transistor. See also **field-effect transistor.**

Mössbauer effect The emission of a gamma ray

without loss of energy by an excited nucleus strongly bound within a crystalline lattice. The entire crystal absorbs the recoil, so that the emitting nucleus is practically recoilless. The emitted quantum has the same energy as the quantum originally absorbed and is available to adjacent nuclei for further absorption and emission.

mountain A topographic elevation rising more than 300 m above country level or 600 m above sea level.

Mpc Megaparsec.

mRNA Messenger RNA.

MSL Mean sea level.

Mt Mount, mountain.

MTL Mean tide level.

mud A mixture of water and of mineral particles smaller than 62.5 μm.

mudstone An indurated mud.

mud volcano A small conical structure constructed by escaping marsh gases.

multiplet 1. A set of close energy levels and corresponding spectral lines resulting from magnetic interaction between the orbital and spin momenta of the valence electrons. 2. A set of particles sharing most but not all properties (e.g. the three pions).

multiplicand A number that is to be multiplied by another number (the multiplier).

multiplier The number that multiplies the multiplicand.

multivalent Having a valence of 3 or more.

muon (μ) A lepton. Charge = $e^{\pm}$; rest mass = 0.11342892 u; $t_{1/2}$ = $2.19709 \cdot 10^{-6}$ s; decay = $e^{\pm}\nu\bar{\nu}$. See **Elementary particles***.

muon magnetic moment (μ_{μ}) Magnetic moment of the muon, equal to $4.4904514 \cdot 10^{-26}$ J T^{-1}.

muon rest mass (m_{μ}) See **muon.**

muscovite The white mica, $KAl_2(AlSi_3O_{10})(OH)_2$.

muskeg A growing Sphagnum bog.

mutation. 1. A change in the DNA structure of a living organism. 2. An error in chromosome duplication.

mV Millivolt.

MW Megawatt (= 10^6 watts).

MWL Mean water level.

Mx Maxwell.

m.y. Million years.

mylo- Prefix meaning *grinding.*

mylonite A microbrecciated rock exhibiting shearing flow structures, a product of intense shear metamorphism.

mylonitization The transformation of a rock into a mylonite by intense shear metamorphism.

myrmekite An intergrowth of oligoclase and quartz.

N

ν 1. Frequency. 2. Kinematic viscosity. 3. Neutrino.

n 1. Index of refraction. 2. Neutron. 3. Principal quantum number.

n- Nano-.

N 1. Newton. 2. Normal concentration or normality. 3. North-seeking pole of a magnetic dipole.

N_A Avogadro number ($= 6.022136 \cdot 10^{23}$).

nabla See **del**.

nacre The iridescent inner layer of oyster and other molluscan shells, consisting of aragonitic lamellae interlayered with organic matter.

NAD Nicotinamide adenine dinucleotide, a coenzyme important in respiration and other metabolic processes. It consists of two ribose sugars linked to each other via two phosphate groups, with a nicotinamide attached to one side and an adenine to the other.

NADH Reduced NAD.

nadir The point on the celestial sphere directly opposite the zenith.

NADP Nicotinamide adenine dinucleotide phosphate, a coenzyme important in respiration and other metabolic processes. It consists of NAD with an additional phosphate group.

NADPH Reduced NADP.

nanno- Common misspelling for **nano-**.

nannofossil Common misspelling for **nanofossil**.

nannoplankton Common misspelling for **nanoplankton**.

nano- 1. Prefix meaning *10^{-9}*. Cf. **atto-, femto-, micro-, pico-**. 2. Prefix meaning *small*.

nanofossil A collective term for discoasters and coccoliths, calcitic elements produced by unicellular marine algae of the phylum Haptophyta, kingdom Protoctista.

nanogram (ng) 10^{-9} g.

nanometer (nm) 10^{-9} m.

nanoplankton Plankton consisting of organisms ranging from 5 to 50 μm, smaller than microplankton but larger than ultraplankton.

nanosecond (ns) 10^{-9} s.

NAP Nonarboreal pollen.

naphthalene $C_{10}H_8$, forming a double ring structure. Mol. mass $= 128.173$. See **naphthalene ring**.

naphthalene ring A system of two hexagonal, planar rings attached to each other. Positions are numbered as shown in the illustration. Position 1,3 and 5,8 are termed α, positions 2,3 and 6,7 are termed β, and positions 9 and 10 are termed γ.

Napierian logarithm The exponent needed to express a number as a power of *e*. Syn. **natural logarithm**. See *e*.

nappe *(Geometry)* A surface generated by a ray fixed at its endpoint and with a point along its length describing either a circle *(circular nappe)* or an ellipse *(elliptical nappe)*. Two nappes with a common axis and joined at the vertex form a *cone*. See **cone**. *(Geology)* A detached recumbent fold.

native element A nongaseous element occurring as such in a mineral deposit.

natron Hydrated sodium carbonate, $Na_2CO_3 \cdot 10H_2O$.

natron lake See **soda lake**.

natural forces The four interactions among particles. **1. strong force** Between quarks or hadrons, mediated by gluons; strength increases with distance, leading to confinement within 10^{-15} m; typ-

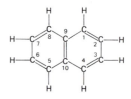

Naphthalene double-ring structure. The numbers identify the positions of the carbon atoms.

ical decay rates $= 10^{-23}$ to 10^{-21} s. **2. electromagnetic force** Between charges, mediated by photons; range, infinite; typical decay rates $= 10^{-19}$ to 10^{-16} s; strength $= 10^{-2}$ to 10^{-3} of strong interaction. **3. weak force** Between baryons and leptons, mediated by gauge bosons; range $= 10^{-18}$ m; typical decay rates $<10^{-10}$ s; strength $= 10^{-6}$ of strong interaction. **4. gravitational force** Between masses, mediated by gravitons; range, infinite; typical decay rates ?; strength $= 10^{-39}$ of strong interaction.

natural logarithm (ln) The exponent needed to express a number as a power of *e*. Syn. **Napierian logarithm.** See *e*. Cf. **logarithm.**

natural number See **numbers.**

natural remanent magnetization (NRM) The magnetization of a rock or mineral that was produced at the time of formation or deposition by the ambient magnetic field.

natural selection The selection by the environment of the individuals, populations, or species most fit for the prevailing environmental conditions.

nautical mile (mi) A nonmetric unit of length used in navigation, originally defined as equal to 1° of latitude at 45°N along the Greenwich meridian. Redefined as equal to 1852.0 m exactly.

neanic Referring to a youthful growth stage, following the nepionic stage and preceding the ephebic stage.

neap tide The lowest tide, occurring when the Moon is at quadrature.

necrocoenosis An assemblage of dead organisms and organic remains representative of the original population. Cf. **thanatocoenosis.**

Néel point The temperature above which antiferromagnetic ordering is not possible and the substance becomes paramagnetic.

Néel temperature Syn. **Néel point.**

negative 1. Referring to a number smaller than zero. 2. Referring to the charge of the electron. 3. Referring to a S magnetic pole. 4. Referring to sinistral rotation. Cf. **positive.**

negative feedback A control procedure by which a portion of the output is fed back 180° out of phase with the input, resulting in reduction of noise and distortion and in stabilization of the output. Cf. **positive feedback.**

negative pole 1. The terminal of a battery exhibiting electron excess. 2. The S or south-seeking pole of a magnetic needle. Cf. **positive pole.**

negatron An alternate name for electron.

nekton Collective name for free-swimming aquatic animals.

neontology The study of extant species, as contrasting *biology,* which is the study of living organisms, and *paleontology,* which is the study of extinct species.

neostratotype A stratotype designed after the holostratotype has been destroyed or become unusable.

neoteny 1. Arrested development resulting from the retention of youthful characteristics through the adult stage. Syn. **paedomorphism, paeodmorphosis.** 2. Early sexual development. Syn. **paedogenesis.**

neotype A single specimen designated as the type specimen of a species or subspecies when the holotype and other specimens in the original collection have been lost or destroyed.

neper (Np) A unit of attenuation in transmission-line engineering. See **attenuation.**

nepheline A feldspathoid, $(Na,K)AlSiO_4$.

nepheline syenite A plutonic rock consisting mainly of alkali feldspar and nepheline.

nephelinite A plutonic rock consisting mainly of nepheline and clinopyroxene.

nepheloid layer A deep oceanic layer, especially along boundary currents, reaching a thickness of several hundred meters above bottom and containing significant amounts (50–100 μg/liter) of suspended clay-size particles.

nepionic The growth stage in molluscan shells following the embryonic stage and preceding the neanic stage.

Neptune The eighth planet from the Sun. Mean distance from the Sun $= 30.10957$ AU. Sidereal period $= 164.79$ y; sidereal rotational period at equator $= 15.8$ h. Equatorial radius $= 24,750$ km. Mass $= 103.0 \cdot 10^{24}$ kg; mean density $= 1.65$ g/cm^3. Internal structure (estimated): Ni-Fe and silicate core [radius $= 8500$ (?) km], a 10,000 (?) km thick mantle consisting of liquid CH_4, NH_3, and H_2O, and a 5500 (?) km thick atmosphere consisting of H_2 (90%) and He (10%). Magnetic field not yet measured. Surface temperature $= 57$ K. Two satellites, Triton (radius = 1600 km, mass = $3.4 \cdot 10^{22}$ (?) kg, density = 2 (?) and Nereid (radius = 150 (?) km, mass = $2.8 \cdot 10^{19}$ (?) kg, density = 2 (?) g/cm^3). See **Planets—atmospheres***, **Planets—physical data***, **Satellites***.

neptunism The classical theory of Abraham Werner (1750–1817) according to which all rocks are crystallized or deposited from seawater. Cf. **plutonism.**

neritic Defining the coastal environment between low tide and the edge of the continental shelf.

nesosilicates See **silicates.**

Neumann lines Fine lines produced by mild shock in kamacite (α-iron) and revealed by etching. Especially common in hexahedrites (iron meteorites).

neuston Plankton supported on the water surface by the water's surface tension.

neutral current The absence of charge exchange in interaction processes mediated by the Z^0 gauge boson. Cf. **charged current.**

neutrino (ν) A stable lepton with zero (?) mass, zero electric charge, and spin 1/2. There are 3 forms of neutrinos, each with its antineutrino: the electron neutrino (ν_e), the muon neutrino (ν_μ), and the tauon neutrino (ν_τ). See **Elementary particles***, **solar neutrino flux, solar neutrino unit.**

neutron (n) An unstable baryon with charge = 0 and mass = 1.00866490 u. It is an integral constituent of all atomic nuclei except that of ^{1}H. In the free state it decays by β^- (τ = 914 ± 6 s = 15.2 m; $t_{1/2}$ = 633 ± 6 s = 10.5 m) into a proton. See **Elementary particles***.

neutron activation analysis A method of elemental analysis by which stable nuclides are bombarded by neutrons, transformed into radioactive isotopes, and thus detected.

neutron capture The capture of thermal neutrons by atomic nuclei.

neutron-capture cross-section The apparent cross section of a nucleus from the point of view of an incoming thermal neutron. Representative neutron-capture cross-section values (in barns): ^{1}H = 0.333, ^{4}He = 0 (lowest), ^{10}B = 3836, ^{12}C = 0.0035, ^{28}Si = 0.177, ^{40}Ca = 0.41, ^{56}Fe = 2.6, ^{58}Ni = 4.6, ^{113}Cd = 19,900, ^{135}Xe = 2,600,000 (highest), ^{235}U = 583.2, ^{238}U = 2.68, ^{239}Pu = 742, ^{244}Pu = 1.7, ^{254}Cf = 100.

neutron Compton wavelength ($\lambda_{C,n}$) Compton shift of incident x-rays and γ-rays in collision with free neutrons with photon scattering angle of 90°. It is a length characteristic of the neutron.

$$\lambda_{C,n} = h/m_n c$$

where h = Planck's constant, m_n = rest mass of the neutron, c = speed of light. It is equal to $1.3195911 \cdot 10^{-15}$ m. Cf. **Compton effect.**

neutron rest mass (m_n) The mass of a neutron at rest or traveling at negligible speed with respect to the speed of light. It is equal to 1.00866490 u.

neutron star Gravitationally collapsed object with mass between the Chandrasekhar limit and 3 solar masses, consisting mostly of neutrons. Density = 10^{14} g/cm^3; radius = 5–15 km. Cf. **black hole.**

névé The perennial snow cover on the head of a glacier.

newton (N) The SI and MKS unit of force, equal to the force needed to accelerate the mass of 1 kg by 1 m/s^2. 1 N = 10^5 dynes.

Newtonian fluid A turbulence-free fluid in which the rate of change in strain at a point in the fluid is proportional to the applied stress. The proportionality factor is the coefficient of viscosity.

Newton's first law "A body in an inertial frame remains at rest or moves at constant velocity along a straight line." This law defines inertial frame.

Newton's law of gravitation "Two masses attract each other with a force proportional to the product of their masses and inversely proportional to the square of their distances."

$$F = Gm_1 m_2 / r^2$$

where $G = Fr^2 / m_1 m_2$ = universal gravitational constant, f = force, m_1 is a mass, m_2 is a second mass, r = distance between the centers of mass of the two masses.

Newton's second law "The acceleration of a body in an inertial frame is equal to the applied force per unit of mass."

$$a = F/m$$

where a = acceleration, F = force, m = mass.

Newton's third law "To each action there is an equal and opposite reaction." This law establishes the principle of conservation of momentum.

N-galaxy A type of spiral galaxy with a very bright nucleus and broadened spectral lines. N-galaxies are more active than Seyfert galaxies, but less active than BL Lacertae objects.

ng Nanogram.

NGC The *New General Catalogue of Nebulae and Stars,* published in 1888 and originally listing 7840 nebulae (most of which were later recognized as galaxies or star clusters).

niche The environmental locus of a species or a population.

nick See **knickpoint.**

nickel-cadmium battery A secondary cell with a $NiO-Ni(OH)_2$ anode, a cadmium cathode, and an alkaline electrolytic solution. emf = 1.2 V.

nickpoint See **knickpoint.**

nicol prism Either the polarizing or the analyzing prism in a petrographic microscope.

nicotinamide The amide of nicotinic acid, $C_5H_4NCONH_2$.

nicotinic acid Niacin, C_5H_4NCOOH.

nit (nt) A unit of luminance, equal to 1 cd/m^2.

niter KNO_3. Syn. **saltpeter.**

nm Nanometer.

NMR Nuclear magnetic resonance.

n.n. Nomen nudum.

noble gases The gases He, Ne, Ar, Kr, Xe, and Rn, belonging to group 18 of the Periodic Table of the Elements and characterized by a full valence shell ($1s^2$ for He, ns^2np^6 for all others).

node *(Physics, Oceanography)* A point, line, or surface in a standing wave where amplitude is zero. *(Astronomy)* Either one of the two points at which the orbit of a celestial body intersects the ecliptic or the celestial equator on the celestial sphere. The **ascending node** is the point at which the celestial body, in moving along its orbit, crosses the ecliptic or the celestial equator and passes from the southern to the northern hemisphere. The **descending node** is the point at which the celestial body crosses the ecliptic or the celestial equator and passes from the northern to the southern hemisphere.

nomen nudum (n.n.) Latin for *naked name,* a name given to a species or subspecies without formal description or illustration.

nonarboreal pollen (NAP) Pollen produced by grasses.

nonconformity A major unconformity between igneous or metamorphic basement and the overlying sedimentary rocks.

nonconservative elements Elements in a solution whose concentrations do not remain constant with time. See **conservative elements.**

noninertial frame A frame of reference with respect to which an inertial mass experiences a force.

nonmetal Any of the electronegative elements on the right side of the Periodic Table of the Elements. Nonmetals are only 24% of all naturally-occurring elements. Cf. **metal.**

non-Newtonian fluid A fluid in which the rate of shear is not proportional to the applied stress.

nonrelativistic 1. Referring to a phenomenon for which the effects of special and general relativity are insignificant. 2. Referring to a particle traveling at a speed much smaller than the speed of light, so that relativistic effects are not detectable.

norite A plutonic rock consisting of labradorite and orthopyroxene.

norm *(Mathematics)* The absolute value of a vector, defined as the square root of the scalar square of the vector. The norm of two vectors is the square root of their scalar product. The norm of a vectorial sum is equal or smaller than the sum of the norms. *(Geology)* The theoretical mineral composition of a rock based on bulk chemical analysis.

normal concentration (N) A concentration characterized by the number of gram equivalents of solute per liter of solution. 1N = 1 gram equivalent, 2N = 2 gram equivalent, etc.

normal distribution A bell-shaped distribution of values symmetrical about the mean. It is given by

$$y = [1/\sigma(2\pi)^{1/2}]e^{-(1/2)[(x-\mu)/\sigma]^2}$$

where σ = standard deviation, μ = mean of distribution.

normal fault A tensional fault with a slanted fault plane along which one side has moved downward with respect to the other.

normal magnetization Magnetization produced at a time of normal polarity.

normality The number of gram equivalents of solute per liter of solution.

normal polarity The polarity of the geomagnetic field prevailing during the present (Brunhes) polarity epoch. Cf. **reversed polarity.**

normal solution A solution containing 1 gram equivalent of solute per liter of solution.

normative mineral A mineral whose presence in a

rock is indicated (but not necessarily proved) by bulk chemical analysis.

north magnetic pole The site in the northern hemisphere where magnetic inclination is 90°. Present location: 77.3°N, 101.8°W. It has moved to this location over the past 150 years from a location at 70°N, 97°W in 1831. Cf. **dip pole, geomagnetic pole, south magnetic pole.**

nova Usually a white dwarf in a binary system that receives an increasing load of hydrogen from a companion star until ignition results in a sudden (days) increase in luminosity (by some 10 magnitudes) accompanied by expulsion of hydrogen gas. A slow (months for the "fast" novae, years for the "slow" novae) decrease in luminosity follows until the original conditions are re-established.

(n,p) *Neutron in, proton out,* a symbolism describing a nuclear reaction in which a nucleus absorbs a neutron and emits a proton.

Np Neper.

npn transistor A transistor with a *p*-type base between an *n*-type emitter and an *n*-type collector. Cf. **pnp transistor.**

N pole North-seeking magnetic pole, i.e. the end of a magnetic dipole that points toward the north magnetic pole. Cf. **magnetic polarity.**

NRM Natural remanent magnetization.

ns Nanosecond ($= 10^{-9}$ s)

nt Nit.

n-type conduction Electrical conduction by free electrons in a semiconductor. Cf. **p-type conduction.**

n-type semiconductor An extrinsic semiconductor in which the majority carriers are electrons. Cf. **p-type semiconductor.** See **semiconductor.**

nuclear binding energy See **binding energy.**

nuclear chain reaction A set of successive nuclear reactions that are self-sustaining because the energy liberated by each reaction is sufficient to initiate one or more similar reactions in adjacent nuclei. See **fission.**

nuclear fission See **fission.**

nuclear force The strong force that binds together nucleons in atomic nuclei, resulting from interaction between quarks confined to different nucleons. It is approximately 10^6 stronger than the chemical bond binding atoms into molecules. See **strong force.**

nuclear magnetic resonance (NMR) The absorption of electromagnetic waves of characteristic frequencies by atomic nuclei.

nuclear magneton (μ_N) A unit of magnetic moment expressing the magnetic moment of baryons and nuclei.

$$\mu_N = eh/4\pi m_p$$
$$= 5.050787 \cdot 10^{-27} \text{ J/T}$$

where e = electron charge, h = Planck's constant, m_p = mass or proton.

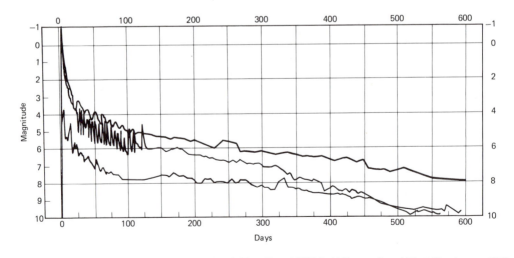

Novae light curves. Nova Aquilae 1918 (upper curve), Nova Persei 1901 (middle curve), and Nova Geminorum 1912 (lower curve) (all "fast" novae). (Jones 1961, p. 367, Fig. 109)

nucleic acid A group of organic compounds that includes DNA and RNA.

nucleon Either the proton or the neutron.

nucleon number The number of nucleons in a nucleus. It is equal to the mass number (A).

nucleoprotein Any of the class of proteins associated with nucleic acids.

nucleoside A segment of a nucleic acid molecule, consisting of a base attached to a pentose sugar.

nucleosome A repetitive segment of DNA consisting of 140 DNA base pairs folded around 8 histone molecules and a linker DNA to which a histone molecule is attached, linking the nucleosome with the next one. See **chromosome.**

nucleosynthesis The formation of atomic nuclei during the Big Bang, in stellar cores, or during supernova explosions. See **element formation.**

nucleotide A segment of a nucleic acid molecule, consisting of a pentose sugar with a base and a phosphate group attached to it. Molecular masses for nucleotides common to DNA and RNA (add

15.999 u for RNA) are: adenosine monophosphate (AMP, $C_{10}H_{14}N_5O_7P$), 347.224 u; cytidine monophosphate (CMP, $C_9H_{14}N_3O_8P$), 323.199 u; guanosine monophosphate (GMP, $C_{10}H_{14}N_5O_8P$), 363.224 u. Molecular mass of thymidine monophosphate (TMP, $C_{10}H_{15}N_2O_8P$) (DNA only), 322.211 u; of uridine monophosphate (UMP, $C_9H_{13}N_2O_9P$) (RNA only), 324.184 u.

nucleus *(Physics)* The central part of the atom, consisting of nucleons (protons and neutrons, except ^1H which consists of a single proton), where most of the mass of the atom is concentrated. Radius = 10^{-15} m; density = 10^{14} g/cm^3. *(Astronomy)* The central portion of a cometary head, consisting of a mixture of frozen gases (H_2O, CO_2, HCN, CH_3CN, CN, CO, etc.) and 25% silicate and metal particles. Density = 1.1 g/cm^3; radius = 1–10 km (?). *(Biology)* The inner part of a cell, containing the chromosomes and bound by the nuclear membrane.

nuclide Any of the set of atoms characterized by a specific number of protons and of neutrons in the nucleus and by a specific nuclear energy level. See **isomer.**

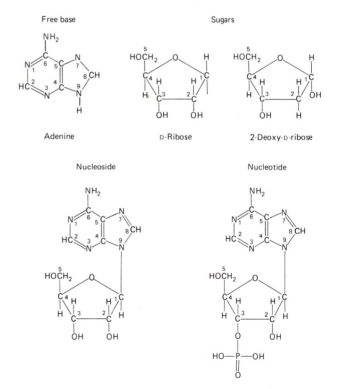

Nucleic acids. An example (adenine) of base linkages.

nuée ardente A cloud of hot water vapor and other gases (mainly CO_2 and SO_2) containing volcanic ash and pyroclastics. Its relatively high density allows it to roll downslope at high speed (>50 km/h) and for a considerable distance.

numbers The major classes of numbers are: **1. natural numbers** The set of positive integers. **2. integers** The set of positive and negative integers plus zero. **3. rational numbers** The set of numbers that can be expressed as a ratio of integers. **4. irrational numbers** The set of number that cannot be expressed as a ratio of integers. **5. real numbers** The set of rational and irrational numbers. **6. imaginary numbers** The set of real numbers each multiplied by $(-1)^{1/2}$. **7. complex numbers** The set of numbers consisting of a real number plus an imaginary number. **8. hypercomplex numbers (quaternions)** The set of numbers consisting of complex numbers generated in three dimensions. **9. cardinal numbers** A set of numbers without regard to their order. **10. ordinal numbers** A set of numbers characterized by their order.

nunatak An isolated bedrock proturberance extending above the surface of a glacier.

Nusselt number (N_{Nu}) A dimensionless number relating the heat transfer across a solid surface in contact with a fluid to the heat conductivity of the fluid.

$$N_{Nu} = hL/k$$

where h = heat transfer coefficient, L = characteristic dimension of the surface, k = thermal conductivity of the fluid.

nutation A slight, periodic but irregular motion of the Earth's axis superimposed upon the precessional motion, caused by the changing attractions of the Moon and the Sun. Amplitude = ± 9 seconds of arc from the mean position; period = 18.61 tropical years.

O

o- Ortho- *(Chemistry)*.

obduction The overthrusting of oceanic crust on the leading margin of a continental plate.

object wave See **hologram.**

oblate Characterizing an ellipsoid of revolution flattened along its axis of rotation. Flattening (f)

$$f = (a - c)/a$$

where a = equatorial diameter, c = polar diamter ($a > c$). See **ellipsoid of revolution.** Cf. **prolate.**

oblateness The degree of flattening of an ellipsoid of revolution or of a celestial or other body.

oblique fault See **oblique-slip fault.**

oblique-slif fault A fault in which slip has both a strike and a dip component.

obsidian A dark-colored, compact volcanic glass.

occluded front The intersection with the ground of a sloping surface separating a cold air mass from a less cold air mass, with warm air above both. Cf. **cold front, stationary front, warm front.** **1. occluded cold front** A front resulting when a colder air mass underrides a less cold air mass, with warm air above both. **2. occluded warm front** A front resulting when a cold air mass overrides a colder air mass, with warm air above both.

occultation The complete or partial obscuration of a celestial body by the Moon or a planet.

ocean An expanse of seawater $>10^6$ km in surface. The major oceans are (area and mean depth in parenthesis): Pacific ($166.241 \cdot 10^6$ km^2, 4188 m); Atlantic ($86.577 \cdot 10^6$ km^2, 3736 m); Indian ($73.427 \cdot 10^6$ km^2, 3872 m); Arctic ($12.257 \cdot 10^6$ km^2, 1117 m). Total area of oceans and seas = $362.033 \cdot 10^6$ km^2; mean depth = 3729 m; volume = $1349 \cdot 10^6$ km^3; mean depth of the oceanic basins = 4500 m.

ocher A pulverulent iron oxide, most commonly limonite (yellow ocher) or hematite (red ocher).

octahedrite Any of the iron meteorites consisting of kamacite, taenite, and 6.5–16% Ni, characterized by the presence of Widmanstätten figures. Oc-

tahedrites represent 75.4% of the iron meteorites, or 4.3% of all meteorites. See **Meteorites*.**

octet 1. The set of eight valence electrons in a valence subshell, providing the least chemically reactive electronic structure. 2. The eight baryons in the SU(3) symmetry (n, p, Λ^0, Σ^-, Σ^0, Σ^+, Ξ^-, Ξ^0).

octupole Two electric or magnetic quadrupoles with opposite polarities and separated by a small distance.

OD Ordnance Datum.

odd-even nucleus Defining a nucleus with an odd number of protons and an even number of neutrons.

odd-odd nucleus Defining a nucleus with an odd number of protons and an odd number of neutrons.

Oe Oersted.

oersted (Oe) The CGS$_{emu}$ unit of magnetic field intensity. It is the intensity of the magnetic field at the center of a one-turn coil with radius = 1 cm through which a current of $1/2\pi$ abampere flows. It is also the intensity of the magnetic field that exerts the force of 1 dyne on a unit magnetic pole. 1 oersted = $1000/4\pi$ ampere/meter.

offlap The mode of succession of sedimentary layers deposited by a regressive sea, each layer covering less of the previously deposited layer. Cf. **onlap.**

ohm (Ω) The SI, MKS, and MKSΩ unit of electric resistance, equal to the resistance of a conductor in which a potential difference of 1 volt produces a current of 1 ampere.

$$\Omega = V/A$$

where V = volt, A = ampere.

Ohm's Law The law relating electric current, resistance, and voltage.

$$I = V/R$$

where I = electrical current, V = voltage, R = resistance.

-oid Suffix meaning *looking like, resembling.*

oligo- Prefix meaning *few, little.*

oligoclase A plagioclase, $Ab_{90}An_{10}-Ab_{70}An_{30}$.

oligomicit Defining a lake that overturns only occasionally.

oligotrophic Referring to a body of water with a deficiency of plant nutrients. Cf. **eutrophic.**

oligotypic Referring to an assemblage consisting of a few species usually represented by a large number of individuals.

olistho- Prefix meaning *slippery.*

olisthostrome A chaotic, heterogeneous sedimentary mass emplaced by subacqueous sliding on a slippery surface.

olistostrome Common misspelling for **olisthostrome.**

olivine An olive-green mineral, $(Mg,Fe)_2SiO_4$, the most common component of mafic and ultramafic rocks.

Olympus Mons The highest volcano on Mars, about 600 km across at the base and rising 25 km above the surrounding territory. Estimated age = $200 \cdot 10^6$ y.

omega (Ω^-) The heaviest baryon in the SU(3) symmetry decuplet. Mass 1672.45 MeV = 1.795434 u. See **Elementary particles*.**

omegatron A miniature mass spectrograph used for static analysis of gases in evacuated chambers.

omphacite A high-pressure, high-density clinopyroxene, $(Ca,Na)(Mg,Fe,Al)(SiO_3)_2$.

onlap The mode of succession of sedimentary layers deposited by a transgressive sea, each layer extending further inshore than the preceding one. Cf. **offlap.**

ontogeny The development of an organism from protocell or zygote to adult.

onyx A variety of chalcedony with parallel, colored bands.

oo- Prefix meaning *egg.*

oolite A sedimentary rock consisting of cemented ooliths.

oolith A spheroidal body, commonly 0.5–1 mm across, consisting of concentric layers of aragonite formed in warm, shallow, turbulent seawater.

Oort cloud The cloud of 10^{12} (?) comets surrounding the solar system to a distance of perhaps 10^5 AU = 2500 solar system radii = 1.6 l.y.

ooze A deep-sea sediment consisting of >30%

$CaCO_3$ as foraminiferal shells and coccoliths and 30% or less of deep-sea clays. See **deep-sea sediments.**

opal Hydrated silica, $SiO_2 \cdot nH_2O$.

open cluster Any of the 1000+ clusters of 15–300+ young (Population I) stars, 6.5–50 l.y. across, common in the Galaxy. Examples are the Pleiades and the Hyades. Mean age of open cluster stars = $60 \cdot 10^6$ y.

operations research The application of objective, quantitative methods to decision-making processes.

ophiolite A low-grade metamorphic assemblage of mafic and ultramafic rocks, including peridotite, gabbro, basalt, and serpentine.

ophitic Referring to a structure characteristic of diabase in which plagioclase crystals are embedded in pyroxene.

opposition The time when the Sun and an outer planet or the Moon lie at the opposite sides of the Earth. The Moon is at opposition each full Moon and during lunar eclipses.

optical binary Two independent stars that appear to form a binary system because of perspective.

optical depth (α) A measure of the opacity of a medium to the transmission of electromagnetic radiation from a source O (or reference point O) to a point x.

$$I_x = I_0 e^{-\alpha x}$$

where I_x = radiation at point x, I_0 = radiation at point O, α = absorption coefficient, x = distance between I_0 and I_x. Cf. **attenuation.**

optic angle The angle between the two optic axes of a biaxial crystal.

optic axis The optical axis normal to the plane of the circular section of the indicatrix.

orbicular Describing an igneous rock texture exhibiting orbicules.

orbicule A spheroidal arrangement of crystals, up to several centimeters across, in an igneous rock.

orbital Any of the quantum states of an electron characterized by a specific set of n, l, and m_l quantum numbers. See **quantum number, shell, subshell.**

orbital angular momentum (l) The angular momentum of an orbiting electron. It is characterized by the quantum number l, which may assume the values 0, 1, 2, 3, 4, 5, ... $(n - 1; n =$ principal quantum number), corresponding to the letters s, p, d, f, g, h,.... See s, p, d, f.

orbital elements The six parameters needed to specify the position and path of an orbiting celestial body in a two-body system with known masses. They are: 1, length a of semimajor axis; 2, eccentricity $e = c/a$, where c = semidistance between the two foci, a = length of semimajor axis; 3, inclination i of orbital plane on plane of ecliptic; 4, longitude Ω of ascending node (=angular distance from vernal equinox to ascending node); 5, orbital orientation ω = angular distance from ascending node to periapsis measured along the direction of motion; 6, time t of periapsis passage. If the sum of the masses is unknown, the orbital period T is also needed to specify the orbits.

order *(Physics, Chemistry)* The state of zero entropy. *(Biology)* A rank in taxonomic classification below class and above family.

order of reaction The sum of the exponents of the concentrations in a rate equation. First-order reaction: only one atomic or molecular species participates (e.g. decomposition of a compound). Second-order reaction, two atomic or molecular species participate. Third-order reaction: three atomic or molecular species participate.

ordinary chondrite A common type of stony meteorite consisting of olvine, pyroxene, and plagioclase chondrules in a microcrystalline matrix of the same minerals. Ordinary chondrites represent 78.9% of the stony meteorites or 67.6% of all meteorites. See **Meteorites***.

Ordnance Datum (OD) The mean sea level at Newlyn, Cornwall, England, to which elevations in British Ordnance Survey maps are referred.

ore The rock from which economically important metals can be extracted.

ore mineral The portion of an ore containing the metallic mineral, as contrasted with the gangue.

organelle An organized structure within a cell, including nucleus, mitochondria, ribosomes, lysosomes, and plastids.

organometallic An organic compound in which a hydrogen atom is replaced by a metallic atom.

oriental emerald Green-colored gem corundum, Al_2O_3. See **Gems***.

oriental ruby Red-colored gem corundum, Al_2O_3. See **Gems***.

oriental sapphire Blue-colored gem corundum, Al_2O_3. See **Gems***.

oriental topaz Yellow-colored gem corundum, Al_2O_3. See **Gems***.

orogen A crustal belt that has been deformed into a mountain range.

orogene See **orogen**.

orogenesis See **orogeny**.

orogeny The formation of mountain ranges.

orographic Referring to a mountain or a mountain system.

ortho- Prefix meaning *straight. (Chemistry)* Prefix indicating the neighboring 1,2 positions in the benzene ring. Cf. **meta-, para-** *(Chemistry)*.

orthoclase A common rock-forming alkali-feldspar, $KAlSi_3O_8$.

orthogenesis The unidirectional evolution of a character within a lineage.

orthogeosyncline The geosyncline complex consisting of a volcanic eugeosyncline seaward and a nonvolcanic miogeosyncline landward.

orthohelium The states of the helium atom in which the spins of the two electrons are parallel. Cf. **parahelium**.

orthohydrogen The state of a hydrogen molecule in which the two protons have total spin equal to 1. It is an energy level higher than that in which the two protons have total spin equal to zero. Cf. **parahydrogen**.

orthopyroxene Any of the pyroxenes crystallizing in the orthorhombic system (e.g. bronzite, enstatite, hypersthene). Cf. **clinopyroxene**.

orthoquartzite A quartzite containing more than 90% quartz.

orthorhombic One of the 6 crystal systems, characterized by three mutually perpendicular axes of different unit length.

oscillation A cyclic change in a given parameter. **1. damped oscillation** An oscillation in the course of which energy of the oscillating system is either lost to the outside or degraded to other forms of energy such as heat. **2. forced oscillation** An oscillation induced on a system by an external driving force. **3. free oscillation** An oscillation of a system not constrained by external conditions.

oscillator A system capable of oscillating. E.g. an LC circuit.

oscilloscope A test instrument using a cathode-ray tube with a fluorescent screen to exhibit the values and waveforms of electrical quantities changing through time.

osculating orbit The truly elliptical orbit of a planetary body in the absence of perturbation by other planetary bodies.

osmosis The diffusion of a liquid or gaseous substance through a membrane until the concentration is the same on both sides.

osmotic pressure (Π) The pressure exerted by a pure solvent on a membrane separating it from the solution.

ostracum The outer, prismatic, calcareous layer of a molluscan shell.

Ostwald law The law relating the degree of ionization α to concentration in a dilute electrolytic solution.

$$K_\Lambda = C\alpha^2/(1 - \alpha)$$

where K_Λ = ionization constant, C = molar concentration, $\alpha = \Lambda_d/\Lambda_0$ = degree of ionization, Λ_d = equivalent conductance at dilution d, Λ_0 = equivalent conductance at infinite dilution.

Ostwald process A process to produce nitric acid. Ammonia is oxidized to NO, which is oxidized to NO_2, which is reacted with H_2O to produce HNO_3.

outer planets Jupiter, Saturn, Uranus, Neptune, and Pluto, with orbits beyond the asteroidal belt.

overprint The superposition of a more recent set of chemical or structural features on an older one.

overthrust A low-angle, large displacement fault.

overturned Describing a fold, one limb of which has been rotated beyond the vertical. Syn. **inverted, reversed.**

oxbow A hairpin meander leaving only a narrow land neck betwen the two branches.

oxidation The loss of one or more electrons by an atom or molecule. Cf. **reduction.**

oxidation-reduction potential The oxidative or reductive potential of an element or molecule with

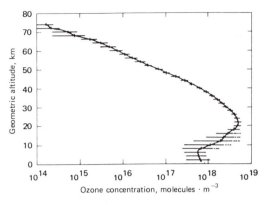

Ozone. The concentration of ozone as a function of altitude in the atmosphere (mid-latitudes). (U.S. Standard Atmosphere, 1976, p. 29, Fig. 31)

respect to the potential of the standard electrode taken as 0.

ozocerite A naturally occurring, dark brown to black paraffin.

ozone layer The layer in the lower stratosphere, centered at altitudes ranging from 30 km at the equator to 18 km at the poles, where ozone concentration is much higher (10^{-5} to 10^{-6}) than in the lower troposphere (10^{-8}). Ozone (O_3) is formed by reaction between molecular and monoatomic oxygen in the presence of any other molecule M (needed to conserve momentum):

$$O_2 + O + M \rightarrow O_3 + M$$

Atomic oxygen is produced during the day by photodissociation of O_2:

$$O_2 + h\nu \rightarrow 2O$$

O_3 is destroyed primarily by reaction with NO:

$$O_3 + NO \rightarrow O_2 + NO_2$$

and, secondarily, by photodissociation during the day:

$$O_3 + h\nu \rightarrow O_2 + O$$

See **atmosphere.**

P

π Symbol for the ratio of the circumference of a circle to its diameter. It is equal to 3.141 592 653 589 793. . . .

Π Osmotic pressure.

p 1. Momentum. 2. Pressure. 3. Principal (see *s, p, d, f*). 4. Proton.

p- Para- *(Chemistry).*

P 1. Parity. 2. Permeance. 3. Poise 4. Primary or pressure wave (see **P wave**).

Pa Pascal.

pack ice A broad expanse of sea ice. Pack ice forms over open polar seas by freezing of seawater. As the freezing point of freshwater is higher (0°C) than that of seawater (−1.872°C), pack ice has lower salinity than seawater (5‰ at low freezing rate, increasing to 10‰ for higher freezing rates). Pack ice averages 3 m in thickness.

packing fraction The ratio $(M - A)/A$, where M = nuclear mass in u, A = mass number.

paedo- Prefix meaning *child*.

paedogenesis See **neoteny.**

paedomorphism See **neoteny.**

paedomorphosis See **neoteny.**

pahoehoe *(Hawaiian)* Basaltic lava surface exhibiting a ropy surface.

paired electron 1. An electron paired with another electron of opposite spin occupying the same orbital. 2. One of two electrons that form a covalent bond between two atoms. Cf. **unpaired.**

pair production The simultaneous production of an electron and a positron by a photon with energy $\geqslant 1.022$ MeV ($= 2m_ec^2$, where m_e = rest mass of electron, c = speed of light), passing through the strong electrostatic field surrounding an atomic nucleus (the nucleus is needed to conserve momentum).

palaeo- See **paleo-.**

palagonite A mixture of smectite, phillipsite, and Fe–Mn oxides resulting from the alteration of basaltic glass.

palagonitization Formation of palagonite.

paleo- Prefix meaning *old, ancient.*

paleothyology The study of ancient storms as revealed by sedimentological features.

Paleozoic The geologic era following the Proterozoic and preceding the Mesozoic. It ranges from 590 to 248 million years ago and is subdivided into the following periods (age of boundaries in million years): 590/Cambrian/505/Ordovician/438/Silurian/408/Devonian/360/Carboniferous/286/Permian/248.

pali A high cliff (Hawaii).

palin- Prefix meaning *again, anew.*

palimpsest (Literally, *scraped again,* as of old manuscripts written on vellum that were scraped clean of previous writing for reuse.) Referring to a younger metamorphic texture or structure superimposed upon an older one; relict sediments on the continental shelf; or a new drainage pattern superimposed upon an older one.

palingenesis *(Geology)* Formation of new magma by melting of previously formed igneous rocks. *(Biology)* Recapitulation of phylogeny during ontogeny.

palinspastic (Literally, *stretched again.*) Referring to a geological cross section in which folded rocks have been stretched to their original geometry at time of formation.

palisade A basaltic cliff.

pallasite Any of the group of stony-iron meteorites consisting of olivine (65 vol. %), Fe–Ni (30 vol.%), troilite (2.3 vol. %), schreibersite (1.2 vol. %), and chromite (0.4 vol. %). Pallasites are 33% of all stony-iron meteorites or 0.5% of all meteorites. See **Meteorites*.**

paludal Referring to a marsh.

palustrine Referring to a marsh.

palygorskite A clay mineral with the chain structure $Mg_2(Al,Fe)_2(Si_2O_5)_4(OH)_2 \cdot 4H_2O$. Syn. **attapulgite.**

palynology The study of pollen and spores.

palynomorph Any of the resistant organic structures (pollen, spores, dinoflagellate cysts, acritarchs, etc.) remaining after acid treatment of vegetable remains.

Pangea The supercontinent that existed in Late Paleozoic time and included most of the continental lithosphere of the Earth.

Panthalassa The world ocean that surrounded Pangea.

para- Prefix meaning *beside, derived from, by.* *(Chemistry)* Prefix indicating the opposite 1,4 positions in the benzene ring. Cf. **meta-, ortho-,** *(Chemistry)*

parabola See **conic sections.**

parachor (P) A quantity proportional to molecular volume.

$$P = M\gamma^{1/4}/(\rho - \rho_0)$$

where M = molecular mass, γ = surface tension, ρ = density of the liquid, ρ_0 = density of the vapor in equilibrium with the liquid.

paraconformity An unconformity between successive, parallel beds representing a significant time interval of nondeposition.

paragenesis The successive formation of associated minerals in the process of mineralization.

paragenetic Referring to paragenesis.

parageosyncline A geosyncline within a craton.

parahelium The states of the helium atom in which the spins of the two electrons are antiparallel. Cf. **orthohelium.**

parahydrogen The state of a hydrogen molecule in which the two protons have total spin equal to zero. It is an energy level lower than that in which the two protons have total spin equal to 1. Cf. **orthohydrogen.**

paralectotype Any of the syntypes other than the lectotype.

paralic Referring to the marginal marine environment where conditions are brackish or alternating beween marine and continental.

paralimnion The marginal environment of a lake, extending from the shore to the maximum depth at which vegetation can root.

parallax The parallax of an object O is the angle $A\widehat{O}B$ where A and B are two different points of observation separated by the base AB. **1. diurnal par-**

allax Parallax of a celestial body resulting from the change in the observer's position due to the rotation of the Earth. The equatorial diameter is the maximum length of the terrestrial base. **2. annual parallax** One half of the angle by which a nearer star appears to be displaced with respect to the backdrop of distant stars as the Earth moves 180° along its orbit. It is equal to the angle by which 1 AU is subtended at the distance of the star. **3. secular parallax** Apparent and continuously increasing displacement of stars resulting from the motion of the Sun through space.

parallel Any of the circles on a sphere, parallel to the equator.

parallel circuit A circuit in which all components have their positive terminals connected to a common positive line, and their negative terminals connected to a common negative line.

parallel evolution The evolution, not necessarily contemporaneous, of different lineages into similar morphotypes because of the development of similar environmental conditions.

paramagnetic Possessing the property of paramagnetism.

paramagnetism The property of a substance whose atoms or molecules have permanent magnetic moments that can be oriented parallel to each other by the application of a magnetic field. The magnetic energy induced in a paramagnetic substance is much lower than its thermal energy at room temperature, so that the induced orientation of the atoms or molecules is lost upon removal of the applied field.

parasitic *(Biology)* Referring to an organism living by parasitism. *(Geology)* Referring to a secondary volcanic cone on the side of a larger one.

parasitism The mode of existence of an organism that derives its food and/or other advantages by close association with another organism. Cf. **symbiosis.**

parastratotype A secondary stratotype to supplement or clarify the holostratotype.

paratype A specimen, other than the holotype, used to supplement the description of a species or subspecies.

Parazoa A subkingdom of lower animals that includes the phyla Placozoa, Porifera, and Archaeocyatha. See **Taxonomy*.**

parity (P) A number describing the symmetry of a system with respect to reflection of all three space coordinates through the origin. $P = +1$ for a symmetric system; $P = -1$ for an antisymmetric system.

Parkes process A process to extract silver from lead by dissolving it in molten zinc.

parsec (pc) Parallax-second, the distance at which 1 AU subtends 1 second of arc. 1 pc = $(360 \cdot 60 \cdot 60)/2\pi$ AU = 206,264.806 AU = 3.261633 l.y. = $30.856772 \cdot 10^{12}$ km.

partial derivative See **calculus.**

parton A pointlike constituent of a nucleon evidenced by collision effects with high-energy leptons. A parton is identifiable with a quark or a gluon.

pascal (Pa) A unit of pressure equal to 1 newton/m^2 or 10 dyn/cm^2.

Pascal's principle "Pressure applied at any point in a confined fluid is distributed undiminished throughout the fluid at all points and to the confining surface."

Paschen series A series of lines in the infrared region of the spectrum of hydrogen, representing transitions beweeen $n > 3$ and $n = 3$ energy levels, where n is the principal quantum number. Energies range from 0.6612 to 1.5113 eV; corresponding wavelengths range from 1.8751 to 0.82036 μm. Cf. **energy level.**

passive margin A continental margin that is moving away from a spreading axis. Cf. **active margin.**

patch reef A small, mound-like reef in a lagoon.

Pauli exclusion principle "No two fermions of the same kind and belonging to the same system (atom, molecule, or such a larger, internally bound system as a metal bar) can exist in the same quantum state as specified by the set of quantum numbers, but must differ in at least one quantum number."

pc Parsec.

PDB Peedee belemnite. An isotopic standard for oxygen and carbon, consisting of a single, ground specimen of *Belemnitella americana* from the Peedee formation of South Carolina. $^{18}O/^{16}O$ of PDB calcite = $^{18}O/^{16}O$ of SMOW + 30.86‰ ; $^{18}O/^{16}O$ in CO_2 from PDB calcite reacted at 25°C with H_3PO_4 = $^{18}O/^{16}O$ of CO_2 gas equilibrated at 25°C with SMOW + 0.22‰ .

PDR Precision depth recorder, a precision echosounder for continuous recording of seafloor depth.

pearl A concretionary, spherical or subspherical body consisting of concentric aragonitic laminae alternating with thin organic layers, produced by various marine and freshwater molluscan species.

peat A compacted, porous mass of vegetable matter that has undergone early diagenesis toward carbonization.

peat bog A bog in which peat has accumulated.

peat coal Diagenized vegetable matter intermediate between peat and lignite.

peat moss Peat formed from moss (especially Sphagnum).

pebble A rock fragment ranging in size from 4 to 64 mm.

peculiar motion Motion of a star relative to neighboring stars.

pedalfer A classic name for a soil rich in Fe oxides.

pediment The slightly upward concave, gently sloping accumulation of sediment at the foot of a mountain range.

pedion An open crystal structure consisting of a single face with no symmetric equivalent.

pedo- See **paedo-.**

pedocal A classic name for a soil exhibiting a concentration of carbonates.

pedogenesis The formation of soil.

pedosphere The totality of soils on Earth.

pegmatite A very coarse igneous rock, usually of granitic composition, consisting of crystals more than 1 cm across, formed in the pegmatitic stage of magma crystallization.

pegmatitic stage The stage of pegmatite formation at the end of magmatic crystallization when residual fluids are enriched in volatiles. Cr. **pneumatolytic stage.**

pelagic Marine, oceanic.

Peléean cloud. See **nuée ardente.**

pelecypod Any of the molluscs belonging to the class Bivalvia.

pelite A sedimentary rock consisting of clay and/or clay-size carbonate particles. Cf. **lutite.**

pelitic Referring to pelite.

pellet A small, rounded or ellipsoidal accretion of mud or clay-size carbonate particles.

pelletoid Referring to a sediment rich in pellets.

pelmicrite A limestone consisting of pellets and recrystallized carbonate mud (micrite).

pelsparite A limestone consisting of pellets and of carbonate mud recrystallized more coarsely (sparite, crystal size >10 μm) than micrite.

pendulum (simple) Period T of small-amplitude oscillation:

$$T = 2\pi(L/g)^{1/2}$$

where L = length of pendulum, g = gravitational acceleration.

pene- Prefix meaning *almost.*

penecontemporaneous Formed at about the same time.

peneplain A land surface eroded to almost a plain.

peneplanation The process of peneplain formation.

penumbra 1. The region of an object that is not totally protected from a light source by an intervening opaque body, such as the portion of the Earth's surface where a solar eclipse is partial. 2. The less dark area of a sunspot surrounding the darker core (umbra). Cf. **umbra.**

peptide bond A bond between adjacent amino acids formed by reaction between the $-NH_2$ group of one and the $-COOH$ group of the other, with the release of one H_2O molecule.

per- Prefix meaning *through* (as in *permeate*) or *more* (as in *peroxide*).

peralkaline Defining an igneous rock with Na + K > Al (as oxides).

peraluminous Referring to an igneous rock with Na + K < Al (as oxides).

peri- Prefix meaning *around, near.*

periapsis The apsis closest to the center of gravity of an orbiting body. Syn. **perihelion** for the solar planets. Cf. **apoapsis, apsides.**

periastron The point in a planetary orbit, or in the orbit of a secondary in a binary system of stars, closest to the star around which the planet or the secondary is revolving. Syn. **perihelion** for solar planets or other objects in circumsolar orbits. Cf. **apastron.**

periclase The mineral MgO.

peridotite A plutonic rock consisting mainly of olivine and pyroxene.

perigee The orbital point of the Moon or an artificial circumterrestrial satellite closest to the Earth. Cf. **apogee.**

periglacial 1. Referring to an area adjacent to a glacier or an ice sheet. 2. Referring to processes taking place in an area adjacent to a glacier or an ice sheet.

perihelion The orbital point of a planet, comet, or asteroid closest to the Sun.

perihelion distance (q) The distance between the Sun and the perihelion of an orbiting body or the vertex of a parabolic orbit.

period *(Physics)* The duration of 1 cycle in a cyclic phenomenon.

$$T = 2\pi/\omega$$
$$= 1/\nu$$

where T = period, ω = angular frequency, ν = frequency. *(Chemistry)* Any of the rows of the Periodic Table of the Elements. *(Geology)* A division of geologic time longer than an epoch but shorter than an era, during which the rocks of a system are formed.

Periodic Table of the Elements* An arrangement of the elements in rows (periods) and columns (groups), exhibiting increase in the number of electrons in the valence shell along the rows and increase in number of shells down column. Syn. **Mendeleev's Table.**

periostracum The thin conchiolin layer covering the outer surface of a molluscan or brachiopod shell.

perlite A volcanic glass that cracked during cooling because of contraction, forming small spherules.

perlitic Referring to the structure of perlite.

permafrost The permanently frozen surface of the Earth in high latitude regions.

permeability (μ) *(Physics)* The effect of a medium on a magnetic field. **1. absolute permeability** (μ) The ratio of magnetic induction **B** to the strength of the applied magnetic field **H**:

$$\mu = B/H$$

It is expressed in henry/meter (SI). The permeability of vacuum (μ_0) is equal to $4\pi \cdot 10^{-7}$ henry/meter. **2. relative permeability** (μ_r) The ratio of the absolute permeability of a material to that of vacuum:

$$\mu_r = \mu/\mu_0$$

Relative permeability is ~ 1 for most substances but it may reach as high as several thousand for ferromagnetic materials within specific ranges of **H**. *(Geology)* The capacity of a rock to allow fluids through, depending on the number, geometry, size, and interconnections of the pores. Permeability in sediments ranges around 1 to 5 darcys; in sedimentary rocks in ranges from 1 darcy to 0.01 millidarcy (chert). See **darcy**.

permeability constant (μ_0) The ratio of magnetic induction to magnetic field strength in vacuo:

$$\mu_0 = \mathbf{B/H}$$
$$= 4\pi \cdot 10^{-7} \text{ henry/meter}$$
$$= 12.5663706144 \ldots \text{ henry/meter}$$

permeability of free space See **permeability constant**.

permeability of vacuum Syn. **permeability constant**.

permeance (P) The reciprocal of reluctance:

$$P = \Phi/mmf$$

where P = permeance, Φ = magnetic flux, mmf = magnetomotive force. It is expressed in henry (SI).

permittivity (ϵ) The effect of a substance on an electric field.

$$\epsilon = \kappa\epsilon_0$$
$$= \sigma/E$$

where ϵ = permittivity of substance, κ = dielectric constant of substance, ϵ_0 = permittivity constant, σ = charge density on opposite, parallel plates, E = electric field intensity across substance beween plates.

permittivity constant (ϵ_0) The inverse of the constant needed in the SI, MKS, and CGS$_{esu}$ systems to reduce to unity the force between two unit electric charges at a unit distance.

$$F = (1/\epsilon_0)(qq/4\pi r^2)$$

from which

$\epsilon_0 = qq/F4\pi r^2$
$= 1/4\pi$ statC^2/dyn cm^2 (CGS$_{esu}$)
$= 1/\mu_0 c^2$ (SI)
$= 8.854187817 \ldots \cdot 10^{-12}$ C^2/N m^2 or F/m (SI)

In the preceding, F = force, q = charge, r = distance between the two charges, μ_0 = permeability constant, c = speed of light.

permittivity of free space Syn. **permittivity constant**.

permittivity of vacuum Syn. **permittivity constant**.

Permocarboniferous The time encompassing the Carboniferous and Permian periods, ranging from 360 to 248 million years B.P. See **Geological time scale***.

permutation A specific arrangement of the elements of a set or of a portion of a set. Cf. **combination**.
1. permutations of *n* elements:

$$P = n!$$

2. **permutation of *n* elements *k* at a time without repetition:**

$$P(n,k) = n!/(n - k)!$$

3. **permutation of *n* elements *k* at a time with repetition:**

$$P_r(n,k) = n^k$$

perovskite The mineral CaTiO$_3$.

perovskitite A dense (zero pressure $\rho = 4.2$ g/cm^3) igneous rock consisting of minerals with the perovskite structure and believed to be the major constituent of the Earth's mantle below 600 km of depth. Cf. **garnetite**.

Perseids A major meteor shower. See **meteor shower**.

perthite An alkali feldspar with parallel laminae of microcline and albite, resulting from exsolution. See **cryptoperthite, microperthite**.

perthitic Describing the structure exhibited by perthite.

petrifaction The fossilization process transforming organic matter into calcium carbonate, silica, or other minerals.

petro- Prefix meaning *stone*.

petrofabric See **fabric**.

petrogenesis The formation of rocks, especially that of igneous and metamorphic rocks.

petroglyph A rock carving.

petrography The description and classification of rocks.

petroleum fraction Any of the hydrocarbon fractions of increasing molecular mass obtained from the fractional distillation of crude oil at increasing temperature. See **Petroleum fractions***.

petrology The study of the formation and evolution of igneous and metamorphic rocks.

Pfund series A series of lines in the infrared region of the spectrum of hydrogen, representing transitions between $n > 5$ and $n = 5$ energy levels, where n is the principal quantum number. Energies range from 0.1662 to 0.5441 eV; corresponding wavelengths range from 7.4578 to 2.2788 μm.

pH A measure of the hydrogen-ion activity of a chemical system = hydrogen-ion concentration for dilute solutions.

$$pH = -\log_{10} [H^+]$$

where $[H^+]$ = hydrogen-ion concentration in moles/liter. See **pH scale***.

phage Any of the viruses that infect bacteria. Syn. **bacteriophage.**

phaneritic Syn. **macrocrystalline.**

Phanerozoic The geologic time since the appearance of abundant Metazoa with exoskeleta, i.e. from the beginning of the Cambrian (590 m.y. B.P.) to the present. Cf. **Cryptozoic.**

phase *(Physics)* The portion of a cycle in a periodic phenomenon that has passed a reference point (symbol ϕ).

$$Q = Q_{max} \sin (\omega t + \phi)$$

where Q = a sinusoidally varying quantity; Q_{max} = maximum value achieved by Q during a cycle; ω = angular frequency, t = time, ϕ = phase angle. Two waves are in phase when $\phi = n2\pi$, where n is an integer (which includes 0); they are on opposite phase when $\phi = n\pi$. *(Chemistry)* A homogeneous portion of a chemical system characterized by specific physical or chemical characteristics.

phase angle (ϕ) *(Physics)* The angle between the rotating vectors generating two waves of the same frequency. *(Electricity)* The angle between voltage and current in ac circuits. *(Astronomy)* The angle Sun-Moon-Earth or Sun-Planet-Earth.

phase constant The imaginary component of the propagation constant. See **propagation constant.**

phase diagram A graph showing the boundaries of the stability fields of the various phases of a system as functions of temperature, pressure, and composition.

phase difference (ϕ) The difference in phase between voltage and current in ac circuits:

$$\phi = \tan^{-1} (X_L - X_C)/R$$
$$= X/R$$

where $X_L = 2\pi fL$ = inductive reactance; $X_C = 1/2\pi fC$ = capacitive reactance; f = frequency; R = resistance; X = reactance.

phase equilibrium The conditions of temperature, pressure, and composition under which the different phases of a system can coexist at equilibrium.

phase rule The equation

$$F = C + 2 - P$$

describing the relationship between the degrees of freedom F (= number of independent variables), the number of components C, and the number of phases P in a heterogeneous system at equilibrium.

phase space A $2n$-dimension space, one for each generalized coordinate and one for each of the conjugate momenta (n = degrees of freedom).

phase transition The change of a substance from one phase to another.

phase velocity The velocity at which the phase of a wave is traveling. E.g. sea waves (the water is not traveling, the wave is).

phasor A vector rotating counterclockwise around an origin O on a plane, representing a sinusoidally varying quantity. The vector's length represents the magnitude and its angle with the x axis, measured counterclockwise, represents the phase.

phe Phenyl.

phenoclast A conspicuous rock fragment in a sedimentary rock.

phenocryst A conspicuous crystal in a porphyritic matrix.

phenols Aryl hydroxides. E.g. phenol, C_6H_5OH.

phenotype The physical appearance of an organism resulting from interaction of genotype with environment.

phenyl The ring radical $-C_6H_5$.

phi (ϕ) The negative logarithm in base 2 of the diameter of a particle. Thus a particle 1/64 mm in size has a phi value of 6, and a pebble 64 mm across has a phi value of -6.

phi grade scale A size scale for particles based on phi units.

phillipsite A zeolite mineral, $(K_2Na_2Ca)Al_2Si_4O_{12} \cdot$ 4–5H_2O.

phi unit The constant geometric interval of 1/2

between successive particle sizes in both the phi and Wentworth grade scales.

phloem The food-conducting tissue of vascular plants.

phon A unit of sound loudness, equivalent to the smallest audible difference in loudness. A difference of 10 phons doubles subjective loudness.

phonolite The microcrystalline, effusive equivalent of nepheline syenite, characterized by alkali feldspar, a feldspathoid, and mafic minerals.

phonon The quantum of acoustic energy in an elastic medium, equal to $h\nu$, where h = Planck's constant, ν = acoustic wave frequency.

phosphate rock A sedimentary rock with a concentration of phosphatic minerals sufficient for economic usage.

phospholipid A family of amphipathic molecular species structurally based on glycerol, with a phosphate polar head and an alkyl acid, nonpolar tail. Phospholipids are important constituents of organic membranes.

phosphorescence Luminescence persisting after removal of the exciting source. Atoms are raised to a metastable state, or electrons are transferred to crystal defect loci, by an exciting process. Light is produced when the thermal energy of the system subsequently returns the atoms to their ground state and recombines the electrons with their carriers.

phosphorite A sedimentary rock with phosphate minerals [fluorapatite, $Ca_5(PO_4)_3F$; carbonate-fluorapatite or francolite, $Ca_5(PO_4CO_3)_3F$] in sufficient concentration to be of economic interest.

phot A unit of illumination, equal to 1 lumen/cm^2 = 10^4 lux.

photic zone See **euphotic zone.**

photocell See **photoelectric cell.**

photoconductivity An increase of electrical or electron-hole conduction exhibited by many nonmetal and metalloid solids, such as alkali and Ag halides, Ge and Si crystals, polymers, etc., when excited by electromagnetic radiation.

photoelectric cell An electronic device capable of developing an electrical output proportional to incident light.

photoelectric effect The release of electrons from a metal surface by the action of photons of appropriate frequency. The maximum kinetic energy K_{max} of the emitted electrons is given by Einstein's photoelectricity equation:

$$K_{max} = h\nu - W_0$$

where h = Planck's constant, ν = frequency of incident light, W_0 = work function of the given metal surface. Work functions of the elements range from 2.14 eV for cesium to 5.9 eV for selenium.

photolysis The dissociation of molecules by photons of appropriate energies.

photomultiplier tube A vacuum tube with a series of dynodes. Electrons released at the photocathode hit successive dynodes that are maintained at increasingly higher positive voltages, liberating increasing numbers of electrons. A cascade is thus developed, the magnitude of which is proportional to the energy of the original photon, the number of dynodes, and the magnitude of the accelerating voltages. Common types include 6–16 stages with 75–150 V difference between one stage and the next, a secondary electron yield of 3–5 electrons per impinging electron, and a transit time for the electrons of 10–100 ns. Materials for the dynode surfaces include BeO and Cs_3Sb, characterized by low work functions, or Cs_2O and GaP with negative electron affinity. A 10-stage dynode with a yield of 4 secondary electrons per impinging electron will produce $4^{10} = 10^6$ electrons at the anode from a single electron at the photocathode. The electron current is led into a 100 kΩ to 100 MΩ resistor and the voltage developed across it is amplified and measured.

photon (γ) The quantum of electromagnetic radiation. Mass = 0; spin = -1; energy = $h\nu$; momentum = $h\nu/c$; equivalent inertial mass = $h\nu/c^2$. See **Elementary particles*.**

photonuclear reaction A nuclear reaction (emission of a neutron or a proton, fission, isomeric transition, etc.) induced by γ-radiation of appropriate energy (>5 MeV). E.g. the reaction $^9Be(\gamma,n)^8Be$.

photosphere The 500-km-thick surface layer of the Sun, forming the Sun's visible surface. Temperature ranges from 8000 K at the bottom to 4000 K at the surface (average = 6000 K), density from 10^{-6} to 10^{-8} g/cm^3. The photosphere exhibits a granular surface and sunspots.

photosynthesis 1. The synthesis of chemical compounds by the action of light. 2. The production, mediated by chlorophylls and enzymes, of (mainly) carbohydrates and oxygen from CO_2 and water. The basic reaction is

$$CO_2 + H_2O + h\nu \rightarrow [CH_2O] + O_2$$
$$- 4.86 \text{ eV} (= 112 \text{ kcal/mol})$$

where $[CH_2O]$ indicates *carbohydrate*. Light ionizes chlorophyll. The energy of the liberated electrons is used to form ATP and NADP. In the more primitive Photosystem I, based on chlorophyll a and utilizing red light ($\lambda > 0.68$ μm), chlorophyll regains the lost electrons from the return of low-energy electrons. This system is used by the Cyanobacteria. In the more advanced Photosystem II, based on a mixture of chlorophyll a, b, c, or d and utilizing light with wavelength shorter than 0.68 μm, chlorophyll regains the lost electrons from water molecules at the chlorophyll site (2 electrons/molecule). The de-electronized water molecule splits into H^+ and OH^- radicals, with H^+ going to form NADPH and OH^- combining to form H_2O and O_2. Energy from ATP is used to synthesize carbohydrates from CO_2 and H_2O, a set of reactions that does not need light. Photosystems I and II combined are used by the Prochlorophyta and all higher photosynthesizers. Plants utilize only 2% of incident solar radiation, limiting factors being the ambient concentration of CO_2 and the concentration and intrinsic reaction rates of the enzymes involved. See **quantasome.**

photovisual magnitude See **magnitude.**

photovoltaic effect The generation of a voltage in a system energized by visible light or other electromagnetic radiation. See **solar cell.**

phreatic Pertaining to groundwater.

phreatic water Groundwater.

phyletic See **phylogenetic.**

phyletic evolution Evolution along a lineage.

phyletic gradualism Evolution by slow change along a lineage.

phyllite A metamorphic rock of grade intermediate between slate and schist.

phyllosilicate See **silicate.**

phylogenetic Pertaining to phylogeny.

phylogeny The evolutionary development of a group of organisms.

phylum The taxonomic rank above class and below kingdom.

physical libration The slight variation in the Moon's rotation caused by the attraction of the Earth on the 1.09-km-high bulge of the Moon that points toward the Earth. It results in a longitudinal displacement of $\pm 0.02°$ with a period of 1 year and a latitudinal displacement of $\pm 0.04°$ with a period of 6 years.

physiography The description of the morphology of the Earth's surface.

physiotope An area of uniform physicochemical conditions.

phyto- Prefix meaning *plant.*

phytocoenosis Plant community.

phytolith A microscopic mineral body secreted by a plant, usually consisting of opaline silica ($SiO_2 \cdot nH_2O$) or calcium oxalate ($CaC_2O_4 \cdot H_2O$).

phytoplankton Plant plankton. Cf. **zooplankton.**

pi bond (π) A covalent bond in which electron density is maximum above and below an axis joining the two nuclei.

pico- Prefix meaning 10^{-12}. Cf. **atto-, femto-, micro-, nano-.**

picrite A hypabyssal rock consisting mainly of olivine with pyroxene and biotite.

picritic Referring to a picrite.

piedmont A gently sloping area at the foot of a mountain or a mountain range.

piezoelectricity 1. The generation of a voltage in a crystal through the application of mechanical stress. 2. The generation of mechanical stress on a crystal through the application of a voltage.

pigeonite A clinopyroxene, $(Ca,Mg,Fe^{2+}) \cdot (Mg,Fe^{2+})(SiO_3)_2$.

pig iron The impure iron produced by a blast furnace and shaped into blocks or *pigs*. It contains up to 4% C, up to 2% Si, some P, and traces of S. Pig iron is purified to produce steel. See **steel.**

pillow basalt Basalt extruded under water and displaying pillow structure.

pillow structure The surface appearance of an igneous rock extruded under water, resembling a layer of closely spaced pillows.

pi meson See **pion.**

pinacoid An open crystal form consisting of two parallel faces. Cf. **pedion.**

pinacoidal class A crystal class in the triclinic system with the center as the only element of symmetry.

pingo An ice diapir, often tens of meters high, covered with soil or rock debris.

pion A triplet of nonstrange mesons: $\pi^\pm$, mass = 0.1498304 u, $t_{1/2}$ = $2.6030 \cdot 10^{-8}$ s; π^0, mass = 0.1448876 u, $t_{1/2}$ = $0.83 \cdot 10^{-16}$ s. Decay products: $\pi^+ \rightarrow \mu^+ + \nu_\mu$; $\pi^- \rightarrow \mu^- + \nu_\mu$; $\pi^0 \rightarrow 2\gamma$ (98.802%) or $\pi^0 \rightarrow e^+ + e^- + \gamma$ (1.198%). See **Elementary particles*, Hadrons—quark structure***.

Pirani gauge A device that measures vacuum from the resistance of a wire heated by an electric current.

pisolite 1. A sedimentary rock, usually carbonate, consisting of pisoliths. 2. A sedimentary rock consisting of accretionary lapilli.

pisolith An accretionary, spherical or subspherical body, usually calcitic or aragonitic, 1 to 10 mm in size.

pisolitic Referring to a pisolite.

pitch The physiological response to sound frequency. It is measured in mels.

pitchblende Massive uraninite (UO_2).

pK The negative logarithm of the ionization constant.
$$pK = -\log_{10} K$$
where K = ionization constant.

placer A beach, alluvial, or fluvial deposit of one or more heavy minerals.

plagioclase Any of the triclinic Na–Ca aluminosilicates, ranging in composition from 100% albite (Ab, $NaAlSi_3O_8$) to 100% anorthite (An, $CaAl_2Si_2O_8$), including

albite	90–100% Ab, 0–10% An
oligoclase	70–90% Ab, 10–30% An
andesine	50–70% Ab, 30–50% An
labradorite	30–50% Ab, 50–70% An
bytownite	10–30% Ab, 70–90% An
anorthite	0–10% Ab, 90–100% An

planck The SI and MKS unit of action (energy/frequency), equal to 1 joule/hertz.

Planck energy *(E_P)* The energy
$$kT = (hc^5/2\pi G)^{1/2}$$
$$= 1.221 \cdot 10^{28} \text{ eV}$$
$$= 1.311 \cdot 10^{19} \text{ u}$$
where k = Boltzmann constant, T = absolute temperature, h = Planck's constant, c = speed of light, G = gravitational constant.

Planck era The time from cosmological time t = 0 to cosmological time t = $5.390 \cdot 10^{-44}$ s, during which quantum gravity was dominant. Cf. **Planck time.**

Planckian The Planck era, ranging from cosmological time t = 0 to cosmological time t = $5.390 \cdot 10^{-44}$ s. It is followed by the Gamowian.

Planck length *(l_P)* The length
$$l_P = (Gh/2\pi c^3)^{1/2}$$
$$= 1.616 \cdot 10^{-35} \text{ m}$$
where G = gravitational constant, h = Planck's constant, c = speed of light.

Planck mass *(m_P)* The mass of a particle whose reduced Compton wavelength equals the Planck length:
$$m_P = (hc/2\pi G)^{1/2}$$
$$= 1.311 \cdot 10^{19} \text{ u}$$
$$= 2.177 \cdot 10^{-8} \text{ kg}$$
where h = Planck's constant, c = speed of light, G = gravitational constant.

Planck's constant *(h)* The quantum of action.
$$h = E/\nu$$
$$= 6.626075 \cdot 10^{-34} \text{ J s}$$
$$= 4.135692 \cdot 10^{-15} \text{ eV s}$$
where E = energy, ν = frequency.

Planck's distribution law See **Planck's radiancy law.**

Planck's formula See **Planck's radiancy law.**

Planck's law See **Planck's radiancy law.**

Planck's radiancy law A law giving blackbody radiancy as a function of wavelength:
$$I_\lambda = (2\pi hc^2/\lambda^5)/(e^{hc/\lambda kT} - 1)$$
where I_λ = blackbody radiancy at wavelength λ, c = speed of light, h = Planck's constant, λ = wavelength, k = Boltzmann constant, T = absolute temperature.

Planck's second See **Planck time.**

Planck temperature *(T_P)* From Planck energy E_P = kT, T_P = Planck energy divided by Boltzmann constant k. It is equal to $1.417 \cdot 10^{32}$ K.

Planck time *(t_P)* Planck length divided by the speed of light.
$$t_P = (Gh/2\pi c^5)^{1/2}$$
$$= 5.390 \cdot 10^{-44} \text{ s}$$
where G = gravitational constant, h = Planck's constant, c = speed of light.

planet A celestial body orbiting around a star and having a mass insufficient to initiate and sustain nuclear reactions in its core. See **Planets—physical data.**

plane table A surveying instrument consisting of a board with an alidade mounted on a tripod.

planetary alignment Approximate alignment of the outer planets occurring every 178 years. The last such alignment occurred in 1981–1982.

planetary system A system of planets accompanying a star. Stars derive from the gravitational collapse of dense interstellar clouds, which consist of atomic, ionic, and molecular species as well as of μm-size particles of predominantly Fe–Ni metal and Fe–Mg silicates (see **interstellar cloud**). Depending upon the initial conditions of the parent cloud (mass, density, internal energy, turbulence, local angular momenta and the magnitude of their resultant, etc.) collapse may produce a double or multiple star system or a single star with a complement of planets. In the latter case, collapse includes the formation of a planetary ring, to which angular momentum is transferred from the central body via Alfvén waves, and an early episode of high luminosity that dissipates much of the energy of accumulation and sweeps gases from the inner to the outer regions of the ring. Condensation of Fe may precede that of the Fe–Mg silicates at the higher temperatures and hydrogen pressures prevailing in the inner region of the planetary ring, while the reverse may obtain in the outer region (cf. **Planetary system formation—condensation sequence***). Clumping of μm-size dust particles leads to meter-size bodies which aggregate to form planetesimals ranging in size from <1 km to 100 km. These are concentrated in specific orbital bands within the ring. Sweeping of the smaller planetesimals by the larger ones within each band leads to the formation of the planets. The formation of the Sun and its planetary system occurred within $\sim 1 \cdot 10^6$ y after the last episode of nucleogenesis in the solar neighborhood, as indicated by the occurrence in meteorites of anomalous isotopic abundances for ^{26}Mg (from ^{26}Al, $t_{1/2} = 720,000$ y), ^{107}Ag (from ^{107}Pd, $t_{1/2} = 6.5 \cdot 10^6$ y), and ^{129}Xe (from ^{129}I, $t_{1/2} = 15.7 \cdot 10^6$ y). Such anomalies would not be present had not the parent nuclides been trapped within meteoritic matter before undergoing decay. Internal differentiation of the planets followed their final accumulation, each developing a core, mantle, and crust. Additional heat was released by this process (2500 J/g for the Earth) which, together with the heat of accumulation and that produced by the decay of long-lived radioactive nuclides (see **Geothermal energy***), has maintained the mantles of the larger planets

($\geqslant$Venus) in a convective mode to this day. The atmospheres of the inner planets are secondary, having been developed by degassing of the interior during differentiation, while those of the outer planets are primary. Satellites and ring systems are believed to have originated from circumplanetary rings of planetesimals in the same way the planets originated from circumsolar rings. The four outermost satellites of Jupiter (Ananke, Carme, Pasiphae, and Sinope) and the outermost satellite of Saturn (Phoebe), all with highly inclined orbits and retrograde revolutions, as well as Triton, may have originated as independent planetoids that were then captured. See **Planets—atmospheres***, **Planets—physical data***, satellite, Satellites***.

planetary wave A major atmospheric wave of long wavelength, significant amplitude, and westward motion.

planetary wind Any of the major wind systems of the Earth.

planetesimal Any of the large number of small bodies (<100 km across) consisting of silicate and metal microcrystals embedded in ices, thought to exist in the early stage of the development of a planetary system.

planetoid A small planetary body not otherwise characterized.

planispiral Having the shape of a spiral coiled along a plane. Cf. **trochoidal.**

plankton Collective name for the organisms that are freely floating, but not actively swimming, in marine or freshwater bodies. Cf. **phytoplankton, zooplankton.**

planosol A soil on flat or almost flat topography consisting of a leached layer underlain by hardpan.

Plantae One of the 5 kingdoms. It includes all higher plants. See **Taxonomy***.

plasma A neutral, highly ionized gas with the free electrons balancing the charges of the positive ions.

plastid Any of the self-replicating organelles in plant cells containing chlorophyll (chloroplasts), other pigments (chromoplasts), or no pigments (leukoplasts, converting glucose to starch). Plastids average 5 μm across and have a variety of shapes. Plastid DNA, like bacterial DNA, is bare of proteins. Plastids may have originated as procaryota symbiotic with early eucaryotes. See **chloroplast.**

PLATE 165 **PNEUMATOLYSIS**

plate Any of the major slabs of the terrestrial lithosphere, 100–150 km thick, resting on the asthenosphere. The major plates are: Eurasian, African, Indian, Pacific, North American, South American, Nazca, and Antarctic. Minor plates are Arabian, Somali, Philippines, Juan de Fuca, Cocos, and Caribbean.

plate margin The line of contact between adjacent plates. The principal plate margin types are: **1. spreading margin** (e.g. Mid-Atlantic Ridge, separating the American plates from the Eurasian and African plates); **2. subduction margin** (e.g. the Peru-Chile trench, separating the Nazca and South American plates); **3. collision margin** (e.g. the Himalayas, separating the Indian and Eurasian plates); and **4. transform margin** (e.g. the San Andreas fault, separating the Pacific and N. American plates).

plateau A flat and broad land expanse elevated with respect to the surrounding territory.

plateau basalt An accumulation of semi-horizontal basaltic flows produced by fissure eruptions, forming a plateau.

plate tectonics The study of the motion and interactions of the lithospheric plates through time.

Platonic year The length of the general precessional period of the Earth, equal to 25,800 y.

platykurtic Defining a distribution less peaked than normal. Cf. **leptokurtic.**

playa A flat, dry lake bed surface (SW United States).

Pleistocene The first period of the Quaternary sub-era, ranging from $1.6 \cdot 10^6$ to 10,000 y B.P. It is followed by the Holocene. See **Geological time scale*.**

Pleistogene The most recent geological sub-era, consisting of the Pleistocene ($1.6 \cdot 10^6$ to 10,000 y B.P.) and the Holocene (10,000 y B.P. to the present). Syn. **Quaternary.**

pleniglacial A major pulse of glaciation, accumulating thick (>1 km) ice sheets on the continents at high to middle latitudes and decreasing sea level by 100 ± 30 m.

pleochroic Referring to a mineral exhibiting pleochroism.

pleochroic halo A series of concentric, colored spherical surfaces in a crystal, up to 50 μm in radius, appearing as rings in thin section. They are produced by radiation damage by α particles of different energies emitted by heavy radionuclides usually contained in zircon, sphene, or apatite inclusions within the crystal.

pleochroism The ability of anisotropic crystals to absorb different wavelengths and, therefore, to exhibit different colors, in different directions.

Plinian eruption An explosive volcanic eruption producing a large, tall plume of volcanic ash. It was first described by Pliny the Younger in A.D. 79 (Vesuvius).

plumbago Graphite.

plume A rising mass of hot air, water, or rock.

plunge The inclination of the hinge of a fold with respect to the horizontal.

plunging fold A fold whose hinge is not horizontal.

Pluto The ninth and outermost planet of the solar system. Mean distance from the Sun = 39.44 AU = 5.467 light hours. Sidereal period = 248.5 y. Orbital eccentricity = 0.248; orbital inclination to ecliptic = 17.2°. Sidereal rotational period = 6.3874 d. Radius = 1145 km. Mass = $1.15 \cdot 10^{22}$ kg. Density = 1.84 g/cm³. Internal structure (estimated): Fe–Ni and silicate core [radius = 350 (?) km] and a 1150 (?) km thick mantle of frozen gases. Surface temperature = 42 K. Atmospheric pressure = 0.1 mb; atmospheric gases, $CH_4 \sim$ 100%, noble gases. One satellite, Charon (radius = 642 km; mass = $4.87 \cdot 10^{20}$ kg; density = 1.84 g/cm³). See **Planets—atmospheres*, Planets—physical data*, Satellites*.**

pluton An igneous rock body formed at depth.

plutonic Referring to an igneous rock crystallized at depth. Cf. **hypabyssal.**

plutonic metamorphism Metamorphism at depth and, therefore, at high temperature and pressure.

plutonism 1. The formation of plutons. 2. The classical theory of James Hutton (1726–1797) according to which the Earth was formed by solidification of molten magma. Cf. **neptunism.**

pluvial 1. Defining a climate with high rainfall. 2. Referring to a time interval of high pluviosity in the African Pleistocene.

pneumatocyst A gas-filled spherical structure that keeps Sargassum and other brown algae afloat.

pneumatolysis Crystallization of minerals or alteration of rocks by gases derived from solidifying magma.

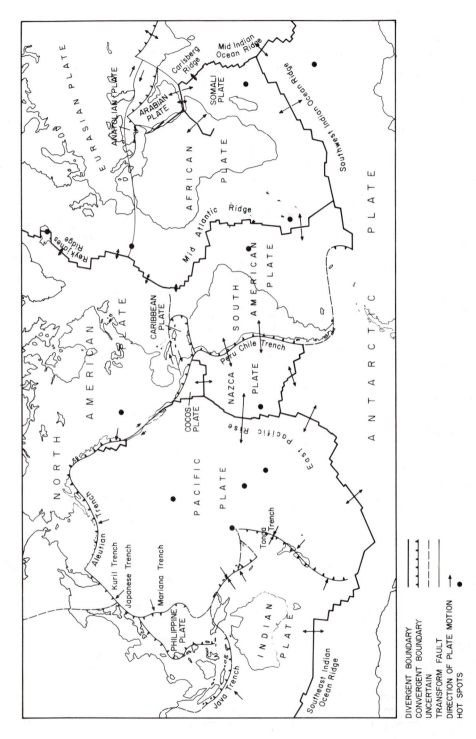

Plate tectonics. Major plates, plate boundaries, and major hot spots. The arrows indicate direction of plate motion and their length in mm × 1.4 gives plate motion velocity in cm/y.

pneumatolytic 1. Referring to pneumatolysis. 2. Formed by pneumatolysis.

pneumatolytic stage A late stage of petrogenesis, with crystallization from residual fluids enriched in gases, following the pegmatitic stage and preceding the hydrothermal stage.

***pn* junction** A thin region between a *p*-type and an *n*-type semiconducting material across which a potential barrier exists. See **semiconductor.**

***pnp* transistor** A transistor with an *n*-type base between a *p*-type emitter and a *p*-type collector. Cf. ***npn* transistor.**

podsol A soil consisting of a top layer rich in organic matter, followed below by a leached, gray layer and an illuvial horizon enriched in Fe and Al oxides and in organic matter.

Pogson ratio The ratio of 2.512 between successive units of apparent celestial magnitude. Cf. **magnitude.**

pOH The negative logarithm of the OH^- ion concentration in an aqueous solution. $pOH = -\log_{10} [OH^-]$, where the brackets signify concentration.

poikilitic A crystalline texture consisting of small crystals dispersed within a larger one.

poikilo- Prefix meaning *varied, variegated, intricate.*

poikilothermic Referring to a cold-blooded animal, i.e. to an animal whose body temperature varies with the ambient temperature.

point defect A crystal defect involving a single point within a crystal.

poise (P) The CGS unit of dynamic viscosity, equal to 1 dyn s/cm^2 = 0.1 Pa·s.

poiseuille The SI and MKS unit of dynamic viscosity, equal to 1 N·s/m^2 = 1 Pa·s.

Poisson distribution A frequency distribution for random events of increasing unlikelihood per unit of measurement, when occurrence in nonoverlapping measurement intervals is independent of preceding occurrences. Examples are the number of alpha particles emitted by a radioactive source per unit of time, the number of earthquakes per unit of time, or the number of diamonds per unit volume of placer sand. In each case, numbers increasingly greater from the mean λ represent increasingly unlikely events. The probability $P(k)$ for a number k to occur is given by the equation

$$P(k) = e^{-\lambda}\lambda^k/k!$$

Poisson ratio (σ) The ratio of lateral contraction to longitudinal extension for a bar under stress parallel to its length.

polar *(Physics)* Having two opposite ends distinguished by usually opposite or different characteristics. *(Geography)* Defining the regions located, or the phenomena occurring, within the polar circles.

polar cell Either of the two atmospheric circulation cells centered at the terrestrial poles. Warmer air rises at the boundary betwen the Ferrel and polar cells; it travels poleward on a course of 45° (northern hemisphere) or 235° (southern hemisphere); it sinks at the poles; and it radiates out on a course of 225° (northern hemisphere) or 315° (southern hemisphere).

polar circle Either the Arctic Circle or the Antarctic Circle, now at 66°33′32″ lat N or S (respectively), fixed by the 23°26′28″ angle between the Earth's axis and the normal to the plane of the Earth's orbit. It is the parallel bounding the polar cap over which there is a 24-hour period of darkness at the winter solstice and a 24-hour period of solar light at the summer solstice.

polar coordinate system A system of coordinates on a plane to identify the position of any point on the plane by its distance from the origin of the horizontal x axis and the counterclockwise angle made with the positive direction of the x axis by the ray connecting the origin to the point. Cf. **Cartesian coordinate system.**

Polaris The brightest star in Ursa Minor and the present North Pole star, located 1.0° from the north celestial pole.

polarity epoch The period of time during which the Earth's magnetic field maintains its polarity. Polarity epochs range in duration from less than 100,000 years to 35 million years (mid-Cretaceous). The present epoch of normal polarity, called Brunhes, began 730,000 y ago. See **polarity event.**

polarity event A short (<10,000 y) period of opposite polarity within a given polarity epoch. Syn. **magnetic event.** See **polarity epoch.**

polarization (electrical) 1. The production of a relative displacement of positive and negative charges in a body by the application of an electric field. 2. Electric dipole moment per unit volume.

polarization (optical) The constraining of the electric vector of electromagnetic radiation to a specified direction or behavior. 1. **linear polariza-**

Age (m. y.)

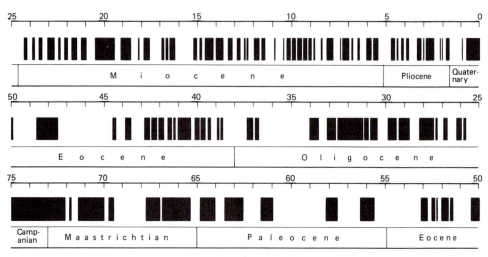

Polarity time scale. Normal polarity in black, reversed polarity in white. (Data from Harland et al. 1982)

tion The electric vector oscillates normally to the direction of propagation on the plane containing that direction. Intensity I of the polarized light:

$$I = I_{max} \cos^2\theta \quad \text{(Law of Malus)}$$

where I_{max} = maximum intensity of the transmitted light, θ = angle of polarization. **2. circular polarization** The electric vector rotates about the direction of propagation. **3. elliptical polarization** The electric vector rotates about the direction of propagation while changing its amplitude, with the change having a period identical to the period of rotation.

polarizing angle See **Brewster's angle.**

polar molecule A molecule having a permanent dipole moment.

polar wandering The motion of the geomagnetic poles with respect to the frame of reference fixed on a given lithospheric plate.

polder A tract of land reclaimed from the sea (Holland).

pole of inaccessibility The point on the Antarctic ice cap most remote from the ocean, approximately 82.0°S and 55.0°E.

poly- Prefix meaning *many.*

polygenetic 1. Consisting of more than one material. 2. Resulting from more than one process.

polymer A chain of repetitive monomers.

polymerization The binding of monomers into a polymer.

polymictic 1. Describing a lake that is continuously mixing. 2. Describing a clastic sediment or rock consisting of different minerals.

polymorph A crystal form of a polymorphic substance.

polymorphic 1. Referring to a substance that exhibits polymorphism. Cf. **allotropic.** 2. Referring to a species that exhibits polymorphism.

polymorphism 1. Property of a substance to crystallize in more than one form. Cf. **allotropy.** 2. Property of a species that exhibits different forms other than those related to sexual differentiation.

polynya An expanse of open water within an ice-covered water body.

polypeptide chain A chain of amino acids linked to each other by peptide bonds. See **alpha helix.**

polyprotic Referring to an acid that can donate more than one proton per molecule. E.g. sulfuric acid, H_2SO_4.

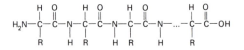

Polypeptide chain. R = amino acid side-group.

polysaccharides A family of carbohydrates containing 18 or more carbon atoms. E.g. cellulose $(C_6H_{10}O_5)_{3000-6000}$; glycogen, $(C_6H_{10}O_5)_{6000-120,000}$; starch, $(C_6H_{10}O_5)_{100-6000}$.

polytypic Referring to a species exhibiting significant geographic variation.

pond An enclosed body of freshwater larger than a pool but smaller than a lake.

pool A very small, enclosed body of water.

population I The population of younger $(10^7-10^8$ y), metal-rich stars in irregular galaxies and in the gas-rich arms of spiral galaxies. The youngest population I stars contain ~ 100 times more metals than the oldest population II stars. See **population II.**

population II The population of older $(\sim 10^{10}$ y) stars in the core of spiral galaxies, in globular clusters, and in elliptical galaxies. The oldest population II stars contain ~ 100 times less metals than the youngest population I stars. See **population I.**

p orbital The orbital of an atomic electron characterized by an orbital angular momentum quantum number $l = 1$. See **s, p, d, f.**

porcelain A white, compact translucent ceramic made by firing pure kaolin.

porosity The volume percent of pores in a solid. Intergranular porosity of common clastic rocks ranges from 10% to 35%. Spheres of identical sizes have porosity of 47.64% with cubic packing, and of ~ 26% with rhombohedral packing.

porphin ring $H_2N_4H_{12}C_{20}$, a heterocyclic ring consisting of 4 pyrrole rings united by $-CH=$ (methene) groups, capable of holding a metallic ion in its center. It is the core of chlorophyll (Mg^{2+} in center), of cytochrome c and hemoglobin (Fe^{2+} in center), and of hemocyanin (Cu^{2+} in center).

porphyrin ring See **porphin ring.**

porphyrins A group of organic compounds consisting of a porphin ring with side chains of various radicals ($-CH_3$, $-CH=CH_2$, $-CH_2COOH$, $-CH_2-CH_2COOH$, etc.). The porphin ring may chelate a metal ion and form such porphyrins as chlorophyll, hemoglobin, hemocyanin, and cytochrome c. See **porphin ring.**

porphyritic Having the texture of porphyry.

porphyry Any igneous rock consisting of phenocrysts in a groundmass of microcrysts.

position Of a body moving along axis x with uniformly accelerated motion:
$$x = x_0 + v_{x0}t + \tfrac{1}{2}a_x t^2$$
where x_0 = initial position, v_{x0} = initial velocity, t = time, a = acceleration. See **acceleration, velocity.**

positive 1. Referring to any number greater than zero. 2. Referring to the electrical charge of sign opposite that of the electron. 3. Referring to a N magnetic pole. 4. Referring to dextral rotation. Cf. **negative.**

positive feedback Feedback in which a portion of the output is fed back in phase with the input, resulting in signal amplification.

positive pole 1. The terminal of a battery exhibiting electron deficiency. 2. the N (north-seeking) pole of the magnetic needle or other magnetic dipole. Cf. **negative pole.**

positron (e^+, β^+) A positive electron, the antiparticle of the electron.

positronium The bound state of an electron and a positron, with half life of $1.39 \cdot 10^{-7}$ s (parallel spins, decaying into 3 γ) or $1.25 \cdot 10^{-10}$ s (antiparallel spins, decaying into 2 γ).

postglacial The time, or referring to the time, since the end of the last ice age, approximately 10,000 years ago. Cf. **Holocene, Recent.**

postorogenic Referring to a geologic processs or event occurring soon after an orogenic phase.

potamic Referring to a river. Syn. **fluviatile.**

potash Potassium carbonate (K_2CO_3) or hydroxide (KOH).

potash lake An alkali lake rich in potash and other salts.

potassium-argon dating method A method of absolute dating based on the decay by K-capture of ^{40}K ($t_{1/2} = 1.277 \cdot 10^9$ y) to ^{40}Ar.

potential 1. A scalar quantity involving energy as a function of position in a field. 2. The work required to bring a unit quantity from infinity to a specific position in the quantity's field.

potential energy See **energy.**

potential temperature 1. The temperature of a mass of air brought adiabatically to the pressure of 1 bar. 2. The temperature of a mass of deep marine or lacustrine water brought adiabatically to the surface.

potentiometer 1. An instrument to measure emf using a reference potential, a voltage divider, and a null meter. 2. A voltage divider yielding a continuous voltage range. It consists of a resistor with a sliding contact.

Potsdam gravity See **gravity standard.**

pound (lb) 1. A nonmetric unit of force. See **poundal, pound force.** 2. A nonmetric unit of mass. See **pound mass.**

poundal (pdl) A nonmetric unit of force, equal to the force that will impart an acceleration of 1 ft/s^2 to 1 pound mass. It is equal to 0.138254954376 newtons (exactly).

pound force (lbf) A nonmetric unit of force equal to the gravitational force experienced by a pound-mass when $g = 32.174049$ ft/s^2 (standard g). It is equal to 4.4482216152605 newtons (exactly).

pound mass (lb) 1. A nonmetric unit of mass equal to 0.45359237 kg (exactly). 2. A nonmetric unit of force. See **pound force.**

power (P) 1. The ratio of work to time:

$$P = W/t$$
$$P = dW/dt$$

where P = power, W = work, t = time. *(Electricity)* 1. *Direct current:*

$$P = IV$$

where P = power, I = current, V = potential. 2. *Alternating current:*

$$P_A = V_E I_E \cos \theta$$

where P_A = average power, V_E = effective voltage, I_E = effective current, θ = phase angle between voltage and current. *(Optics)* See **diopter.**

power factor The ratio of the average to the apparent power in an ac circuit.

power spectrum A spectrum showing the distribution of the intensity of a periodic phenomenon as a function of frequency.

Poynting-Robertson effect The loss of orbital momentum by a small (<1 mm) dust particle orbiting the Sun. Spiraling time:

$$t = 7.0 \cdot 10^6 r\rho aq$$

where t = time (years), r = radius of particle (cm), ρ = density of particle (g/cm^3), a and q = semi-major axis and perihelion distance, respectively, of initial orbit (AU). Comets continuously replenish lost dust.

Poynting vector (S) A vector describing the electromagnetic energy transported by a planar electromagnetic wave across a surface normal to the wave.

$$S = E \times H = (E \times B)/\mu_0$$

where S = Poynting vector, E = electric field, H = magnetic field intensity, B = magnetic induction, μ_0 = permeability constant.

ppb Parts per billion.

ppm Parts per million.

ppt Parts per thousand.

Prandtl number 1. A dimensionless number (Pr_M) describing diffusion in flowing systems:

$$Pr_M = \mu/\rho D$$

where μ = dynamic viscosity, ρ = density, D = diffusivity. 2. A dimensionless number (N_{Pr}) used in the study of convection:

$$N_{Pr} = \mu C_p/k$$

where μ = dynamic viscosity, C_p = specific heat at constant pressure, k = thermal conductivity.

prasinite A greenschist with approximately equal amounts of hornblende, chlorite, and epidote.

Precambrian The time following the Gamowian and preceding the Cambrian. It ranges from $4.7 \cdot 10^9$ y B.P. to $590 \cdot 10^6$ B.P. See **Geological time scale*.**

precession The conical motion of the axis of a body rotating about its center when subjected to a torque tending to alter the axis' direction in space.

precession of the equinoxes Westward motion of the nodes (equinoxes) of the ecliptic on the celestial sphere due to the precessional motions of the Earth's axis and the Earth's orbit caused by torques applied by the Moon, the Sun, the planets, and by a relativistic effect. The lunisolar torque on the Earth's equatorial bulge causes the Earth's axis to process clockwise as seen from the North ("luni-solar precession") by $(50.4001 + 0.0049T)''/y = 50.4044''/y$ (1987), describing a cone with an apical angle of $46°52'54.98''$ (1987). The relativistic effect and the torques applied by the other planets cause the Earth's orbit as a whole to process counterclockwise (as seen from the North). The relativistic effect (= $0.0192''/y$) reduces precession to $50.3852''/y$ ("geodetic precession"), and the planetary torques further reduce it [by $(0.1248 - 0.0189T)''/y = 0.1084''/y$ (1987)] to = $50.2768''/y$ (1987) ("general precession"). The general precessional period is thus $360°/50.2768'' = 25,777.297$ y (1987). Because of the opposite sense of the axial and orbital precessions, the seasonal precessional period (e.g. northern summer solstice at perihelion to next northern summer solstice at perihelion) is shorter than the general precessional period; it is equal to sidereal year/(anomalistic year − tropical year) = 20,943.791 y (1987). Because of the time terms and various uncertainties principally related to the angular momentum of the Earth (which depends upon its internal mass distribution), the general precessional period should be rounded to 25,800 y and the seasonal precessional period to 21,000 y. (In the preceding, T = centuries from A.D. 1900.0.) See **nutation.**

precipitate An insoluble substance formed by a chemical reaction in an aqueous solution.

precipitation *(Chemistry)* The formation of a precipitate. *(Meteorology)* Water that falls out of the atmosphere either as liquid or solid. The amount of precipitation is expressed as the thickness of the liquid water precipitated.

precision The amount of consistency in repeated measurements, expressed by the number of decimal places. Cf. **accuracy.**

pressure (*p*) Force per unit of surface:
$$p = dF/dS$$
where p = pressure, F = force, S = surface. Pressure in open vessel or body of water:
$$p = p_0 + \rho g h$$
where p = pressure, p_0 = pressure at open surface, ρ = density of fluid in vessel or body of water, g = gravitational acceleration, h = depth below surface.

primärrumpf Primary swell, a broad swell of the Earth's crust rising slowly enough to be continuously eroded and, therefore, maintaining a mature topography.

primary *(Astronomy)* The celestial body nearest the center of mass of a system around which one or more secondary bodies also orbit. *(Physics)* Referring to any of the radionuclides having a sufficiently long half life to be still present in nature in measurable amount. See **primary radionuclide.** *(Seismology)* Defining a compressional acoustic wave. See **P wave.**

Primary A term introduced by Giovanni Arduino (1714–1795) for the igneous rocks forming the core of the Alps. Arduino called *secondary* the sedimentary rocks draping the Alps and *tertiary* the loose sediments below. The terms *Primary* and *Secondary* have been replaced by the terms *Paleozoic* and *Mesozoic,* respectively. The term *Tertiary* has been reassigned to the pre-Quaternary Cenozoic. Arduino's *tertiary* has been replaced by

the term *Quaternary* introduced by Desnoyers in 1829. Cf. **Geological time scale*.**

primary battery A battery consisting of one or more primary cells.

primary cell A nonreversible and, therefore, nonrechargeable electrolytic cell.

primary color Any of the set of colors (e.g. red, yellow, and blue) that may be variously combined to form a wide range of other colors.

primary optic axis Any of the four directions of equal wave-front velocity in biaxial crystals. Cf. **optic axis, secondary optic axis.**

primary radionuclide Any of the radionuclides having a sufficiently long half life to be still present in nature in measurable amount. Examples are ^{40}K, ^{87}Rb, ^{147}Sm, ^{176}Lu, ^{187}Re, ^{190}Pt, ^{232}Th, ^{235}U, and ^{238}U. Cf. **induced radionuclide, secondary radionuclide.** See **Isotope chart*.**

primary wave See **P wave.**

primordial lead Lead that has not undergone changes in the relative abundances of its isotopes by the addition of radiogenic lead from the decay of ^{238}U, ^{235}U, and ^{232}Th. An example is the lead in the troilite phase of meteorites ($^{206}Pb/^{204}Pb$ = 9.307; $^{207}Pb/^{204}Pb$ = 10.294; $^{208}Pb/^{204}Pb$ = 29.476).

principal focus The point to which parallel rays crossing an optical system converge.

principal planes Two planes normal to the axis of an optical system on its opposite sides, forming images on each other with magnification = 1.

principal point The intersection of the axis of an optical system with a principal plane.

principal quantum number (*n*) The quantum number that identifies the shell to which an atomic electron belongs. It can assume only positive integer values (in $h/2\pi$ units), ranging from 1 for the innermost or K shell up. Cf. **energy level.**

principle of equivalence "An inertial frame having a uniform gravitational field and a second frame uniformly accelerated with respect to the first, with no gravitational field of its own, are equivalent." Therefore, gravitational mass = inertial mass. Experiments carried out within the two frames give identical results.

prism A polyhedron with congruent polygonal bases and parallelograms as sides.

prismatic layer The middle layer of a molluscan shell, between the periostracum outside and the lamellar layer inside.

probability The ratio of the number of occurrences of a given event to the total number of possible occurrences. It ranges from 0 (impossible) to 1 (certain).

Procaryota One of the two superkingdoms of life on Earth, consisting of single-celled or colonial organisms with cells lacking nuclear membrane and organelles. Cf. **Eucaryota.** See **Taxonomy*.**

procaryote 1. Referring to a cell without nuclear membrane or organelles. 2. Any of the single-celled or colonial organisms with cells lacking nuclear membrane and organelles. Cf. **eucaryote.**

prodelta The submerged portion of the delta beyond the delta front.

proglacial Immediately in front of a glacier or an ice sheet.

progradation The seaward advance of sediment deposition.

prograde Referring to the counterclockwise motion of a planet or other celestial body around the Sun, or of a satellite around its planet, as seen from the north. Cf. **retrograde.**

projectile Equation of trajectory:

$$z = x(\tan \vartheta_0) - gx^2/2(v_0 \cos \vartheta_0)^2$$

where z = vertical distance, x = horizontal distance, ϑ_0 = initial inclination; g = gravitational acceleration; v_0 = initial velocity.

prokaryote See procaryote.

prolate Defining an ellipsoid of revolution elongated along its axis of rotation. Cf. **oblate.** See **ellipsoid of revolution.**

proloculus The first chamber of a foraminiferal shell.

propagation coefficient (γ) A coefficient rating a line transmitting an ac current.

$$\gamma = \alpha + j\beta$$

where α = attenuation coefficient of the current with distance from source, $j = (-1)^{1/2}$, β = phase constant, measured in radians per unit length and representing the voltage or current phase lag at a point x downline with respect to the phase at the source or other upline reference point (lag = βx). See **attenuation.**

proper motion (μ) Apparent motion of a star on the celestial sphere, resulting from its peculiar motion and the motion of the Sun with respect to it.

proportionality factor A constant or a variable expression relating two quantities.

prosthetic group A nonpeptide attachment to a protein molecule.

protactinium-thorium dating method A method of absolute dating of deep-sea sediments based on the changing ^{231}Pa/^{230}Th ratio ($t_{1/2}$ = 60,100 y).

protein A chain of amino acids with prosthetic groups (conjugated proteins) or without (simple proteins). The number of amino acids in a protein chain ranges from as little as 8 to as large as 7000, corresponding to a molecular mass range of 880 to 770,000 u (average amino acid molecular mass = 110 u). Common proteins consist of 130 to 630 amino acids, corresponding to a molecular mass range of 14,000 to 70,000 u. Protein molecules are arranged (secondary structure) in the α-helix (a spiral chain with positive helicity, stabilized by intrachain H bonds) or β structure (two or more parallel or antiparallel, extended chains stabilized by interchain H bonds). These structures may be folded over (tertiary structure) and two or more chains may associate (quaternary structure). The length of a fully stretched common protein chain is in the range of 0.15 to 0.30 μm.

Proterozoic The most recent Precambrian era, from the appearance of stromatolites ($2.7 \cdot 10^9$ y B.P.) to the beginning of the Cambrian ($590 \cdot 10^6$ y B.P.). It follows the Archean.

Protista The set of all single-celled eucaryota.

protium The common isotope of hydrogen, ^{1}H.

protoconch The earliest portion of a gastropod or cephalopod shell.

Protoctista One of the 5 kingdoms, including the Protista and their immediate evolutionary, multicellular descendants (multicellular algae, slime molds, etc.). Cf. **Protista.** See **Taxonomy*.**

protodolomite A disordered crystal of dolomitic composition. Cf. **dolomite.**

proton (p) A stable (?) baryon. Rest mass = 1.00727623 u = 938.223 MeV. m_p/m_e = 1836.1527, where m_p = rest mass of proton, m_e = rest mass of electron.

proton Compton wavelength ($\lambda_{C,p}$) Compton shift of incident x-rays and γ-rays in collision with protons with photon scattering angle of 90°. It is a length characteristic of the proton:

$$\lambda_{C,p} = h/m_p c$$
$$= 1.3214100 \cdot 10^{-15} \text{ m}$$

where h = Planck's constant, m_p = rest mass of proton, c = speed of light.

proton magnetic moment (μ_p) $\mu_p = 1.4106076 \cdot 10^{-26}$ J T^{-1}.

proton magnetometer An instrument that measures the absolute value of the magnetic flux but not its direction. It is used mainly at sea and airborne. It consists of a bottle with water (or other H-containing fluid) surrounded by a coil. The magnetic axes of the spinning protons, normally randomly oriented, become aligned when a DC current is made to flow through the coil. When the current is interrupted, the axes precess around the direction of the Earth's magnetic field on their way to resume random orientation (unless of course the induced field should happen to be parallel to the Earth's field). This process takes about 1 second. The frequency of precession (Larmor frequency) is proportional to the ambient field B:

$$B = 23.4874 \, \nu$$

where B = ambient field in nanotesla or γ, ν = frequency in hertz. For $B = 50,000 \ nT$, ν is ~ 2000 Hz. Precision is $<0.5 \ \gamma$, limited by the accuracy with which the proton gyromagnetic ratio is known.

proton-proton chain A series of nuclear reactions involving H, He, Li, Be, and B, and resulting in the synthesis of ^{4}He from ^{1}H. Depending upon the pathway, the energy produced is 26.203, 25.934, or 22.735 MeV per ^{4}He synthesized, equivalent to 5.5–$6.3 \cdot 10^{11}$ joules per gram of ^{1}H consumed. Together with the carbon cycle, the proton-proton chain is responsible for the production of thermonuclear energy in the cores of Main Sequence stars. See **carbon cycle, Stars—energy production***.

Protophyta The set of single-celled plants.

protoplasm All living matter inside the cell wall, including nucleus and cytoplasm.

prototype See **archetype**.

Protozoa The set of all single-celled animals.

protozoan Any of the single-celled animals belonging to the Protozoa.

Proxima Centauri The star closest (4.26 l.y.) to the Sun. It is the fainter component of the triple α Centauri. The two brighter components, α Centauri A and B, are a visual binary with masses similar to that of the Sun and an orbital period of 80.1 years.

psammite Syn. **arenite**.

psammitic Referring to a psammite or arenite.

psammon The fauna living within loose sand.

psammophilic Referring to an organism living within loose sand.

psammophyte A plant adapted to living in sand or sandy soil.

psephite Syn. **rudite**.

psephitic Referring to a psephite or rudite.

pseudo- Prefix meaning *false*.

pseudoforce Any force experienced by an inertial mass with respect to a noninertial frame. Syn. **inertial force**.

pseudomorph A mineral, developed by substitution or alteration, that reproduces the crystal form of the preceding mineral.

pseudopodium A protoplasmic projection in protozoa used for locomotion and/or feeding.

PSI Pound per square inch, a nonmetric unit of pressure equal to 6894.757 Pa = 0.06804596 atm.

psychro- Prefix meaning *cold*.

psychrophyte A plant adapted to a cold environment.

psychrosphere The totality of cold environments on Earth, including high latitudes, high altitudes, and the deep ocean. Cf. **thermosphere** (def. 1).

pteropod Any of the pelagic gastropods belonging to the order Pteropoda. Pteropods form light aragonitic shells.

pteropod ooze A deep-sea sediment consisting of at least 30% pteropod shells.

p-type conduction Electrical conduction by holes in a semiconductor. Cf. **n-type conduction**.

p-type semiconductor An extrinsic semiconductor in which the majority carriers are holes. Cf. **n-type semiconductor**. See **semiconductor**.

pulsar A rapidly rotating neutron star emitting electromagnetic radiation in regular pulses related to its rotational period. Known periods range from 1.557 806 449 059 ms for pulsar PSR1937+21 (the fastest rotating pulsar known) to 33.1 ms for pulsar 0531+21 (the Crab pulsar) and to 4.308 s for pulsar 1845−19 (the slowest rotating pulsar known). With the exception of pulsar PSR1937+21, pulsar periods increase with time by amounts ranging up to $422.69 \cdot 10^{-15}$ s/s (Crab pulsar). See **glitch**.

pulse-height analyser An instrument counting the pulses falling within a given interval of measurement.

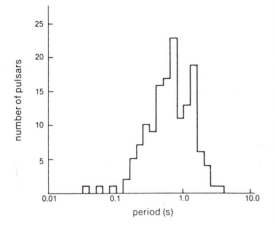

Pulsars. Pulsar period distribution for 149 pulsars. (Bowers and Deeming 1984, p. 303, Fig. 16.11)

pumice A vesicular, glassy volcanic ejectum usually of rhyolitic composition.

pumpellyite-prehnite facies A low-temperature (150–300°C), intermediate pressure (4–6 kbar) metamorphic facies characterized by the presence of pumpellyite and prehnite.

punctuated equilibrium A model of evolution according to which lineages exhibit long periods of stasis interrupted by short intervals during which evolutionary change is rapid.

pure water Water containing no salts or other dissolved substances.

purine 1. $C_5H_4N_4$ (mol. mass = 120.114), consisting of a double heterocyclic ring structure. 2. Referring to the double heterocyclic ring structure characteristic of purine.

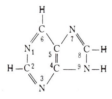

Purine. (King and Stansfield 1985, p. 320).

P wave A primary or pressure body wave with alternating compression and expansion in the direction of propagation. Cf. **S wave**. See **Seismic waves***.

pycno- Prefix meaning *dense, solid.*

pycnocline A sharp density gradient in a stratified fluid.

pycnometer An instrument consisting of a bulb of precisely known volume, used to weigh a specific volume of a liquid in order to determine its density.

pyrimidine $C_4H_4N_2$ (mol. mass = 80.089), consisting of a single heterocyclic ring structure.

pyrite Iron sulfide, FeS_2.

pyroclast Any of the fragments, regardless of size, ejected during a volcanic eruption.

pyroclastic Referring to a volcanic ejectum.

pyrolite A name for the undifferentiated mantle rock (equivalent to a mixture of 2/3 peridotite and 1/3 basalt) from which the uppermost mantle and crust differentiated.

pyrope The Mg–Al end member of the garnet family, $Mg_3Al_2Si_3O_{12}$. See **garnet.**

pyroxenes A group of dark, Ca–Mg–Fe silicates crystallizing either in the orthorhombic system (orthopyroxenes) or the monoclinic system (clinopyroxenes). Pyroxenes are major components of igneous and metamorphic rocks. Common orthopyroxenes:

enstatite	$MgSiO_3$
hypersthene	$(Mg,Fe)SiO_3$

Common clinopyroxenes:

augite	$(Ca,Na)(Mg,Fe,Al)\cdot$
	$[(Si,Al)O_3]_2$
clinoenstatite	$MgSiO_3$
clinohypersthene	$(Mg,Fe)SiO_3$
diopside	$CaMg(SiO_3)_2$
hedenbergite	$CaFe(SiO_3)_2$

pyroxenite An ultramafic, plutonic rock consisting mainly of pyroxene.

pyrrole $(CH)_4NH$, a component of the porphin ring structure. See **porphin ring.**

Q

q Perihelion distance.

Q 1. Aphelion. 2. Electric charge. 3. Heat. 4. Q factor.

QCD Quantum chromodynamics.

QED Quantum electrodynamics.

Q.E.D. *Quod erat demonstrandum,* Latin for *which was to be demonstrated.*

Q factor (Q) Quality factor, a measure of the ability of a cyclic system to store energy.

$$Q = 2\pi E/\Delta E$$

where E = maximum energy stored/cycle, ΔE = energy dissipated/cycle.
(Electricity)

$Q = X_L/R$	(RL circuits)
$Q = X_C/R$	(RC circuits)
$Q_0 = 2\pi f_0 L/R$	
$= 1/2\pi f_0 CR$	(RLC series circuits)

where X_L = inductive reactance, X_C = capacitative reactance, Q_0 = Q factor at resonance, f_0 = resonant frequency, L = inductance, C = capacitance, R = resistance.
(Seismology)

$$Q = \omega/2c\alpha$$

where ω = angular frequency, c = phase velocity, α = attenuation coefficient.

Q-mode factor analysis Factor analysis concerned with relationships among items. Cf. **R-mode factor analysis.**

QSO Quasi-stellar object. Syn. **quasar.**

quadrant 1/4th of the circumference of a circle.

Quadrantids A major meteor shower. See **meteor shower.**

quadrature The position of the Moon or of a superior planet when the angle Sun-Earth-Moon or Sun-Earth-planet is 90°.

quadrupole Two adjacent electric or magnetic dipoles with inverse polarity so as to practically cancel each other's field beyond some distance.

quadrupole mass spectrometer A mass spectrometer with four parallel rods between which the ions pass. An alternating potential superimposed on a steady potential between pairs of rods lets through only ions of a specific mass.

quagma A hypothetical quark-gluon plasma with density of $1-2 \cdot 10^{15}$ g/cm^3 (4–8 times greater than the density of nuclear matter) that may have been the stable state of matter for a brief interval following the initiation of the Big Bang.

quagmire A soft bog.

quality factor (Q) See **Q factor.**

quantasome The smallest organized unit capable of storing light quanta as ATP bonds in the process of photosynthesis. A quantasome consists of 230 chlorophyll molecules (160 chlorophyll a, 70 chlorophyll b), 48 carotenoid molecules, 700 molecules of other lipids, and protein. Total molecular mass = 1,920,000 u.

quantum chromodynamics (QCD) The theory of the strong quark-quark, quark-gluon, and gluon-gluon interaction arising from the color properties of quarks and gluons and accounting for the strong (color) force.

quantum electrodynamics (QED) The quantum theory of electromagnetic radiation and its interaction with charged particles.

quantum field theory The quantum theory of wave fields, including gravitational, electromagnetic, acoustic, etc., and incorporating the basic quantum of action.

quantum gravity The quantum theory of gravitation, not yet formulated.

quantum jump The change of a system from one stationary state to another by emission or absorption of discrete units (quanta) of energy.

quantum mechanics The modern theory of matter and electromagnetic radiation and their interactions.

quantum number Any of the numbers needed to characterize the state of an atomic or subatomic system. The four quantum numbers needed to identify the energy state of atomic electrons are (with values in units of $h/2\pi$ in parenthesis): the principal quantum number n ($n \geqslant 1$); the orbital angular momentum quantum number l ($l < n$); the magnetic orbital angular momentum quantum number (or simply magnetic quantum number) m_l (m_l = any integer between $-l$ and $+l$; thus, $l = 0$, $m_l = 0$; $l = 1$, $m_l = -1, 0, +1$; $l = 2$, $m_l = -2, -1, 0 +1, +2$; $l = 3$, $m_l = -3, -2, -1, 0 +1, +2, +3$); and the magnetic spin angular momentum quantum number (or simply spin quantum number) m_s ($m_s = \pm 1/2$).

quantum of action (h) See **Planck's constant.**

quantum state A state characterized by a specific set of quantum numbers.

quark A fundamental particle forming all hadrons, with charge $-1/3$ or $+2/3$ of the electron charge, baryon number $1/3$, spin $1/2$, strangeness 0 or -1, and charm 0 or $+1$. Baryons consist of 3 quarks and mesons of quark-antiquark pairs. Quarks come in 6 "flavors" [up (u), charm (c), and top (t), each with a $+2/3$ charge; down (d), strange (s), and bottom (b), each with a $-1/3$ charge], each of which comes in three "colors" (red, green, blue). The total number of quarks is thus 18. Antiquarks have reversed charge, baryon number, strangeness, and charm. The terms for flavors and colors are used to identify quark properties and bear no relationship to their common meanings.

quartz Crystalline silica, SiO_2. See **silica.**

quartz clock A clock in which a quartz crystal is inserted in an oscillating electric circuit having a frequency similar to the natural frequency of the quartz crystal. The latter regulates the former. See **quartz oscillator.**

quartz diorite A plutonic rock with more quartz than diorite. It grades into granodiorite with increasing alkali feldspar content.

quartzite A recrystallized or metamorphosed clean silica sandstone.

quartz lamp A mercury-vapor lamp with quartz glass that transmits most of the ultraviolet radiation.

quartz oscillator An oscillating electric circuit that includes a quartz crystal. Because of its piezoelectric properties, the quartz crystal provides a sharply tuned frequency.

quasar Any of the compact extragalactic objects of stellar appearance but with large (>0.4) redshift and emitting, if at cosmological distance, amounts of power (radio, infrared, optical, and x-ray) ranging from 10^{38} to 10^{41+} W (cf. $3.8 \cdot 10^{26}$ W for the Sun, $6.8 \cdot 10^{37}$ W for the Galaxy). Most distant

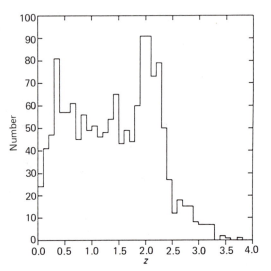

Quasars. Abundance as a function of redshift parameter z. The sharp drop at $z = 2.5$ is probably due to inadequate detection of quasars with $z > 2.5$. (Hewitt and Burbidge 1980, p. 65, Fig. 2)

quasar: 0046–293, $15.4 \cdot 10^9$ l.y. away, with redshift parameter $z = 3.80$ and recessional speed of 276,820 km/s.

Quaternary The most recent geological sub-era. It ranges from $1.6 \cdot 10^6$ y B.P. to the present and includes the Pleistocene ($1.6 \cdot 10^6$ to 10,000 y B.P.) and the Holocene (10,000 y B.P. to the present). See **Geological time scale*.**

quaternary compound A molecule containing four different types of atoms.

quaternary system A chemical system consisting of four different chemical components.

quaternion See **numbers.**

quebrada A gorge.

quenching Rapid chilling.

q.v. *Quod vide,* Latin for *which see, which you should see.*

Q value The gain in kinetic energy (= loss in rest mass) in nuclear bombardment processes.

$$Q = K_B + K_b - K_a - K_A$$
$$= (m_a + m_A - m_B - m_b)c^2$$

where a = bombarding particle, A = target nucleus, B = residual nucleus, b = product particle, K = kinetic energy, m = mass, c = speed of light.

Q wave See **Love wave.**

R

ρ 1. Density (mass/volume). 2. Resistivity. 3. Volumetric electric charge density.

R 1. Cosmic scale factor. 2. Electric resistance. 3. Gas constant. 4. Radical.

$\Re$ Reluctance.

R$_\infty$ Rydberg constant.

RA Right ascension.

race A population with physical or behavioral characteristics that are discernible, but not as marked as in a subspecies, and that exhibit intergradations with populations adjacent in space or time.

racemic (dl) Referring to a mixture of dextro- (d) and laevo- (l) rotatory molecules of a given compound.

racemization The transformation of a d or l compound into a racemic mixture.

rad 1. Radian. 2. The standard unit of absorbed radioactive dosage, equal to the absorption of 100 ergs per gram of matter.

radar Radio detection and ranging, a system using short (1–100 cm) radio waves in short (1 μs), high-power (10–10,000 kW) pulses to measure the distance of an object from the two-way travel time of the wave.

radian The SI unit of plane angle, defined as that angle that subtends a segment on the circumference of a circle equal in length to the radius. 1 radian = $180°/\pi$ = 57°17′44.8″.

radiancy Electromagnetic energy emitted by a surface per unit area per unit time.

$$R = \sigma T^4$$

where R = radiancy, σ = Stefan-Boltzmann constant, T = absolute temperature. See **Stefan-Boltzmann constant, Stefan-Boltzmann law.**

radiant *(Physics)* Emitting electromagnetic radiation. *(Astronomy)* The point on the celestial sphere from which the meteors in a meteor shower appear to originate.

radiant flux density Radiant power per unit surface. Syn. **irradiance.**

radiant power The energy of electromagnetic radiation impinging on a surface or passing through it per unit time.

radiation *(Physics)* The emission and propagation of electromagnetic waves or of particles. *(Biology)* The dispersal of a population of organisms to different locations and environments, leading to evolutionary divergence.

radiation balance The balance between incoming solar radiation and the outgoing terrestrial backradiation.

radiation damage The damage done to a crystal lattice by the decay of radioactive isotopes by fission or by α emission. The flying apart of the two major fragments in ^{238}U fission create tunnels in the crystal lattice about 10 μm in length but only a few Å in width. These can be made visible by etching the polished crystal surface with suitable acids. Alpha particles of varying energies produce concentric spheres, up to 50 μm in radius, exhibiting radiation damage. See **fission track, fission track dating, pleochroic halo.**

radical 1. An atom or a molecule with at least one unpaired electron. 2. A group of atoms behaving as a single atom in chemical reactions.

radioactive decay The decay of a radionuclide into another by α, $\beta^\pm$, K-capture, fission, or isomeric transition. Radioactive decay involves energies much higher than those commonly available in planetary processes. As a result, radioactive decay is independent of environmental conditions and the decay rate is constant:

$$-dN/dt = \lambda N$$

where dN = number of atoms decaying during the timer interval dt, λ = decay constant, N = number of atoms present. The number of atoms of the original radionuclide remaining after time t is given by

$$N = N_0 e^{-\lambda t}$$

where N = number remaining, N_0 = original number, λ = decay constant, t = time. See **decay constant.**

radioactive equilibrium A state of equilibrium

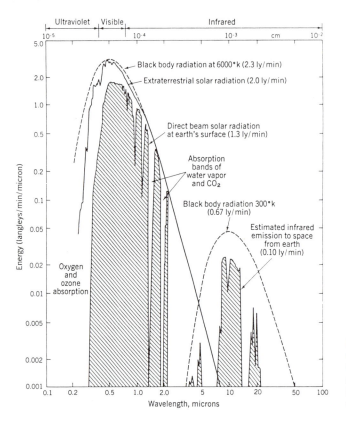

Radiation balance. Solar spectrum to the left, terrestrial spectrum (backradiation) to the right. Blackbody spectra and absorption bands as shown. (Strahler 1971, p. 199, Fig. 13.1; based on Sellers 1965, p. 20, Fig. 6)

achieved between daughter and parent radionuclide in a given radioactive decay series in which their relative abundances are proportional to their half lives or inversely proportional to their decay constants. E.g.: ^{222}Ra abundance/^{238}U abundance $= t_{1/2}(^{222}\text{Ra})/t_{1/2}(^{238}\text{U}) = \lambda(^{238}\text{U})/\lambda(^{226}\text{Ra}) = 3.58 \cdot 10^{-7}$.

radioactivity The emission of particles and/or radiation during the process of radioactive decay.

radiocarbon The radioactive (β^- decay) isotope ^{14}C, with $t_{1/2} = 5730$ y.

radiocarbon dating method See **carbon-14 dating method.**

radio frequency (RF) The frequency band of communication, ranging from $\sim 10^4$ to $\sim 3 \cdot 10^{11}$ Hz. See **Electromagnetic spectrum*.**

radiogenic Referring to a nuclide produced by the decay of a radionuclide.

Radiolaria A class of planktonic marine protozoa belonging to the phylum Actinopoda, kingdom Protoctista, with tests consisting of opaline silica [except for the Acantharia, which have tests con-

sisting of celestite (SrSO$_4$), and the Phaeodaria, which have tests consisting of a mixture of silica and organic compounds]. Radiolaria live mainly in the upper 200 m of the water column, but living specimens have been found at depths as great as 3000 m.

radiolarian 1. A protozoan belonging to the Radiolaria. 2. Pertaining to the Radiolaria.

radiolarian ooze A deep-sea deposit consisting of more than 30% radiolarian tests. Five percent of the Pacific floor and 0.5% of the Indian Ocean floor are covered with radiolarian ooze. This type of sediment is now absent from the Atlantic, but it was present during the last deglaciation (from 15,000 to 10,000 y B.P.).

radiolarite Recrystallized radiolarian ooze.

radiometric age An age determined by radiometric dating. See absolute dating method.

radiometric dating Age determination based on the radioactive decay of unstable nuclides. See **absolute dating method.**

radionuclide A radioactive nuclide.

raised beach A beach raised above sea level by uplift or by eustatic lowering of sea level.

raised reef A reef raised above sea level by uplift or by eustatic lowering of sea level.

RAM Random-access memory. A semiconductor device allowing input and retrieval directly to and from individual cells, each storing one bit of information. Single chips may contain 64K, 128K, or 256K individual cells. Cf. **ROM.**

Raman effect The scattering of light passing through a transparent medium, due to interaction of light with the rotational and vibrational motions of the molecules of the medium.

rand A long, low, rock ridge (South Africa).

Raoult's law "The partial vapor pressure of a solvent is equal to its vapor pressure in the pure state times its mole fraction."

$$p = p_o[N/(n + N)]$$

where p = partial vapor pressure of solvent; p_o = vapor pressure of pure solvent; N = number of moles of solvent; n = number of moles of solute(s).

rapakivi A granite or quartz monzonite characterized by orthoclase with plagioclase overgrowth.

rare-earth elements (REE) 1. The oxides of the lanthanides. 2. The oxides of Sc, Y, La, Ac, the lanthanides, and the actinides, belonging to group 3 of the Periodic Table of the Elements.

rare earths See **rare-earth elements.**

rate constant The proportionality constant defining the rate of a chemical reaction.

rate process A process in which the derivatives of the variables with respect to time depend on the value(s) of the variable(s) at the given time.

rational face A crystal face whose intercepts on the XYZ axes bear a simple numerical ratio to each other.

rationalized unit Any of the units in a system of units (such as the SI), where the factors 2π (circular symmetry) or 4π (spherical symmetry) are included in the definition. E.g. Coulomb's constant.

ray 1. The normal to the wavefront of an electromagnetic or acoustic wave. 2. A thin beam of light. 3. A geometric line with one endpoint (by comparison, a line has no endpoints and a segment has two endpoints).

Rayleigh criterion A criterion holding that the images of two points may be considered resolved if the principal diffraction maximum of one falls exactly on the first diffraction minimum of the other.

Rayleigh wave A seismic surface wave characterized by retrograde, ellipsoidal motion at the free surface.

Rb/Sr dating method See **rubidium-strontium dating method.**

RC circuit A circuit containing resistance and capacitance in series.

reactance (X) The imaginary part of impedance in an ac circuit. It is expressed in ohms.
1. capacitative reactance (X_C):

$$X_C = 1/2\pi f C$$

where X_C = capacitative reactance, $2\pi f = \omega$ = angular frequency, f = frequency, C = capacitance.
2. inductive reactance (X_L):

$$X_L = 2\pi f L$$

where X_C = inductive reactance, $2\pi f = \omega$ = angular frequency, f = frequency, L = inductance.
3. inductive-capacitative reactance (X_{LC}):

$$X_{LC} = 2\pi f L - 1/2\pi f C$$

where L = inductance, C = capacitance, $2\pi f = \omega$ = angular frequency, f = frequency.

reaction series See **Bowen's reaction series.**

reactive power The product $V_e I_e \sin \theta$, where V_e = effective voltage, I_e = effective current, θ = phase angle between voltage and current.

reactor A system consisting of a core with fissionable material (^{235}U or ^{239}Pu) capable of sustaining a nuclear reaction. The heat created is used to produce steam to run turbogenerators. Light-water-cooled power reactors use UO_2 with ^{235}U enriched by several percent. Spontaneous fission of ^{238}U produces neutrons (2.5 neutrons/fission) to fission ^{235}U. The high-energy (~ 2 MeV), high-speed (20,000 km/s) fission neutrons are slowed down to thermal speeds (a few km/s) by moderators of low atomic mass (water, heavy water, Be, C, hydrocarbons, etc.). Efficiency (electricity generated/heat produced) = 0.3 (30%). Nuclear reactors for power production deliver 500–1000 MW of electricity.

recapitulation The theory according to which ontogeny recapitulates phylogeny. Syn. **palingenesis.**

Recent See **Holocene.**

rechargeable battery See **storage battery.**

reciprocal lattice A lattice of points along axes normal to those of the crystalline lattice, located at distances from the center reciprocal to those of the crystal planes.

rectifier A circuit component (e.g. a diode) allowing flow of current in only one direction.

recumbent fold A fold whose axial plane is 90° or more from the vertical.

recurrence horizon A horizon within a peat bog deposit marking a period of nondeposition. Syn. **Grenzhorizont.**

red algae The algae of the phylum Rhodophyta, characterized by the pigment phycoerythryn. E.g. Lithothamnion.

red beds Clastic deposits with sand, silt, and clay particles coated with hematite, characteristic of warm, arid environments.

red clay A deep-sea deposit consisting of water- and wind-transported quartz and clays, clays from the alteration of volcanic glasses, and Fe–Mn oxides. It contains $<30\%$ $CaCO_3$ and covers 38% of the ocean floor.

red giant A star in a post-main sequence evolutionary stage characterized by a large, extended envelope and a compact He, C, or O core. Diameter $= 10$–100 solar diameters; surface temperature $= 1000$–3000 K; mean density $= 10^{-3}$ to 10^{-6} g/cm^3.

red noise Noise in which the amplitudes of the lower frequencies are higher than those of the other frequencies. Cf. **blue noise, white noise.**

redox Reduction-oxidation.

redox potential See **oxidation-reduction potential.**

red shift The apparent increase in the wavelength of light as seen by an observer when the distance between source and observer increases. A special case of the Doppler shift. Cf. **blue shift.** See **Doppler shift, redshift parameter.**

redshift parameter (z) The relative increase in wavelength of light from a source receding from an observer:

$$z = \frac{\Delta\lambda}{\lambda} = (\lambda_r/\lambda) - 1$$

where λ = wavelength of light emitted, λ_r = wavelength of light received.
Recessional velocity:

$$v_r = c \cdot [(z + 1)^2 - 1] / [(z + 1)^2 + 1]$$

where v_r = recessional velocity, c = velocity of light, z = redshift parameter. See **Doppler shift.**

Red Spot See **Great Red Spot.**

red tide An expanse of seawater containing a bloom of dinoflagellates. See **bloom.**

reduction The gain of one or more electrons by an atom or a molecule. Cf. **oxidation.**

reduzates Sediments deposited under reducing conditions.

REE Rare-earth elements.

reef A submarine mound or ridge constructed of rock debris or formed by $CaCO_3$-depositing marine organisms.

reef complex The ensemble of fore reef, reef flat, and back reef.

reef flat A shallow platform consisting of dead reef and covered with reef debris extending behind the fore reef.

reef tract A reef system.

reference wave See **hologram.**

refraction The change in direction of propagation of a wave passing from one medium to another one in which the wave velocity is different, or passing through a medium with a changing refractive index.

refractive index See **index of refraction.**

refugium A restricted area in which a flora or fauna can survive protracted adverse conditions affecting the surrounding territory.

reg A desert area from which the finer sediments have been removed by wind action, leaving a cover of coarse gravel. Cf. **hammada.**

regional metamorphism Metamorphism affecting a broad area.

regional metasomatism Metasomatism affecting a broad area.

regolith Common misspelling for **rhegolith** (q.v.).

regression *(Mathematics)* The relationship between two or more variables. *(Geology)* The withdrawal of the sea from land.

regression analysis A statistical method to quantify the relationship between two or more variables.

regressive Referring to phenomena or deposits related to marine regression.

rejuvenated Referring to geological features reactivated by renewed action of the pertinent geological agents (erosion, faulting, etc.).

relative density (ρ) The density of a substance relative to that of another, specified substance, usually pure water at 3.98°C (temperature of maximum density for pure water) and 1 atm for solids and liquids, or air at STP for gases. The density of pure water under these conditions is equal to 1.000000 g/cm³. Cf. **absolute density, density, Water—density***. Syn. **specific gravity**.

relative humidity The percent ratio of the amount of water vapor in an air mass to the amount that would saturate that air mass at the same temperature.

relative permeability (μ_r) The ratio of the permeability μ of a substance to the permeability μ_0 of vacuum:

$$\mu_r = \mu/\mu_0$$

See **permeability**.

relative permittivity (ϵ_r) The ratio of the permittivity ϵ of a substance to the permittivity ϵ_0 of vacuum:

$$\epsilon_r = \epsilon/\epsilon_0$$

See **dielectric constant, permittivity**.

relativistic 1. Referring to a particle traveling at a speed approaching that of light, for which relativistic effects are appreciable. 2. Referring to a frame moving at a speed approaching that of light, within which relativistic effects become appreciable.

relativity The theory developed by Albert Einstein between 1905 and 1916, based on the principle that the speed of light in vacuo is invariant. **1. special relativity** (1905) Physical phenomena yield identical observations in nonrotating reference systems in uniform rectilinear motion relative to each other. **2. general relativity** (1909–1916) An extension of special relativity including gravitation as an effect of the curvature of the space-time continuum.

relaxation time The time required by a system to return to its original state after cessation of a stimulus.

relict Referring to a geological feature (crystal, sediment, landscape, etc.) remaining from a preceding time when conditions were different.

reluctance ($\Re$) A measure of the resistance of a substance or magnetic circuit to magnetic flux.

$$\Re = mmf/\Phi$$
$$= l/\mu A$$

where mmf = magnetomotive force, Φ = magnetic flux, l = length, μ = permeability, A = cross section. It is expressed in henry^{-1}. Cf. **permeance**.

rem Roentgen-equivalent-man. The amount of radiation producing the same damage to humans as 1 roentgen of high-voltage x-rays.

remanent magnetization The magnetization acquired by a rock or sediment at the time of formation.

remanié Reworked.

repetition The duplication of a stratigraphic sequence by recumbent folding.

residence time The average amount of time a given substance spends as part of a given system. Residence time is proportional to the size of the system and inversely proportional to the flux of the substance through the system.

resistance (R) The opposition of a circuit to the flow of electricity. It is expressed in ohms.
1. *dc circuits:*

$$R = V/I$$

where V = voltage, I = current
2. *ac R circuits:*

$$R = V_M/I_M$$

where V_M = maximum voltage, I_M = maximum current. Cf. **conductance**.
3. *ac RLC circuits:* The real part of the complex impedance Z:

$$|Z| = [(R^2 + (\omega L - 1/\omega C)^2]^{1/2}$$

where $\omega = 2\pi f$ = angular frequency in radians/second; L = impedance, C = capacitance. Cf. **reactance**.

resistates Sediments formed by minerals resistant to weathering.

resistivity (ρ) The specific resistance of a material to the flow of electric current.

$$\kappa = RA/l$$
$$= 1/\sigma$$

where R = resistance, A = cross-section of conductor, l = length of conductor, σ = conductivity. Resistivity is measured in ohms·m (Ω m). It ranges from $1.6 \cdot 10^{-8}$ to $185 \cdot 10^{-8}$ for metals, from 10^{-6} to 10^6 for semiconductors, and from 10^6 to 10^{15} for insulators. Examples: Ag = $1.586 \cdot 10^{-8}$ (20°C) (lowest among the elements), Cu = $1.678 \cdot 10^{-8}$ (20°C), Mn = $185 \cdot 10^{-8}$ (25°C), Si = $3-4 \cdot 10^{-2}$ (0°C), yellow S = $2 \cdot 10^{15}$ (20°C) (highest among the elements). Cf. **conductivity**.

resistor An electrical component providing resistance to the flow of current.

resonance *(Physics)* The enhanced response of a system having a natural oscillation frequency when excited by a similar frequency from outside.

(Chemistry) A bond hybrid between two or more Lewis structures.

respiration The process of oxidation of organic molecules within a cell leading to the formation of ATP and the release of energy.

rest mass The mass of a particle at rest in an inertial frame.

retrograde motion 1. Referring to the clockwise motion of a planet or other celestial body around the Sun, or of a satellite around its planet, as seen from the North. Cf. **prograde**. 2. Referring to the apparent reversal in the motions of the outer planets along their orbits due to the motion of the Earth along its orbit. Cf. **epicycle**.

reverse bias See **bias**.

reversed See **overturned**.

reversed fault A compressional fault with a slanted fault plane along which one block tends to override the other.

reversed magnetization Magnetization produced at a time of reversed polarity.

reversed polarity The polarity of the geomagnetic field opposite that prevailing during the present (Brunhes) polarity epoch. Cf. **normal polarity**.

reversible Referring to a process that dissipates no energy.

revolution *(Astronomy)* 1. The motion of a body about a center external to the body. 2. The orbiting of a celestial body around another. *(Geology)* An intense period of orogenesis.

reworked A sediment or a fossil that has been transported and redeposited into a younger formation.

Reynolds number (N_{Re}) The ratio of inertia force to viscous force in a fluid in motion.

$$N_{Re} = \rho v l / \mu$$

where ρ = density, v = velocity, l = characteristic length of the system, μ = viscosity.

RF Radio frequency.

rhabd- Prefix meaning *rod*.

rheg- Prefix meaning *carpet*.

rhegolith Unconsolidated, fragmented cover of rock materials. Commonly misspelled *regolith*.

rheidity Fluidity.

rheo- Prefix meaning *flow*.

rheology The study of flow.

rheostat A variable electrical resistor.

rhizo- Prefix meaning *root*.

rhombochasm A rhombic gap in the Earth's crust, caused by tension in two different directions.

rhombohedral A subsystem of the hexagonal crystal system, characterized by having a rhombohedral unit cell.

rhombohedron A parallelepiped with six identical rhombic faces.

rhumb line A line crossing successive meridians at a constant angle. Except for the equator, it spirals toward the pole and is represented by a straight line on a Mercator projection. Syn. **loxodrome**. See **Mercator projection**.

rhyolite The extrusive equivalent of granite, consisting of quartz and orthoclase phenocrysts in a glassy groundmass.

rhythmite A unit in a rhythmic succession of sedimentary beds.

ria A drowned river valley.

ribonucleic acid See **RNA**.

ribose A pentose sugar, $C_5H_{10}O_5$.

ribosome An organelle, about 250 Å across, present in all animal and plant cells, where protein synthesis occurs. Mass of ribosome ≈ 2.5–$4.5 \cdot 10^6$ u. Number of ribosomes/cell $\sim 10^4$ (procaryota), $\sim 10^6$–10^7 (eucaryota).

Richter scale A scale of earthquake intensity. The magnitude of an earthquake on the Richter scale is given by the equation

$$M = \log_{10}(A/T)$$

where M = magnitude, A = maximum amplitude of ground motion in μm registered by a standard Wood-Anderson short-period seismometer at the standard distance of 100 km (special tables are used to reduce observations to this standard distance), and T = dominant wave period in seconds. The smallest earthquakes have a Richter magnitude of -2 or -3, and the largest ones have a magnitude of up to 9. Magnitude is related to energy dissipated by the equation

$$\log_{10} E = 5.24 + 1.44\, M$$

where E = energy in joules, M = Richter magnitude.

rift A major graben, caused by doming produced by a deep-seated spreading line.

rift valley A valley developed along a rift.

right ascension (Ra, α) Longitudinal coordinate in the equatorial coordinate system. See **coordinate system (celestial).**

right-hand rules A set of unnecessary "rules" to describe the vectorial product

$$F = il \times B$$

where F = force, i = current (either consisting of positive ions or taken as having a sense opposite that of electron flow), l = length of current-carrying conductor, B = external magnetic field. 1. (Direction of magnetic field created by a *positive* charged particle moving in a straight line or a *positive* current flowing through a straight conductor.) If the fingers of the right hand are wrapped around the path of the particle or around the conductor and if the thumb is extended in the direction of motion of the particle or the current flow, the fingers will indicate the path of the circular magnetic field created. 2. (Force on a *negative* charged particle or on a conductor carrying a *negative* current in an external magnetic field.) If the thumb, first, and second finger of the right hand are extended at 90° to each other, and if the second finger indicates the direction of motion of the charged particle or of the negative current flowing through the conductor, and if the first finger indicates the direction of the external magnetic field, the thumb indicates the direction of the force experienced by the particle or the conductor. See **left-hand rules.**

right-lateral fault A fault in which a block appears to have moved to the right when viewed from the opposite block. Cf. **left-lateral fault.**

rill 1. A streamlet. 2. A lunar surface structure resembling the dry bed of a streamlet.

ringer A thin-bedded, fine-grained, cemented sandstone that "rings" when hit with a hammer.

riparian Pertaining to the bank of a river.

rip current A strong, narrow return current of water piled onshore by waves and made convergent by coastal morphology.

ripple mark A ribbed surface structure on loose sand or silt caused by water motion.

*R***-mode factor analysis** Factor analysis concerned with relationships among variables. Cf. **Q-mode factor analysis.**

rms Root mean square.

RNA Ribonucleic acid, consisting of a ribose sugar-phosphate chain with one of four bases (adenine, cytosine, guanine, uracil) attached to each sugar. RNA is coded by DNA in the nucleus and carries the genetic information to the ribosomes where protein synthesis takes place. See **messenger RNA, transfer RNA.** Cf. **DNA.**

Roche limit The minimum distance between a larger celestial body and a smaller one orbiting around it at which tidal effects from the larger body fail to overcome the self-gravitation of the smaller body, thus making possible its accumulation and continued existence. The Roche limit is $2.4R$, where R = radius of the larger body, for two bodies of zero tensile strength and identical density.

roche moutonnée A bare rock surface sculptured by an overriding glacier or ice sheet so as to resemble a flock of sheep as seen from above.

rock An aggregate of one or more minerals formed by crystallization, solidification, sedimentation, or precipitation.

rock crystal Quartz crystal.

rock flour The very fine product of bedrock abrasion by rocks embedded in the underside of a glacier or ice sheet.

rock salt Coarsely crystallized NaCl.

rock-stratigraphic unit See **lithostratigraphic unit.**

roentgen The amount of x-ray or gamma radiation that produces, in 1 cc of dry air at 0°C and 760 mmHg, ions carrying 1 statcoulomb of electricity of either sign.

ROM Read-only memory. A semiconductor device allowing direct retrieval ("read only") from (but no input to) individual cells, each storing one bit of information. Single chips may contain 64K, 128K, or 256K individual cells. Retrieval time is in the order of $0.1–1$ μs.

room temperature 20°C (68°F; U.S.) or 15.5°C (60°F; U.K.).

root mean square (rms) The square root of the arithmetic mean of the squares of a set of numbers.

root-mean-square current See **effective current.**

root-mean-square voltage See **effective voltage.**

rosette A crystalline aggregate of barite, marcasite, or pyrite in sedimentary rocks, resembling a rose.

rot Rotation. See **curl.**

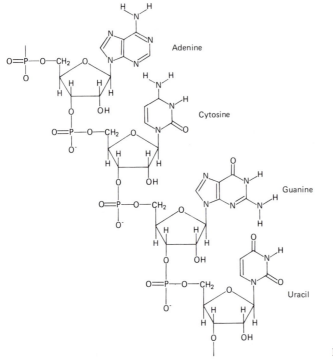

RNA segment.

rotation 1. The spinning of a body about an internal axis of symmetry. 2. See **curl.**

rotational energy *(Classical mechanics)* The kinetic energy E_k of a rotating body:

$$E_k = \tfrac{1}{2}I\omega^2$$

where I = rotational inertia = $\int r^2 dm$ (r = shortest distance between axis of rotation and mass increment dm), ω = angular velocity. *(Physical chemistry)* The component of the total energy of a diatomic or polyatomic molecule related to its rotational motion. For a diatomic molecule:

$$E_{rot} = J(J + 1)h^2/8\pi^2I$$

where E_{rot} = rotational energy, J = inner quantum number, h = Planck's constant, I moment of inertia of the molecule.

rotational inertia See **moment of inertia.**

roundness The ratio of the average radius of curvature of a sedimentary particle to the radius of the maximum inscribed sphere.

***r*-process** The rapid capture, by iron group-elements, of free neutrons released by a supernova explosion, and the consequent, rapid formation of all heavier elements, including the transuranic ones, by neutron capture and β^- decay. See **element formation.**

RR Lyrae stars Old population II giant periodic variable stars occurring mainly in globular clusters. The period is usually <1 day.

rubidium-strontium dating method An absolute dating method based on the β^- decay of ^{87}Rb ($t_{1/2}$ = 48·10^9 y) to ^{87}Sr (stable). See **isochron.**

ruby Red-colored gem corundum (Al_2O_3). Cf. **oriental ruby. See Gems*.**

rudaceous Referring to the texture of a rudite.

rudite A sedimentary rock consisting of fragments larger than coarse sand (i.e. >2 mm across). Cf. **arenite, lutite, siltite.**

runnel 1. A rivulet. 2. The channel excavated by a rivulet.

runoff The amount of water directly returned to the ocean by stream flow.

rupestral Referring to a high, rocky cliff or area.

rust A mixture of iron oxides, hydroxides, and carbonates.

rutile The mineral TiO_2.

Rydberg constant (R_∞) An atomic constant describing the energy binding the electron to the atomic nucleus (assumed to have infinite mass).

$$R_\infty = 2\pi^2 me^4/h^3c$$

where m = rest mass of electron, e = electron charge, c = speed of light, h = Planck's constant. It is equal to $10,973,731.534$ m^{-1}.

S

σ 1. Electric conductivity. 2. Neutron-capture cross section. 3. Poisson ratio. 4. Standard deviation. 5. Stefan-Boltzmann constant. 6. Surface electric charge density.

Σ Summation.

s 1. Second. 2. Sharp (see **s, p, d, f**).

S 1. Entropy. 2. Secondary or shear (seismic waves). 3. Siemens. 4. South-seeking pole of a magnetic dipole. 5. Strangeness number.

s_A Atomic second.

s_E Ephemeris second.

Sabellaria A genus of encrusting worms belonging to the order Sedentaria, class Polychaeta, phylum Annelida.

sabellarid reef A mound of encrusting sabellarid worms.

sabkha A supratidal, coastal environment with evaporites and eolian deposits, characteristic of arid coastlines.

saccharoidal Referring to a microcrystalline texture resembling sugar.

salada A dry salt-lake bed.

sal ammoniac Ammonium chloride, NH_4Cl.

salcrete A hard crust along a shoreline consisting of sand grains cemented by salt crystals.

salina A salt flat where halite deposits can be found.

salinity The amount of dissolved salts in marine or continental waters, with Br and I represented as Cl. It is expressed in g/kg or ppt. See **chlorinity.**

salt 1. Any of the products of reactions between acids and bases. A salt contains the anion of the acid and the cation of the base. 2. Sodium chloride, NaCl.

saltation The transport of sand grains by moving water along the bottom through a series of progressive jumps as the particles hit other particles resting on the bottom.

salt dome A salt diapir, usually round in cross section, 2–3 km across and 5–10 km high, originating in a deep salt bed and piercing overlying sediments.

salt lake A lake in a region where evaporation exceeds precipitation and runoff, with no outlet, and with water enriched in salts. The excess major ions include Cl^- and Na^+ (e.g. Great Salt Lake, Utah) and, in addition, SO_4^{2-} (e.g. Caspian Sea) or CO_3^{2-} (e.g. Lake Laach, Germany), or all four (e.g. Mono Lake, California).

salt marsh A coastal marsh periodically flooded with seawater.

salt pan A small depression in which water evaporates leaving a salt deposit.

saltpeter KNO_3. Syn. **niter.**

salt weathering The fragmentation of a rock by the crystallization of salts in its pores.

samarium-neodymium dating method An absolute dating method based on the α decay of ^{147}Sm ($t_{1/2} = 106 \cdot 10^9$ y) to ^{143}Nd (stable). The ratio ^{147}Sm/^{143}Nd is plotted vs. ^{143}Nd/^{144}Nd to obtain an isochron. See **isochron.**

sammelkristallization A recrystallization process by which smaller crystals are overgrown by larger crystals and incorporated into their structure.

sand A sediment with particle size ranging from 1/16 (0.0625) mm to 2 mm.

sand line A wire line used to lower and raise tools through a drill hole.

sandshale A sedimentary deposit consisting of alternating sandstone and shale beds.

sandstone A sedimentary rock consisting of sand-size particles in a finer matrix. The particles are variously cemented by carbonates, silica, and/or iron oxides.

sand wave A submarine accumulation of sand in parallel ridges, caused by bottom currents.

sanidine A high-temperature, disordered form of orthoclase ($KAl Si_3O_8$) with some Na substituting for K.

saponite A Mg-rich clay mineral of the montmorillonite group.

sapphire Blue-colored gem corundum (Al_2O_3). Cf. **oriental sapphire.** See **Gems*.**

sapro- A prefix meaning *rotten, putrid.*

sapropel A putrescent accumulation of algal and other vegetable matter decaying under anaerobic conditions.

sapropelite A sapropelic coal.

saprophyte A plant living on decaying organic matter.

Sargassum A floating brown alga belonging to the class Cyclosporeae, phylum Phaeophyta.

saros The period of 223 synodic lunar months = 6585.32 days = 18.030 y for the recurrence of a particular sequence of solar and lunar eclipses.

satellite Any of the bodies orbiting a planet. The solar system includes 54+ satellites, distributed among the planets as follows: Mercury, 0; Venus, 0; Earth, 1; Mars, 2; Jupiter, 16; Saturn, 17; Uranus, 15; Neptune, 2; Pluto, 1. Masses range from $2.0 \cdot 10^{15}$ kg (Deimos of Mars) to $1.490 \cdot 10^{23}$ kg (Ganymede of Jupiter). Densities range from 3 to 3.5 for the Moon, Io, and Europa, and from 1 to 2 for most other satellites. See **Satellites*.**

satin spar A white, translucent variety of gypsum.

saturated *(Chemistry)* 1. Referring to a solution incapable of holding additional solute. 2. Referring to a carbon compound in which all C to C bonds are single. *(Geology)* Referring to an igneous rock with quartz in its norm.

Saturn The sixth planet from the Sun. Mean distance from the Sun = 9.554747 AU. Sidereal period = 29.4577 y; sidereal rotational period at equator = 10.233 h. Equatorial radius = 60,268 km; polar radius = 54,364 km. Mass = $568.8 \cdot 10^{24}$ kg. Mean density = 0.67 g/cm³. Internal structure (estimated): Fe–Ni and silicate core with radius = 16,000 km and a 12,000-km-thick mantle of metallic H. Magnetic field = 0.2 gauss. Atmosphere thickness = 32,000 km; surface atmospheric pressure $\gg$ 100 bar. Gases in atmosphere, H_2 = 90%, He = 10%. Seventeen satellites, the largest of which is Titan [radius = 2575 km; internal structure (estimated): Ni–Fe and silicate core 1720 km in radius and an 840-km-thick mantle of frozen gases; atmosphere thickness = 300 km; surface atmospheric pressure = 1.6 bar; gases in atmosphere, N_2 = 96%, CH_4 = 2%, other gases = 2%].

Ring system <200 m thick, consisting of debris up to meter size and extending from close to the equatorial surface of the planet to 480,000 km away. See **Planets—atmospheres*, Planets—physical data*, Satellites*.**

saussurite A rock consisting mainly of albite and zoisite derived from the alteration of plagioclase in basalts and gabbros.

saussuritization A metamorphic or deuteric process by which saussurite is produced in basalts and gabbros.

savanna An open, grassy expanse with shrubs and scarce trees in the semiarid regions of Africa.

sb Stilb.

scabland A basaltic plateau rapidly eroded during the ice ages, exhibiting deep, dry channels.

scaglia A fine-grained pelagic limestone characterized by conchoidal fracture (Upper Cretaceous-Lower Tertiary, Italy).

scalar A quantity that has magnitude but no direction. Cf. **vector.**

scalar product (·) The scalar product of two vectorial quantities. Syn. **dot product.** Cf. **vectorial product.**

scalenohedron A closed crystal form bound by scalene triangles.

scanning electron microscope (SEM) A microscope in which a beam of electrons a few angstrom across is made to scan the surface of a sample. The intensity of the secondary electrons thus generated produces a signal that is fed to a cathode-ray tube screen or to a photographic plate. Cf. **electron microscope.**

scattering layer See **deep scattering layer.**

schiefer A laminated rock, such as a shale, slate, or schist.

schist A metamorphosed shale with foliated texture largely due to muscovite crystals.

schistosity Foliation, as exhibited by schists.

schlieren Tabular bodies, 10 cm to 20 m thick, in igneous rocks having different color index from that of the surrounding rock.

schreibersite The mineral $(Fe,Ni)_3P$, common in iron meteorites.

Schroedinger equation A partial differential

equation describing the wave function of a non-relativistic particle:

$$(h^2/8\pi^2m)(\partial^2\Psi/\partial x^2 + \partial^2\Psi/\partial y^2 + \partial^2\Psi/\partial z^2) - V\Psi = (h/2\pi i)(\partial\Psi/\partial t)$$

where h = Planck's constant, m = mass of particle, Ψ = wave function, V = potential energy of particle, $i = (-1)^{1/2}$.

Schulze's solution A macerating solution consisting of a saturated aqueous solution of $KClO_3$ and varying amounts of HNO_3.

Schwarzschild radius The radius of the event horizon of a nonrotating black hole:

$$R_s = 2GM/c^2$$

where R_s = Schwarzschild radius, G = gravitational constant, M = mass of object, c = speed of light.

scintillation The conversion of radioactive energy into light by scintillators.

scintillator A solid or liquid substance capable of emitting a light pulse of short decay time (10–$250 \cdot 10^{-9}$ s) following absorption of an ionizing particle. Solid scintillators for β^- particles include sodium iodide (NaI) and anthracene (a polycyclic aromatic hydrocarbon, $C_{14}H_{10}$); liquid scintillators include 2,5-diphenyloxazole (PPO) plus 1,4-*bis*-(5-phenyloxazol-2-yl)-benzene (POPOP) dissolved in toluene or xylene to which an appropriate liquid containing the radioactive substance is added. Light emission is detected with a photomultiplier tube. For α particles the scintillator is a thin layer of Ag-activated zinc sulfide (ZnS) coated on the envelope of the photomultiplier tube.

scissor fault A fault with a pivotal point on either side of which there is an increasing offset with increasing distance along the strike.

scoria A vesicular lava surface or lava fragment.

scoriaceous Referring to the texture of scoria.

SCR Silicon-controlled rectifier.

scree An accumulation of loose rock fragments at the base of a cliff.

screw dislocation A crystal defect of increasing amplitude along a crystal plane from inside out. As the crystal grows, the defect appears to rotate around a central axis normal to the crystal plane.

sea 1. A secondary body of saltwater compared to the major oceans. Seas with surface $>1\cdot10^6$ km^2 are (surface and average depth in parenthesis): South China Sea ($2.795\cdot10^6$ km^2, 1437 m), Caribbean Sea ($2.515\cdot10^6$ km^2, 2575 m), Mediterranean

Sea ($2.510\cdot10^6$ km^2, 1502 m), Bering Sea ($2.261\cdot10^6$ km^2, 1492 m), Gulf of Mexico ($1.507\cdot10^6$ km^2, 1614 m), Sea of Okhotsk ($1.392\cdot10^6$ km^2, 973 m), Sea of Japan ($1.013\cdot10^6$ km^2, 1667 m). 2. A major, inland body of saltwater. Inland seas with surface $>50\cdot10^3$ km^2 are: Caspian Sea ($370.8\cdot10^3$ km^2, av. depth = 1025 m), and the Aral Sea ($64.5\cdot10^3$ km^2, av. depth = 67 m).

sea ice Ice resulting from the freezing of seawater. See **pack ice.**

sea level See **mean sea level.**

sealevel datum A reference mean sea level for expressing elevations and depths.

seamount An elevation of 1000 m or more above the ocean floor, regardless of how it originated.

seaquake An earthquake below the ocean floor.

seasonality The differential seasonal behavior of a phenomenon, organism, or group of organisms.

seawater The water of the ocean or sea (also written *sea water*). Concentration of major ions (g/kg) for average salinity of 35.0‰ : Cl^- = 19.353; Na^+ = 10.775; SO_4^{2-} = 2.712; Mg^{2+} = 1.295; Ca^{2+} = 0.412; K^+ = 0.400; HCO_3^- = 0.145; Br^- = 0.067; Sr^{2+} = 0.008; B^{3+} = 0.0046; F^- = 0.0013. Estimated number of H_2O molecules within the structure domain of a single Na^+ ion: 52 (5°C), 34 (20°C), 21 (50°C).

sec Secant.

secant See **trigonometric function.**

second (s) The SI, MKS, and CGS unit of time. See **atomic second, ephemeris second.**

secondary A celestial body orbiting a primary.

secondary cell See **storage cell.**

secondary electrons Electrons emitted by a target as a result of the impact of particles or waves of sufficient energy.

secondary optic axis Any of the four directions of equal ray velocity from the center of the indicatrix of a biaxial crystal. Cf. **primary optic axis.**

secondary radionuclide Any of the relatively short-lived radionuclides that occur naturally because they are continuously produced as part of the decay chain of primary radionuclides. Examples are ^{234}U, ^{231}Pa, ^{230}Th, ^{226}Ra, ^{222}Rn, and ^{210}Pb. Cf. **induced radionuclide, primary radionuclide.** See **Isotope chart*.**

secondary rock A rock consisting of particles derived from pre-existing rocks.

secondary structure A structure acquired by a rock subsequent to its formation and emplacement.

secondary wave See **S wave.**

second law of thermodynamics "The change in entropy of a reversible system is equal to the heat absorbed by the system divided by the absolute temperature of the system."

$$dS = dQ/T$$

where S = entropy, Q = heat absorbed by the system, T = absolute temperature. See **entropy.**

second order reaction A chemical reaction proceeding at a rate proportional to the product of the concentrations of two reactants or to the square of the concentration of a single reactant. See **order of reaction.**

secular Having a long time-range.

sedimentary rock A rock formed by consolidation and cementation of subaqueous or subaerial sediments or by precipitation from marine or freshwater. See **Sedimentary rocks*.**

sediment load The amount of solid matter carried in suspension by moving water.

Seebeck effect The development of an emf (up to several tens of millivolts) caused by a temperature difference between two junctions of different metals in the same circuit. See **thermocouple.**

segregation *(Geology)* Concentration of specific crystals in specific parts of a magma during cooling and crystallization. *(Biology)* The separation of alleles and homologous chromosomes during meiosis.

seiche A standing wave in a lake or bay.

seif A major dune elongated in the wind direction, reaching up to 200 m in height and 300 km in length.

seismic Pertaining to abrupt, internal motions of the Earth, either natural or artificial.

seismic reflection The study of the shallower internal structure of the solid Earth by generating acoustic waves and registering their reflection from subsurface layers.

seismic refraction The study of the shallower internal structure of the solid Earth by generating acoustic waves and registering their return at increasing distances from the acoustic source, thus detecting waves that have been transmitted along deeper layers of greater elasticity and rigidity. See **critical distance.**

seismic wave An elastic wave transmitted through the body or along an interface of the Earth (including ocean and atmosphere). See **Seismic waves*.**

seismograph An inertial system to detect vibrations of the solid Earth.

selection rules Rules that pertain to the changes, in quantum numbers of a quantum mechanical system, to effect with appreciable probability a transition between two states. If the probability is too low, the transition is termed *forbidden.*

selenite Macrocrystalline gypsum.

selenium cell A photocell using selenium as an electron emitter. Selenium has a high (5.9 eV) work function.

selenology The study of the Moon.

self-induction The appearance of emf in a coil when the current in the coil changes.

SEM Scanning electron microscope.

semi Prefix meaning *one half.*

semiconductor A solid crystalline material with electric conductivity ranging from 10^{-6} to 10^6 siemens·m, intermediate between that of insulators (10^{-4} to 10^{-6} S·m) and that of metals (10^6 to 10^8 S·m). Semiconductors consisting of semiconducting elements such as Si and Ge, which belong to group 14 of the Periodic Table (4 electrons in valence shell), are called *intrinsic.* Contrary to conductors, the highest electronic energy band (valence band) in semiconductors is separated from the conduction band by an energy gap (1.09 eV for Si, 0.72 eV for Ge at room temperature) which, however, is not as wide as that in insulators. When electrons in the valence band are sufficiently energized to cross the energy gap and enter the conduction band, the vacancies left in the valence band, called holes, act as positive particles. Under the influence of an applied voltage, electrons and holes move in opposite directions. Electron conduction is enhanced by *doping* the semiconductor with a minute amount (10^{-6}) of an element from group 15 of the Periodic Table of the Elements (especially P and As), thus introducing one essentially free electron per atom of the doping substance and creating an *n-type* semiconductor. Hole induction is enhanced by doping the semiconductor with an equally minute amount (10^{-6}) of an element from group 13 of the Periodic Table (especially Al or Ga) with 3 electrons in the valence shell, thus introducing one hole per atom of the doping substance and creating a *p-type* semiconductor. Doped semiconductors are called *extrinsic*. A *pn* junction is the area of contact between a *p*-type and an *n*-type semiconductor. Electrons from the *n*-type region in the immediate vicinity of the junction move toward the *p*-type region leaving behind a thin, electron-depleted (and, therefore, positive) area; analogously, holes from the *p*-type region moving toward the *n*-region leave behind a thin, hole-depleted (and therefore negative) area. These two areas form a potential barrier preventing further motion of the electrons from the *n*-type region to the *p*-type region and of holes in the opposite direction. This barrier is reduced or eliminated by the application of a forward bias (increased potential difference between the *n*- and *p*-regions by the application of an external voltage), resulting in current flow.

semiconjugate axis The portion of a conjugate axis between the vertex of a hyperbola and either asymptote.

semidiurnal Referring to a phenomenon occurring twice a day.

semimajor axis (a) The distance between the center of an ellipse through one of the foci to the perimeter.

semiminor axis (b) The distance between the center of an ellipse and the periphery normally to the major axis.

senescence The process of aging.

senescent *(Geology)* Referring to an almost completely eroded landscape. Cf. **mature** (def. 2). *(Biology)* Referring to an aging organism.

sensu lato (s.l.) Latin for *in a broad sense, broadly speaking.*

sensu stricto (s.s.) Latin for *in a strict sense, strictly speaking.*

sepiolite A very light, porous, hydrated Mg silicate, $Mg_4(Si_2O_5)_3 \cdot (OH)_2 \cdot 6H_2O$. Cf. **meerschaum.**

septum Partition.

sequestering The removal of a metal ion from a system by chelation.

serac Any of the broken ice ridges that form when the bed of a glacier suddenly steepens.

sere A temporary ecologic community occurring as part of a series of such communities succeeding each other in the same area.

series A chronostratigraphic unit below system and above stage in rank, formed during a geologic epoch.

series circuit A circuit in which each element has its negative terminal connected to the positive terminal of the next element.

serpentine A hydrated Mg–Fe silicate, $(Mg, Fe)_3Si_2O_5(OH)_4$, resulting from alteration of olivine and other Mg–Fe silicates.

serpentinite A rock consisting mainly of the mineral serpentine.

serpentinization The process of alteration of olivine and other Mg–Fe silicates resulting in the formation of serpentine.

serpulid Any of the worms belonging to the family Serpulidae, order Sedentaria, class Polychaeta, phylum Annelida, capable of building contorted tubes encrusting submarine surfaces.

serpulid reef A small reef consisting largely of serpulid tubes.

servomechanism A feedback mechanism used to control mechanical systems.

sesqui- Prefix meaning *one and one half times.*

sessile Referring to a plant or animal attached to a substrate since early development.

sextant A navigational instrument to measure the angular elevation of the Sun or other celestial body above the horizon. It consists of a telescopic sight to view the horizon, a pivoting 60° sector graduated into 120°, a mirror (the *index glass*) to reflect the light of the Sun or other celestial body into a half-silvered mirror (the *horizon glass*) in front of the telescopic sight. By pivoting the graduated sector, the image of the Sun or other celestial body is superimposed on the horizon and the angle is read. Cf. **astrolabe.**

Seyfert galaxy A type of spiral galaxy with a bright, very active, turbulent nucleus emitting 10^4 times more energy than normal spiral galaxies. Spectral lines are broadened indicating turbulence in the source. Cf. **N-galaxy.**

shadow matter A hypothetical form of matter arising from superstring theories and capable of interacting with ordinary matter (quarks and leptons) only through gravitational interaction. This form of matter is hypothesized to have come into existence during the Planck era (the first $5.390 \cdot 10^{-44}$ s). It would be detectable today only by its gravitational effect on regions of space larger than the solar system.

shale A sedimentary rock consisting of indurated silt and clay in thin layers.

shallow earthquake See **earthquake.**

shallow-focus earthquake See **shallow earthquake.**

shaly Having the texture of shale.

shared Referring to an electron pair shared by two adjacent atoms in the formation of a covalent bond.

shear A deformation along a plane or set of planes tangential to the force applied.

shear modulus The ratio of shear stress to angle of deformation expressed in radians. Cf. **Young's modulus.**

shear strength The internal resistance of a body to shear stress.

shear stress Stress on a body caused by a force tangential to one of its surfaces.

shear wave See **S wave.**

shelf break See **shelf edge.**

shelf edge The edge of the continental shelf at an average depth of 130 m below mean sea level.

shell *(Physics)* Any of the principal quantum states of an electron, identified by the principal quantum number n. See **orbital, quantum number, subshell.** *(Biology)* The hard, calcareous, siliceous chitinous, or chitinophosphatic outer covering of an invertebrate animal.

shield The inner portion of a craton, consisting of exposed, ancient rock basement. See **craton.**

shield volcano A broad volcano built by lava flows of low viscosity (e.g. the Mauna Loa and Mauna Kea, Hawaii).

shoal A shallow bottom.

shock metamorphism Metamorphism produced by shock waves, as in meteoritic impacts.

shock wave 1. A large-amplitude acoustic wave caused by supersonic motion of a body in a medium. 2. An acoustic wave traveling at supersonic speed through a body.

shoestring A long, narrow body of sediment, usually sand or sandstone, embedded in silt, clay, or shale, and representing channel fill, a beach deposit, or a bar.

shore The narrow strip of land immediately adjacent to a body of water.

shoreface The permanently submerged zone below the low-water line where wave action is strong.

shoreline The changing borderline between land and water.

shott See **chott.**

shower A cascade of particles produced in the upper atmosphere by the impact of high-energy galactic protons with nuclei of atmospheric gases. A giant shower originating from a single particle with an energy of 10^{20} eV will produce millions of particles and cover more than 1 km^2 of ground at sea level.

SI Système International d'Unités, the International System of Units based on the meter, kilogram, second, ampere, kelvin, candela, and mole. See **Units*.**

sial Silica-aluminum, indicating the upper portion of the continental crust above the sima, largely consisting of aluminosilicates. Cf. **sima.**

sialic Pertaining to sial.

sibling species Species that, as such, are genetically isolated but are nevertheless morphologically similar.

sidereal Stellar.

sidereal day The time interval between successive passages of the vernal equinox across a given celestial meridian. It is divided into 24 hours and is equal to $86,164.09055 + 0.0015T$ s_E (where T = tropical centuries from 1900.0) = $86,164.09186$ s (1987) = 23h 56m 4.0918s (1987). It starts at sidereal noon.

sidereal hour 1/24th of a sidereal day = 59m 50.170s (1987).

sidereal month The time the Moon takes to make a revolution around the Earth with respect to the backdrop of the fixed stars. It is equal to $27.32166140 + 0.00000016T$ d_E, where T = tropical centuries from 1900.0.

sidereal noon The moment when the vernal equinox crosses the local meridian.

sidereal period (P) The time taken by a planet or satellite to complete a revolution around its primary, with distant stars as a frame of reference. For the inferior planets:

$$1/P = 1/P_e + 1/P_{syn}$$

where P_e = sidereal period of the Earth, P_{syn} = synodic period of the planet or satellite. For the superior planets:

$$1/P = 1/P_e - 1/P_{syn}$$

Cf. **synodic period.**

sidereal time The time based on the rotation of the Earth with respect to the fixed stars.

sidereal year The time required for the longitude of a distant star to increase by 360°. It is equal to $(31,558,149.984 + 0.010T)$ s_E or $(365.25636556 + 0.000000011T)$ d_E, where T = tropical centuries from 1900.0.

siderite 1. Iron carbonate, $FeCO_3$. 2. Any of the iron meteorites, consisting of Fe–Ni alloys with 5–30% Ni. See **Meteorites***.

siderolite See **stony-iron meteorite.**

sideromelane Basaltic glass.

siderophile Describing an element having affinity for the metallic rather than for the sulfide or silicate phases. **Cf. chalcophile.**

siemens (S) The SI and MKS unit of conductance, equal to the conductance between two points of a conductor that allows the passage of 1 ampere when the potential difference between them is 1 volt. The siemens is the reciprocal of the ohm, and was formerly called *mho.*

sigma *(Physics)* **(Σ)** A triplet of strangeness-1 baryons. See **Elementary particles***. *(Mathematics)* **(σ)** Standard deviation.

sigma bond (σ) A covalent bond with high electronic density between the two nuclei and exhibiting symmetry about the axis joining the two nuclei.

silex Latin for *silica.*

silica A polymorphic mineral, SiO_2. Phase changes at 1 atm of pressure: low-quartz/573°C/high-quartz/867°C/tridymite/1470°C/cristobalite/1713°C/liquid. High-pressure phases are coesite (density = 2.915) and stishovite (density = 4.28 g/cm³).

silica gel A colloidal combination of silica particles and water. It can absorb water from the air and be dehydrated by heating.

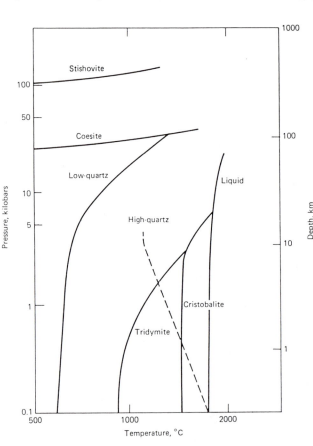

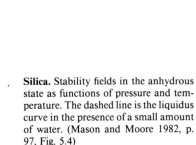

Silica. Stability fields in the anhydrous state as functions of pressure and temperature. The dashed line is the liquidus curve in the presence of a small amount of water. (Mason and Moore 1982, p. 97, Fig. 5.4)

silica glass 1. Glass made of pure silica. 2. The mineral lechatelierite.

silica sand Sand consisting of pure silica grains.

silicate Any of the minerals containing SiO_4 tetrahedra as part of their crystal structure. There are six families: **1. neososilicates** $(SiO_4)^{4-}$; Si/O = 0.25. Independent tetrahedra. E.g. olivine. **2. sorosilicates** $(Si_2O_7)^{6-}$; Si/O = 0.29. Tetrahedra pairs sharing an oxygen. E.g. epidote. **3. cyclosilicates** $(Si_3O_9)^{6-}$, $(Si_4O_{12})^{-6}$, $(Si_6O_{18})^{6-}$; Si/O = 0.33. Rings of tetrahedra each sharing two oxygens. E.g. beryl. **4. inosilicates** $(SiO_3)_x^{2-}$, $(Si_4O_{11})_x^{6-}$, Si/O = 0.33 or 0.36. Chains of tetrahedra each sharing 2 or 3 oxygens. E.g. augite. **5. phyllosilicates** $(Si_2O_5)_x^{2-}$; Si/O = 0.40. Sheets of tetrahedra each sharing 3 oxygens. E.g. kaolinite. **6. tectosilicates** SiO_2; Si/O = 0.50. Three-dimensional framework of tetrahedra each sharing 4 oxygens. E.g. quartz.

siliceous ooze Either diatom or radiolarian ooze, deep-sea sediments that contain at least 30% siliceous skeletal elements. Siliceous oozes cover 15% of the ocean floor. See **deep-sea sediments.**

silicified wood Wood that has been transformed into chalcedony in such a way that the original cell structure is preserved.

silicoflagellate Any of the unicellular flagellate algae of the class Silicoflagellata, phylum Chrysophyta, kingdom Protoctista, with an exoskeleton of opaline silica.

silicon-controlled rectifier (SCR) A *pnpn* device in which the base *n* adjacent to the anode (the first *p*) functions also as emitter for the *npn* section, and the base *p* adjacent to the cathode (the second *n*) functions both as collector for the first section *(pnp)* and gate for the second section *(npn)*. Conduction occurs only when the gate terminal is activated by an appropriate signal but it continues after removal of the signal until anode voltage is reduced, removed, or reversed. Silicon-controlled rectifiers are used as relays, switches, or rectifiers. The silicon-controlled rectifier is the solid-state equivalent of the thyratron.

sill 1. A sheet-like intrusion of an igneous rock parallel to the bedding plane or foliation of the country rock. Sills range from centimeters to meters in thickness. Cf. **dike.** 2. A shallow submarine ridge separating two basins.

silt A sediment with particle sizes ranging from 1/16 (0.0625) mm to 1/256 (~0.0039) mm.

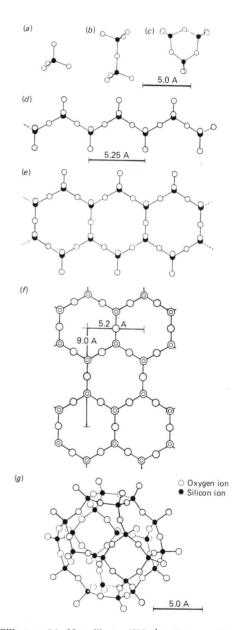

Silicates. (a) Nesosilicate $(SiO_4)^{4-}$; (b) sorosilicate $(Si_2O_7)^{6-}$; (c) cyclosilicate $(Si_3O_9)^{-6}$, $(Si_4O_{12})^{-6}$, or $(Si_6O_{18})^{6-}$; (d) inosilicate $(SiO_3)^{2-}$, showing chain structure of pyroxenes; (e) inosilicate $(Si_4O_{11})^{6-}$, showing structure of amphiboles; (f) phyllosilicate $(Si_2O_5)^{2-}$, showing extended sheets; (g) tectosilicate, showing three-dimensional structure of lazurite. A = angstrom. (Bragg 1955, p. 134–135, Fig. 82)

siltite A sedimentary rock consisting of silt-size particles. Cf. **arenite, lutite, rudite.**

siltstone A massive siltite.

sima Silica-magnesium, indicating the oceanic crust and the lower portion of the continental crust below the sial, largely consiting of Mg-rich silicates. Cf. **sial.**

simatic Referring to sima.

sin Sine.

sine See **trigonometric function.**

singularity A point where density becomes infinite and, as a consequence, space and time are infinitely distorted.

sinistral Counterclockwise on a plane or having negative helicity in space. Cf. **dextral.**

sinistral coiling See **coiling direction.**

sinistral fault Syn. **left-lateral fault.**

sinkhole A funnel-like depression in a limestone surface, caused by solution of the limestone by CO_2-rich vadose waters.

sinter A siliceous or carbonate encrustation precipitated from mineral waters, especially hot springs.

sixth power law "The power of river water to transport suspended sediment is proportional to the sixth power of its velocity."

skeleton 1. The hard support for the flesh of an organism (endoskeleton). 2. The hard, external cover produced by an organism for protection of its softer parts (exoskeleton).

skewness (*Sk*) A measure of the asymmetry of a frequency distribution.

$$Sk = (\Sigma\ x^3 f(x)/N)/\sigma^3$$

where $x = x - \bar{x}$ = distance of class interval x from mean $\bar{x}$, $f(x)$ = frequency of items in class interval x, σ = standard deviation.

skin effect The tendency of ac electricity, especially at RF frequencies, to flow near the surface of a conductor. Skin thickness (the distance from the surface at which current density has decreased to $1/e$) is, for copper, 0.3 mm at 60 Hz, 0.06 mm at 1 kHz, 0.02 mm at 1 MHz.

s.l. *Sensu lato.*

SLAC Stanford Linear Accelerator Center (Palo Alto, California).

slack water Water between tidal ebb and flow that has come to almost a standstill.

sleugh See **slough.**

slew See **slough.**

slickenside A smooth, polished, and finely striated rock surface produced by friction along a fault plane.

slide mark A groove on a sediment surface produced by sediments or objects sliding over it.

sliding stone An isolated stone from the margins of a playa that has slid across it for some distance (up to 300 m). Sliding is by wind force when evaporating water under the stone acts as a lubricant.

slip The relative displacement across a fault.

slope failure Downslope movement of rock or sediment when the angle of repose is exceeded.

slough 1. A small marsh. 2. A small body of water in a marsh.

slue See **slough.**

small circle A circle formed by the intersection of a plane with a spherical surface when the plane does not intersect the center of the sphere.

smaller Foraminifera The Foraminifera other than those belonging to the larger Foraminifera. See **larger Foraminifera.**

smectite See **montmorillonite.**

smoky quartz A variety of quartz colored brown-gray by dispersed Al^{3+}.

SMOW Standard Mean Ocean Water. Central Pacific water from a depth of about 1500 m used as an isotopic standard for hydrogen and oxygen. See **PDB.**

Snell's law A law relating the angles of incidence and refraction to the refractive indices of two media:

$$n_2/n_1 = \sin\vartheta_1/\sin\vartheta_2$$

where n_1 = refractive index of medium 1, n_2 = refractive index of medium 2, ϑ_1 = angle of incidence through medium 1, ϑ_2 = angle of refraction through medium 2.

snowline The line above which snow persists across the summer.

SNR Supernova remnant.

SNU Solar neutrino unit.

soapstone See steatite.

soda   Sodium carbonate, Na_2CO_3.

soda lake   An alkali lake rich in sodium salts.

sodic feldspar   See **sodium feldspar.**

sodic plagioclase   See **sodium feldspar.**

sodium feldspar   The mineral albite ($NaAlSi_3O_8$) or a feldspar containing mainly albite. Cf. **plagioclase.**

soffione   A natural steam vent.

soft   Referring to radiation whose particles or photons have low energy and, therefore, do not readily penetrate matter.

soft rock   Sedimentary rock.

soft-rock geology   The geology and petrology of sedimentary rocks. Cf. **hard-rock geology.**

soft water   Water containing less than 60 mg/liter of dissolved carbonates expressed as $CaCO_3$. Cf. **hard water.**

soil   1. The surface layer rich in organic matter produced by plants and where plants can grow. 2. The loose rock material covering bedrock. Syn. **rhegolith.**

sol   A liquid colloidal dispersion.

solar cell   A *pn* junction within a silicon wafer that converts solar light directly into electricity. Solar photons increase the concentration of both holes in the *p* region and electrons in the *n* region, resulting in an increased potential difference across the junction (0.6 V at saturation with full sunlight). Maximum power output is 110 W/m^2 and efficiency is about 11% (full sunlight at sea level $\sim$1000 W/m^2). See **hole, photovoltaic effect, semiconductor.**

solar constant   The mean amount of solar radia-

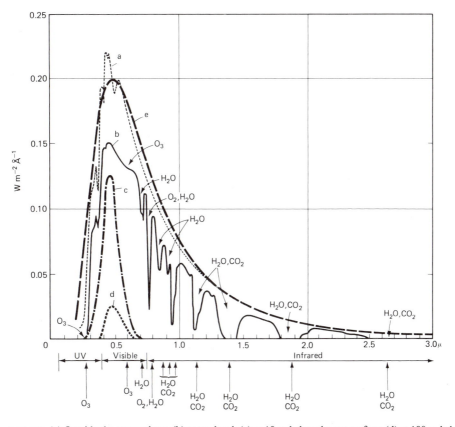

Solar spectrum. (a) Outside the atmosphere; (b) at sea level; (c) at 10 m below the sea surface; (d) at 100 m below the sea surface; (e) blackbody spectrum at 5900 K. Absorption by atmospheric gaseous molecules as shown. (From Dietrich et al. 1975, p. 167, Fig. 4.10)

tion per unit of surface normal to the Sun rays at the mean distance of the Earth from the Sun. It is equal to 1.950 cal cm^{-2} min^{-1} = 1360 W/m^2. Full sunlight at sealevel ~ 1000 W/m^2.

solar neutrino flux Neutrino flux received by the Earth from the Sun. It is equal to 2.10 ± 0.26 SNU or about one-third of what theory predicts.

solar neutrino unit (SNU) Solar neutrino flux capable of producing the reaction

$$\nu + {}^{37}Cl \rightarrow {}^{37}Ar + e^-$$

at a rate of 10^{-36}/s at the mean distance of the Earth from the Sun. The ^{37}Ar produced is detected by its decay to ^{37}Cl (K capture, $t_{1/2} = 35.04$ d).

solar system A gravitationally bound system consisting of the Sun (mass = $1.9891 \cdot 10^{30}$ kg), 9 planets (total mass = $2.670 \cdot 10^{27}$ kg), 54+ satellites (total mass = $0.720 \cdot 10^{24}$ kg), and a large number of minor planets and meteoroids (total mass = $1.8 \cdot 10^{21}$ kg) and of comets (total mass $\approx 10^{26}$ kg). Total mass of solar system = $1.9919 \cdot 10^{30}$ kg (99.86% contributed by the Sun). Mean density = 10^{-23} g/cm^3. Total angular momentum = $3.1643 \cdot 10^{43}$ kg m^2 s^{-1} (Sun = 3.100%, Mercury = 0.028%; Venus = 0.057%; Earth = 0.082%; Mars = 0.011%; Jupiter = 59.622%; Saturn = 24.155%; Uranus = 5.234%; Neptune = 7.752%; Pluto = 0.0001%). The solar system revolves around a barycenter that can be as far as $1.5 \cdot 10^6$ km (2.15 solar radii) from the center of the Sun at times of planetary alignment. Total rotational energy = $3.1 \cdot 10^{25}$ J. The solar system is located 30,000 l.y. from the center of the Galaxy near the inner edge of one of the spiral arms; it revolves around the galactic center at a speed of 250 km/s; and it oscillates across the galactic plane with a total excursion of about 20 l.y. (present location: 10 l.y. north of the galactic plane) and a period of $33 \pm 3 \cdot 10^6$ y. The solar system completes one revolution around the center of the Galaxy in $225 \cdot 10^6$ y. It presently moves relative to the background radiation at a speed of about 358 km/s toward the Virgo cluster coordinates ($l^{II} = 283$, $b^{II} = +75$). See also **apex.**

solar wind Supersonic flow of ionized atoms (mainly H and He) from the Sun through the solar system and beyond. In the region of the Earth the ion density is 1–30 ions/cm^3. Ion velocities range from 300 to 700 km/s, corresponding to travel times from the Sun ranging from 5.7 to 2.5 days.

sole mark A convex, elongated protrusion on the underside of a graded bed representing the filling, by the bed's bottom sediments, of a depression on the surface of the underlying bed.

solenoid A coil of insulated wire that produces an axial magnetic field when a current flows through it.

solfatara A sulfurous fumarole.

solid angle The angle subtending a surface from a point not part of the surface. The angle formed at the vertex of a cone or a pyramid. Solid angles are measured in steradians. The solid angle subtended by the center of a sphere is equal to 4π steradians.

solidus The line (in a binary system), surface (in a ternary system), or volume (in a quaternary system) separating the solid from the solid-liquid equilibrium domain in a temperature vs. composition diagram. Cf. **liquidus.**

solifluction The downslope creep of water-logged rhegolith.

solitary coral A noncolonial coral.

solitary wave A single wave crest progressing along the surface of a medium. It is produced by a single, unidirectional pulse by a solid surface normally to the medium's surface.

soliton A solitary wave.

solstice Either of the two times in the year when the Sun is highest (summer solstice) or lowest (winter solstice) at the local noon.

solubility product The product of the concentrations of ions of a compound in a saturated solution, each raised to the number of times the element appears in the chemical formula of the compound. Cf. **ion product.**

solvus The line (in a binary system), surface (in a ternary system), or volume (in a quaternary system) separating a solid solution domain from the domain of two or more phases that may be produced by exsolution. Cf. **liquidus, solidus.**

sonar Sound navigation and ranging. An acoustic system used to detect submerged objects, structures, or the ocean floor by echo ranging.

sone The unit of loudness, based on the physiological response to sound loudness. As the human ear is more sensitive to higher frequencies, a higher frequency sound appears louder than a lower frequency sound having the same sound-pressure level. See **sound pressure.**

sonobuoy A buoy, teetered or free, used in seismic refraction at sea to gather data and relay them by means of a transmitter.

***s* orbital** The orbital of an atomic electron characterized by orbital angular momentum quantum number $l = 0$. See **s, p, d, f.**

sorosilicate See **silicate.**

sorption A collective name that includes both absorption and adsorption.

sorting The separation by size, shape, or density, of sedimentary particles by a transporting agent.

soufrière A volcanic crater or vent system emitting sulfurous gases.

sound A wave disturbance in the density and pressure of a solid, liquid, or gas, or in the elastic strain of a solid. As a wave phenomenon, sound is characterized by amplitude (maximum absolute value), wavelength λ, frequency $\nu = c/\lambda$ (where c = velocity of sound), period $T = 1/\nu$, and phase. Characteristic frequencies (hertz, 20°C if in air): free oscillations of the Earth = $1–3.5 \cdot 10^{-3}$ (λ = 10,000–20,000 km); earthquake P waves ~ 1 ($\lambda \sim$ 4.8 km); A_4 (piano) = 440 (λ = 0.78 m); male speech range = 100–9000 (λ = 3.43 m to 3.81 cm); female speech range = 150–10,000 (λ = 2.29 m to 3.43 cm); audible frequency range = 16–20,000 (λ = 21.4 m to 1.71 cm); highest sound frequency produced = $6 \cdot 10^8$ (λ = 0.57 μm $\sim$ wavelength of visible light); pressure difference of loudest sound tolerable by human ear = 280 μbar = 11 μm displacement; pressure difference in faintest audible sound = $2 \cdot 10^{-4}$ μbar = $8 \cdot 10^{-6}$ μm displacement $\sim 1/10$ of the atomic radius. Examples of sound velocities (km/s): air (20°C) = 0.343; freshwater (20°C) = 1.403; seawater (25°C, 35‰ salinity) = 1.535; common surface rocks = 2–6; iron = 5.950; diamond = 18.1. See **P wave, S wave, Seismic waves*, Sound*.**

sound level The sound pressure level at a given point in a sound field. Examples of sound levels: threshold of pain, 120 db; loud scream, 90 db; loud conversation, 70 db; quiet conversation, 40 db; whisper, 20 db; threshold of hearing, 0. See **Sound*.**

sound pressure The instantaneous pressure change in a medium when a sound wave passes through it. The sound pressure level in decibels is equal to $20 \log P/P_{ref}$, where P = sound pressure in microbars, P_{ref} = reference sound pressure = $2 \cdot 10^{-4}$ μbar.

source rock The sedimentary rock in which organic matter has been transformed into hydrocarbons.

south magnetic pole 1. See **magnetic polarity.** 2. The site in the southern hemisphere where mag-

netic inclination is 90°. Present location: 65.8°S, 139.0°E. Cf. **dip pole, geomagnetic pole, north magnetic pole.**

sp. Species (sing.). Cf. **spp.**

space group Any of the 230 possible combinations of symmetry elements with one of the 14 Bravais lattices as lattice of translation. See **symmetry elements.**

space lattice A repetitive, geometrical array of points in space.

spacetime The four-dimensional space-time continuum.

spallation The breakup of a nucleus into three or more fragments due to bombardment by energetic particles. Cf. **fission.**

spar Any transparent or translucent, cleavable, nonmetallic crystal.

sparite Crystalline limestone matrix with crystal sizes >4 μm. Cf. **micrite.**

sparry Pertaining to spar.

sparry calcite See **sparite.**

s, p, d, f Sharp, principal, diffuse, fundamental: the first four values (0, 1, 2, and 3) of the orbital angular momentum quantum number l.

special relativity See **relativity.**

speciation The evolution of a new species.

species *(Biology)* "*Species sunt tot quot ab initio produxit infinitum Ens*" (Linnaeus 1738). "Species are groups of actually or potentially interbreeding natural populations" (Mayr 1942). *(Physics, Chemistry)* A specific type of particle, nuclide, nucleus, atom, molecule, or ion.

specific Referring to a quantity expressed per unit of mass, volume, weight, density, etc.

specific gravity Syn. **relative density.**

specific heat The ratio of the heat added to a system to the resulting temperature rise.

specific humidity The mass of water vapor per unit mass of humid air.

specific rotation The power of optically active substances to rotate plane-polarized light. It is measured in degrees/mm (solids) or in degrees/dm for solutions.

specific volume The volume of a substance per unit mass. It is the reciprocal of density.

spectral classification A classification of stars according to spectral type and surface temperature. Spectral types and effective temperatures (K): O5 = 40,000, BO = 28,000, B5 = 15,500, AO = 9900, A5 = 8500, FO = 7400, F5 = 6580, GO = 6030, G5 = 5520, KO = 4900, K5 = 4130, MO = 3480, M5 = 2800, M8 = 2400. Cf. **Hertzsprung-Russell diagram, Main Sequence.**

spectrophotometer An instrument that measures the intensity of light at different wave lengths.

speed The magnitude of the velocity vector.

speleo- Prefix meaning *cave.*

speleothem A deposit precipitated in a cave from percolating water. Included are stalactites, stalagmites, and crusts.

spessartine A garnet, $Mn_3Al_2Si_3O_{12}$.

spessartite A lamprophyre consisting of hornblende or clinopyroxene phenocrysts in a groundmass of Na plagioclase.

Sphagnum A moss contributing importantly to bog peat.

sphalerite The mineral ZnS.

sphene The mineral $CaTiO(SiO_4)$.

spheno- Prefix meaning *wedge.*

sphenochasm A triangular opening in the crust resulting from the relative rotation of two plates.

sphenolith An igneous intrusion shaped like a wedge.

sphere The locus of all points equidistant from a given point chosen as center, given by the equation (center at origin):

$$x^2 + y^2 + z^2 = 1$$

Volume $V = \frac{4}{3}\pi r^3$; surface $S = dV/dr = 4\pi r^2$.

spheroid 1. Any surface resembling a sphere. 2. See **ellipsoid of revolution.**

spheroidal Pertaining to a spheroid.

spherule A small sphere.

spherulite A spherical or spheroidal mineral mass with radial internal structure.

spicule *(Astronomy)* Any of the narrow (1000 km across), hot (10,000–20,000 K) jets of gas rising from the lower chromosphere to the inner corona of the Sun at speeds of 20–30 km/s. Spicules last only 5–10 minutes before falling back and reforming. *(Biology)* Any of the small calcareous or siliceous structures secreted by many invertebrates and used for tissue support.

spilite A submarine basalt, often with pillow structure, altered by contact with seawater into a greenstone with characteristic low-grade metamorphic minerals (albite, chlorite, actinolite, sphene, epidote, calcite, prehnite).

spin The rotation of a body about its axis expressed as the angular momentum vector. The spin of elementary particles is expressed in units of $h/2\pi$, where h = Planck's constant. Fermions have half-integer spin and bosons have integer spin. See **helicity.**

spinel A Mg–Al oxide, $MgAl_2O_4$, often colored by dispersed Cr^{2+}. See **Gems*.**

spinel structure A closely packed cubic structure characteristic of high-pressure, high-temperature phases.

spinifex Referring to the texture of elongated, interwoven olivine crystals in komatiites, probably formed by quenching.

spin quantum number See **magnetic spin angular momentum quantum number.**

spiral galaxy See **galaxy.**

S pole South-seeking magnetic pole, i.e. the end of a magnetic dipole that points toward the south magnetic pole. Cf. **magnetic polarity.**

spongin A scleroprotein, forming skeletal tissue in sponges.

spontaneous fission A mode of radioactive decay of ^{244}Pu ($t_{1/2} = 25 \cdot 10^9$ y vs. $80.8 \cdot 10^6$ for α decay) and ^{238}U ($t_{1/2} = 10.1 \cdot 10^{15}$ y vs. $4.468 \cdot 10^9$ y for α decay) in which the nucleus splits into two major fragments and several smaller ones, including free neutrons (2.5 neutrons/fission).

spore A unicellular, haploid body resistant to unfavorable conditions and capable of developing into a gametophyte when conditions are favorable.

sporophyte The diploid phase in the life cycle of a plant, arising from a zygote and producing spores (which are haploid). In plants higher than algae the sporophyte is the dominant phase. See **gametophyte.**

sporopollenin The extremely resistant substance of which the spore capsule is made.

spp. Species (plural). Cf. **sp.**

SP$\overline{P}$S Super Proton-Antiproton Synchrotron (CERN, Geneva). See **collider.**

***s*-process** The slow process of formation of the heavier elements by addition, to existing nuclei, of thermal neutrons followed by β^- decay. The free neutrons are primarily produced by the following reactions:

$$^{12}C(p,\gamma)^{13}N(\epsilon,\gamma)^{13}C(\alpha,n)^{16}O$$
$$^{14}N(\alpha,\gamma)^{18}F(\beta^+\nu)^{18}O(\alpha,\gamma)^{22}Ne(\alpha,n)^{25}Mg$$
$$^{18}O(\alpha,n)^{21}Ne(\alpha,n)^{24}Mg$$

where (x,y) means x *in*, y *out*. See **element formation.**

spruit A small, intermittent stream (South Africa).

SPS Super Proton Synchrotron (CERN, Geneva). See **collider.**

sr Steradian.

s.s. *Sensu stricto.*

SSC Superconducting Super Collider, a proposed giant collider with a radius of 24 km (United States).

St Stoke.

stadion A Greek measure of length equal to 189.6 m. See **Measures*.**

stadium A Roman measure of length equal to 184.4 m. See **Measures*.**

stage A chronostratigraphic unit below series and above substage in rank, consisting of rocks deposited during an age.

staghorn coral The coral *Acropora cervicornis.*

stagnum A marshy pond.

stalactite A columnar or conical spelothem hanging from the roof of a cave.

stalagmite A columnar or conical speleothem arising from the floor of a cave.

standard absolute entropy (S^0) Entropy of a substance at 25°C, 1 atm.

standard atmosphere A model atmosphere having a temperature of 15°C and a pressure of 1013.250 mb at sea level, with an average gradient of temperature, pressure, and density.

standard deviation (σ) The square root of the square of the average difference between the values of a normally distributed quantity and its mean.

$$\sigma = (\Sigma\ x_i^2/N)^{1/2}$$

where x_i = deviation of value i from the mean, N = number of values. For normal distributions, the percentage of the values falling within given σ intervals from the mean are as follows: $\pm 1\sigma$ = 68.27%; $\pm 2\sigma$ = $\pm 95.45\%$; $\pm 3\sigma$ = 99.73%; $\pm 4\sigma$ = 99.994%; $\pm 5\sigma$ = 99.99994%.

standard electrode A half cell used for measuring electrode potentials. See **hydrogen electrode.**

standard error The variability of a given parameter when obtained from a set of samples independently and randomly drawn from the same universe.

standard free energy of formation (ΔG_f^0) The free energy change at 25°C, 1 atm, in forming 1 mole of a substance from its constituent elements also at 25°C, 1 atm.

standard g (g_0) The Earth's gravitational acceleration at sea level at 45° lat as measured in 1888, used as a conventional gravity standard for many applications, including the definition of atm as a unit of pressure [1 atm = pressure of a 760-mm-high column of Hg with density of 13.5951 g/cm^3 and subject to a gravitational attraction of g_0. It is equal to 1.013250 bar = 101,325.0 Pa (exactly)]. g_0 = 980.665 gal (exactly).

standard heat of formation The heat needed at constant pressure to form one mole of a substance at 25°C, 1 atm, from its elements also at 25°C, 1 atm.

standard model A modern theory of matter and radiation than includes the electroweak theory and quantum chromodynamics. According to this model, all matter and radiation consists of combinations of 18 colored quarks [6 flavored quarks (u, c, t, d, s, b) each existing in 3 different colors (red, green, and blue)], 6 leptons (electron, muon, tauon and their neutrinos), and 12 force carriers (photon, 8 colored gluons, 3 weak bosons). The standard model successfully describes all interactions (except gravitational) down to a scale of 10^{-18} m.

standard pressure The pressure of 1 atmosphere = 760 mmHg (exactly) = 1.013250 bar (exactly) = 101325.0 Pa (exactly).

standard section A stratigraphic section as complete as possible for a given time interval, representing a given region.

standard solution A solution containing a known amount of a solute, used as a standard.

standard state The state of a pure substance at 1 atmosphere and at a specified temperature (usually 298.15 K = 25.0°C).

standard temperature The temperature of 273.15 K = 0°C.

standard time Mean solar time at any of the 24 internationally agreed-upon time zones into which the Earth's surface is divided.

standard volume The volume occupied by 1 mole of gas at 0°C of temperature at 760 mmHg of pressure. It is equal to 22.41410 liters for an ideal gas.

standing wave The wave resulting from the superposition of an incident wave and its reflected wave. It forms when the distance between origin and reflector is an integral number of half wavelengths.

$$y = 2y_{max} \sin nx \cos \omega t$$

where y = amplitude, y_{max} = maximum amplitude, n = wave number = $2\pi/\lambda$ (λ = wavelength), ω = angular frequency, t = time. Amplitude y is maximum when $nx = k(\pi/2)$ and k = 1, 3, 5 . . . (antinodes); it is minimum when $nx = k\pi$ and k = 1, 2, 3 . . . (nodes).

stannous Pertaining to tin.

star *(Astronomy)* A celestial body of sufficient mass and density to initiate and sustain nuclear reactions in its core. The relative abundances of spectral types from B (blue) to M (red) are: B = 3%, A = 27%, F = 10%, G = 16%, K = 37%, M = 7%. See **spectral classification.** *(Physics)* A group of tracks radiating from the same point, exhibited by a nuclear emulsion plate or by a photograph of a cloud or bubble chamber process. It is produced by spallation or by successive radioactive decays of a single nuclide.

star formation Stars derive from the collapse of dark clouds, which consist of atomic, ionic, and molecular species as well as solid particles (see **dark cloud, Molecules–interstellar*, Molecules–interstellar, relative abundances*).** If the mass of the cloud is greater than the Jeans mass, self-collapse may occur. Free-fall accumulation is estimated to take $\sim 1 \cdot 10^6$ y, with retardation (accompanied by dissipation of some of the accumulating heat) produced by the dynamics of the cloud's magnetic field. Collapse may be triggered by shock waves from nearby supernovae or by density waves. Frequency of star formation in the Galaxy ≈ 1 solar mass per million year, representing $\sim 10\%$ of the mass of the parent cloud. When temperature in the core of the protostar reaches values $>10^6$ K, nuclear reactions are initiated and the protostar becomes a star.

Stark effect The splitting, shifting, or broadening of spectral lines in the presence of an electric field.

starquake A quake in the interior of a stellar body. See **glitch.**

stat Prefix identifying units in the CGS_{esu} system.

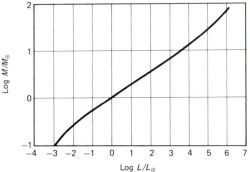

Stars: mass-luminosity relation. M = mass of star; $M_\odot$ = mass of Sun; L = luminosity of star; $L_\odot$ = luminosity of Sun. White dwarfs excluded. (Data from Allen 1976, p. 209)

statA Statampere.

statampere (statA) The CGS_{esu} unit of electric current.

$$\begin{aligned} statA &= statC/s \\ &= aA/c \\ &= 10 \ A/c \\ &= 3.335641 \cdot 10^{-10} \ A \end{aligned}$$

where c = speed of light in cm/s.

statC Statcoulomb.

statcoulomb (statC) The unit of electric charge in the CGS_{esu} system, equal to the charge that exerts the force of 1 dyne on an equal charge at a distance of 1 cm.

$$\begin{aligned} statC &= aC/c \\ &= 10 \ C/c \\ &= 3.335641 \cdot 10^{-10} \ C \end{aligned}$$

where c = speed of light in cm/s.

state function A function dependent upon the state of a system.

statF Statfarad.

statfarad (statF) The CGS_{esu} unit of capacitance, defined as the capacitance of a capacitor that exhibits the potential difference of 1 stavolt between its plates when each is charged with 1 statcoulomb of opposite electricity.

$$\begin{aligned} statF &= statC/statV \\ &= 10 \ c^{-1} \ C/10^{-8} \ c \ V \\ &= 1.11265 \cdot 10^{-12} \ F \end{aligned}$$

where c = speed of light in cm/s.

Star Chart. (Strahler 1971, p. 17, Fig. 1.18)

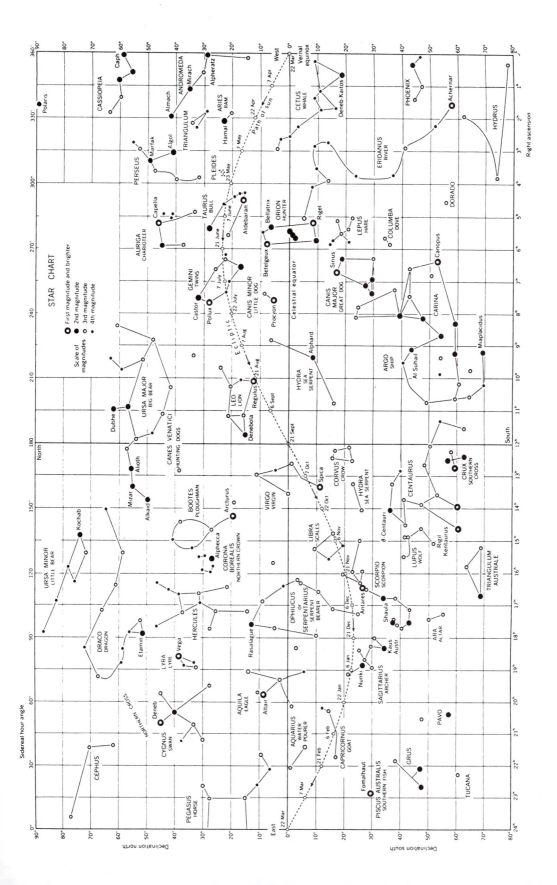

statH Stathenry

stathenry (statH) The CGS_{esu} unit of inductance and permeance, defined as the self or mutual inductance of a closed circuit in which an emf of 1 statvolt is produced when the current changes uniformly at the rate of 1 statampere/second.

$$statH = statV/statA\ s^{-1}$$
$$= 10^{-8}\ c\ V\ s/10\ A\ c^{-1}$$
$$= 8.98755 \cdot 10^{11}\ H$$

where c = speed of light in cm/s.

static limit The surface surrounding a black hole *(event horizon)*, so named because an observer cannot remain at rest on or within it, but must revolve around the black hole in the same direction as the black hole's rotation.

stationary front A front that does not move. See **front** *(Meteorology)*. Cf. **cold front, occluded front, warm front.**

stationary satellite A geosynchronous satellite 36,000 km above the equator, orbiting the Earth with the same period as the rotational period of the Earth.

stationary wave See **standing wave.**

statistical mechanics The study of the physical properties of systems consisting of a large number of particles.

statohm (statΩ) The CGS_{esu} unit of electric resistance.

$$statΩ = statV/statA$$
$$= 2.99792458 \cdot 10^2/3.335641 \cdot 10^{-10}$$
$$= 8.987551 \cdot 10^{11}\ Ω.$$

statute mile (mi) A nonmetric measure of length, equal to 5280 feet = 1760 yards = 1609.344 m (exactly).

statV Statvolt.

statvolt (statV) The CGS_{esu} unit of electromotive force or potential difference.

$$statV = aV/c$$
$$= 10^{-8}\ c\ V$$
$$= 2.99792458 \cdot 10^2\ V$$

where c = speed of light in cm/s.

statΩ Statohm.

steatite A metamorphic rock consisting mainly of talc $[Mg_3Si_4O_{10} \cdot (OH)_2]$, produced by late metamorphism of ultramafic rocks following serpentinization. Syn. **soapstone.**

steatitization The metamorphic process producing steatite.

steel An alloy of Fe and C, with only 0.2–1.5% C. Stainless steel contains 12–30% Cr.

Stefan-Boltzmann constant (σ) The proportionality constant relating the radiancy of a blackbody radiator to absolute temperature.

$$R = σT^4$$

where R = radiancy, T = absolute temperature. It is equal to $5.6705 \cdot 10^{-8}\ W\ m^{-2}\ K^{-4}$.

Stefan-Boltzmann law "The radiancy of a blackbody is proportional to the fourth power of the absolute temperature of the body."

$$R = σT^4$$

where R = radiancy, $σ$ = Stefan-Boltzmann constant, T = absolute temperature.

steinkern A hardened sediment filling of the internal cavity of molluscan shell.

stellarator A tube twisted as a nonintersecting figure 8 with external coils creating magnetic fields parallel to the walls of the tube. It is used to confine a plasma and to prevent it from touching the walls.

steno- Prefix meaning *narrow.*

stenobathic Describing an organism with a narrow depth-range tolerance.

stenohaline Describing an organism with a narrow salinity-range tolerance.

stenothermal Describing an organism with a narrow temperature-range tolerance.

steppe A treeless, grassy, semiarid expanse.

steradian (sr) The SI unit for solid angles, defined as that solid angle with vertex at the center of a sphere that subtends on the surface of the sphere an area equal to the square of the radius of the sphere.

stereo- Prefix meaning *double.*

stereoisomer Any of two or more chemical compounds that have the same composition and atomic bonding but different arrangements of their atoms. Cf. **structural isomer.**

stereonet An equal-angle coordinate system for projecting crystallographic and structural parameters. Syn. **Wulff net.**

stilb (sb) A unit of luminance, equal to 1 cd/cm^2.

Stirling cycle A thermodynamic cycle consisting of two isothermal and two isochoric phases.

stishovite A high-pressure (>100 kb), high-density (4.28 g/cm^3) phase of quartz. See **silica.**

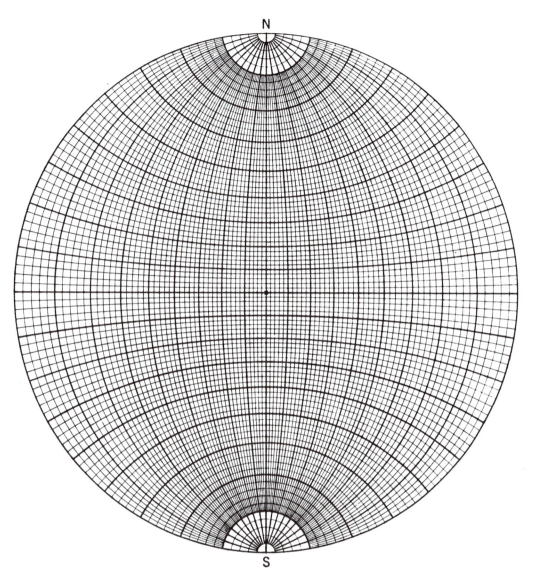

Stereonet.

stochastic Pertaining to random processes or to random variables.

stoichiometric Referring to a chemical system or compound *with the elements measured,* i.e. with the components in exact proportions.

stoichiometry The relative proportions of reactants and products in a chemical reaction.

stoke (St) A unit of kinematic viscosity, equal to 1 poise/g/cm^3.

Stokes' law A law giving the velocity V of a small sphere accelerated through a fluid.

$$V = 2ar^2(\rho_s - \rho_m)/9\eta$$

where a = acceleration, r = radius of the sphere, ρ_s = density of the sphere, ρ_m = density of the medium, η = coefficient of viscosity of the medium.

stony-iron meteorites A family of meteorites consisting mainly of olivine, pyroxene, and Fe–Ni alloys. Stony-iron meteorites form 1.5% of all meteorites, and are divided into mesosiderites (60%, consisting of pyroxene, plagioclase, and Fe–Ni alloys), pallasites (33%, consisting of olivine and Fi–Ni alloys), and others (7%, consisting of pyroxene, olivine, and Fe–Ni alloys). See **Meteorites***.

stony meteorites A family of meteorites consisting mainly of Mg–Fe silicates and plagioclase. Stony meteorites form 92.8% of all meteorites and are divided into chondrites (92.4%) and achondrites (7.6%). See **achondrite, chondrite, Meteorites***.

stoping The mining of a steeply inclined or vertical vein by excavation in a series of steps.

storage battery A battery consisting of storage cells.

storage cell An electrolytic cell that can be recharged by a current opposite that of the discharge current.

storm beach A ridge of coarse rubble piled up at the inner edge of a beach or behind it by storm waves.

storm surge The wind-driven onshore surge of seawater during a storm.

stoss Referring to the upstream side of an obstruction (hill, etc.) within a moving glacier or ice sheet.

STP Standard temperature and pressure, equal to 273.15 K = 0.00°C and 1 atmosphere (760 mmHg).

straat A trough between dunes (South Africa).

strain The response of a deformable body to stress. A total of 9 possible strains can affect an anisotropic body, 3 tensile, 3 shear, and 3 volumetric.

strand Syn. **shore**.

strandline See **shoreline**.

strangeness A hadronic quantum number conserved in strong and electromagnetic interactions, and responsible for the slow decay of massive particles of nonzero strangeness.

strangeness number (S) A hadronic quantum number equal to the hypercharge minus the baryon number. $S = 0$ for the π and η mesons, the proton, and the neutron; $S = 1$ or -1 for the

kaon; $S = -1$ for the Λ and Σ hyperons; $S = -2$ for the Ξ hyperon; $S = -3$ for the Ω hyperon. Antiparticles have strangeness number and charge of opposite sign.

strange particle A hadron with a strangeness number different from 0.

strange quark The s quark, with baryon number 1/3, charge $-1/3$, and strangeness number -1, or its antiquark, with reversed charge and strangeness.

stratiform Having the shape of a layer or stratum.

stratigraphic section A sequence of sedimentary rocks.

stratigraphic trap A trap for hydrocarbons resulting from lateral lithological changes within a sedimentary formation.

stratigraphy The study of the succession of sedimentary rocks and their fossil content.

stratizone The stratigraphic interval represented by a biozone.

stratosphere The atmospheric layer above the troposphere and below the mesosphere, extending from 10–16 km to 50 km of altitude in the standard atmosphere. See **atmosphere**.

stratotype The type section representative of a given stratigraphic name.

stratovolcano A stratified volcanic cone consisting of alternating layers of lava and pyroclastics.

stratum A layer. Pl. *strata*.

streak plate A plate of unglazed porcelain (hardness = 7) used for mineral identification by observing the color of a mineral streak on it.

stream A body of water running within a bed, regardless of size.

streamer A string of hydrophones towed behind a ship.

stream load The amount of sediment carried by a stream along its bed, in suspension, and in solution.

stress An external force acting on the surface of a body at any angle except zero.

stress-strain curve A curve showing the relationship between stress applied to a body and the resulting strain.

strike The direction of the line representing the intersection of an inclined plane with the horizontal.

strike-slip fault A fault with the movement parallel to the fault's strike. Syn. **lateral fault.** Cf. **dip-slip fault.**

string Any of the string-like space defects predicted by some GUTs to have formed before cosmological time $t = 10^{-35}$ s. Strings are said to have acted as nuclei for matter accretion during and after the inflationary period, eventually leading to the formation of galaxies. Cf. **inflation.**

stringer A long and thin ore body.

stromatolite A fossil algal structure, consisting of laminae rich in carbon and iron oxides alternating with laminae consisting largely of inorganic siliceous or calcareous sediment particles. Cf. **algal mat.**

strombolian eruption A volcanic eruption characterized by the emission of jets of basaltic lava from the central crater.

strong acid An acid that is completely ionized in water solution. Cf. **weak acid.**

strong base A base that is completely ionized in water solution. Cf. **weak base.**

strong force The color force carried by gluons in the strong interaction between quarks within hadrons and between quarks in adjacent hadrons. Effective range = 10^{-15} m; strength $\sim 10^2$ to 10^3 times greater than the electromagnetic force, 10^6 greater than the weak force, 10^{39} times greater than the gravitational force. See **color force, gluon, natural forces, strong interaction.**

strong interaction The interaction between quarks, gluons, and hadrons mediated by gluons. See **color froce, gluon, natural forces, strong force.**

strontium-90 A radioactive isotope of strontium produced by the fission of uranium and plutonium isotopes. ^{90}Sr decays (β^-, $t_{1/2}$ = 28.6 y) to ^{90}Y, which in turn decays (β^-, $t_{1/2}$ = 64.1 h) to ^{90}Zr (stable).

structural isomer See **isomer** *(Chemistry).*

structural trap A trap for hydrocarbons resulting from tectonic deformation.

stylolite A finely indented solution plane within a limestone body.

SU(3) The group of special unitary transformations in three dimensions of a multiplet of elementary particles.

SU(5) The group of special unitary transformations in five dimensions that combines fermions (leptons and quarks) in multiplets within which the transformations quarks-leptons and quarks-

antiquarks become possible through mediation by massive (10^{14} GeV) X and Y bosons. SU(5) explains the fractional quark charge and the charge identity between electron and proton.

subage The time during which the rocks forming a substage are deposited.

subalkaline Defining an igneous rock without feldspathoids and with a lower alkali/silica ratio than alkaline rocks.

subarkose A sandstone intermediate in composition between arkose and quartzarenite.

subarkosic wacke A graywacke rich in feldspar.

subduction The plunging of an oceanic plate at a steep (60°) angle under an adjacent plate bearing a continent at its leading margin.

subduction zone The belt along which subduction occurs. Cf. **Benioff zone.**

subgraywacke A sedimentary rock intermediate between graywacke and quartzarenite.

subhedral Referring to a subhedron.

subhedron A crystal only partly bound by its own rational faces, the rest being bound by adjacent crystal faces.

sublimation The direct passage of a substance from the solid to the gaseous phase, or vice versa.

sublittoral See **neritic.**

submarine canyon A valley deeply incised on the continental terrace by subaerial or submarine erosion, or by a combination of the two.

submarine fan An accumulation of sediments derived from the continent in front of a large river mouth or at the foot of a submarine canyon.

submarine valley See **submarine canyon.**

submarine weathering See **halmyrolysis.**

subshell Any of the quantum states of an electron characterized by a specific set of n and l quantum numbers. See **orbital, quantum number, shell.**

subsidence Crustal sinking.

subsidiary quantum number Syn. **orbital angular momentum quantum number.** See **quantum number.**

subspecies Any population within a species that can be identified by one or more characteristics.

substage A portion of a stage represented by sedimentary rocks deposited during a subage.

Suess effect The dilution of natural $^{14}CO_2$ in the atmosphere by the addition of CO_2 from the burning of fossil fuels.

sulfide layer A hypothetical layer rich in sulfides at the surface of the outer core. Syn. **chalcosphere.**

sulfur bacteria 1. Unpigmented, autotrophic bacteria that draw energy from the oxidation of H_2 and S:

$$2H_2S + O_2 = 2H_2O + 2S$$

$$2S + 3O_2 + 2H_2O = 2H_2SO_4$$

2. Pigmented photosynthesizing bacteria using H_2S instead of H_2O:

$$CO_2 + 2H_2S = [CH_2O] + H_2O + 2S$$

$$2S + 3O_2 + 2H_2O = 2H_2SO_4$$

The sulfuric acid is then neutralized to gypsum:

$$H_2SO_4 + CaCO_3 = CaSO_4 + H_2O + CO_2$$

Sun The central body of the solar system. Radius = $695,990 \pm 70$ km. Mass = $1.9891 \cdot 10^{30}$ kg. Mean density = 1.409 g/cm^3. The Sun has a layered structure, consisting of core (0 to 170,000 km), radiative layer (170,000 to 590,000 km), convective layer (590,000 to 695,500 km), and photosphere (695,500 to 695,990 km). Above the photosphere lie the chromosphere, extending to an altitude of 2500 km, and the corona, extending to more than 10^7 km. Core temperature = $15 \cdot 10^6$ K; core density = 160 g/cm^3; core composition = 38% H, 62% He. Composition of radiative layer, convective layer, and photosphere = 72% H, 27% He, 1% heavier elements. Effective temperature of photosphere = 5770 K. Power emitted = $3.826 \cdot 10^{26}$ watts = $4.257 \cdot 10^6$ tonnes/s. Apparent visual magnitude = -26.74; absolute visual magnitude = $+4.83$. Spectral type = G2 (yellow dwarf). Dipole magnetic field = 1–2 gauss (?); magnetic field in sunspots = 1000–4000 gauss. Sidereal period at equator = 25.53 days; sidereal period at poles = 36.61 days. Inclination of solar equator to ecliptic = $7°15'$. Velocity relative to near stars = 19.7 km/s toward the solar apex. Mean distance from Earth = 1 AU = 149,597,870.7 km = 8.316746 light minutes = 499.004784 light seconds. The Sun orbits around the barycenter of the solar system. See **solar system.**

sunspot Any of the large (mean radius = 15,000 km) areas on the solar surface consisting of a darker and cooler (4240 K) center (umbra) and a warmer (5680 K), less dark periphery (penumbra), characterized by a strong magnetic field (1000–4000 gauss). Sunspot activity exhibits an approximate 11-year cycle during which sunspots move from mid-latitudes toward the equator. Individual sunspots last days to weeks.

sunspot cycle See **sunspot.**

sunstone An oliooclase crystal with dispersed hematitic flakes that reflect a golden light.

superconductivity The property of many metals and alloys to exhibit vanishing resistivity at temperatures close to the absolute zero.

supercooling The cooling of a substance below its freezing point but without actual freezing, resulting in a metastable fluid state.

superfluidity The frictionless state of liquid ^{4}He at temperature below 2.172 K or of liquid ^{3}He at temperatures below 0.00093 K, both at zero pressure.

superforce The single force that, according to theory, was the only force during the first 10^{-32} s of our universe, giving origin to the four natural forces as the universe expanded and temperature decreased. Cf. **inflation, natural forces.**

supergene A near-surface mineral deposit formed by descending solutions.

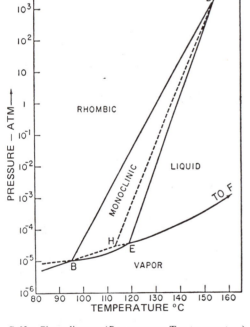

Sulfur. Phase diagram (P = pressure; T = temperature). B: $P = 0.01$ mmHg, $T = 95.5°C$; D: $P = 1290$ atm., $T = 155°C$; E: $P = 0.025$ mmHg, $T = 119.3°C$; F (critical point): $P = 204$ atm., $T = 1041°C$. Dotted lines separate metastable phases. H = rhombic-liquid-vapor eutectic. (Moore 1950, p. 108, Fig. 5.3)

supergiant A massive, large, highly luminous, red or blue star. Examples of red supergiants are Antares in Scorpio (radius = 140 solar radii, distance = 425 l.y.) and Betelgeuse in Orion (radius = 400 solar radii, distance = 650 l.y.); an example of a blue supergiant is Rigel in Orion (radius = 20 solar radii, distance 815 l.y.).

superheating The heating of a substance above its boiling point without actual boiling, resulting in a metastable fluid state.

superimposed stream A stream that has eroded through a surface structural pattern and has maintained its course while eroding through an underlying, different structural pattern.

superior planet Any of the planets (Mars, Jupiter, Saturn, Uranus, Neptune, or Pluto) with orbits at a distance from the Sun greater than the distance of the Earth's orbit.

supernova A star in the process of exploding. There are 5 types of supernovae, of which Type I (seen in both elliptical and spiral galaxies) and Type II (seen in the arms of spiral galaxies) are the most common. Type I supernovae, which reach an absolute magnitude of -19, may originate from white dwarfs that have accumulated sufficient matter to exceed the Chandrasekhar limit. Type II supernovae, which reach an absolute magnitude of about -17, originate from the collapse of the core of supergiants with masses >8 solar masses, and

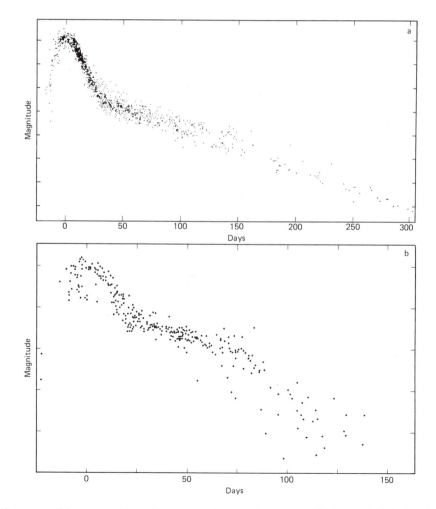

Supernovae light curves. (a) Type I supernovae; (b) type II supernovae. (Barbon et al. 1974, 1974a)

result in the formation of a neutron star or black hole and in the dispersal of most (90%) of the star's matter into space. The light curves of Type I exhibit a rapid rise and an equally rapid decline, followed by a slower decline; those of Type II show a rapid rise and an early rapid decline, followed by a plateau and by a second phase of rapid decline. The light curves of Type I supernovae are more uniform than those of Type II. Frequency of supernovae ≈ 0.025/y/galaxy.

supernova remnant (SNR) An expanding shell of gas produced by a supernova explosion (145 are known in the Galaxy).

supersaturation A state in which the concentration of a solute in a solvent exceeds saturation.

superstring theory Any of the set of field theories based on supersymmetry and the evolution of strings as the universe expanded to a radius equal to the Planck length ($1.616 \cdot 10^{-35}$ m). Superstring theories attempt to unify color, electroweak, and gravitational forces. Cf. **string, supersymmetry.**

supersymmetry (Susy) A GUT theory relating fermions to bosons. It requires that each elementary particle be associated with a supersymmetric partner (quark/squark, lepton/slepton, photon/photino, gluon/gluino, W/Wino, Z/Zino, graviton/gravitino).

supracrustal Referring to rock layers overlying basement rock.

surface tension (γ) The force binding to each other the surface molecules of a liquid. Most liquids in contact with air at room temperature have surface tensions ranging between 15 and 75 dyn/cm. Examples (dyn/cm): ethyl alcohol (0°C) = 24.05; chloroform (20°C) = 27.14; benzene (20°C) = 28.85; glycerol (20°C) = 63.4; water (18°C) = 73.05.

surface wave A seismic wave propagating along a surface. Typical periods of surface waves = 10–100 s; typical velocities = 3.5–4.2 km/s.

surf base The depth at which surf begins to break, equal to about 1/4 of the wavelength.

surf beat The grouping of a series of waves of higher and lower amplitude observed along many coastlines, caused by the superposition of waves and swell.

surf zone The zone between offshore surf break and upshore limit of water uprush.

susceptance (B) The imaginary part of admittance. It is expressed in siemens. See **admittance.**

susceptibility (χ) The capacity of a substance to be affected by an electric or a magnetic field.
1. electric susceptibility (χ_e) The polarizability of a dielectric in the presence of an electric field.

$$\chi_e = \mathbf{P}/\epsilon_0 \mathbf{E}$$

where $\mathbf{P}$ = polarization, ϵ_0 = permittivity constant, $\mathbf{E}$ = electric field strength.
2. magnetic susceptibility (χ_m) The ratio of magnetization of a material to the strength of the applied magnetic field.

$$\chi_m = \mathbf{M}/\mathbf{H}$$
$$= (\mu_r - \mu_0)/\mu_0$$

where $\mathbf{M}$ = magnetization, $\mathbf{H}$ = applied magnetic field, μ_r = relative permeability, μ_0 = permeability constant.

suspensoid A system of colloidal particles, sufficiently small to be kept in suspension by Brownian motion.

Susy Supersymmetry.

sverdrup A unit of ocean current transport, equal to 10^6 m^3/s.

swale A depressed, marshy tract of land.

swamp A broad tract of land covered with water and vegetation, with sufficient exchange of water to prevent the accumulation of peat.

S wave A secondary or shear body wave with motion transversal to the direction of propagation. S waves commonly propagate at 55–65% of the speed of P waves. Cf. **P wave.** See **Seismic waves*.**

swell A broad wave characterized by long-period (30–300 s), long wavelength (several hundred meters), and appreciable height (a few meters), formed by major storms at sea and traveling at high speed (tens of kilometers per hour) out of the storm area and across the ocean for as much as 10,000 km without appreciable loss of energy.

syenite A plutonic rock consisting mainly of K-feldspar, with secondary amounts of alkali plagioclase, hornblende, and biotite.

sylvite The mineral KCl.

symbiont A member of symbiotic pair.

symbiosis A relationship between two organisms or two species that is beneficial to both, to similar or different degrees.

symbiotic Pertaining to symbiosis.

symbolic logic A formal treatment of logics using symbols to represent quantities and their relationships.

symmetry The property of a physical system enabling it to remain invariant under specific transformations in space, time, or a mathematical field. Space translation symmetry results in the conservation of momentum; time translation symmetry results in the conservation of energy; and rotational symmetry results in the conservation of angular momentum. Symmetry in gauge transformations results in the conservation of charge, baryon number, and lepton number. See **SU(3), SU(5)**.

symmetry elements The 22 elements by which the symmetry of crystalline structures can be described. These are: 1 center of symmetry; 4 rotation axes (2-, 3-, 4-, and 6-fold), 4 retroinversion axes (2-, 3-, 4-, and 6-fold, plus inversion); 11 helicoidal axes (1 binary, 2 ternary, 3 quaternary, and 5 senary); 1 plane; and 1 glide plane. There are 230 combinations (space groups) of these elements that give unique structural motifs.

sympatric 1. Referring to a species living in the same area as another one. 2. Referring to a process occurring in the same area as another one.

syn- Prefix meaning *with*.

synchrocyclotron A frequency-modulated cyclotron used to accelerate protons, deuterons, alpha particles, and heavier ions, in which the frequency of the accelerating voltage is modulated to keep in phase with the frequency of the spiraling particles.

synchrotron A device for accelerating electrons or protons in closed orbits by a constant-RF potential and a varying magnetic field.

synchrotron radiation Electromagnetic radiation emitted by particles accelerated to relativistic speeds in a magnetic field.

syncline A concave fold. Cf. **anticlyne**.

synclinorium A complex of parallel synclines.

syndiagenesis The chemical and physical alterations of sediments during deposition and shortly after. Syn. **early diagenesis**.

syndiagenetic Pertaining to syndiagenesis.

syneresis The spontaneous dehydration of a gel with aging.

synergism The combined action of two or more agents, the total effect of which is greater than the sum of the individual effects taken separately.

syngenetic Referring to a process occurring at the time of formation of an object, organism, or rock formation.

synodic Pertaining to the conjunction of celestial bodies.

synodic month The time intervals between successive New Moons, equal to $29.5305882 + 0.00000016T$ d_E, where T = centuries from 1900.0.

synodic period (P_{syn}) The average time interval between successive returns of a planet or the Moon to the same position with respect to the Earth and the Sun. For inferior planets:

$$P_{syn} = 1/P_p - 1/P_e$$

where P_p = sidereal period of the planet or the Moon, P_e = sidereal period of the Earth. For superior planets:

$$1/P_{syn} = 1/P_e - 1P_p$$

synonym The name of a taxon that is deemed to be identical to that of an already named taxon.

syntexis The melting of pre-existing rock. Cf. **anatexis**.

syntype 1. Any of the specimens accompanying the holotype in the original outcrop or collection. 2. Any of the specimens in an original collection from which no holotype was designated.

system A chronostratigraphic unit below erathem and above series in rank, consisting of rocks deposited during a period.

Système International d'Unités See **SI**.

syzygy Alignment of the Sun, Earth, and either the Moon or a planet, occurring when the Moon or the planet are at conjunction or at opposition.

T

τ 1. Hour angle. 2. Mean life. 3. Shear stress.

t 1. Temperature (Celsius). 2. Time.

T 1. Absolute temperature. 2. Period. 3. Tesla.

$t_{1/2}$ Half life.

t_A Atomic time.

t_E Ephemeris time.

T_{eff} Effective temperature.

t_U Universal time (cf. **UT**).

tachylite See **sideromelane.**

tachyon A hypothetical particle, consistent with special relativity, which can exist only at speeds greater than the speed of light.

taconite A cherty rock with laminae enriched in iron oxides and, more rarely, iron carbonates.

taenite γ-Ni–Fe (face-centered cubic) alloy (up to 65 wt. % Ni) in iron meteorites. Cf. **kamacite.**

TAI Temps Atomique International (International Atomic Time).

taiga A forested, circumarctic area bound by tundra to the north and by steppe to the south.

talc A hydrated magnesium silicate,

$$Mg_3SiO_4O_{10}(OH)_2.$$

talc schist A schist rich in talc with muscovite and quartz.

talus A rock debris accumulating at the base of a mountainside or a cliff. Syn. **scree.**

tan Tangent.

tangent See **trigonometric function.**

taphocoenosis A community of fossils buried together by transport and sedimentation, not representative of a specific biocoenosis. Cf. **thanatocoenosis.**

taphrogeosyncline A geosyncline developed within a crustal rift bound by normal faults. Cf. **aulacogen.**

tauon (τ). An unstable lepton. Charge = $e^\pm$; rest mass = 1.91540 u; $t_{1/2}$ = $3.4 \cdot 10^{-13}$ s; decay = $\mu^\pm \nu \bar{\nu}$, $e^\pm \nu \bar{\nu}$, etc. See **Elementary particles*.**

tautonym A name used to designate both genus and species (and occasionally also subspecies). The species thus designated becomes the type species for the genus.

taxis Orientation or movement of an organism in response to a stimulus.

taxon 1. Any of the formal groupings of organisms (kingdom, phylum, class, order, family, genus, species, subspecies). 2. The organisms classified within any one of such groups.

taxonomy The science dedicated to the formal classification of living and fossil organisms. See **Taxonomy*.**

technicolor 1. A theory according to which the Higgs particle is composite and its components are held together by a new force (analogous to the color force) called technicolor. 2. The force holding together the components of the Higgs particle (presumed range $\sim 10^{-20}$ m).

tecto- Prefix meaning *to construct.*

tectogenesis See **orogeny.**

tectonic Pertaining to tectonics.

tectonic breccia A breccia resulting from friction between two crustal blocks affected by tectonic motions.

tectonics The study of the relative motions of crustal blocks.

tectonism The motions of crustal blocks.

tectosilicates See **silicate.**

tektite Any of the silica-rich, fused bodies, centimeters across, occurring in specific strewn fields of specific ages (Australasian, $0.7 \cdot 10^6$ y; Ivory Coast, $1.3 \cdot 10^6$ y; Czechoslovakia, $14.8 \cdot 10^6$ y; Texas, $35 \cdot 10^6$ y) and apparently formed by cometary impacts on Earth. Microtektites ($<$1 mm across) related to the falls (except the Czechoslovakian one) have been found in deep-sea sediments (Australasian, western Pacific-Indian Ocean; Ivory Coast, eastern equatorial Atlantic; Texas, Caribbean-equatorial Pacific). Average chemical composition (mass %): $SiO_2 = 75.6$, $Al_2O_3 = 13.0$, $FeO + Fe_2O_3 = 4.1$, $Na_2O + K_2O = 3.5$, $MgO = 1.7$, $CaO = 1.4$, $TiO_2 = 0.8$, $H_2O = 0.005$.

teleseism An earthquake distant from a given seismological observatory.

tell A midden (North Africa, Middle East).

telluric Pertaining to the Earth.

temperature The level of the internal kinetic energy of a system of particles or a body.

Temps Atomique International (TAI) See **International Atomic Time.**

tephra Volcanic ash and related pyroclastics.

tephrite The extrusive equivalent of theralite, consisting of Ca plagioclase, a feldspathoid, and augite.

tera- Prefix meaning 10^{12}.

terminal moraine The end moraine of a glacier or ice sheet.

terminus The lower margin of a glacier.

ternary system A chemical system consisting of 3 different chemical components.

terra 1. Earth, land *(Latin)*. 2. A lunar highland (pl. *terrae*). See **highlands.**

terra rossa Red earth *(Italian)*, a red soil, rich in Fe and Al oxides and hydroxides, residual from the weathering of limestone.

terrigenous Referring to a sediment derived from land.

Tertiary The Cenozoic subera encompassing the Cenozoic time before the Quaternary. Periods (boundary ages in million years): 65/Paleocene/ 54.9/Eocene/38.0/Oligocene/24.6/Miocene/5.1/ Pliocene/1.6. See **Geological time scale*.**

tesla (T) The SI unit of magnetic flux density (= magnetic induction), equal to 1 weber/m^2. Cf. **gauss.**

test The outer covering of an organism, including shell, regardless of composition. Cf. **shell** *(Biology).*

Tethys The ancient sea separating Gondwana from Laurasia.

tetragonal One of the 6 crystal systems, charac-

terized by 3 perpendicular symmetry axes, two of which have the same length.

tetravalent Referring to a valence of 4.

Tevatron The Fermilab accelerator at Batavia, west of Chicago, capable of accelerating protons to 1 TeV (1 teraelectronvolt = 10^{12} eV).

thalassa *(Greek)* Sea.

thallophytes The algae, characterized by the zygote not developing into an embryo and by the absence of vascular tissue. Cf. **embryophytes.**

thalweg The midstream line where the two sides of a valley intersect.

thanatocoenosis. A death assemblage representative of the biocoenosis. Cf. **taphocoenosis.**

theodolite A survey instrument consisting of a horizontal circle on a tripod serving as a base to a vertical circle to which a telescope is attached. It is used to measure horizontal and vertical angles.

theralite A gabbro rich in feldspathoids.

thermal conductivity See **conductivity.**

thermal cross section The cross section of a thermal neutron. It is expressed in barns.

thermal equator The line connecting the points at the ocean surface where air temperature has the highest yearly average.

thermal neutron A low-energy (av. 0.025 eV = 17°C) neutron active in neutron-capture events.

thermal noise Noise produced in conductors and circuits by the thermal agitation of conducting electrons.

thermal radiation Electromagnetic radiation emitted by all solids, liquids, or gases at temperatures $T > 0$ K.

thermistor A ceramic semiconductor whose resistivity decreases (nonlinearly) with increasing temperature.

thermocline The layer below the surface of a lake or the ocean where temperature decrease is steepest.

thermocouple Two strips of different metals (Chromel-Alumel, Fe-constantan, Cu-constantan, or Pt-Pt with either 10% or 13% Rh) joined at each end. An emf (up to several tens of millivolts) develops when the two ends are kept at different temperatures (Seebeck effect). It is used to measure temperature by holding one end at a reference temperature.

thermodynamic cycle A cyclic process involving reversible conversions among different forms of energy.

thermodynamics The study of energy (esp. heat) transfer within or between systems.

thermodynamic temperature scale The absolute temperature scale, starting at the absolute zero and measured in kelvins.

thermohaline circulation The density-driven vertical circulation of the ocean.

thermoionic emission The emission of electrons by a heated cathode in a vacuum tube.

thermoluminescence The emission of light by electrons in a solid substance that have been energized to higher energy levels by ambient radioactivity as they return to the ground state when the substance is heated or otherwise excited.

thermonuclear reaction The fusion of lighter atomic nuclei into heavier ones, with release of energy. See **fusion**.

thermopause The boundary between thermosphere and exosphere at 650 km of altitude, where the mean free path of atmospheric gas molecules equals the altitude (critical level).

thermophilic Referring to an organism adapted to high temperatures.

thermoplastic Referring to a substance that is softened by heat.

thermoremanent magnetization Magnetization acquired by a substance while cooling from above the Curie point to room temperature in the ambient magnetic field.

thermosetting Referring to a substance that is not softened by heat.

thermosphere 1. The totality of warm environments on Earth, including low altitudes at low latitudes, and the low-latitude surface layer of the ocean. Cf. **psychrosphere**. 2. The outer layer of the terrestrial atmosphere (85–650 km of altitude), bound by the thermopause. See **atmosphere**.

Thévenin's theorem "Any linear active circuit seen from two defined terminals A and B is equivalent to a voltage source V_{oc} in series with a resistance R_{Th}, where V_{oc} is the open-circuit voltage between A and B and R_{Th} is the equivalent resistance (Thévenin resistance) between A and B when all voltage sources in the circuit are left open and all current sources are shorted out."

thio- Prefix meaning *sulfur*.

thioalcohols See **thiols**.

thioethers A family of organic compounds consisting of two radicals (alkyl or aryl) joined by a sulfur atom.

thiols A family of organic compounds formed by an alkyl radical and an $-SH$ group. They are the sulfur equivalent of alcohols.

third law of thermodynamics "A crystal has zero entropy at the absolute zero (0 K)."

$$\text{Lim}_{T \to 0} \Delta S_T = 0$$

where T = absolute temperature, S = entropy.

thixotropic Exhibiting thixotropy.

thixotropy The ability of a mixture of clay-size particles and water to form H bonds, thus producing a relatively rigid structure.

thol- Prefix meaning *dome-like*.

tholeiite A subalkaline basalt enriched in iron.

tholeiitic basalt See **tholeiite**.

tholoid A broad volcanic dome.

tholus A dome-like elevation, especially on Mars.

threshold temperature 1. The temperature above which, but not below which, a given phenomenon or process can take place. 2. The temperature above which the reaction energy $\rightleftarrows$ matter is reversible. It is equal to mc^2/k (where m = mass of particle, c = speed of light, k = Boltzmann constant). See **element formation**.

throw The vertical component of the total slip of a fault.

thrust The amount of overriding of one structure over another.

thrust fault A reverse fault with a dip of 45° or less.

thylacoid The basic chlorophyll-containing structural unit of chloroplasts. A thylacoid is a flattened, folded membrane stacked in 2–50 layers to form a granum, a few tens of which form a chloroplast.

thymidine $C_{10}H_{14}N_2O_5$ (mol. mass = 242.232), a nucleoside consisting of thymine linked to a ribose sugar.

thymine A nucleic acid base, $C_5H_6N_2O_2$ (mol. mass = 126.115 u).

thyratron A hot-cathode, gas-filled triode in which a high positive potential at the anode is neutralized by a high negative potential at the grid. Discharge occurs when the grid potential is reduced. Argon or argon-mercury mixtures are commonly used. Deionization following discharge requires about 1 msec (recovery time). Hydrogen gas is used for high-speed applications (recovery time = 1 μsec). Thyratrons have been used as relays, rectifiers, and counters for radioactive particles. The thyratron is now replaced in many applications by its solid-state counterpart, the silicon-controlled rectifier.

tidal basin A coastal basin flooded or partly flooded at each high tide.

tidal bore See **bore.**

tidal flat A low-lying flatland covered with water during each high tide.

tidal force A stress caused on a celestial body by the gravitational attraction of another one, the attraction being greatest at the near side of the body and least at the opposite side. See **tide-generating force.**

tidal marsh A coastal marsh invaded by seawater during each high tide.

tidal pool A pool of water left in a depression on a coastline by the retreating tide.

tidal wave Incorrect name for *storm surge* or *tsunami.*

tide The periodic change in the level of the ocean and other large water bodies in response to the gravitational attraction of the Moon and, to a lesser degree, of the Sun. See **tide-generating force.**

tide-generating force 1. The differential force produced on different parts of a celestial body by the differential gravitational attraction by a neighboring body. 2. The deviation, from point to point on the Earth's surface, of the gravitational attraction of the Moon and of the Sun with respect to the gravitational attraction on the center of the Earth. This deviation is inversely proportional to the cube of the distance, while the attraction on the center is inversely proportional to the square of the distance. As a result, the tide-generating force produced by the Sun is 2.17 times smaller than that produced by the Moon, while the gravitational attraction of the Sun on the center of the Earth is 179 times greater.

tidewater Water brought in by the tide.

tiger-eye A gem variety of quartz.

till An unsorted, unconsolidated, chaotic glacial deposit.

tillite A lithified till.

till sheet A sheet of till deposited by an ice sheet.

timberline See **tree line.**

time See **atomic time, ephemeris time, International Atomic Time, sidereal time, universal time.**

time constant 1. The time it takes for a quantity to increase from 0 to $1 - 1/e$ (or to 63.2%) of its final value, if the rise varies with time t as $1 - e^{-kt}$ (k = constant). 2. The time it takes for a quantity to decrease in $1/e$ (or to 36.8%) of its initial value if the decrease varies with time t as e^{-kt} (k = constant).

time of flight The time it takes for an accelerated particle to travel a specific distance, as from source to collector or between two detectors.

time-stratigraphic unit See **chronostratigraphic unit.**

time-transgressive See **diachronous.**

time zone Any of the 24 internationally agreed-upon zones, about 15° of longitude wide, in which the Earth is divided. Starting with time zone Z, bisected by the Greenwich Meridian (see **Z, ZULU**) the time zones are lettered A to M eastward of zone Z, and N to Y westward of it. The boundary between time zones M and Y is the date line across which one passes from a day to the following day (westward) or to the preceding day (eastward).

Titan See **Saturn.**

titanaugite The mineral $Ca(Mg,Fe,Ti)(Si,Al)_2O_6$.

Titius-Bode law A numerical relationship yielding the distance r of the planets and the asteroidal belt from the Sun in terms of Astronomical Units (AU).

$$r = 0.4 + 0.3 \cdot 2^n \text{ AU}$$

where n may assume the values of $-\infty$ (Mercury), 0 (Venus), or the integers 1 to 8 (Earth to Pluto, including the asteroidal belt). The distances predicted for the planets up to and including Uranus are quite accurate, but those predicted for Neptune and Pluto are 29% and 95% (respectively) too large.

titration The measurement of the concentration of a substance in a solution by the addition of an increasing amount of an appropriate reactant until the reaction is completed, as indicated by color change or electric measurement.

tjaele Frozen ground, not necessarily permanently.

tokamak A device for confining plasma within a toroidal chamber by the application of strong magnetic fields.

tombolo A sand bar connecting an offshore island with the mainland.

ton 1. A metric unit of mass equal to 1000 kg. 2. A nonmetric unit of weight equal to 2000 lb (short ton). 3. A nonmetric unit of weight equal to 2240 lb (long ton). Cf. **tonne.**

tonalite A plutonic, granitoid rock rich in plagioclase and poor in alkali feldspar.

tonne The metric ton, equal to 1000 kg. Cf. **ton.**

tool mark An imprint on the soft surface of a graded bed left by an object carried by a turbidity current.

topaz An orthorhombic mineral, $Al_2SiO_4 \cdot (OH,F)_2$, white or yellowish in color. Cf. **oriental topaz. See Gems*.**

topographic correction A correction to a gravity measurement to account for the topography in the vicinity of a station.

topology The study of geometric spaces invariant under deformation.

topotype A specimen collected at the same location from which the holotype was collected.

topset bed Any of the almost horizontal beds deposited on the surface of an advancing delta.

tornado A funnel-shaped air vortex descending from a cloud, with peripheral winds reaching more than 200 km/h and a tip that may or may not touch ground. Tornadoes touching ground commonly cut a destructive path ~50 m wide and 5 km long. Tornadoes move at speeds averaging 50–60 km/h.

toroid Shaped like a torus.

toroidal See **toroid.**

torque A rotational vector applied to a point P, equal to the vectorial product $\mathbf{D} \times \mathbf{F}$, where $\mathbf{F}$ = force, $\mathbf{D}$ = distance between point P and the point of application of the force.

torr 1/760 of 1 atmosphere, equal to 133.32237 Pa = 1 mmHg (within $1 \cdot 10^{-7}$).

torrent A seasonal mountain stream.

torus The surface generated by a circle of radius r, normal to the plane of a second circle of radius $>2r$, when its center is moving along the entire perimeter of the second circle.

touchstone A velvet-black, cherty stone for testing the purity of gold and silver by the color of their streaks.

tourmaline The mineral $(Na,Ca)(Li,Mg,Al) \cdot (Al,Fe,Mn)_6(BO_3)_3Si_6O_{18} \cdot (OH)_4$. See **Gems*, Minerals*.**

trace element An element dispersed in a given substance in amounts smaller than ~1%.

trace fossil A mark or imprint left on a sediment surface by the activity of an animal. Syn. **ichnofossil.**

Tracheophyta A grade of the Kingdom Plantae, characterized by the sporophyte being the major plant body and by the presence of vascular tissue. Cf. **Bryophyta.**

trachyte The extrusive equivalent of syenite.

trades See **trade winds.**

trade winds The low-latitude, low-altitude return flow of the Hadley cells on either side of the equator. Direction of flow is from the northeast in the northern hemisphere, from the southeast in the southern hemisphere. See **Hadley cell.**

transceiver An instrument that both transmits and receives radio waves.

transducer Any device transforming an input into an output of a different form.

transfer function The mathematical relationship between output and input in a control system.

transfer RNA (tRNA) The anticodon-bearing RNA molecule that carries specific amino acids to the ribosomal site where protein synthesis occurs. tRNA anticodons are matched to mRNA codons so that the amino acid sequence results in a protein of the prescribed structure.

transformer A device that transfers electric energy from one circuit to another. It consists of two coils inductively coupled.

transform fault A lateral fault offsetting a spreading axis.

transgression 1. Rigorously, either an ingression or a regression of the sea. 2. Loosely, a marine ingression.

transistor An electronic device with two junctions and three terminals (emitter, base, and collector) consisting of either an *npn* system or a *pnp* system. Depending upon the voltages and signals applied to or across the terminals and the characteristics of the circuitry, transistors can operate as voltage, current or power amplifiers or regulators, or as switches.

transit Passage of an inferior planet across the solar disc.

transition element Any of the elements which, in the ground state, have a partially filled (<10 electrons) *d* subshell. Included are elements 21-28, 39-45, 57-78, and 89+. See **Elements—electronic structure***.

transmittance (T) The ratio of the radiant power *I* transmitted by a body to the radiant power I_0 incident on it:

$$T = I/I_0$$

transversal wave A wave in which the parameter involved changes perpendicularly to the direction of wave propagation. Cf. **longitudinal wave**. See **S wave**.

transverse fault A fault striking a structural trend at an angle.

transverse wave See **transversal wave.**

traveling wave A wave that transports energy from one place to another, in contrast to a standing wave.

traveltime The time taken by a seismic wave to travel from source to target or to receiver.

travertine A dense, microcrystalline calcitic or araragonitic precipitate from a hot spring or from surface or groundwater. Cf. **tufa.**

tree line The line across a mountainside above which trees do not grow.

trench A narrow, elongated, deep depression of the sea floor. Major trenches are (max. depth in m): Aleutian (7679), Amirante (9074), Banda (7440), Bougainville (8940), Cayman (7093), Diamantina (8230), Japan (9810), Java (7450), Kermadec (10,047), Kuril-Kamchatka (9750), Marianas (10,915, Challenger Deep), Middle America (6662), New Britain (8245), New Hebrides (9165), Palau (8054), Peru-Chile (8055), Philippine (10,030), Puerto Rico (9200), Ryukyu (7507), South Sandwich (8264), Tonga (10,882), Yap (8527).

triangulation A method for determining the distance *d* of a target C from a point *A* by measuring

the length of a base line $AB = b$ and the two angles α and β formed by the two lines joining the target to the two ends of the baseline, and by resolving for the length of the line *h* from C to *b* normally to *b*.

$$h = b/(\cot \alpha + \cot \beta)$$
$$d = h/\sin \alpha$$

tribe A grouping of petrologic clans. See **clan, family.**

tribo- Prefix meaning *friction.*

tributary A stream or a glacier emptying into a larger one.

triclinic One of the 6 crystal systems, characterized by three axes of unit length different from each other and intersecting at angles different from 90°.

tridymite A high-temperature polymorph of quartz, stable between 867°C and 1470°C at atmospheric pressure and in the absence of water or impurities. See **silica.**

trigonal A subsystem of the hexagonal crystal system, characterized by one axis of threefold symmetry.

trigonometric function Any of the functions related to an angle on a unit circle. **1. sine (sin)** The ordinate of the endpoint of an arc on a unit circle, starting at 3 o'clock and drawn counterclockwise. **2. cosine (cos)** The abscissa of the endpoint of an arc on a unit circle, starting at 3 o'clock and drawn counterclockwise. **3. tangent (tan)** The ratio sine/cosine. **4. cotangent (cot)** The ratio cosine/sine. **5. secant (sec)** The ratio 1/cosine. **6. cosecant (csc)** The ratio 1/sine. **7. versine (vers)** The difference 1 − cos. **8. coversine (covers)** The difference 1 − sin.

triple-alpha process (3α process) The collision of 3 alpha particles to form a ^{12}C nucleus. It occurs in H-depleted stellar cores at temperatures of 10^8 K.

triple point The temperature and pressure at which the three phases of a system (solid, liquid, gas) can coexist.

triplet 1. The electronic state of an atom or molecule having total spin angular momentum quantum number of 1. The total angular momentum quantum number *J* can thus assume the three values $L + 1$, L, and $L − 1$, where L = orbital angular momentum vector. 2. A set of three particles sharing most but not all properties, such as the three pions.

tripoli A pulverulent weathering product of chert or siliceous limestone.

tritium The isotope ^{3}H. $t_{1/2} = 12.33$ y.

tritium dating method The absolute dating of subterranean water, deeper ocean water, and other H-bearing substances previously in exchange with the atmosphere, by the residual amount of tritium present.

triton The nucleus of tritium.

TRM Thermoremanent magnetization.

tRNA Transfer RNA.

trochoidal Having the shape of a widening helicoidal spiral.

troctolite A gabbro consisting mainly of Ca plagioclase and olivine.

troilite The mineral FeS, common in meteorites.

Trojan Any of the minor planets belonging to the two groups that occupy Lagrangian points 4 and 5 in the orbit of Jupiter, leading and trailing the planet by 60°. See **Lagrangian points.**

trona The mineral $Na_2CO_3 \cdot NaHCO_3 \cdot 2H_2O$, common in dry salt-lake beds.

trophic Pertaining to nutrition.

tropic Either the tropic of Cancer (23°26′28″ N) or the tropic of Capricorn (23°26′28″ S), at latitudes fixed by the present angle of 23°26′28″ be-

tween the Earth's rotational axis and the normal to the plane of the Earth's orbit. The tropics are the loci where the Sun stands vertical at local noon during the summer solstice.

tropical century One hundred tropical years.

tropical year The time interval between successive vernal equinoxes. It is equal to $(31,556,925.9747 - 0.530T)$ s_E or $(365.24219878 - 0.0000006167)$ d_E, where T = centuries from the year 1900.0. The correction is required because of the 11.2″/century secular deceleration caused by the lunar tides. Between 1986 and 2002 the tropical year is remaining equal to 31,556,925.47 s_E or 365.242193 d_E within the last decimal figure.

tropical year 1900 The primary unit of ephemeris time. It is the time the Earth required to complete a revolution around the Sun beginning at the vernal equinox of the year 1900. It is equal to 31,556,925.9747 s_E or 365.24219878 d_E.

tropic of Cancer See **tropic.**

tropic of Capricorn See **tropic.**

tropism The turning of a sessile organism in response to a stimulus.

tropopause The boundary surface separating the troposphere below from the stratosphere above. Its altitude increases from 10 km at the poles to 16 km at the equator. See **atmosphere.**

troposphere The lowest layer of the atmosphere,

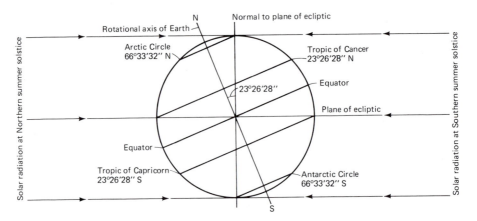

Tropics. The latitude of the tropics is determined by the inclination of the Earth's axis from the normal to the plane of the ecliptic. As a result of this inclination, the Sun's rays are perpendicular to the Earth's surface at the tropic at local noon, summer solstice. At the same time, the polar cap (the area bound by the polar circle) in the same hemisphere receives maximum illumination. At local noon, winter solstice, the Sun's rays are inclined 46°52′56″ to the vertical at the tropic, while the polar cap receives no solar light.

separated from the stratosphere by the tropopause. Its thickness ranges from 10 km at the poles to 16 km at the equator. See **atmosphere.**

trough An elongated depression on the sea floor, broader and shallower than a trench.

trunk glacier The main glacier into which tributary glaciers flow.

trunk stream The main stream into which tributary streams empty.

Tschermak molecule The hypothetical silicate molecule $(Mg,Fe)O \cdot (Al,Fe)O_3 \cdot SiO_2$. This molecule becomes the Ca-Tschermak molecule if $(Mg,Fe)O$ is replaced by CaO.

tsunami A long-period (3 min to 3 hr), long-wavelength (up to 200 km), low-amplitude (centimeters to decimeters away from the source and terminus), high-speed (600–800 km/h) surface seawave produced by a seaquake. Funneling of seawater by coastline morphology may cause local water uprush in excess of 20 m.

T Tauri stars Rapidly rotating, irregularly variable stars exemplified by T Tauri and characterized by strong infrared radiation from a surrounding dust and gas cloud. T Tauri stars may represent young stars in the process of evolving into Main Sequence stars and perhaps developing a planetary system. Age $\sim 1 \cdot 10^6$ y.

tufa A less dense variety of travertine.

tuff A pyroclastic deposit consisting of pyroclasts and volcanic ash.

tundra The treeless, high-latitude belt of vegetation north of the taiga.

Tunguska event The collision of a small $(5 \cdot 10^7$ kg) comet with the terrestrial atmosphere, creating an explosion at an altitude of 8.5 km above the valley of the Podkamennaya Tunguska River (a tributary of the Lena) in NE Siberia at 7.17 A.M. on June 30, 1908. No crater was formed but the explosion, releasing an energy of $5 \cdot 10^{16}$ joules (~ 1.2 megatons), flattened trees radially to a distance of 15 km from ground zero (but trees directly under ground zero were left standing). Magnetite with up to 8.5% Ni and glassy silicate spherules of extraterrestrial origin were found in soil samples at the site.

tunnel effect The ability of a particle to cross a potential energy barrier greater than its own total energy.

tunneling The crossing by a particle of a potential barrier greater than its own total energy.

turbidite A graded-bedded clastic subaqueous layer deposited by a turbidity current.

turbidity current Downslope subaqueous movement of sediment-laden water.

turn One complete loop.

turquoise The mineral $CaAl_6(PO_4)_4(OH)_8 \cdot 4H_2O$. See **Gems*.**

twin crystal A crystal exhibiting twinning.

twinning The occurrence of two or more parts of the same crystal having different lattice orientations related to each other by simple symmetry operations. Twinning may result from crystal growth, polymorphic transformation, or mechanical stress.

twin axis The axis of symmetry of a twin pair superimposing one twin on the other by rotation.

twin plane The common interface joining twin crystals.

two-pi Pertaining to all directions on a plane radiating from a point on that plane. The expression refers to the plane angle (2π) subtended by the center of a circle. Cf. **four-pi.** See **rationalized unit.**

two-way travel time The time an acoustic signal takes to travel from source to reflector and back. See **travel time.**

Tycho The youngest among the larger lunar craters. Age $= 270 \cdot 10^6$ y.

Tycho's star The galactic supernova that flared on November 11, 1572 and that was studied by Tycho Brahe.

type locality The location of a type section.

type section The originally described section of a given stratigraphic unit.

type specimen The single specimen upon which a taxon is based, i.e. the holotype, or the neotype, or the lectotype.

typhoon A tropical cyclone in the western tropical Pacific, equivalent to the hurricane of the western tropical Atlantic. See **hurricane.**

U

u 1. Atomic mass unit, defined as 1/12th of the mass of the neutral atom of ^{12}C. Cf. **amu, atomic mass unit.** 2. Velocity.

U Internal energy.

UBV. Ultraviolet-Blue-Visual, the system of photoelectric stellar magnitudes (V is for *visual,* meaning *yellowish-green* which is the color band to which the human eye is most sensitive). See **magnitude.**

Udden-Wentworth scale See **Wentworth grade scale.**

ultrabasic Defining a rock with <40% SiO_2, regardless of mineralogical composition. Cf. **acid** *(Geology),* **basic** *(Geology),* **intermediate** (def. 1).

ultramafic Defining a rock with 90% or more mafic minerals.

ultrametamorphism Metamorphism at the highest temperature and pressure possible without complete melting of the rock.

ultraplankton Plankton smaller than 5 μm, smaller than nanoplankton.

ultrasonic Referring to a frequency greater than the highest audible frequency of 20,000 Hz.

ultrastructure A tissue or skeletal structure visible only by electron microscopy.

umbilicus An external depression at the intersection of the coiling axis of a shell with the last whorl.

umbra 1. The area of complete shadow of a body being eclipsed. 2. The darker, cooler (4240 K) center of a sunspot surrounded by the less dark, warmer (5680 K) penumbra.

uncertainty principle "The product of the uncertainties Δp in momentum p and Δr in position r of a particle is equal to, or greater than, Planck's constant divided by 2π."

$$\Delta p \, \Delta r \geqslant h/2\pi$$

unconfined aquifer An aquifer with a free surface.

unconfined groundwater Groundwater with a free surface (water table).

unconformity A break in the sedimentary record. There are four major types of unconformity. **1. nonconformity** The rocks below are massive igneous or metamorphic rocks separated from the sedimentary rocks above by an erosional surface. **2. angular unconformity** Bedding of the sedimentary rocks below is at an angle with respect to that of the sedimentary rocks above, indicating a period of diastrophism and erosion. **3. disconformity** Bedding of the sedimentary rocks below is parallel to that of the sedimentary rocks above, but a conspicuous erosional surface separates the two. **4. paraconformity** Bedding of the sedimentary rocks below is parallel to that of the sedimentary rocks above, but the two are separated by a period of nondeposition. Cf. **diastem.**

underclay A clay layer below a coal seam, representing the soil upon which the coal-forming plants grew.

undersaturated Referring to an igneous rock containing low-Si minerals (feldspathoids, pyroxene, olivine).

underthrust A fault block that has been thrust under an opposite block at low angle.

undulatory extinction Wavy extinction exhibited by strained minerals as the petrographic microscope stage is rotated.

undulipodium Any of the cilia or flagella in eucaryotic cells, consisting of a bundle of nine peripheral microtubules and two axial ones originating from a basal body (kinetosome). Kinetosomes are self-duplicating organelles suggesting that undulipodia may have originated from spirochetes symbiotic with early eucaryotes.

uniaxial Referring to a crystal having only one optic axis.

unified field theory Any of the theories attempting to combine color, weak, electromagnetic, and gravitational fields into a single system.

uniformitarianism The concept that the nature of natural processes is invariant with time.

unipolar transistor A transistor using only majority carriers for conduction. Cf. **bipolar transistor.**

unitary transformation A linear transformation in vector space that conserves norms and scalar products.

unit cell The smallest portion of a crystal lattice containing all atomic or molecular species characteristic of that crystal.

unit magnetic pole The unit of magnetism in the CGS system, defined as that quantity of magnetism that, when placed at the distance of 1 cm from a similar quantity, attracts it or repels it with a force of 1 dyne.

$$F = \mathfrak{M}\mathfrak{M}/r^2$$

where F = force, $\mathfrak{M}$ = unit magnetic pole, r = distance between the two unit magnetic poles. Cf. **oersted.**

unit pole See **unit magnetic pole.**

univalent Referring to an atom or molecule with a single valence.

univalve Referring to an organism forming only one valve (e.g. a gastropod). Cf. **bivalve.**

universal stage A device on the stage of an optical microscope to allow the rotation of a mineral grain or thin section in all three directions in space.

universal time (t_U, U_T) Mean solar time counted from midnight at the Greenwich meridian. Identical to International Atomic Time on 1958 January 0d 0h 0m 0s. See **Greenwich Mean Time.**

universe *(Cosmology)* The totality of what exists, both visible and invisible. Physical parameters of the observable universe (all uncertain values): radius = $16.6 \cdot 10^9$ light years (Hubble distance); age = $16.6 \cdot 10^9$ y (Hubble time); mass = $1.3 \cdot 10^{52}$ kg; density = $0.9 \cdot 10^{-30}$ g/cm³; rotation (?) = 10^{-13} rad y⁻¹ (?); angular momentum (?) = $1.2 \cdot 10^{84}$ kg m² s⁻¹ (?); total number of nucleons = $8.0 \cdot 10^{78}$; total number of photons = $6 \cdot 10^{87}$. Cf. **cosmos, inflation, metagalaxy.** *(Statistics)* A population of items from which a sample is drawn.

universe, evolution of The present universe appears to have originated $16.6 \cdot 10^9$ years ago (Hubble time, an uncertain value) and to have expanded ever since. Three possibilites are recognized, depending upon the density parameter Ω. 1. $\Omega > 1$: space is positively curved (spherical), expansion will stop, the universe will collapse over itself into a cosmocrunch from which a new expansion may arise. 2. $\Omega = 1$: space is flat (Euclidean), the universe will continue expanding forever at a decreasing rate approaching zero at infinity

(parabolic expansion). 3. $\Omega < 1$: space is negatively curved (hyperbolic), the universe will keep expanding forever at a decreasing rate toward at an asymptotically constant rate (hyperbolic expansion). If $\Omega > 1$, the universe is closed; if $\Omega < 1$, the universe is open. At present $\Omega = 0.15$ (very uncertain value).

unpaired electron An electron for which there is no other electron with the same energy but opposite spin in the same atom.

unsaturated *(Chemistry)* 1. Referring to a solution capable of containing additional solute. 2. Referring to a carbon compound having one or more double or triple C to C bonds. *(Geology)* Referring to a mineral that does not form in the presence of free silica (e.g. feldspathoids, olivine).

unshared Referring to an electron not partaking in the formation of a covalent bond.

unsorted A sediment whose elements have not been sorted by the transporting medium. Cf. **sorting.**

unstable *(Physics)* Referring to a nuclide that undergoes radioactive decay. *(Geology)* Referring to a mineral or a rock poorly resistant to weathering.

Upper Referring to the upper portion of a chronological or chronostratigraphic unit. Cf. **Lower, Middle.**

upper mantle The layer between the Mohorovičić discontinuity and the seismic wave velocity discontinuity at 670 km of depth.

uracil A nucleic acid base characteristic of RNA, $C_4H_4N_2O_2$ (mol. mass = 112.088).

uraninite The mineral UO_2.

uranium-lead dating method A method of absolute dating based on the decay of ^{238}U ($t_{1/2}$ = $4.468 \cdot 10^9$ y) to ^{206}Pb, or on the decay of ^{235}U ($t_{1/2}$ = $704 \cdot 10^6$ y) to ^{207}Pb. Cf. **lead-uranium age.**

uranium-thorium disequilibrium dating method Any of a set of absolute dating methods based on the disequilibrium of the daughter products in the U-Pb or Th-Pb decay series. Examples are the $^{234}U/^{238}U$ and $^{230}Th/^{234}U$ ratios in aragonitic carbonates, and the $^{231}Pa/^{230}Th$ ratio in deep-sea sediments.

Uranus The seventh planet from the Sun. Mean distance from the Sun = 19.21814 AU. Sidereal period = 84.0139 y. Inclination of equator to orbital plane 97.92°. Sidereal rotational period =

17.24 h (retrograde). Equatorial radius = 25,400 km. Mass = $86.9 \cdot 10^{24}$ kg. Mean density = 1.31 g/cm^3. Internal structure (estimated): Fe–Ni and silicate core [radius = 8000 (?) km] and a 10,000-km-thick mantle consisting of liquid CH$_4$, NH$_3$, and H$_2$O. Atmosphere thickness = 8000 (?) km. Gases in atmosphere: H$_2$ = 90%, He = 10%. Magnetic field = 0.25 gauss, inclined 55° with respect to the axis of rotation. Surface temperature = 57 K. Fifteen satellites, the largest of which are Titania (radius = 805 km, density = 1.6 g/cm^3) and Oberon (radius = 775 km, density = 1.5 g/cm^3). Ring system, <10 km to 100 km thick, extending from 37,000 to 51,160 km above the equatorial surface. See **Planets*, Satellites***.

urea NH$_2 \cdot$CO$\cdot$NH$_2$.

USGS United States Geological Survey.

U-shaped valley A glacially eroded valley with a broad floor and ripid sides. Cf. **V-shaped valley**.

UT Universal Time.

UV Ultraviolet.

uvarovite A garnet, Ca$_3$Cr$_2$Si$_3$O$_{12}$.

V

v Velocity.

V 1. Electric potential. 2. Potential energy. 3. Volt. 4. Voltage. 5. Volume.

v_P P wave velocity.

v_S S wave velocity.

vadose Referring to the water above the permanent water table.

vale A broad, gently sloping valley.

valence The capacity of an atom or a molecule to form bonds with other atoms or molecules. The bonds are formed by the transfer or sharing of electrons residing in the outer (valence) shell.

valence band The highest electronic energy band in a semiconductor or insulator, separated from the conduction band by an energy gap. Conduction can occur only when electrons in the valence band are sufficiently energized (5.4 eV for C, 1.107 eV for Si, 0.67 eV for Ge at room temperature) to cross the energy gap and enter the conduction band. The vacancies left in the valence band, called *holes,* act as positive particles. Under the influence of an applied voltage or of the photovoltaic effect, electrons and holes move in opposite directions, thus giving rise to an electric current.

valence bond The bond formed by the overlap of two atomic orbitals, each containing an unpaired electron.

valence shell The outer electronic shell containing the electron(s) participating in the formation of chemical bonds.

valence-shell electron-pair repulsion theory (VSEPR) A theory on the shape of a covalent molecule based on the repulsion among bonding and nonbonding electron pairs in the valence shell of the central atom.

valley train A long strip of glacial outwash deposited by a glacial river beyond the terminus of a glacier.

Van Allen belts Two concentric toroidal belts around the Earth with axes coinciding with the Earth's magnetic axis and containing protons and electrons. The inner one, centered at 3200 km of altitude, consists of high-energy protons (>15 MeV) and electrons (>1.5 MeV) of probable galactic origin. The outer one, centered around 25,000 km of altitude, consists of low-energy protons (~ 0.2 MeV) and electrons (~ 0.4 MeV) of probable solar origin.

van de Graaff accelerator A van de Graaff generator with an evacuated conduit through which particles are accelerated.

van de Graaff generator A high-voltage generator consisting of a rubberized belt running between two drives, one at ground potential and the other insulated and surrounded by a large metal sphere. Electrical charges are fed to the belt from metal whiskers at 10,000 to 50,000 volts near the ground drive, and removed at the other end to be deposited on the metal sphere. Potentials as high as 10^7 volts are routinely produced.

van der Waals equation A modification of the equation of state for ideal gases to make it applicable to real gases by accounting for the finite size of atoms and molecules and the attractive forces among them.

$$(p + n^2a/V^2)(V - nb) = nRT$$

where p = pressure, n = number of moles, a = constant characteristic of each gas and independent of temperature and pressure; V = volume of the gas; b = volume occupied by 1 mole of atoms or molecules; R = gas constant; T = temperature.

van der Waals force The weak (<0.5 eV) mutual attraction arising when atoms or molecules mutually distort their charge distribution and induce opposite dipole moments on each other.

van't Hoff's law "The osmotic pressure of a solution equals that of the solute if it were an ideal gas occupying the same volume."

$$\Pi = nRT/V$$

where Π = osmotic pressure, n = number of moles of solute, R = gas constant, T = absolute temperature, V = volume of solution.

vapor A gas below its critical temperature, capable of being liquified by pressure alone. Cf. **gas.**

variability See **coefficient of variability.**

variable A quantity whose change determines a change in another quantity. See **function.**

variable star Any of the stars that change their outputs, either periodically or nonperiodically. Included are stars whose output appears to change because of eclipsing produced by a companion. See **Cepheid, RR Lyrae star, T Tauri.**

variance The square of the standard deviation.

varve A pair of thin sediment layers deposited during one year, one richer in organic matter and the other poorer. In glacial and periglacial lakes, the organic-rich layer is deposited during the winter when the streams feeding the lake are frozen and sediment input is reduced to zero; the organic-poor layer is deposited from spring to fall when the streams are flowing and sediment input is resumed.

vector A quantity that has both magnitude and direction.

vectorial product (×) The vectorial product of two vectors. The magnitude of the product vector is given by the area of the parallelogram constructed using the two vectors as sides; its direction is normal to the plane of the two vectors and in the direction of an advancing screw whose rotation by an angle $<180°$ superimposes the multiplicand on the multiplier. Syn. **cross product.** Cf. **scalar product.**

Vela pulsar Pulsar 0833-45, 1600 l.y. distant in Vela, that originated from a supernova explosion about 11,000 y ago.

velocity (v) The first derivative of position vs. time.

$$\mathbf{v} = d\mathbf{r}/dt$$

where $\mathbf{v}$ = velocity, $\mathbf{r}$ = displacement vector, t = time. Cf. **acceleration, jerk.**

velocity of recession The recessional velocity v_r of a distant celestial body.

$$v_r = c\,[(z + 1)^2 - 1]\,/\,[(z + 1)^2 + 1]$$

where c = speed of light in vacuo, z = redshift parameter.

ventifact A stone shaped by wind in a desertic area. Syn. **glyptolith.**

Venturi meter A device that measures the velocity of a fluid in a pipe. The difference in pressure in the pipe and at a constricted section of the pipe is proportional to the flow rate.

Venus The second planet from the Sun. Mean distance from the Sun = 0.723332 AU. Sidereal period = 0.61521 tropical years = 224.701 d. Sidereal rotational period = 243.01 (retrograde). Equatorial radius R_{eq} = 6051.4 km. Mass = $4.871 \cdot 10^{24}$ kg. Mean density = 5.243 g/cm^3. Internal structure (estimated): Fe–Ni core (radius = 0.43 R_{eq}), silicate mantle. Magnetic field = $<0.5 \cdot 10^{-4}$ gauss. Surface temperature = 700 to 760 K (average = 730 K). Atmospheric pressure = 90 bar. Gases in atmosphere: CO_2 = 96.0%, N_2 = 3.5%, SO_2 = 0.01%. Ar = 0.01%. No satellites. See **Planets*.**

verde antico A dark green ornamental rock consisting of serpentine with calcitic veins.

vermetid reef A small intertidal reef consisting of cemented tubes of the gastropod species *Vermetus nigrans.*

vermicular quartz A wormlike intergrowth of quartz and oligoclase. Cf. **myrmekite.**

vermiculite A group of hydrated Mg–Fe aluminosilicates derived from the weathering of Mg–Fe-bearing micas, especially biotite and phlogopite. It is one of the clay minerals.

vernacular Referring to the common name of a given taxon as opposed to the scientific Latin name.

vernal equinox (♈) The spring equinox. See **First Point of Aries.**

verrucano A cemented conglomerate with quartz pebbles.

vers Versine.

versed sine Versine.

versine See **trigonometric function.**

vertical circle A great circle on the celestial sphere passing through an observer's zenith.

vibrational energy The part of the total energy of a diatomic or polyatomic molecule produced by its vibration. For a diatomic molecule

$$E_{vib} = h\nu_0\,(v + 1/2)$$

where h = Planck's constant, ν_0 = fundamental frequency, v = vibrational quantum number = 0, 1, 2, 3, ... For v = 0, $E_{vib} = \frac{1}{2}h\nu_0$ = zero-point energy, possessed by all vibrating systems at 0 K.

vibrational level An energy level of a diatomic or polyatomic molecule related to a particular value of its vibrational energy.

virgation 1. The divergence of a group of parallel folds. 2. The splitting of a fault into diverging branches.

virial See **virial of a system.**

virial coefficients See **virial equation.**

virial equation An empirical equation of state with terms above those for an ideal gas.

$$pV = RT + Ap + Bp^2 + Cp^3 + \cdots$$

where p = pressure, V = volume, R = gas constant, T = absolute temperature. The virial coefficients $A, B, C, \ldots$ are functions of temperature to be determined empirically. The behavior of any real gas may be determined exactly by the application of a sufficient number of terms.

virial of a system The average "strength" or energy of a system of inert particles taken over a long period of time.

$$E_{av} = -\tfrac{1}{2}(\overline{\Sigma \mathbf{F}_i \cdot \mathbf{r}_i})$$

where E_{av} = average energy, $\mathbf{F}_i$ = attractive (+) or repulsive (−) force experienced by particle i, and $\mathbf{r}_i$ = position vector of particle i.

virion A complete virus particle.

viroid Any of the naked segments of intracellular RNA causing diseases especially in plants. Molecular mass = 10^4 u.

virtual Nonreal.

virtual particle A particle that mediates the transmission of force between real particles. A virtual particle can exist only for a time Δt such that

$$\Delta E \cdot \Delta t \leqslant h/2\pi,$$

where E = energy of particle, h = Planck's constant.

virus Any of the submicroscopic (20–400 nm in length, $1-150 \cdot 10^6$ in molecular mass) pathogens consisting of a nucleic acid core surrounded by a protein sheath. Viruses lack independent metabolism and can only reproduce inside living cells.

viscosity The tangential force

$$F = \eta \, A(dv/dz)$$

needed to move a planar surface of area A along its plane when separated from a similar surface by a distance z filled with a fluid having a coefficient of viscosity η. Characteristic viscosities (poises): air (20°C) = $1.86 \cdot 10^{-6}$; water (20°C) = 0.010; ethanol (20°C) = 0.012; mercury (20°C) = 0.015; olive oil (20°C) = 0.84; glycerol (20°C) = 14.90; glass at melting temperature = 10^3; glass at working temperature = 10^7; pitch (15°C) = 10^{10} g; glass at annealing temperature = 10^{13}; glacier ice = 10^{13}; halite (20°C) = 10^{15} to 10^{17}; alabaster (20°C) = $10^{16}-10^{18}$; Earth's mantle (from glacial rebound) = 10^{21}; glass (20°C) = 10^{22}; marble (20°C) = 10^{22}.

viscous magnetization Secondary magnetization acquired by a mineral or a rock after formation or deposition, caused by the influence of the ambient magnetic field during protracted time.

viscous remanent magnetization (VRM) See **viscous magnetization.**

visual binary A binary system the two components of which are sufficiently apart to be seen as separate stars either with the naked eye or with a telescope.

void ratio The ratio of the volume of void space to the volume of solid matter in a porous or vesicular material.

volcanic dust Pyroclastic particles <62.5 μm in diameter. Cf. **ash (volcanic).**

volcanoclastic Referring to a clastic deposit produced by a volcano.

volcanogenic Referring to igneous or sedimentary rocks produced by volcanic activity.

volt (V) The SI and MKS unit of electromotive force or potential difference, defined as the potential difference needed to develop the energy of 1 joule per coulomb of charge or the power of 1 watt per ampere of current.

$$V = J/C$$
$$= W/A$$

voltaic cell A primary electrolytic cell with two electrodes consisting of different metals.

volt-ampere The SI and MKS unit of apparent power, equal to effective voltage times effective current.

vorticity A vector equal to the curl of the flow velocity.

VRM Viscous remanent magnetization.

V-shaped valley A valley excavated by running water and, therefore, exhibiting a V-shaped cross section. Cf. **U-shaped valley.**

vug A small cavity in a rock, usually lined with crystals.

W

W 1. Energy. 2. Watt. 3. Work.

$W^{\pm}$ The weak bosons, with mass = 86.7 ± 2.9 u, which carry the charged current in weak interaction processes. See **Elementary particles***.

wacke A poorly sorted sandstone rich in unstable mineral grains and rock fragments in an abundant matrix of silt and clay. Cf. **graywacke.**

wad (Dutch) A tidal flat. Pl. *wadden.*

wadden See **wad.**

wadi *(Arabic)* The flat-bottomed, steep-sided bed of an episodic torrent in an arid region. Syn. **arroyo.**

Wallace line The line between Bali and Lombok and northward through the Strait of Makasar between Borneo and Celebes, separating the Eurasian flora and fauna to the northwest from the Australian flora and fauna to the southeast.

water H_2O. Mol. mass = 18.0153 u; density = 0.9998396 g/cm^3 (0°C), = 1.000000 g/cm^3 (3.98°C, the temperature of highest density); melting point = 0.00°C (1 atm); triple point = 0.01°C (vapor pressure = 4.57 mmHg); boiling point = 100°C (1 atm); critical temperature = 374.15°C; viscosity at 0°C = 1.7916 cP, at 25°C = 0.8903 cP, at 100°C = 0.2820 cP; surface tension at 25°C = 7.214·10^{-2} N/m; dielectric constant at 25°C = 77.738; dissociation constant at 25°C, 1 atm = 1.0·10^{-14}. See **Water density*, Water—physical properties*.**

water gas A mixture of CO and H_2 obtained by passing hot (1000°C) steam over glowing coke.

water mass A large body of seawater identifiable from its temperature and salinity.

water molecule (H_2O) Bond length = 0.95718·10^{-10} m; bond angle = 104.523°.

water of crystallization See **water of hydration.**

water of hydration Water as H_2O molecules bound to ions in a hydrate. It can be removed by heating.

watershed 1. The line where water parts along the crest of a mountain ridge or a chain. 2. The drainage basin of a stream or river.

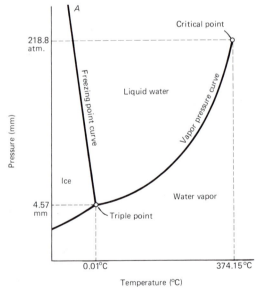

Water. Phase diagram. (Krauskopf 1979, p. 289, Fig. 13.1)

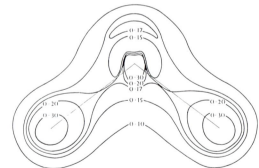

The water molecule. Electron density. Bond length = 0.95718·10^{-10} m; bond angle = 104.523°; root-mean-square vibration amplitude in the ground vibrational state = 0.067·10^{-10} m or 7% of the bond length. (Eisenberg and Kauzman 1969, p. 26, Fig. 1.6)

water table The free surface of groundwater.

watt (W) The SI and MKS unit of power, equal to 1 joule/second.

watt-hour (Whr) A practical unit of electrical energy, equal to 3600 joules.

wave A cyclic phenomenon propagating in space, characterized by amplitude, velocity, frequency or wavelength, and phase.

wavelength	λ	$= c/\nu$
velocity	c	$= \lambda\nu$
frequency	ν	$= \omega/2\pi$
		$= 1/T$
period	T	$= 1/\nu$
		$= 2\pi/\omega$
wave number	n	$= 1/\lambda$ (or $2\pi/\lambda$)

where λ = wave length; c = velocity; ν = frequency; ω = angular frequency; T = period.

wave function (Ψ) A function the square of which is a measure of the probability of a particle occurring within a specified volume of space.

waveguide A metallic conduit along which electromagnetic waves (especially microwaves) can be transmitted.

wavelength (λ) The shortest distance in space between two points on or in a wave train that are consecutive and have the same phase angle.

wave mechanics A branch of quantum mechanics dealing with the wavelike properties of particles.

wave number The number of waves per unit of length. See **wave.**

wave packet A well-defined disturbance in a train of waves resulting in a short-lived amplitude disturbance due to interference.

wave train A series of waves produced by the same agent.

wave vector A vector directed in the direction of propagation of a wave, with a magnitude equal to the wave number.

wave vector space The space defined by surfaces normal to the wave vectors in a crystal system.

wave velocity See **phase velocity.**

wavy extinction See **undulatory extinction.**

Wb Weber.

weak acid An acid that is only partly ionized in an aqueous solution. Cf. **strong acid.**

weak base A base that is only partly ionized in an aqueous solution. Cf. **strong base.**

weak force The force carried by the $W^{\pm}$ and Z^{o} bosons. It is effective to distances $<10^{-18}$ m. It is 10^4 times weaker than the electromagnetic force and 10^6 times weaker than the strong force. See **natural forces.**

weak interaction The interaction involving hadrons and leptons, resulting in the β^- decay of the neutron, and the decays of the muon, the tauon, the charged pion, and other mesons. It is mediated by the $W^{\pm}$ and Z^{o} particles. Cf. **natural forces, weak force.**

weathering The process of rock and mineral disintegration by physical, chemical, or biochemical agents.

weber (Wb) The SI and MKS unit of magnetic flux. It is defined as that magnetic flux that, when threading through a circuit of 1 turn, produces in it an emf of 1 volt as it is reduced to zero in 1 second at a uniform rate.

welded tuff A tuff rich in volcanic ash that has become compacted by aggregation of the glass matrix under the relatively high residual pressure and under compaction.

Wentworth grade scale A logarithmic grade scale on base 2 for particle size classification. The following size grades are recognized (boundary sizes in mm): boulder/256/cobble/64/pebble/2/sand/0.625/silt/0.004/clay.

Wentworth scale See **Wentworth grade scale.**

Weston cell A standard cell used for accurate voltage calibration, with a cadmium cathode, a mercury anode, and a saturated cadmium sulfate solution as electrolyte. emf = 1.018636 volts at 20°C. Cf. **Daniell cell, Leclanché cell.**

westward drift The westward motion of the nondipole magnetic field of the Earth, amounting to an average of 0.2° of longitude per year.

west wind The wind blowing from the west at middle northern and southern latitudes, representing the ground flow of the Ferrel cell. See **Ferrel cell.**

wetlands A collective name for permanently or seasonally wet habitats, including swamps, sloughs, marshes, tidal flats, stagna, ponds, bogs, and pools.

Wheatstone bridge An arrangement of known resistances to measure an unknown resistance.

white dwarf A gravitationally collapsed star with mass smaller than the Chandrasekhar limit, derived from the core of an evolving red giant. Radius = 5000 to 10,000 km; density = $5 \cdot 10^5$ g/cm^3; absolute magnitude = +10 to +15.

white ice Ice rich in air bubbles.

white noise Noise in which the amplitudes of all frequencies are similar. Cf. **blue noise, red noise.**

whiting A milky area in seawater caused by resuspension of fine carbonate mud.

whole-rock analysis Geochemical analysis in which bulk rock is used rather than individual crystal concentrates.

Widmanstätten figures A triangular pattern of laths of kamacite bordered by taenite, made visible by polishing and etching the surface of an octahedrite.

Wiechert-Gutenberg discontinuity See **Gutenberg discontinuity.**

Wien's constant The quantity 2897.8 μm·K, which is the product of absolute temperature times the wavelength (in μm) at which blackbody emission reaches its maximum power.

Wien's displacement law A law stating that the wavelength at which blackbody emission reaches its maximum power is equal to 2897.8 divided by the absolute temperature of the blackbody.

$$\lambda_{max} = 2897.8/T$$

where λ_{max} = wavelength of radiative maximum, T = absolute temperature.

wildflysch A chaotic flysch deposit.

Wilson cloud chamber See **cloud chamber.**

Wolf-Rayet star A very massive ($\sim$50 solar masses), very hot (T = 50,000 K), very luminous, highly turbulent star in the process of ejecting much mass. Wolf-Rayet stars may be supermassive stars in the stage preceding supernova explosion.

wollastonite The mineral $CaSiO_3$.

work (W) The change in kinetic energy of a body or system under the influence of a force $\mathbf{F}$:

$$W = \int_a^b \mathbf{F} \cdot d\mathbf{r}$$

where $\mathbf{r}$ = displacement.

work function The minimum amount of energy needed to remove an electron from a clean metal

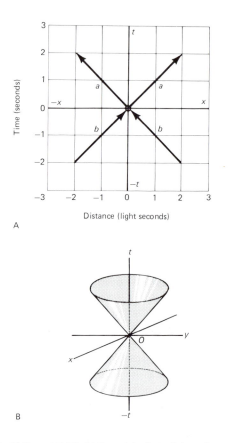

World lines. (A) World lines: (a) of two light pulses radiating simultaneously in vacuo from 0 (representing the present) along the $-x$ and $+x$ directions; and (b) of two light pulses simultaneously reaching 0 from the $-x$ and $+x$ directions. Either one of lines (a) also represents the world radius of the light wave-front sphere radiating isotropically from 0; and either one of lines (b) also represents the world radius of the sphere encompassing light reaching 0. (B) The light cone, representing light radiating from 0 in all directions along the xy plane (upper half) as well as light reaching 0 at any one given instant from all directions along the xy plane (lower half). The regions in the plane $x \pm t$ between the world lines and the $\pm x$ axis (A) or outside the light cone (B) are forbidden in the sense that speeds greater than the speed of light in vacuo would be required for motions to be represented in these regions. Events, i.e., phenomena of short (theoretically infinitesimal) duration, are represented by points in the 4-dimensional space-time continuum. 0 = present; t = future time; $-t$ = past time.

surface in vacuo. The values for pure elements range from 2.14 eV (cesium) to 5.9 eV (selenium).

world line A line within four-dimensional space-time representing the motion of a particle or body. The world line of a beam of light from origin 0 along the x axis with time t represented in units of length (e.g. meters) along the z (vertical) axis is a straight line forming an angle of 45° with either axis. The world lines of a continuous set of light beams radiating away as well as toward origin 0 along the horizontal x-y plane form a light cone described by $\pm(x^2 + y^2 + z^2)^{1/2} = t$ (where the $+$ sign indicates the future and the $-$ sign indicates the past). The space not bound by the cone is forbidden as it would represent motions at speeds greater than the speed of light.

Wulff net See **stereonet.**

wüstite The mineral FeO.

X

χ_ε Electric susceptibility.

χ_μ Magnetic susceptibility.

X Reactance.

$\mathbf{X_C}$ Capacitative reactance.

$\mathbf{X_L}$ Inductive reactance.

$\mathbf{X_{LC}}$ Inductive-capacitative reactance.

xenocryst A crystal of origin different from that of the rock in which it is enclosed.

xenolith A foreign inclusion in an igneous rock.

xenomorphic Referring to a crystal texture produced by crystals growing against each other and thus unable to develop rational crystal faces.

xenotopic Referring to a sedimentary rock formed by precipitation or recrystallization in which the crystals are anhedral.

xeric Dry.

xerophilic Referring to an organism that prefers dry conditions.

xerophyte A plant adapted to dry conditions.

xerophytic Referring to a plant adapted to dry conditions.

Xi (Ξ) A pair of strangeness-2 baryons. See **Elementary particles***.

XPS See **x-ray photoelectron spectroscopy**.

x-ray diffraction The scattering of x-rays by crystal lattices.

x-ray fluorescence 1. Emission by a substance of characteristic x-ray lines when exposed to x-rays. 2. An analytical method whereby the x-ray fluorescence lines emitted by a substance when excited by x-rays are diffracted by a crystal with known lattice spacings, and the elements in the substance and their abundances are identified from the spectral lines and their intensities.

x-ray photoelectron spectroscopy (XPS, ESCA) An analytical method by which a specimen is irradiated with monochromatic x-rays. The energies of the resulting photoelectrons yield the elements present in the specimen and their abundances.

x-rays An electromagnetic spectrum band ranging in wavelength from $3 \cdot 10^{-10}$ to $3 \cdot 10^{-13}$ m (approximately). See **Electromagnetic spectrum***.

Y

y Year.

Y 1. Admittance. 2. Young's modulus.

yard A nonmetric unit of length, equal to 0.9144 m (exactly).

year A period of time variously defined as follows: **1. anomalistic year** The time interval between successive passages of the Earth at perihelion. It is equal to $365.25964134 + 0.00000304T$ $d_E = 31,558,433.240$ s_E (1987). See **anomalistic year**. **2. calendar year** Syn. Gregorian year. **3. eclipse year** The time interval between successive passages of the Sun through the same node of the Moon's orbit. It is equal to $346.620031 + 0.000032T$ $d_E = 29,947,973.083$ s_E (1987). There are 19.0 lunar eclipses in a saros. **4. Gregorian year** The time interval of 365.2425 d $= 31,556,952$ s. **5. Julian year** The time interval of 365.25 d $= 31,557,600$ s. **6. Platonic year** The general precessional period, equal to 25,800 y. **7. sidereal year** The time required for the longitude of a distant star to increase by 360°. It is equal to $365.25636556 + 0.00000011T$ $d_E = 365.25636566$ d_E (1987) $= 31,558,149.993$ s_E (1987). **8. tropical year** The time interval between successive vernal equinoxes. It is equal to $365.24219878 - 0.00000616T$ $d_E = 365.24219342$ d_E (1987) $= 31,556,925.511$ s_E (1987). In the preceding, $T =$ centuries since 1900.0. See **tropical year.**

yellow-green algae Single-celled or filamentous, predominantly freshwater algae belonging to the phylum Xanthophyta. Cf. **golden-yellow algae.**

yield *(Nuclear reactions)* 1. The ratio of the mass + energy of the particle(s) produced in a nuclear reaction to that of the incident particle(s). 2. The total energy released by a nuclear explosion. *(Secondary electrons)* The ratio of emitted electrons to incident electrons in the electron bombardment of a target.

yield strength The stress under which a material begins to deviate from proportionality between stress and strain.

Young's modulus The ratio of tensional stress to the resulting strain.

youthful Defining a landscape that has undergone little erosion.

Z

ƺ Zenith distance.

z Redshift parameter.

Z 1. ac impedance. 2. Atomic number. 3. Valence. 4. ZULU.

Z⁰ Weak boson with 0 charge that carries the neutral current in weak interaction (mass = 99.7 ± 1.7 u). See **Elementary particles***.

Zeeman effect The splitting of spectral lines emitted by atoms or molecules in a static magnetic field due to interactions between the electron magnetic moments and the applied field.

zenith The point on the celestial sphere vertically above the observer.

zenith distance (ƺ) Coaltitude, the angular distance of a celestial body from the observer's zenith.

zeolite Any of the hydrous Na, K, Ca aluminosilicates characteristic of low-pressure, low-temperature metamorphism.

zeolite facies A low-temperature (100–250°C), low-pressure (1–4 kbar) metamorphic facies characterized by the presence of zeolites.

zephyr The west wind.

zero-point energy The vibrational energy retained by molecules and crystals at 0 K. See **vibrational energy.**

zero-point entropy The entropy that a disordered crystal retains at 0 K.

zero-point vibration The vibrational motions in a crystal at 0 K. See **vibrational energy.**

zigzag fold A kink fold with unequal limbs.

zircon The mineral $ZrSiO_4$.

zodiac A band on the celestial sphere extending to about 8° on either side of the ecliptic and containing the 12 zodiac constellations of antiquity (Aries, Taurus, Gemini, Cancer, Leo, Virgo, Libra, Scorpius, Sagittarius, Capricornus, Aquarius, Pisces). These constellations represent mainly animals, hence the name zodiac (from ζῷον = animal).

zodiacal light The faint, cone-shaped glow tapering from the horizon upward along the plane of the ecliptic that can be seen after sunset and before sunrise especially at low latitudes. It is caused by the scattering of solar light by interplanetary dust particles, which are concentrated along the plane of the ecliptic.

zoisite A mineral of the epidote group, $Ca_2Al_3Si_3O_{12}(OH)$.

zoo- Prefix meaning *animal.*

zooid A semi-independent member of a colony, such as a coral or a bryozoan.

zooplankton Animal plankton.

ZULU Greenwich mean time, so called because the time line bisected by the Greenwich meridian bears the letter Z. See **time zone.**

Zürich relative sunspot number (R) An index of sunspot activity:

$$R = k(f + 10g)$$

where k = constant related to the quality of a given observer and the equipment used; f = total number of sunspot forming a group; g = number of sunspot groups.

zwitterion A complex ion carrying a positive and a negative charge in different sites. E.g. internally neutralized glycine, $^+NH_3-CH(H)-COO^-$.

zygo- Prefix meaning *joined, united.*

zygote 1. The fertilized egg before cleavage. 2. A diploid organism produced by the union of two gametes.

Ω

ω 1. Angular frequency. 2. Angular velocity. 3. Argument of perihelion.

Ω Ohm.

Ω^- Omega particle, a strangeness-3 baryon. See **Elementary particles***.

TABLES

$t_{1/2} =$ half-life; $t_0 =$ age of the solar system $= 4.7 \cdot 10^9$ y.

Method	$t_{1/2}$ (years)	Effective range (y B.P.)
^{3}H	12.33	10^0-10^2
^{14}C	5,730	$10^2-5 \cdot 10^4$
^{40}Ar/^{39}Ar	$1.277 \cdot 10^9$	10^4-t_0
^{40}K/^{40}Ar	$1.277 \cdot 10^9$	10^4-t_0
^{40}K/^{40}Ca	$1.277 \cdot 10^9$	10^6-t_0
^{87}Rb/^{87}Sr	$48 \cdot 10^9$	10^8-t_0
^{147}Sm/^{143}Nd	$106 \cdot 10^9$	$?-t_0$
^{176}Lu/^{176}Hf	$36 \cdot 10^9$	$?-t_0$
^{187}Re/^{187}/Os	$50 \cdot 10^9$	$?-t_0$
^{206}Pb/^{204}Pb	—	10^7-t_0
^{207}Pb/^{206}Pb	—	10^8-t_0
^{230}Th (unsupported)	75,380	$10^4-2 \cdot 10^5$
^{231}Pa (unsupported)	32,760	10^4-10^5
^{231}Pa/^{230}Th	60,100	$10^4-1.5 \cdot 10^5$
^{232}Th/^{208}Pb	$14.05 \cdot 10^9$	10^7-t_0
^{234}U(excess)	245,000	10^4-10^6
^{235}U/^{207}Pb	$704 \cdot 10^6$	10^7-t_0
^{238}U/^{206}Pb	$4.468 \cdot 10^9$	10^7-t_0
He/U + Th	—	10^6-t_0
^{238}U fission track	$10.1 \cdot 10^{15}$	10^0-t_0

From Brownlow 1979, p. 42, Table 42; Wapstra and Audi 1985; Tuli 1985; Faure 1986.

Amino acids

The 20 common amino acids found in proteins. The group in parenthesis is attached to the C of the CH component. me = methyl radical ($-CH_3$); ph = phenyl radical ($-C_6H_5$).

Name	Abbreviation	Chemical formula	Molecular mass (u)
alanine	ala	$NH_2 \cdot CH \cdot COOH \cdot (me)$	89.094
arginine	arg	$NH_2 \cdot CH \cdot COOH \cdot (CH_2CH_2CH_2NHCNHNH_2)$	174.203
asparagine	asn	$NH_2 \cdot CH \cdot COOH \cdot (CH_2CONH_2)$	132.119
aspartic acid	asp	$NH_2 \cdot CH \cdot COOH \cdot (CH_2COOH)$	133.104
cysteine	cys	$NH_2 \cdot CH \cdot COOH \cdot (CH_2 \cdot SH)$	121.160
glutamic acid	glu	$NH_2 \cdot CH \cdot COOH \cdot (CH_2CH_2COOH)$	147.131
glutamine	gln	$NH_2 \cdot CH \cdot COOH \cdot (CH_2CH_2CONH_2)$	146.146
glycine	gly	$NH_2 \cdot CH \cdot COOH \cdot (H)$	75.067
histidine	his	$NH_2 \cdot CH \cdot COOH \cdot (CH_2C_3H_3N_2)$	155.157
isoleucine	ile	$NH_2 \cdot CH \cdot COOH \cdot (CHmeCH_2me)$	131.175
leucine	leu	$NH_2 \cdot CH \cdot COOH \cdot (CH_2CHme_2)$	131.175
lysine	lys	$NH_2 \cdot CH \cdot COOH \cdot (CH_2CH_2CH_2CH_2NH_2)$	146.190
methionine	met	$NH_2 \cdot CH \cdot COOH \cdot (CH_2CH_2Sme)$	149.214
phenylalanine	phe	$NH_2 \cdot CH \cdot COOH \cdot (CH_2ph)$	165.192
proline	pro	$NH \cdot CH \cdot COOH \cdot (CH_2)_3$	115.132
serine	ser	$NH_2 \cdot CH \cdot COOH \cdot (CH_2OH)$	105.094
threonine	thr	$NH_2 \cdot CH \cdot COOH \cdot (CHmeOH)$	119.121
tryptophan	trp	$NH_2 \cdot CH \cdot COOH \cdot [CH_2CCH(C_6H_4)NH]$	204.229
tyrosine	tyr	$NH_2 \cdot CH \cdot COOH \cdot (CH_2phOH)$	182.192
valine	val	$NH_2 \cdot CH \cdot COOH \cdot (CHme_2)$	117.148

Asteroids

The fifteen asteroids with diameters greater than 250 km (a = semimajor axis of orbit; e = eccentricity; i = inclination).

Size rank	Number	Name	Diameter (km)	Orbital elements		
				a (AU)	e	$i(°)$
1	1	Ceres	1025	2.76784	0.07685	10.598
2	2	Pallas	583	2.77315	0.23254	34.800
3	4	Vesta	555	2.36168	0.08967	7.144
4	10	Hygeia	443	3.13822	0.11838	3.835
5	704	Interamnia	338	3.06018	0.15318	17.290
6	511	Davida	335	3.18083	0.17192	15.897
7	65	Cybele	311	3.42839	0.10979	3.553
8	52	Europa	291	3.09517	0.10909	7.465
9	451	Patientia	281	3.06501	0.06769	15.202
10	31	Euphrosyne	270	3.14788	0.22761	26.327
11	15	Eunomia	261	2.64211	0.18776	11.759
12	324	Bamberga	256	2.68524	0.33676	11.169
13	107	Camilla	252	3.48698	0.07462	9.952
14	87	Sylvia	251	3.48295	0.09268	10.879
15	45	Eugenia	250	2.72035	0.08363	6.601

From Bender 1979; Bowell et al. 1979.

Impact craters with diameter >10 km.

Name	Location	Latitude	Longitude	Diameter (km)	Age ($\times 10^6$ y)
Vredefort	South Africa	27°00'S	27°30'E	140	1970 ± 100
Sudbury	Ontario, Canada	46°36'N	81°11'W	140	1840 ± 150
Popigai	Siberia	71°30'N	111°00'E	100	38 ± 9
Puchezh-Katunki	Russia, USSR	57°06'N	43°35'E	80	183 ± 3
Manicouagan	Quebec, Canada	51°23'N	68°42'W	70	210 ± 4
Siljan	Sweden	61°02'N	14°52'E	52	365 ± 7
Kara	Russia, USSR	69°10'N	65°00'E	50	57
Charlevoix	Quebec, Canada	47°32'N	70°18'W	46	360 ± 25
Araguainha Dome	Brazil	16°46'S	52°59'W	40	<250
Carswell	Saskatchewan, Canada	58°27'N	109°30'W	37	485 ± 50
Clearwater Lake West	Quebec, Canada	56°13'N	74°30'W	32	290 ± 20
Clearwater Lake East	Quebec, Canada	56°05'N	74°07'W	22	290 ± 20
Manson	Iowa, United States	42°35'N	94°31'W	32	<70
Slate Island	Ontario, Canada	48°40'N	87°00'W	30	350
Mistastin	Labrador, Canada	55°53'N	63°18'W	28	38 ± 4
Boltysh	Ukraine, USSR	48°45'N	32°10'E	25	100 ± 5
Kamensk	Russia, USSR	48°20'N	40°15'E	25	65
Steen River	Alberta, Canada	59°31'N	117°38'W	25	95 ± 7
Strangways	Northern Territory, Australia	15°12'S	133°35'E	24	150 ± 70
Ries	Germany	48°53'N	10°37'E	24	14.8 ± 0.7
Rochechouart	France	45°49'N	0°50'E	23	160 ± 5
St. Martin	Manitoba, Canada	51°47'N	98°33'W	23	225 ± 40
Gosses Bluff	Northern Territory, Australia	23°50'S	132°19'E	22	130 ± 6
Haughton Dome	NW Territories, Canada	75°22'N	89°40'W	20	15
Karla	Russia, USSR	57°45'N	48°00'E	18	10
Logoisk	Byelorussia, USSR	54°12'N	27°48'E	17	100 ± 20
Kaluga	Russia, USSR	54°30'N	36°15'E	15	360 ± 10
Dellen	Sweden	61°55'N	16°32'E	15	230
Obolon	Russia, USSR	49°30'N	32°55'E	15	160
Janisjärvi	Russia, USSR	61°58'N	30°55'E	14	700
Lappajärvi	Finland	63°09'N	23°42'E	14	<600
Wells Creek	Tennessee, USA	36°23'N	87°40'W	14	200 ± 100
Kentland	Indiana, USA	40°45'N	87°24'W	13	300
Sierra Madera	Texas, USA	30°36'N	102°55'W	13	100
Nicholson Lake	NW Territories, Canada	62°40'N	102°41'W	12.5	<450
Serra de Canghala	Brazil	8°05'S	46°52'W	12	<300
Deep Bay	Saskatchewan, Canada	56°24'N	102°59'W	12	100 ± 50
unnamed oasis	Libya	24°35'N	24°24'E	11.5	<120
Bosumtwi	Ghana	6°32'N	1°25'W	10.5	1.3 ± 0.2
Zhamanshin	Kazakh, USSR	49°00'N	61°00'E	10	~1

From Glass 1982, p. 138–139, Table 5.3.

Astroblemes
recent

Recent (<2 million year) impact craters associated with meteorite fragments (asterisks identify the diameter of the largest crater in the cases of multiple craters produced by the same event).

Name	Location	Latitude	Longitude	Diameter (m)	Age
Barringer	Arizona, USA	35°02′N	111°01′W	1200	50,000 y
Boxhole	Northern Territory, Australia	22°37′S	135°12′E	180*	<2 m.y.
Campo del Cielo	Argentina	27°38′S	61°42′W	90	<2 m.y.
Dalgaranga	Western Australia, Australia	27°43′S	117°05′E	20	25,000 y
Haviland	Kansas, USA	37°37′N	99°05′W	10	<2 m.y.
Henbury	Northern Territory, Australia	24°34′S	133°10′E	150*	<2 m.y.
Kaalijärvi	Estonia, USSR	58°24′N	22°40′E	110*	<2 m.y.
Morask	Poland	52°29′N	16°54′E	100*	<2 m.y.
Odessa	Texas, USA	31°48′N	102°30′W	170*	20,000 y
Sikhote Alin	Siberia, USSR	46°07′N	134°40′E	30*	A.D. 1947
Sobolev	Siberia, USSR	46°18′N	137°52′E	50	<2 m.y.
Wabar	Saudi Arabia	21°30′N	50°28′E	100*	<2 m.y.
Wolf Creek	Western Australia, Australia	19°10′S	127°47′E	850	<2 m.y.

From Glass 1982, p. 138–139, Table 5.3.

Atmosphere
composition

Concentration (by volume) of components at ground level (excluding local pollutants) and their residence times.

Component	Concentration	Residence time
N_2	0.78084	$4 \cdot 10^8$ y for cycling through sediments
O_2	0.20946	6000 y for cycling through biosphere
H_2O	$(4-0.004) \cdot 10^{-2}$	—
Ar	$9.34 \cdot 10^{-3}$	largely accumulating
CO_2	$0.320 \cdot 10^{-3}$	10 y for cycling through biosphere
Ne	$1.818 \cdot 10^{-5}$	largely accumulating
He	$5.24 \cdot 10^{-6}$	$2 \cdot 10^6$ y for escape
CH_4	$1.4 \cdot 10^{-6}$	2.6–8 y
Kr	$1.14 \cdot 10^{-6}$	largely accumulating
H	$5.5 \cdot 10^{-7}$	4–7 y
N_2O	$3.3 \cdot 10^{-7}$	5–50 y
CO	$2-0.6 \cdot 10^{-7}$	0.5 y
Xe	$8.7 \cdot 10^{-8}$	largely accumulating
O_3 (ozone)	$3-1 \cdot 10^{-8}$	—
CH_2O (formaldehyde)	$<1 \cdot 10^{-8}$	—
NH_3 (ammonia)	$20-6 \cdot 10^{-9}$	about 1 day
SO_2	$4-1 \cdot 10^{-9}$	hours to weeks
$NO + NO_2$	$1 \cdot 10^{-9}$	<1 month
CH_3Cl (methyl chloride)	$5 \cdot 10^{-10}$	—
CCl_4 (carbon tetrachloride)	$2.5-1 \cdot 10^{-10}$	—
CF_2Cl_2 (Freon 12)	$2.3 \cdot 10^{-10}$	45–68 y
H_2S	$\leqslant 2 \cdot 10^{-10}$	$\leqslant 1$ day
$CFCl_3$ (Freon 11)	$1.3 \cdot 10^{-10}$	45–68 y

From Holland 1978, p. 251, Table 6.1.

Above 120 km of altitude, the atmospheric parameters given below depend on the phase of the sunspot cycle. The values given refer to the year 1976 (sunspot minimum).

Height (km)	Pressure (mb)	Temperature (°K)	Density (kg/m³)	Mean molecular mass (u)
0	$1.01 \cdot 10^3$	288	$1.23 \cdot 10^0$	28.96
5	$5.40 \cdot 10^2$	256	$7.36 \cdot 10^{-1}$	28.96
10	$2.65 \cdot 10^2$	223	$4.14 \cdot 10^{-1}$	28.96
20	$5.53 \cdot 10^1$	217	$8.89 \cdot 10^{-2}$	28.96
40	$2.87 \cdot 10^0$	250	$4.00 \cdot 10^{-3}$	28.96
60	$2.20 \cdot 10^{-1}$	247	$3.10 \cdot 10^{-4}$	28.96
80	$1.05 \cdot 10^{-2}$	199	$1.85 \cdot 10^{-5}$	28.96
100	$3.20 \cdot 10^{-4}$	195	$5.60 \cdot 10^{-7}$	28.40
150	$4.54 \cdot 10^{-6}$	634	$2.08 \cdot 10^{-9}$	24.10
200	$8.47 \cdot 10^{-7}$	855	$2.54 \cdot 10^{-10}$	21.30
300	$8.77 \cdot 10^{-8}$	976	$1.92 \cdot 10^{-11}$	17.73
400	$1.45 \cdot 10^{-8}$	996	$2.80 \cdot 10^{-12}$	15.98
500	$3.02 \cdot 10^{-9}$	999	$5.21 \cdot 10^{-13}$	14.33
600	$8.21 \cdot 10^{-10}$	1000	$1.14 \cdot 10^{-13}$	11.51

From U.S. Standard Atmosphere 1976, Tables I and II.

Beaufort wind scale

Beaufort scale	Velocity (knots)	Marine term	Description Sea	Description Land
0	<1	calm	sea like mirror	smoke rises vertically
1	1–	light air	gentle ripples (<30 cm high)	smoke drifts slowly
2	4–	light breeze	small waves (<1 m)	gentle leaf rustling
3	7–	gentle breeze	1–1.5 m waves	leaves and twigs in motion
4	11–	moderate breeze	1.5 m waves	small branches moving
5	17–	fresh breeze	2 m waves	small trees waving
6	22–	strong breeze	2.5 m waves, whitecaps everywhere	large branches in motion
7	28–	moderate gale	3 m waves	whole trees swaying
8	34–	fresh gale	3.5 m waves; foam in streaks	twigs broken off trees
9	41–	strong gale	5–6 m waves; strong foam streaks	branches broken off trees
10	48–	whole gale	6–12 m waves; spray	smaller trees uprooted
11	56–	storm	12–15 m waves; strong spray	large trees uprooted
12–17	>64	hurricane	waves >15 m; very strong spray	heavy structural damage

From Schule 1966, p. 790–791, Table 2.

Chelates important to living systems

Name	Formula	Molecular mass (u)
pyrrole ring	$NH(CH)_4$	67.090
porphin	$N_2(NH)_2C_8(CH)_{12}$	310.358
heme	$Fe^{++} \cdot N_4 \cdot C_{16}(CH)_4(CH_3)_4(CHCH_2)_2(CH_2CH_2COOH)_2$	552.502
hematin	$Fe(OH) \cdot N_4 \cdot C_{16}(CH)_4(CH_3)_4(CHCH_2)_2(CH_2CH_2COOH)_2$	633.508
hemoglobin (av.)	$(Fe \cdot S_2 \cdot N_{203}O_{208}C_{738}H_{1166})_4$	$(16330.590)_4 = 65322.360$
hemocyanin (av.)	$(Cu \cdot S_2 \cdot N_{203}O_{208}C_{738}H_{1166})_n$	$(16338.289)_n$
chlorophyll a	$Mg \cdot N_4 \cdot O_5 \cdot C_{55}H_{72}$	893.505
chlorophyll b	$Mg \cdot N_4 \cdot O_6 \cdot C_{55}H_{70}$	907.489
corrin	$H \cdot N_4 \cdot C_{19}H_{21}$	306.410
cyanocobalamin	$Co \cdot N_4 \cdot PO_4 \cdot N_{10}O_{10}C_{63}H_{88}$	1355.384
cytochrome c	$Fe \cdot N_4 \cdot C_8(CH)_4(CH_3)_4(CH_2CH_2COOH)_4(CH_3CH_2)_2 \cdot protein$	$542.591 + protein$

Comets
chemical composition

Organic	$C, C_2, C_3, CH, CN, CO, CS, HCN, CH_3CN$
Inorganic	$H, NH, NH_2, O, OH, S, Si, H_2O$
Metals	Na, K, Ca, V, Cr, Mn, Fe, Co, Ni, Cu
Ions	$C^+, CO^+, CO_2^+, CH^+, CN^+, N_2^+, OH^+, H_2O, Ca^+$
Dust	silicates

Irvine and Hjalmarson 1983, p. 115, Table 1; Whipple 1985, p. 344, Table 1; Weaver et al. 1986.

Error given as standard deviation in last digits.

Quantity	Symbol	Value	Error	SI	CGS	Others
atomic mass unit	u	1.6605402	10	10^{-27} kg	10^{-24} g	—
		931.49432	28	—	—	MeV
Avogadro number	N_A	6.0221367	36	10^{23} mol^{-1}	10^{23} mol^{-1}	—
Bohr magneton	μ_B	9.2740154	31	10^{-24} J T^{-1}	10^{-21} erg G^{-1}	—
Bohr radius	a_0	0.529177249	24	10^{-10} m	10^{-8} cm	—
Boltzmann constant, R/N_A	k	1.380658	12	10^{-23} J K^{-1}	10^{-16} erg K^{-1}	—
		8.617384	72	—	—	10^{-5} eV K^{-1}
electron charge	e	1.60217733	49	10^{-19} C	10^{-20} emu	—
		4.80320680	49	—	10^{-10} esu	—
electron charge/mass ratio	e/m_e	1.75881962	53	10^{11} C kg^{-1}	10^{7} emu g^{-1}	—
		5.27280857	53	—	10^{17} esu g^{-1}	—
electron Compton wavelength	λ_c	2.42631058	22	10^{-12} m	10^{-10} cm	—
electron magnetic moment	μ_e	9.2847701	31	10^{-24} J T^{-1}	10^{-21} erg G^{-1}	—
electron magnetic moment in Bohr magnetons	μ_e/μ_B	1.001159652193	10	—	—	—
electron magnetic moment/proton magnetic moment	μ_e/μ_p	658.21068801	66	—	—	—
electron rest mass	m_e	0.91093897	54	10^{-30} kg	10^{-27} g	—
		5.48579903	13	—	—	10^{-4} u
		0.51099906	15	—	—	MeV
electron rest mass/proton rest mass	m_e/m_p	5.44617013	11	10^{-4}	10^{-4}	—
		1/1836.15270	11	—	—	—
Faraday constant	eN_A	9.6485309	29	10^{4} C mol^{-1}	10^{3} emu mol^{-1}	—
fine structure constant $(e^2/\hbar c)$	α	0.00729735308	49	—	—	—
	α^{-1}	137.0359895	61	—	—	—
gas constant	R	8.314510	70	J mol^{-1} K^{-1}	10^{7} erg mol^{-1} K^{-1}	—
		1.9858866	70	cal mol^{-1} K^{-1}	cal mol^{-1} K^{-1}	—
		82.0578	26	—	cm^3 atm K^{-1}	—
gravitational constant	G	6.67206	8	10^{-11} N m^2 kg^{-2}	10^{-8} dyn cm^2 g^{-2}	—
impedance of vacuum $(\mu_0/\epsilon_0)^{\frac{1}{2}}$	Z_0	3.767303134	—	10^2 Ω	10^2 Ω	—

Error given as standard deviation in last digits.

Quantity	Symbol	Value	Error	SI	CGS	Others
light speed in vacuo	c	299792458	0	m s⁻¹	10^2 cm s⁻¹	—
light speed in vacuo squared	c^2	8987551787368 1764	0	m² s⁻²	10^4 cm² s⁻²	—
magnetic flux quantum	$h/2e$	2.06783461	61	10^{-15} Wb	10^{-7} Mx²	—
molar ideal gas volume (STP)	V_m	22.41410	19	10^{-3} m³ mol⁻¹	10^3 cm³ mol⁻¹	—
muon magnetic moment	μ_μ	4.4904514	15	10^{-26} J T⁻¹	10^{-23} erg G⁻¹	—
muon rest mass	m_μ	1.8835327	11	10^{-28} kg	10^{-25} g	—
		0.113428913	17			u
neutron Compton wavelength (h/m_nc)	$\lambda_{C,n}$	1.31959110	12	10^{-15} m	10^{-13} cm	—
neutron rest mass	m_n	1.6749286	10	10^{-27} kg	10^{-24} g	—
		1.008664904	14			u
		939.56563	28			MeV
nuclear magneton	μ_N	5.0507866	17	10^{-27} J T⁻¹	10^{-24} erg G⁻¹	—
permeability constant $(4\pi \cdot 10^{-7})$	μ_0	12.5663706144...	—	10^{-7} H m⁻¹	—	—
		1	—		(Dimensionless)	—
permittivity constant $(1/\mu_0c^2)$	ϵ_0	8.854187817...	0	10^{-12} F m⁻¹	10^{-10} F cm⁻¹	—
Planck's constant	h	6.6260755	40	10^{-34} J Hz⁻¹	10^{-27} erg Hz⁻¹	—
		4.1356692	12			10^{-15} eV Hz⁻¹
$(h/2\pi)$	$\hbar$	1.0545726	6	10^{-34} J Hz⁻¹	10^{-27} erg Hz⁻¹	—
proton Compton wavelength	$\lambda_{C,p}$	1.32141002	12	10^{-15} m	10^{-13} cm	—
proton magnetic moment	μ_p	1.41060761	47	10^{-26} J T⁻¹	10^{-23} erg G⁻¹	—
proton magnetic moment in Bohr magnetons	μ_p/μ_B	1.521032202	15	10^{-3}	10^{-3}	—
proton mass/electron mass	m_p/m_e	1836.152701	37			—
proton rest mass	m_p	1.6726231	10	10^{-27} kg	10^{-24} g	—
		1.007276470	12			u
		938.2231	28			MeV
Rydberg constant	R_∞	1.0973731534	13	10^7 m⁻¹	10^5 cm⁻¹	—
Rydberg energy	hcR_∞	13.6056981	40			eV
Stefan-Boltzmann constant	σ	5.67051	19	10^{-8} W m⁻² K⁻⁴	10^{-5} erg cm⁻² s⁻¹ K⁻⁴	—

From Cohen and Taylor 1986.

$A = XB$; $B = A/X$; $c = $ speed of light in vacuo in cm/s; asterisk identifies exact values; all values involving c or π are also exact.

A	X	B
abampere	10*	ampere
	1*	emu
	$2.997\ 924\ 58 \cdot 10^{10}\ (= c)$	statampere
abampere-turn	10*	ampere-turn
	$12.566\ 371\ (= 4\pi)$	gilbert
abcoulomb	0.002 777 8	ampere-hour
	10*	coulomb
	$6.241\ 506 \cdot 10^{19}$	electron charge
	$2.997\ 924\ 58 \cdot 10^{10}\ (= c)$	statcoulomb
abfarad	10^{9}*	farad
	10^{15}*	microfarad
	$8.987\ 552 \cdot 10^{20}\ (= c^2)$	statfarad
abhenry	10^{-9}*	henry
	$1.112\ 650 \cdot 10^{-21}\ (= 1/c^2)$	stathenry
abmho	10^{9}*	mho
	$8.987\ 552 \cdot 10^{20}\ (= c^2)$	statmho
abohm	10^{-9}*	ohm
	$1.112\ 650 \cdot 10^{-21}\ (= 1/c^2)$	statohm
abvolt	$3.335\ 641 \cdot 10^{-11}\ (= 1/c)$	statvolt
	10^{-8}*	volt
abweber	1*	maxwell
acre	4046.856 4	m^2
aeon	10^{9}*	year
ampere	0.1*	abampere
	1*	coulomb/s
	$2.997\ 924\ 58 \cdot 10^{9}\ (= c/10)$	statampere
ampere-hour	360*	abcoulomb
	3600*	coulomb
	0.037 311 7	faraday
ampere-turn	0.1	abampere-turn
	$3.767\ 303\ 3 \cdot 10^{10}$	esu
	1.256 637 1	gilbert
angstrom	10^{-8}*	cm
	10^{-10}*	m
	10^{-4}*	μm
are	100*	m^2
astronomical unit	149 597 870.7	km
	8.316 746	light-min
	499.004 784	light-sec
	$1.581\ 284 \cdot 10^{-5}$	light-yr
atmosphere	1.013 250*	bar
	76*	cmHg (0°C)
	$1.013\ 250 \cdot 10^{6}$*	dyne/cm^2
	1033.227	g/cm^2
	1.033 227	kg/cm^2
	10.332 27	mH$_2$0 (3.98°C)
	760*	mmHg (0°C)
	$1.013\ 250 \cdot 10^{5}$*	N/m^2
	$1.013\ 250 \cdot 10^{5}$*	Pa
	760*	torr

Conversion factors *(continued)*

$A = XB$; $B = A/X$; $c =$ speed of light in vacuo in cm/s; asterisk identifies exact values; all values involving c or π are also exact.

A	X	B
atomic mass unit (u)	$931.494\ 32 \cdot 10^6$	eV
	$1.660\ 540\ 2 \cdot 10^{-24}$	g
	$1.660\ 540\ 2 \cdot 10^{-27}$	kg
	$931.494\ 32$	MeV
avogram	$1.660\ 540\ 2 \cdot 10^{-24}$	g
bar	$0.986\ 923\ 3$	atmosphere
	$10^{6}*$	barye
	$75.006\ 2$	cmHg
	$10^{6}*$	dyne/cm^2
	1019.716	g/cm^2
	10197.16	kg/m^2
	$10^{3}*$	millibar
	$750.061\ 7$	mmHg
	$10^{5}*$	Pa
barn	$10^{-24}*$	cm^2
	$10^{-28}*$	m^2
barye	$9.869\ 233 \cdot 10^{-7}$	atmosphere
	$10^{-6}*$	bar
	$1*$	dyn/cm^2
	$0.001\ 019\ 716$	g/cm^2
	$0.001*$	millibar
bel	$10*$	decibel
Btu$_{IT}$	$251.995\ 75$	cal$_{IT}$
	$1055.055\ 852\ 62*$	J
calorie (gram)	$4.185\ 5$	J
calorie$_{IT}$	$4.1868 \cdot 10^{7}*$	erg
	$4.1868*$	J
candela/cm^2 (stilb)	$3.141\ 592\ 653 \ldots (= \pi)$	lambert
carat	$3.086\ 472\ 2$	grain
	$0.2*$	g
Celsius	$1.8*$	Fahrenheit
centigrade	$1*$	Celsius
	$1.8*$	Fahrenheit
centimeter	$10^{8}*$	angstrom
	$0.1*$	decimeter
	$0.393\ 700\ 79$	inch
	$0.01*$	m
	$10^{4}*$	μm
	$10*$	mm
cm/s	$0.036*$	km/hr
circumference	$6.283\ 185\ 306 \ldots (= 2\pi)$	radian
coulomb	$0.1*$	abcoulomb
	$6.241\ 506\ 4 \cdot 10^{18}$	electron charge
	$0.1*$	abcoulomb
	$2.997\ 924\ 58 \cdot 10^{9}\ (= c/10)$	statcoulomb
cubic centimeter	$10^{-15}*$	km^3
	$0.001*$	liter
cubic decimeter	$1*$	liter
cubic foot	$28\ 316.846\ 592*$	cm^3
	$28.316\ 846\ 592*$	liter
cubic inch	$16.387\ 064*$	cm^3

$A = XB$; $B = A/X$; c = speed of light in vacuo in cm/s; asterisk identifies exact values; all values involving c or π are also exact.

A	X	B
cubic kilometer	10^{15}*	cm^3
	10^{9}*	m^3
cubic meter	10^{6}*	cm^3
	10^{-9}*	km^3
	10^{9}*	mm^3
cubic millimeter	10^{-3}*	cm^3
	10^{-9}*	m^3
curie	$3.7 \cdot 10^{10}$*	dps (disintegr./sec.)
day (mean solar)	1.002 737 91	day (sidereal)
day (sidereal)	0.997 269 57	day (mean solar)
decibel	0.115 129 255	neper
decimeter	10*	cm
	0.1*	meter
degree	0.017 453 293	radian
dekameter	10*	m
dyne	10^{-5}*	newton
dyne$\cdot$cm	1*	erg
	10^{-7}*	N$\cdot$m
dyne/cm^2	$9.869\ 233 \cdot 10^{-7}$	atmosphere
	10^{-6}*	bar
	1*	barye
	0.1*	N/m^2
	0.1*	Pa
electron charge	$1.602\ 177\ 3 \cdot 10^{-20}$	abcoulomb
	$1.602\ 177\ 3 \cdot 10^{-19}$	coulomb
	$4.803\ 242\ 4 \cdot 10^{-10}$	statcoulomb
electron volt	1 239.842 5	angstrom
	$1.602\ 177\ 3 \cdot 10^{-12}$	erg
	$2.417\ 969\ 6 \cdot 10^{14}$	hertz
	$1.602\ 177\ 3 \cdot 10^{-19}$	J
	$1.160\ 45 \cdot 10^{4}$	kelvin
	$1.073\ 535\ 4 \cdot 10^{-9}$	u
erg	$2.388\ 459 \cdot 10^{-8}$	cal$_{IT}$
	$6.241\ 506\ \cdot 10^{11}$	eV
	$2.389\ 201 \cdot 10^{-8}$	g-cal
	10^{-7}*	J
	$2.389\ 201 \cdot 10^{-11}$	kcal (g-cal)
	$2.388\ 459 \cdot 10^{-11}$	kcal$_{IT}$
	$6.241\ 506\ \cdot 10^{5}$	MeV
	$6.700\ 531 \cdot 10^{2}$	u
erg/s	10^{-7}*	watt
erg$\cdot$s	$1.509\ 190 \cdot 10^{26}$	Planck's constant
farad	10^{-9}*	abfarad
	1.000 495	farad (Int.)
	10^{6}*	microfarad
	$8.987\ 552 \cdot 10^{11}\ (= 10^{-9}c^2)$	statfarad
farad (Int.)	0.999 505	farad
faraday	96 485.309	coulomb
fathom	182.88*	cm
	6*	foot
	1.828 8*	m

$A = XB$; $B = A/X$; $c =$ speed of light in vacuo in cm/s; asterisk identifies exact values; all values involving c or π are also exact.

A	X	B
femtometer	$10^{-13}*$	cm
	$1*$	fermi
	$10^{-15}*$	m
fermi	$10^{-13}*$	cm
	$10^{-15}*$	m
foot	$30.48*$	cm
	$0.304\ 8*$	m
furlong	$660*$	ft
	$201.168*$	m
	$0.125*$	mile (statute)
	$220*$	yard
gal	$1*$	cm/s^2
	$0.01*$	m/s^2
gallon (British)	4546.090	cm^3
	$4.546\ 090$	liter
	$4*$	quart
gallon (U.S.)	$3785.411\ 784$	cm^3
	$3.785\ 411\ 784$	liter
	$4*$	quart
gamma	$10^{-5}*$	gauss
	$10^{-5}*$	oersted
	$10^{-9}*$	tesla
gauss	$1*$	abtesla
	$0.999\ 670$	gauss (Int.)
	$1*$	line/cm^2
	$1*$	maxwell/cm^2
	$1*$	oersted
	$3.335\ 641 \cdot 10^{-11}\ (= 1/c)$	stattesla
	$10^{-4}*$	tesla
	$10^{-8}*$	weber/cm^2
	$10^{-4}*$	weber/m^2
g-cal: *see* gram calorie		
gilbert	$0.079\ 577\ 472\ (= 1/4\pi)$	abampere-turn
	$0.795\ 774\ 72\ (= 10/4\pi)$	ampere-turn
	$1*$	emu
	$2.997\ 924\ 58 \cdot 10^{10}\ (= c)$	esu
gilbert/cm	$1*$	oersted
grain	$0.064\ 798\ 91$	g
	$64.798\ 91$	mg
gram	$6.022\ 045 \cdot 10^{23}$	avogram
	$5*$	carat
	$10^{-3}*$	kg
	$10^{3}*$	mg
	$10^{-6}*$	ton (metric)
gram calorie	4.1855	J
gram/cm^2	$0.000\ 967\ 841$	atmosphere
	$0.000\ 980\ 665$	bar
	$10*$	kg/m^2
	$0.014\ 223\ 343$	lb/inch^2
	$0.735\ 559\ 2$	mmHg

$A = XB$; $B = A/X$; c = speed of light in vacuo in cm/s; asterisk identifies exact values; all values involving c or π are also exact.

A	X	B
hectare	2.471 053 8	acre
	100*	are
	10^4*	m^2
hectogram	100*	g
	0.1*	kg
hectoliter	100*	liter
hectometer	100*	m
henry	10^9*	abhenry
	10^9*	emu
	$1.112\ 650 \cdot 10^{-12}\ (= 10^9/c^2)$	stathenry
horsepower		
(mechanical)	550.0*	foot-pound/second
	745.700	J/s
	0.745 700	kilowatt
	745.700	watt
hour (mean solar)	1/24*	day (mean solar)
	60*	minute
	3600*	second
	1.002 737 91	hour (sidereal)
hour (sidereal)	0.997 269 58	hour (mean solar)
	59.836 175	minute (mean solar)
	3590.170 5	second (mean solar)
inch	2.54*	cm
	0.083 333 . . .	foot
	0.0254*	m
	1000*	mil
inch of Hg	25.4*	mmHg
joule	0.238 845 9	cal_{IT}
	10^7*	erg
	$6.241\ 506 \cdot 10^{18}$	eV
	0.238 920 1	g-cal
	$2.777\ 78 \cdot 10^{-7}$	kw-hr
	$6.241\ 506 \cdot 10^{12}$	MeV
karat (1/24 gold)	41.667	mg/g
kilocalorie/mole	4.339 28	eV
kilogram	15 432.361	grain
	1000*	g
	35.273 962	ounce (avdp.)
	2.204 622 6	pound (avdp.)
	10^{-3}*	ton (metric)
kilogram/cm^2	0.967 841	atmosphere
	0.980 665	bar
	73.555 914	cmHg (standard)
	14.223 343	lb/$inch^2$
kilogram/m^2	$9.678\ 41 \cdot 10^{-5}$	atmosphere
	$9.806\ 65 \cdot 10^{-5}$	bar
	0.1*	g/cm^2
	0.001 422 334 3	lb/$inch^2$
	0.073 555 914	mmHg (standard)
kilojoule/mole	$1.03642 \cdot 10^{-2}$	eV

$A = XB$; $B = A/X$; c = speed of light in vacuo in cm/s; asterisk identifies exact values; all values involving c or π are also exact.

A	X	B
kilometer	10^{5*}	cm
	1000^*	m
	0.539 956 80	mile (nautical)
	0.621 371 19	mile (statute)
kilometer/hour	27.777 778	cm/s
	0.277 777 78	m/s
	0.539 956 80	mile (nautical)/hr
	0.621 371 19	mile (statute)/hr
kilowatt	3412.141 8	Btu/hr
	$8.598\ 452 \cdot 10^5$	cal_{IT}/hr
	$8.601\ 123 \cdot 10^5$	g-cal/hr
	10^{10*}	erg/s
	1.341 02	horsepower (mechanical)
	$3.6 \cdot 10^{6*}$	J/hr
	1000^*	J/s
kilowatt-hour	3412.141 8	Btu
	$8.598\ 452 \cdot 10^5$	cal_{IT}
	$8.601\ 123 \cdot 10^5$	g-cal
	$3.6 \cdot 10^{6*}$	J
knot	51.444 444	cm/s
	1.687 809 9	ft/s
	1.852^*	km/hr
	0.514 444 444 . . .	m/s
	1^*	mile (nautical)/hr
lambert	0.318 309 886 . . . ($= 1/\pi$)	candela/cm^2
	1^*	lumen/cm^2
league (statute)	$4.828\ 032^*$	km
	3^*	mile (statute)
light year	63 239.727	astronomical unit
	$9.460\ 528\ 4 \cdot 10^{12}$	km
	0.306 594 89	parsec
line	1^*	maxwell
line/cm^2	1^*	gauss
liter	1000^*	cm^3
	0.001^*	m^3
	0.879 877 0	quart (British)
	1.056 688 2	quart (U.S., liquid)
liter/second	2.118 883 5	ft^3/min
	15.850 324	gallon (U.S. liquid)/min
lumen (5550 Å)	0.079 577 472 ($= 1/4\pi$)	candela
	0.001 470 588 2	watt
lumen/cm^2	1^*	lambert
	1^*	phot
lux	1^*	lumen/m^2
maxwell	1^*	abweber
	1^*	gauss/cm^2
	1^*	line
	$3.335\ 641 \cdot 10^{-11}$ ($= 1/c$)	statweber
	10^{-8*}	volt-sec
	10^{-8*}	weber

$A = XB$; $B = A/X$; c = speed of light in vacuo in cm/s; asterisk identifies exact values; all values involving c or π are also exact.

A	X	B
maxwell/cm^2	1*	gauss
	$3.335\ 641 \cdot 10^{-11}\ (= 1/c)$	stattesla
	1*	gauss
megaton	$4.1868 \cdot 10^{15}$*	J
megmho/cm	0.001*	abmho/cm
megohm	10^6*	ohm
	$1.112\ 650 \cdot 10^{-6}\ (= 10^{15}/c^2)$	statohm
meter	10^{10}*	angstrom
	100*	cm
	0.546 806 65	fathom
	3.280 839 9	ft
	0.001*	km
	1000*	mm
meter/second	3.6*	km/hr
	1.943 844 5	knot
	1.943 844 5	mile (geographic)/hr
	2.236 936 3	mile (statute)/hr
metric ton: *see* ton		
MeV: *see* million electron volt		
mho	10^{-9}*	abmho
	1*	ohm^{-1}
	$8.987\ 552 \cdot 10^{11}\ (= 10^{-9}c^2)$	statmho
microfarad	10^{-15}*	abfarad
	10^{-6}*	farad
	$8.987\ 552 \cdot 10^5\ (= 10^{-15}c^2)$	statfarad
microgram	10^{-6}*	g
	0.001*	mg
micrometer	10^4*	angstrom
	10^{-4}*	cm
	10^{-6}*	m
	10^{-3}*	mm
micromicrofarad	10^{-12}*	farad
mile (geographic)	1.000	min of lat. at 45° lat.
	6076.115 5	ft
	1.852*	km
	1852*	m
	1*	nautical (Intern.)
mile (U.S., statute)	5280*	ft
	8*	furlong
	1.609 344*	km
	1609.344*	m
	1760*	yards
mile (U.S., statute)/hr	44.704*	cm/s
	88*	ft/min
	1.466 666 7	ft/sec
	1.609 344*	km/hr
	0.868 976 24	knot
	26.822 4*	m/min
millibar	10^{-3}*	bar
	1000	barye

$A = XB$; $B = A/X$; c = speed of light in vacuo in cm/s; asterisk identifies exact values; all values involving c or π are also exact.

A	X	B
milligal	10^{-3}*	gal
milligram	0.005	carat
	0.015 432 358	grain
	10^{-3}*	g
milligram/liter	1*	ppm (part per million)
milliliter	1*	cm^3
	10^{-3}*	liter
millimeter	0.1*	cm
	0.001*	m
million electron volt	$1.602\ 1892 \cdot 10^{-13}$	J
mmHg	0.001 315 789 5	atmosphere
	0.001 333 223 1	bar
	1333.223 1	dyn/cm^2
	1.359 51	g/cm^2
	13.595 1	kg/m^2
	1*	torr
month (lunar, synodic)	29.530 604 2	day (mean solar)
myriagram	10^4*	g
	10*	kg
nanometer	10*	angstrom
	10^{-7}*	cm
	10^{-9}*	m
	10^{-6}*	mm
neper	8.685 890	decibel
newton	10^5*	dyne
	7.233 013 871	poundal
	0.224 808 943	pound-force
newton-meter	10^7*	dyn-cm
	1*	joule
	0.737 562 15	lb-ft
nit	1*	$candela/m^2$
oersted	$79.577\ 472\ (= 10^3/4\pi)$	ampere/meter
	1*	emu
	$2.997\ 924\ 58 \cdot 10^{10}\ (= c)$	esu
	1*	gilbert/cm
ohm	10^9*	abohm
	$1.112\ 650 \cdot 10^{-12}\ (= 10^9/c^2)$	statohm
ounce (apoth., troy)	480*	grain
	31.103 476 8	gram
	1.097 142 9	ounce (avdp.)
	20*	pennyweight
	24*	scruple
ounce (avdp.)	16*	dram (avdp.)
	28.349 523	g
	0.911 458 3	ounce (apoth., troy)
ounce (U.S. fluid)	29.573 530	cm^3
	8*	dram (U.S. fluid)
parsec	206 264.806	astronomical units
	$30.856\ 776 \cdot 10^{12}$	km
	3.261 633	light year

$A = XB$; $B = A/X$; $c =$ speed of light in vacuo in cm/s; asterisk identifies exact values; all values involving c or π are also exact.

A	X	B
pascal	$9.869\ 233 \cdot 10^{-6}$	atmosphere
	$10^{-5}*$	bar
	$10*$	barye
	$10*$	dyn/cm^2
	$7.500\ 617 \cdot 10^{-3}$	mmHg
	$7.500\ 617 \cdot 10^{-3}$	torr
pint (U.S. liquid)	$473.176\ 48$	cm^3
	$0.473\ 176\ 48$	liter
	$16*$	ounce (U.S. fluid)
	$0.5*$	quart (U.S. fluid)
poise	$1*$	g/cm s
	$0.1*$	pascal-second
pound (avdp.)	$256*$	dram (avdp.)
	$453.592\ 37*$	g
	$0.453\ 592\ 37*$	kg
	$16*$	ounce (avdp.)
	$0.031\ 081$	slug
pound/inch^2	$70.306\ 958$	g/cm^2
	6894.757	Pa
poundal	$13\ 825.495\ 437\ 6*$	dyn
	$1*$	lb-ft/s^2
	$0.138\ 254\ 954\ 376$	newton
quart (U.S. liquid)	$946.352\ 946$	cm^3
	$256*$	dram (U.S. liquid)
	$0.25*$	gallon (U.S. liquid)
	$0.946\ 352\ 946$	liter
	$32*$	ounce (U.S. liquid)
	$2*$	pint (U.S. liquid)
quintal	10^5*	g
	$100*$	kg
radian	$57.295\ 78$	degree
roentgen	$2.58 \cdot 10^{-4}*$	coulomb/kg
slug	$14.593\ 903$	kg
	$32.174\ 0$	pound (avdp.)
statampere	$3.335\ 641 \cdot 10^{-11}\ (= 1/c)$	abampere
	$3.335\ 641 \cdot 10^{-10}\ (= 10/c)$	ampere
statcoulomb	$3.335\ 641 \cdot 10^{-11}\ (= 1/c)$	abcoulomb
	$3.335\ 641 \cdot 10^{-10}\ (= 10/c)$	coulomb
	$2.081\ 942 \cdot 10^9$	electron charge
statfarad	$1.112\ 650 \cdot 10^{-21}\ (= 1/c^2)$	abfarad
	$1.112\ 650 \cdot 10^{-12}\ (= 10^9/c^2)$	farad
stathenry	$8.987\ 552 \cdot 10^{20}\ (= c^2)$	abhenry
	$8.987\ 552 \cdot 10^{11}\ (= 10^{-9}c^2)$	henry
statohm	$8.987\ 552 \cdot 10^{20}\ (= c^2)$	abohm
	$8.987\ 552 \cdot 10^{11}\ (= 10^{-9}c^2)$	ohm
stattesla	$2.997\ 924\ 58 \cdot 10^{10}\ (= c)$	gauss
	$2.997\ 924\ 58 \cdot 10^6\ (= 10^{-4}c)$	tesla
statvolt	$2.997\ 924\ 58 \cdot 10^{10}\ (= c)$	abvolt
	$2.997\ 924\ 58 \cdot 10^2\ (= 10^{-8}c)$	volt

$A = XB$; $B = A/X$; c = speed of light in vacuo in cm/s; asterisk identifies exact values; all values involving c or π are also exact.

A	X	B
statweber	$2.997\ 924\ 58 \cdot 10^{10}\ (= c)$	maxwell
	$2.997\ 924\ 58 \cdot 10^{-2}\ (= 10^{-8}c)$	weber
stere	1^*	m^3
stilb	1^*	candela/cm^2
stoke	1^*	cm^2/s
tesla	10^{4*}	gauss
	10^{4*}	oersted
	1^*	weber/m^2
ton (metric)	10^{6*}	g
	1000^*	kg
ton (of refrig., U.S.)	72 574.8	kcal/day
torr	1^*	mmHg
volt	10^{8*}	abvolt
	$0.003\ 335\ 641\ (= 10^8/c)$	statvolt
watt	3.412 142	Btu/hr
	860.112 29	g-cal/hr
	10^{7*}	erg/s
	1^*	J/s
	0.001^*	kilowatt
watt-hour	3600^*	joule
	0.860 112 29	kcal (g-cal)
	0.859 845 23	kcal$_{IT}$
weber	10^{8*}	abweber
	10^{8*}	line
	10^{8*}	maxwell
	$0.003\ 335\ 641\ (= 10^8/c)$	statweber
	1^*	volt-second
weber/cm^2	10^{8*}	gauss
	10^{8*}	oersted
yard (U.S.)	91.44^*	cm
	0.5^*	fathom
	3^*	ft
	$0.914\ 4^*$	m
year (sidereal) (1900)	365.256 365 56	day (mean solar)
	366.256 78	day (sidereal)
	$31.558\ 149\ 984 \cdot 10^6$	second (ephemeris)
	1.000 038 8	year (tropical) (1900)
year (tropical) (1900)	365.242 198 78	day (mean solar)
	366.242 58	day (sidereal)
	$31.556\ 925\ 974\ 7 \cdot 10^6$	second (ephemeris)
	0.999 961 22	year (sidereal)

From Forsythe 1964; Allen 1976, p. 18–29; Dean 1985, p. 2.11–2.42; Weast 1986, p. F292–F305; other sources.

Radioisotope produced	Source nuclide	Half-life	Production rate (atoms/cm^2/s)		Global inventory
			Troposphere	Total atmosphere	
^{3}H	^{14}N, ^{16}O	12.33 y	$8.4 \cdot 10^{-2}$	0.25	3.5 kg
^{7}Be	^{14}N, ^{16}O	53.3 d	$2.7 \cdot 10^{-2}$	$8.1 \cdot 10^{-2}$	3.2 g
^{10}Be	^{14}N, ^{16}O	$1.6 \cdot 10^6$ y	$1.5 \cdot 10^{-2}$	$4.5 \cdot 10^{-2}$	430 tons
^{14}C	^{14}N	5730 y	1.1	2.5	75 tons
^{22}Na	^{40}Ar	2.602 y	$2.6 \cdot 10^{-5}$	$8.7 \cdot 10^{-5}$	1.8 g
^{24}Na	^{40}Ar	15.02 h	—	—	—
^{26}Al	^{40}Ar	720,000 y	$3.8 \cdot 10^{-5}$	$1.4 \cdot 10^{-4}$	1.1 tons
^{28}Mg	^{40}Ar	20.90 h	—	—	—
^{31}Si	^{40}Ar	2.62 h	—	—	—
^{32}Si	^{40}Ar	105 y	$5.4 \cdot 10^{-5}$	$1.6 \cdot 10^{-4}$	1.4 kg
^{32}P	^{40}Ar	14.26 d	$2.7 \cdot 10^{-4}$	$8.1 \cdot 10^{-4}$	0.4 g
^{33}P	^{40}Ar	25.3 d	$2.2 \cdot 10^{-4}$	$6.8 \cdot 10^{-4}$	0.6 g
^{35}S	^{40}Ar	87.5 d	$4.9 \cdot 10^{-4}$	$1.4 \cdot 10^{-3}$	4.5 g
^{38}S	^{40}Ar	2.84 h	—	—	—
^{34}Cl	^{40}Ar	32.23 m	—	—	—
^{36}Cl	^{40}Ar	301,000 y	$4.0 \cdot 10^{-4}$	$1.1 \cdot 10^{-3}$	15 tons*
^{39}Cl	^{40}Ar	55.6 m	—	—	—
^{37}Ar	^{40}Ar	35.04 d	—	—	—
^{39}Ar	^{40}Ar	269 y	—	23	—
^{81}Kr	^{80}Kr	210,000 y	—	16.2	—

*Including estimate of ^{36}Cl produced by n capture at the Earth's surface.

From Lal 1974, p. 1002, Table 3; Kathren 1984, p. 32, Table 2.5; other sources.

Densities

The density of various substances is given in g/cm³. Unspecified temperature and pressure values may be considered to refer to room temperature (approximately 20°C) and to the pressure of 1 atmosphere.

Substance	Density (g/cm³)	Temperature (°C)	Pressure (atm)
acetone	0.792	20	1
air (gas)			
dry	$1.204 \cdot 10^{-3}$	20	1
ambient, sea level	$1.2250 \cdot 10^{-3}$	15	1
ambient, 10,000 m altitude	$0.41351 \cdot 10^{-3}$	−50	0.261
albite	2.622	—	—
alcohol, ethyl	0.80625	0	—
aluminosilicates	2.6–2.7	—	—
aluminum	2.6989	20	—
ammonia (gas)	$0.7719 \cdot 10^{-3}$	—	—
andalusite	3.144	25	—
anglesite	6.324	25	—
anhydrite	2.963	26	—
aniline	1.584	24	—
anorthite	2.758–2.764	—	—
anthracite	1.4–1.8	—	—
antimony	6.684	25	—
aragonite	2.930	26	—
argon (gas)	$1.7840 \cdot 10^{-3}$	0	1
arsenic	5.727	14	—
atom	0.07–22.5	—	—
augite	3.2–3.5	—	—
barite	4.480	26	—
barium	3.51	20	—
basalt	2.8–3.1	—	—
benzene	0.899	0	—
beryl	2.640	25	—
beryllium	1.85	20	—
biotite	2.8–3.2	—	—
bismuth	9.80	—	—
bone	1.7–2.2	20	—
boron	2.35	—	—
brass			
red	9.75	—	—
yellow	8.4–8.5	—	—
brick	1.4–2.2	—	—
bromine	3.119	20	—
bromoform	1.595	24	—
bronze	8.2–8.7	—	—
butter	0.86–0.87	—	—
cadmium	8.642	—	—
calcite	2.712	26	—
calcium	1.54	—	—
carbon			
amorphous	1.8–2.1	—	—
diamond	3.51	—	—
graphite	2.67	15	—

The density of various substances is given in g/cm³. Unspecified temperature and pressure values may be considered to refer to room temperature (approximately 20°C) and to the pressure of 1 atmosphere.

Substance	Density (g/cm³)	Temperature (°C)	Pressure (atm)
carbon tetrachloride	1.5867	—	—
celestite	3.972	26	—
celluloid	1.4	—	—
cement, Portland	1.5	—	—
cerium	6.657	—	—
cesium	1.8785	15	—
chalk	1.9–2.6	—	—
chlorine (gas)	$3.215 \cdot 10^{-3}$	0	—
chloroform	1.489	20	—
chromium	7.20	28	—
clay minerals	2.5–2.6	—	—
cobalt	8.9	—	—
coesite	2.911	25	—
copper	8.92	—	—
copper–beryllium	8.3–8.7	—	—
cork	0.22–0.26	—	—
corundum	3.988	26	—
cristobalite			
α	2.334	25	—
β	2.194	405	—
cuprite	6.104	26	—
cyanogen (gas)	$2.335 \cdot 10^{-3}$	—	—
diamond	3.5154	25	—
diopside	3.277	—	—
diorite	2.84 ± 0.12	—	—
dolomite	2.866	26 ± 3	—
dysprosium	8.5500	—	—
Earth			
mean	5.518	—	—
inner core	13	4200?	$3.68 \cdot 10^{6}$
electron	$7.9 \cdot 10^{10}$	—	—
enstatite	3.198 ± 0.007	—	—
epidote	3.25–3.50	—	—
erbium	9.006	—	—
ether	0.736	0	—
ethyl alcohol	0.80625	0	—
europium	5.2434	—	—
feldspar	2.6–2.7	—	—
flint	2.63	—	—
fluorine (gas)	$1.69 \cdot 10^{-3}$	15	—
fluorite	3.181	25	—
gabbro	2.98 ± 0.13	—	—
gadolinium	7.9004	—	—
galena	7.597	—	—
gallium	5.904	29.6	—
garnet	3.5–4.3	—	—

The density of various substances is given in g/cm^3. Unspecified temperature and pressure values may be considered to refer to room temperature (approximately 20°C) and to the pressure of 1 atmosphere.

Substance	Density (g/cm^3)	Temperature (°C)	Pressure (atm)
gasoline	0.66–0.69	—	—
germanium	5.35	20	—
glass	2.4–2.8	—	—
glycerin	1.260	0	—
gneiss	2.7–2.9	—	—
gold	18.88	20	—
18 K	14.61	—	—
granite	2.67 ± 0.14	—	—
graphite	2.267	15	—
gypsum	2.317 ± 0.005	—	—
hafnium	13.31	20	—
halite	2.163	25	—
heavy water (D_2O)	1.10469	0	—
hedenbergite	3.55 ± 0.01	—	—
helium			
gas	$0.1785 \cdot 10^{-3}$	0	—
liquid	0.1249	−268.94	1
solid	0.2105	−273.15	50
hematite	5.275	25	—
holmium	8.7947	—	—
hornblende	3.0	—	—
hydrogen			
gas (1 atm)	$0.08988 \cdot 10^{-3}$	0	—
liquid	0.0708	−253	—
solid	0.0706	−262	—
ice			
air-free	0.9174	0	1
glacier	0.88–0.91	0	1
ilmenite	4.786	—	—
inconel	8.1–8.4	—	—
indium	7.30	20	—
iodine	4.93	—	—
iridium	22.421	—	—
iron	7.874	—	—
cast	7.2	—	—
wrought	7.7	—	—
jadeite	3.315 ± 0.020	—	—
krypton (gas)	$3.736 \cdot 10^{-3}$	—	—
kyanite	3.674	25	—
lanthanum			
α	6.1453	—	—
β	6.17	—	—
lead	11.3437	16	—
leucite	2.480	—	—
lignite	1.1–1.4	—	—

The density of various substances is given in g/cm^3. Unspecified temperature and pressure values may be considered to refer to room temperature (approximately 20°C) and to the pressure of 1 atmosphere.

Substance	Density (g/cm^3)	Temperature (°C)	Pressure (atm)
limestone	2.68–2.76	—	—
lithium	0.82	—	—
lutethium	9.8404	—	—
magnesite	3.009	26 ± 3	—
magnesium	1.738	20	—
magnetite	5.200	25	—
manganese	7.42	—	—
marble	2.6–2.8	—	—
mercury	13.5951	0	—
	13.5939	20	—
methane (gas)	$0.560 \cdot 10^{-3}$	—	—
minium	8.9–9.2	—	—
molybdenum	10.22	—	—
monel	8.4–8.8	—	—
muscovite	2.834	27	—
neodymium	7.004	—	—
neon (gas)	$0.89990 \cdot 10^{-3}$	0	1
nepheline	2.623	—	—
neutron	$2.8 \cdot 10^{14}$	—	—
neutron star	10^{14}		
nickel	8.902	25	—
niobium	8.57	—	—
nitric acid	1.5027	25	—
nitrogen (gas)	$1.2506 \cdot 10^{-3}$	—	—
nucleus (atomic)	$2.8 \cdot 10^{14}$	—	—
olive oil	0.918	15	—
olivine	3.2–4.4	—	—
fayalite	4.393	25	1
forsterite	3.214	25	1
orthoclase	2.551 ± 0.008	—	—
osmium	22.48	20	—
oxygen (gas)	$1.429 \cdot 10^{-3}$	0	1
ozone	2.144	0	—
palladium	12.02	20	—
peat	0.6–0.8	—	—
periclase	3.584	25	—
peridotite	3.23 ± 0.06	—	—
perovskite	4.03	—	—
phosphorus			
black	2.70	—	—
red	2.34	—	—
violet	2.36	—	—
yellow	1.82	20	—
platinum	21.45	0	—
platinum-iridium (10% Ir)	21.5	—	—
plutonium	19.84	25	—

The density of various substances is given in g/cm³. Unspecified temperature and pressure values may be considered to refer to room temperature (approximately 20°C) and to the pressure of 1 atmosphere.

Substance	Density (g/cm³)	Temperature (°C)	Pressure (atm)
polonium	9.4	—	—
potassium	0.86	20	—
praseodymium	6.773	—	—
protactinium	15.37	—	—
proton	$2.8 \cdot 10^{14}$	—	—
pumice	0.4–0.9	—	—
pyrite	5.016	—	—
quartz			
α	2.648	25	—
β	2.533	575	—
radium	6.0	20	—
radon (gas)	$9.73 \cdot 10^{-3}$	—	—
rhenium	20.53	—	—
rhodium	12.4	—	—
rubidium	1.532	—	—
ruthenium	12.30	—	—
rutile	4.250	25	—
samarium	7.520	—	—
sandstone	2.14–2.36	—	—
scandium	2.9890	—	—
seawater (35‰ salinity)	1.025	20	1
selenium	4.81	20	—
serpentine	2.50–2.65	—	—
shale	2.6–2.9	—	—
siderite	3.944	26 ± 3	—
silicon	2.33	25	1
sillimanite	3.247	25	—
silver	10.50	20	—
slate	2.6–3.3	—	—
sodium	0.97	—	—
sphalerite	4.088	—	—
spinel	3.582	26	—
starch	1.53	—	—
steel	7.8	—	—
stainless	7.7	—	—
stishovite	4.287	—	—
strontium	2.6	20	—
sugar	1.61	—	—
sulfur			
α	2.07	20	—
β	1.96	—	—
γ	1.92	—	—
sulfuric acid (96–98%)	1.841	—	—
Sun			
mean	1.409	—	—
core	160	—	—

The density of various substances is given in g/cm³. Unspecified temperature and pressure values may be considered to refer to room temperature (approximately 20°C) and to the pressure of 1 atmosphere.

Substance	Density (g/cm³)	Temperature (°C)	Pressure (atm)
tantalum	16.6	—	—
tellurium	6.25	—	—
terbium	8.2294	—	—
thallium	11.85	—	—
thorium	11.7	—	—
thulium	9.3208	—	—
tin			
gray	5.75	—	—
white	7.31	—	—
titanium	4.5	20	—
topaz	3.563	26	—
tridymite	2.265	—	—
	2.192	405	—
troilite	4.830	—	—
tungsten	19.35	20	—
uranium	19.05 ± 0.02	—	—
vanadium	5.96	—	—
water			
liquid, pure	1.000	3.98	1
solid	0.9174	0	1
wollastonite	2.909	—	—
wood			
balsa	0.11–0.14	—	—
ebony	1.11–1.33	—	—
fir	0.48–0.55	—	—
Lignum vitae	1.17–1.33	—	—
oak	0.6–0.9	—	—
pine	0.43–0.67	—	—
poplar	0.35–0.50	—	—
spruce	0.45	—	—
teak, African	0.99	—	—
teak, Indian	0.66–0.88	—	—
xenon (gas)	$5.887 \cdot 10^{-3}$	—	—
ytterbium	6.9654	—	—
yttrium	4.4689	—	—
zinc	7.133	25	—
zircon	4.68	—	—
zirconium	6.506	20	—

From Forsythe 1964; Robie et al. 1966, p. 60–70, Table 5.2; Trent et al. 1972; Strauss and Kaufman 1976; Moses 1978; Weast 1986.

Earth
astronomical and geophysical data

---

a = equatorial radius; c = polar radius; G = gravitational constant; h = altitude in m; M_E = Earth's mass; T = centuries from A.D. 1900.0; ϕ = latitude in degrees.

---

age	$= 4.6 \cdot 10^9$ y
age of earliest fossils	$= 3.4 \cdot 10^9$ y
age of oldest rocks	$= 3.8 \cdot 10^9$ y
angular momentum	$= 5.861 \cdot 10^{33}$ m^2 kg s^{-1}
angular rotational velocity	$= 7.2921157 \cdot 10^{-5}$ rad s^{-1}
anomalistic year	$= 31{,}558{,}433.237$ s$_E$ (1986)
aphelion distance	$= 1.01675104$ AU
artificial satellites	
velocity to attain circular orbit	
(minimum for orbiting)	$= 7.91$ km s^{-1}
velocity to attain parabolic orbit	
(minimum for escape)	$= 11.18$ km s^{-1}
astronomical unit (AU)	$= 149{,}597{,}870.7$ km
	$= 8.31675$ light minutes
	$= 499.004784$ light seconds
atmosphere	
mass	$= 5.136 \cdot 10^{18}$ kg
mean molecular mass	$= 28.964$ u
barycenter of Earth-Moon system (mean)	$= 4640$ km from Earth's center
	$= 1731$ km below Earth's surface
biosphere	
mass (dry)	$= 1.1 \cdot 10^{15}$ kg
yearly production (dry)	$= 4.9 \cdot 10^{13}$ kg
calendar year	$=$ Gregorian year
centripetal acceleration at equator	$= 0.033915$ m s^{-2}
degree of latitude	
length (ϑ = latitude)	$= 111.1334 - 0.5594 \cos 2\phi + 0.012 \cos 4\phi$ km
length at equator	$= 110.5752$ km
length at 45° lat	$= 111.1322$ km
	$= 60.0066$ nautical miles
density	
core	
inner	$= 12.7$ to 13.0 g cm^{-3}
outer	$= 9.9$ to 12.7 g cm^{-3}
crust	
continental	
lower	$= 2.92$ g cm^{-3}
upper	$= 2.72$ g cm^{-3}
mean	$= 2.85$ g cm^{-3}
oceanic	
igneous	$= 2.89$ g cm^{-3}
igneous + sediments	$= 2.85$ g cm^{-3}
mantle	$= 3.3$ to 5.5 g cm^{-3}
mean	$= 5.518 \pm 0.004$ g cm^{-3}
eccentricity of orbit	$= 0.01675104$
eccentricity period	
highest amplitude	$= 413{,}000$ y
second highest	$= 95{,}000$ y
eccentricity range	$= {\sim}0.01$ to ${\sim}0.07$
energy	
rotational	$= 2.137 \cdot 10^{29}$ J
orbital	$= 2.651 \cdot 10^{33}$ J

a = equatorial radius; c = polar radius; G = gravitational constant; h = altitude in m; M_E = Earth's mass; T = centuries from A.D. 1900.0; ϕ = latitude in degrees.

energy of accumulation (work required to dissipate Earth matter against its own gravitational field)	= $2.49 \cdot 10^{32}$ J
ephemeris day (d_E)	= 86,400 s_E
ephemeris second (s_E)	= 1/31,556,925.9747 of tropical year 1900
equator	
length	= 40075.24 km
radius (mean)	a = 6378.139 $\pm$ 0.003 km
equatorial ellipticity	
$(a_{max} - a_{min})/a$	= $1.5 \cdot 10^{-5}$
	= 102 m
equatorial rotational velocity	= 465.10 m s^{-1}
escape velocity	11.18 km s^{-1}
flattening	
$(a - c)/a$	= 0.00335282
	= 1/298.256
geothermal flux	
mean	= $6.1420 \cdot 10^{-2}$ W m^{-2}
	= $1.467 \cdot 10^{-6}$ cal cm^{-2} s^{-1}
total	= $3.13273 \cdot 10^{13}$ W
gravity	
at equator	= 9.7803185 m s^{-2}
at 30° lat	= 9.7932402 m s^{-2}
at 45° lat	= 9.8061907 m s^{-2}
at 60° lat	= 9.8191698 m s^{-2}
at 90° lat	= 9.8321776 m s^{-2}
mean	= 9.8062222 m s^{-2}
gravity standard (g_0)	= 9.860665 m s^{-2}
gravity variation with altitude h	= $0.0003086h$ m s^{-2}
Gregorian year	= 365.2425 d
	= 31,556,952 s
hydrosphere	
atmospheric water	= $1.3 \cdot 10^{16}$ kg
freshwater in lakes and rivers	= $3 \cdot 10^{16}$ kg
ice	= $0.02 \cdot 10^{21}$ kg
mass (total)	= $1.72 \cdot 10^{21}$ kg
ocean	= $1.37 \cdot 10^{21}$ kg
pore water in rocks and sediments	= $0.33 \cdot 10^{21}$ kg
inclination of axis from normal to plane of orbit (= obliquity of ecliptic)	= 23°26′28.0″ (1986)
inclination angle range	= ~21°39′ to ~24°36′
inclination angle period (mean)	= 41,200 y
land	
area	= $148.017 \cdot 10^6$ km^2
mean elevation	= 840 m
magnetic dipole moment	= $7.90 \cdot 10^{22}$ A m^2 (1985)
	= $7.90 \cdot 10^{25}$ gauss cm^3 (1985)
magnetic field (mean)	= 0.5 gauss
mass	
atmosphere	= $5.1 \cdot 10^{18}$ kg
core (total)	= $1.900 \cdot 10^{24}$ kg
crust	= $2.4 \cdot 10^{22}$ kg

a = equatorial radius; c = polar radius; G = gravitational constant; h = altitude in m; M_E = Earth's mass; T = centuries from A.D. 1900.0; ϕ = latitude in degrees.

mass *(continued)*	
inner core	$= 1.2 \cdot 10^{23}$ kg
inner-outer core transition	$= 1.2 \cdot 10^{22}$ kg
mantle	$= 4.052 \cdot 10^{24}$ kg
ocean	$= 1.4 \cdot 10^{21}$ kg
outer core	$= 1.768 \cdot 10^{24}$ kg
total	$M_E = (5.9737 \pm 0.0004) \cdot 10^{24}$ kg
mass × gravitational constant	$GM_E = 3.986005 \cdot 10^{14}$ m^3 s^{-2}
mean density	$= 5.518 \pm 0.004$ g cm^{-3}
mean equatorial ellipticity	
$(a_{max} - a_{min})/a$	$= 1.5 \cdot 10^{-5}$
	$= 102$ m
mean solar day	$= 86,400 + 0.0015T$ s$_E$
	$= 86,400.00129$ (1986)
meridional quadrant	$= 10,002.02$ km
minute of latitude	
length at 45°	$= 1852.20$ m
	~ 1 nautical mile
moment of inertia	
about equatorial axis	$= 0.3295\ M_E\ a^2$
	$= 8.010 \cdot 10^{37}$ kg m^2
about polar axis	$= 0.33078\ M_E\ a^2$
	$= 8.0415 \cdot 10^{37}$ kg m^2
obliquity of the ecliptic	$= 23°27'8.26'' - 46.845''T - 0.0059''T^2 + 0.00181''T^3$
	$= 23°26'28.0''$ (1986)
obliquity period (mean)	$= 40,600$ y
obliquity range	$= \sim 21°39'$ to $\sim 24°36'$
oceans	
area	$= 362.033 \cdot 10^6$ km^2
mean depth	$= 3729$ m
volume	$= 1349.929 \cdot 10^6$ km^3
orbital radius (mean)	$= 149,597,870.7$ km
	$= 1$ AU
orbital velocity (mean)	$= 29.784$ km s^{-1}
perihelion advance	$= 0.1085''$ y^{-1}
perihelion distance	$= 0.98324896$ AU
polar radius	$c = 6356.779$ km
Potsdam gravity standard (g_0)	$= 9.81260$ m s^{-2}
precession	
general	$= 50.2858''$ y^{-1} (1986)
lunisolar	$= 50.4043''$ y^{-1} (1986)
planetary	$= 0.1085''$ y^{-1} (1986)
relativistic (geodetic)	$= 0.0192''$ y^{-1}
precessional angle	$= 46°52'56''$ (1986)
precessional angle periodicity	$= 40,625$ y
precessional angle secular range =	
obliquity range	$= \sim 21°39'$ to $\sim 24°36'$
precessional period	
climatic precession	$= 21,000$ y
general precession	$= 25,800$ y

a = equatorial radius; c = polar radius; G = gravitational constant; h = altitude in m; M_E = Earth's mass; T = centuries from A.D. 1900.0; ϕ = latitude in degrees.

radius
    equatorial (mean) $a = 6378.139 \pm 0.003$ km
    mean $(a^2c)^{1/3} = 6371.03$ km
    polar $c = 6356.779$
    sphere of equivalent volume $= 6370.8$ km
     $= 0.02125$ light seconds

rotational velocity
    angular $= 7.2921157$ rad s^{-1}
     $= 15.041066''$ s^{-1}
    linear at equator $= 465.10$ m s^{-1}
semimajor axis of orbit $= 149{,}597{,}870.7$ km
     $= 1$ AU
semiminor axis of orbit $= 149{,}578{,}480$ km
     $= 0.999870$ AU
sidereal day $= 86{,}164.09055 + 0.0015T$ s$_E^{\bullet}$
     $= 23$h 56m 4.091 s of mean solar time
sidereal year $= 365.25636565$ d$_E$ (1986)
     $= 31{,}558{,}149.993$ s$_E$ (1986)

surface
    area
        land $= 148.017 \cdot 10^6$ km^2
        ocean $= 362.033 \cdot 10^6$ km^2
        total $= 510.0501 \cdot 10^6$ km^2
    mean elevation $= -2430$ m
temperature (mean, surface) $= 288$ K
     $= 15°$C

tidal energy dissipated
    mean tide $= 1.4 \cdot 10^{12}$ J s^{-1}
    spring tide $= 2.6 \cdot 10^{12}$ J s^{-1}
tropical year $= 365.14119348$ d$_E$ (1986)
     $= 31{,}556{,}925.5189$ s$_E$ (1986)
volume $= 1.0831 \cdot 10^{21}$ m^3
     $= 1.0831 \cdot 10^{12}$ km^3

From Schmid and Koch 1972, p. 2.98–2.99; Allen 1976, p. 112–114, 140–141; McQuillin and Ardus 1977, p. 134; Stacey 1977, p. 332–333; Berger 1978, p. 44–45; Garland 1979; Lang 1980, p. 526, Table 57; Turcotte and Schubert 1982, p. 429–430.

Layer			Depth (km)	Temp. (°C)	Pressure (atm)	Density (g/cm³)	Gravity (m/s²)	Velocity (km/s) P waves	Velocity (km/s) S waves	Mass 10²¹ kg	Mass percent
			0		1		9.81				
Crust	lithosphere	oceanic / continental		—	—	2.8	—	6	3.6	24	0.4
			12 / 35	500	2200/9600	2.9 / 3.3	9.84	7.2 / 8.1	4.3 / 4.5		
			65 / 120	1300	38,200	3.4	9.87	8.0	4.4		
Mantle	upper		170	—	—	3.45	9.89	7.8	4.3	1206	20.2
			220	—	72,200	3.5	9.90	8.0	4.4		
			670	2400	242,300	4.4	10.01	10.5	5.9		
	lower		2000	—	—	5.1	—	12.8	6.9	2846	47.6
			2885	3800	1,372,000	5.5 / 9.9	10.69	13.7 / 8.0	7.2 / 0.0		
Core	outer			—	—	—	—	—	—	1768	29.6
			4720	6000	3,067,200	11.9	5.74	9.9	0.0		
	transition		5000	—	—	—	—	10.1	0.0	12	0.2
			5170	6300	3,341,800	12.7	4.36	10.8			
	inner		5750	—	—	12.9	—	—	—	120	2
			6371	6600	3,680,500	13.0	0	11.15			

From Stacey 1977, p. 337–341, Tables G1 and G2; Ringwood 1979, p. 4, Table 1.1; other sources.

The electromagnetic spectrum is continuous and, as a result, the different types of electromagnetic radiation grade into each other. The energy (in eV) is obtained by multiplying the frequency by the value of the Planck's constant in eV ($= 4.14 \cdot 10^{-15}$).

Frequency (Hz)	Wavelength		Name	Typical source
10^{23}	$3 \cdot 10^{-13}$	cm	cosmic gamma rays	supernovae
10^{22}	$3 \cdot 10^{-12}$	cm	gamma rays	unstable atomic nuclei
10^{21}	$3 \cdot 10^{-11}$	cm	gamma rays hard x-rays	unstable atomic nuclei
10^{20}	$3 \cdot 10^{-10}$	cm	hard x-rays	inner atomic shell
10^{19}	$3 \cdot 10^{-9}$	cm	x-rays	electron impact on solids
10^{18}	$3 \cdot 10^{-8}$	cm	soft x-rays	electron impact on solids
10^{17}	$3 \cdot 10^{-7}$	cm	ultraviolet	atoms in discharges
10^{16}	$3 \cdot 10^{-6}$	cm	ultraviolet	atoms in discharges
10^{15}	0.3	μm	visible spectrum	atoms, molecules hot bodies
10^{14}	3	μm	infrared	molecules, hot bodies
10^{13}	30	μm	infrared	molecules, hot bodies
10^{12}	0.3	mm	far infrared	molecules, hot bodies
10^{11}	3	mm	microwaves	communication devices
10^{10}	3	cm	microwaves, radar	communication and detection devices
10^{9}	30	cm	radar	communication and detection devices
10^{8}	3	m	video, FM	television, FM radio
10^{7}	30	m	short-wave	short-wave radio
10^{6}	300	m	AM	AM radio
10^{5}	3	km	long-wave	long-wave radio
10^{4}	30	km	—	induction heating
10^{3}	300	km	—	induction heating
10^{2}	3,000	km	—	rotating electromagnets
10	30,000	km	—	rotating electromagnets
1	300,000	km	—	rotating electromagnets
0	infinite		dc current	batteries

Elementary particles

| Particle Name | Symbol | Mass | | | Mean life (seconds) | Decay | |
		MeV	u	electron = 1		Principal mode(s)	Percent
Classons							
graviton	—	0	0	0	infinite	—	—
photon	γ	0	0	0	infinite	—	—
Gauge Bosons							
—	$W^\pm$	80800	86.7	158,120	—	e^+e^-	—
—	Z^0	92900	99.7	181,800	—	$\mu^+\mu^-$	—
Leptons							
electron	e^-	0.5110034	0.0005485803	1	infinite	—	—
positron	e^+	0.5110034	0.0005485803	1	infinite	—	—
e neutrino	ν_e	0	0	0	infinite	—	—
e antineutrino	$\bar{\nu}_e$	0	0	0	infinite	—	—
muon	$\mu^\pm$	105.65932	0.11342892	206.76833	$2.19709 \cdot 10^{-6}$	$e^\pm \nu \bar{\nu}$	—
μ neutrino	ν_μ	0	0	0	infinite	—	—
μ antineutrino	$\bar{\nu}_\mu$	0	0	0	infinite	—	—
tauon	$\tau^\pm$	1784.2	1.91540	3491.5619	$3.4 \cdot 10^{-13}$	$\mu^\pm \nu \bar{\nu}$	18.5
						$e^\pm \nu \bar{\nu}$	16.2
						etc.	
τ neutrino		0	0	0	infinite	—	—
τ antineutrino		0	0	0	infinite	—	—
Nonstrange Mesons							
pion	$\pi^\pm$	139.5673	0.1498304	273.12401	$2.6030 \cdot 10^{-8}$	$\mu^\pm \nu$	100
	π^0	134.9630	0.1448876	264.1137	$0.83 \cdot 10^{-16}$	$\gamma\gamma$	98.802
						$e^+e^-\gamma$	1.198
eta	η	548.8	0.589156	1074	$0.75 \cdot 10^{-18}$	$\gamma\gamma$	39.0
						$\pi^0\pi^0\pi^0$	31.8
						$\pi^+\pi^-\pi^0$	23.7
						$\pi^+\pi^-\gamma$	4.9
Strange Mesons							
kaon	$K^\pm$	493.667	0.529969	966.0738	$1.2371 \cdot 10^{-8}$	$\mu^\pm \nu$	63.51
						$\pi^\pm\pi^0$	21.17
						etc.	
	$K^0, \bar{K}^0$	497.67	0.534266	973.9074	consists of 50% K_S^0 + 50% K_L^0		
	K_S^0	497.67	0.534266	973.9074	$0.8923 \cdot 10^{-10}$	$\pi^+\pi^-$	68.61
						$\pi^0\pi^0$	31.39
						etc.	
	K_L^0	497.67	0.534266	973.9074	$5.183 \cdot 10^{-8}$	$e^\pm \pi^\pm \nu$	38.7
						$\pi^\pm \mu^\pm \nu$	27.1
						$\pi^0\pi^0\pi^0$	21.5
						$\pi^+\pi^-\pi^0$	12.4
						etc.	
Charmed Nonstrange Mesons							
—	$D^\pm$	1869.4	2.006867	3658.29	$0.92 \cdot 10^{-12}$	$K^0\bar{K}^0 \ldots$	48
						$e^\pm \ldots$	19

| Particle Name | Symbol | Mass | | | Mean life (seconds) | Decay | |
		MeV	u	electron = 1		Principal mode(s)	Percent
Charmed Nonstrange Mesons (*continued*)						K^- ...	16
						K^+ ...	6
						etc.	
—	D^0, D^0	1864.7	2.001821	3649.09	$4.4 \cdot 10^{-13}$	K^- ...	44
						K^0K^0 ...	33
						K^+ ...	8
						etc.	
Charmed Strange Mesons							
—	$F^\pm$	1971	2.11594	3857	$1.9 \cdot 10^{-13}$	$\eta\pi^\pm$	?
						etc.	
Bottom Mesons							
—	$B^\pm$	5270.8	5.6584	10314.6	$1.4 \cdot 10^{-12}$	$D^0\pi^\pm$	4.2
						etc.	
—	B^0	5274.2	5.6620	10321.3		D^0 ...	80
						etc.	
Nonstrange Baryons							
proton	$p^\pm$	938.2796	1.00727647	1836.1515	infinite (?)	—	—
neutron	n^0	939.5731	1.00866490	1838.6827	914 ± 6	$pe^-\bar{\nu}$	100
Strangeness-1 Baryons							
lambda	Λ	1115.60	1.197636	2183.1557	$2.632 \cdot 10^{-10}$	$p\pi^-$	64.2
						$n\pi^0$	35.8
sigma	Σ^+	1189.36	1.276820	2327.4992	$0.800 \cdot 10^{-10}$	$p\pi^0$	51.64
						$n\pi^+$	48.36
	Σ^0	1192.46	1.280148	2333.5657	$5.8 \cdot 10^{-20}$	$\Lambda\gamma$	100
	Σ^-	1197.34	1.285387	2343.1155	$1.482 \cdot 10^{-10}$	$n\pi^-$	100
Strangeness-2 Baryons							
Xi	Ξ^0	1314.9	1.411592	2573.1727	$2.90 \cdot 10^{-10}$	$\Lambda\pi^0$	100
	Ξ^-	1321.32	1.418484	2585.7362	$1.641 \cdot 10^{-10}$	$\Lambda\pi^-$	100
Strangeness-3 Baryons							
omega	Ω^-	1672.45	1.795434	3272.8745	$0.819 \cdot 10^{-10}$	ΛK^-	68.6
						$\Xi^0\pi^-$	23.4
						$\Xi^-\pi^0$	8.0
						etc.	
Nonstrange Charmed Baryons							
—	Λ_c^+	2282.0	2.4498	4465.7	$2.3 \cdot 10^{-13}$	e^+ ...	4.5
						$pK^-\pi^+$	2.2
						etc.	

From Aguilar-Benitez et al. 1984; Wapstra and Audi 1985.

Alphabetical list, symbol, atomic number, and atomic mass of the neutral atom in atomic mass units (u). Atomic mass is based on average isotopic abundance in common terrestrial matter. Accuracy, as indicated by the decimal figures, depends on the variability of the isotopic abundances in different terrestrial substances. The atomic mass value of unstable elements with relatively short half-lives, identified by an asterisk, refers to that of the longest-lived isotope.

Name	Symbol	Atomic number	Atomic mass ($^{12}C = 12.000$)
actinium*	Ac	89	227.027750
aluminum	Al	13	26.981539
americium*	Am	95	243.061375
antimony	Sb	51	121.75
argon	Ar	18	39.9477
arsenic	As	33	74.921594
astatine*	At	85	219.0113
barium	Ba	56	137.327
berkelium*	Bk	97	247.070300
beryllium	Be	4	9.012182
bismuth	Bi	83	208.980374
boron	B	5	10.811
bromine	Br	35	79.904
cadmium	Cd	48	112.41
calcium	Ca	20	40.078
californium*	Cf	98	251.079580
carbon	C	6	12.011
cerium	Ce	58	140.115
cesium	Cs	55	132.90543
chlorine	Cl	17	35.453
chromium	Cr	24	51.9961
cobalt	Co	27	58.933198
copper	Cu	29	63.546
curium*	Cm	96	247.070347
dysprosium	Dy	66	162.498
einsteinium	Es	99	254.088022
erbium	Er	68	167.26
europium	Eu	63	151.965
fermium*	Fm	100	257.09510
fluorine	F	9	18.998403
francium*	Fr	87	223.019733
gadolinium	Gd	64	157.252
gallium	Ga	31	69.723
germanium	Ge	32	72.59
gold	Au	79	196.966543
hafnium	Hf	72	178.49
helium	He	2	4.002602
holmium	Ho	67	164.930319
hydrogen	H	1	1.00794
indium	In	49	114.82
iodine	I	53	126.904473
iridium	Ir	77	192.22
iron	Fe	26	55.847
krypton	Kr	36	83.80
lanthanum	La	57	138.9055

Alphabetical list, symbol, atomic number, and atomic mass of the neutral atom in atomic mass units (u). Atomic mass is based on average isotopic abundance in common terrestrial matter. Accuracy, as indicated by the decimal figures, depends on the variability of the isotopic abundances in different terrestrial substances. The atomic mass value of unstable elements with relatively short half-lives, identified by an asterisk, refers to that of the longest-lived isotope.

Name	Symbol	Atomic number	Atomic mass ($^{12}C = 12.000$)
lawrencium*	Lr	103	260.1053
lead	Pb	82	207.2
lithium	Li	3	6.941
lutethium	Lu	71	174.967
magnesium	Mg	12	24.305
manganese	Mn	25	54.938047
mendelevium*	Md	101	258.0986
mercury	Hg	80	200.59
molybdenum	Mo	42	95.94
neodymium	Nd	60	144.242
neon	Ne	10	20.179
neptunium*	Np	93	237.048168
nickel	Ni	28	58.688
niobium	Nb	41	92.906377
nitrogen	N	7	14.0067
nobelium*	No	102	255.0933
osmium	Os	76	190.2
oxygen	O	8	15.9994
palladium	Pd	46	106.42
phosphorus	P	15	30.973762
platinum	Pt	78	195.08
plutonium*	Pu	94	244.064199
polonium*	Po	84	208.982404
potassium	K	19	39.0983
praseodymium	Pr	59	140.907647
promethium*	Pm	61	144.912743
protactinium*	Pa	91	231.035880
radium*	Ra	88	226.025403
radon*	Rn	86	222.017571
rhenium	Re	75	186.207
rhodium	Rh	45	102.905500
rubidium	Rb	37	85.4678
ruthenium	Ru	44	101.07
samarium	Sm	62	150.36
scandium	Sc	21	44.955910
selenium	Se	34	78.96
silicon	Si	14	28.0855
silver	Ag	47	107.8682
sodium	Na	11	22.989768
strontium	Sr	38	87.62
sulfur	S	16	32.066
tantalum	Ta	73	180.9749
technetium*	Tc	43	97.907215
tellurium	Te	52	127.60
terbium	Tb	65	158.925342

Alphabetical list, symbol, atomic number, and atomic mass of the neutral atom in atomic mass units (u). Atomic mass is based on average isotopic abundance in common terrestrial matter. Accuracy, as indicated by the decimal figures, depends on the variability of the isotopic abundances in different terrestrial substances. The atomic mass value of unstable elements with relatively short half-lives, identified by an asterisk, refers to that of the longest-lived isotope.

Name	Symbol	Atomic number	Atomic mass (^{12}C = 12.000)
thallium	Tl	81	204.383
thorium	Th	90	232.038051
thulium	Tm	69	168.934212
tin	Sn	50	118.710
titanium	Ti	22	47.88
tungsten	W	74	183.85
uranium	U	92	238.0289
vanadium	V	23	50.9415
xenon	Xe	54	131.29
ytterbium	Yb	70	173.04
yttrium	Y	39	88.905849
zinc	Zn	30	65.39
zirconium	Zr	40	91.224

From Walker et al. 1984, p. 59; Tuli 1985; Wapstra and Audi 1985.

Abundances are given as number of atoms per 10^6 atoms of Si. Asterisks identify elements with no isotope sufficiently stable to exist in any appreciable amount.

Element	Abundance	Element	Abundance	Element	Abundance
1 H	27,200,000,000	31 Ga	37.8	61 Pm	*
2 He	2,180,000,000	32 Ge	118	62 Sm	0.261
3 Li	59.7	33 As	6.79	63 Eu	0.0972
4 Be	0.78	34 Se	62.1	64 Gd	0.331
5 B	24	35 Br	11.8	65 Tb	0.0589
6 C	12,100,000	36 Kr	45.3	66 Dy	0.398
7 N	2,480,000	37 Rb	7.09	67 Ho	0.0875
8 O	20,100,000	38 Sr	23.8	68 Er	0.253
9 F	843	39 Y	4.64	69 Tm	0.0386
10 Ne	3,760,000	40 Zr	10.7	70 Yb	0.243
11 Na	57,000	41 Nb	0.71	71 Lu	0.0369
12 Mg	1,075,000	42 Mo	2.52	72 Hf	0.176
13 Al	84,900	43 Tc	*	73 Ta	0.0226
14 Si	1,000,000	44 Ru	1.86	74 W	0.137
15 P	10,400	45 Rh	0.344	75 Re	0.0507
16 S	515,000	46 Pd	1.39	76 Os	0.717
17 Cl	5,240	47 Ag	0.529	77 Ir	0.660
18 Ar	104,000	48 Cd	1.69	78 Pt	1.37
19 K	3,770	49 In	0.184	79 Au	0.186
20 Ca	61,100	50 Sn	3.82	80 Hg	0.52
21 Sc	33.8	51 Sb	0.352	81 Tl	0.184
22 Ti	2,400	52 Te	4.91	82 Pb	3.15
23 V	295	53 I	0.90	83 Bi	0.144
24 Cr	13,400	54 Xe	4.35	84 Po	*
25 Mn	9,510	55 Cs	0.372	85 At	*
26 Fe	900,000	56 Ba	4.36	86 Rn	*
27 Co	2,250	57 La	0.448	97 Fr	*
28 Ni	49,300	58 Ce	1.16	88 Ra	*
29 Cu	514	59 Pr	0.174	89 Ac	*
30 Zn	1,260	60 Nd	0.836	90 Th	0.0335
				91 Pa	*
				92 U	0.0090

Anders and Ebihara 1982, p. 2364, Table 1.

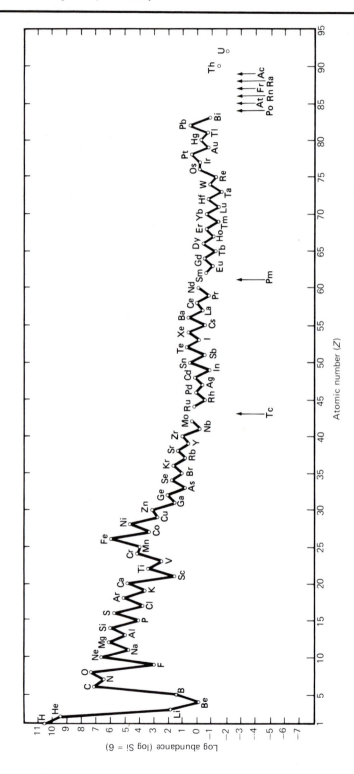

Shell	K	L		M			N				O				P				Q			
n quantum no.	1	2		3			4				5				6				7			
l quantum no.	0	0	1	0	1	2	0	1	2	3	0	1	2	3	0	1	2	3	0	1	2	3
Atomic No. Element																						
1 H	1																					
2 He	2																					
3 Li	2	1																				
4 Be	2	2																				
5 B	2	2	1																			
6 C	2	2	2																			
7 N	2	2	3																			
8 O	2	2	4																			
9 F	2	2	5																			
10 Ne	2	2	6																			
11 Na	2	2	6	1																		
12 Mg	2	2	6	2																		
13 Al	2	2	6	2	1																	
14 Si	2	2	6	2	2																	
15 P	2	2	6	2	3																	
16 S	2	2	6	2	4																	
17 Cl	2	2	6	2	5																	
18 Ar	2	2	6	2	6																	
19 K	2	2	6	2	6		1															
20 Ca	2	2	6	2	6		2															
21 Sc	2	2	6	2	6	1	2															
22 Ti	2	2	6	2	6	2	2															
23 V	2	2	6	2	6	3	2															
24 Cr	2	2	6	2	6	5	1															
25 Mn	2	2	6	2	6	5	2															
26 Fe	2	2	6	2	6	6	2															
27 Co	2	2	6	2	6	7	2															
28 Ni	2	2	6	2	6	8	2															
29 Cu	2	2	6	2	6	10	1															
30 Zn	2	2	6	2	6	10	2															
31 Ga	2	2	6	2	6	10	2	1														
32 Ge	2	2	6	2	6	10	2	2														
33 As	2	2	6	2	6	10	2	3														
34 Se	2	2	6	2	6	10	2	4														
35 Br	2	2	6	2	6	10	2	5														
36 Kr	2	2	6	2	6	10	2	6														
37 Rb	2	2	6	2	6	10	2	6			1											
38 Sr	2	2	6	2	6	10	2	6			2											
39 Y	2	2	6	2	6	10	2	6	1		2											
40 Zr	2	2	6	2	6	10	2	6	2		2											
41 Nb	2	2	6	2	6	10	2	6	4		1											
42 Mo	2	2	6	2	6	10	2	6	5		1											
43 Tc	2	2	6	2	6	10	2	6	6		1											

Shell		K	L		M			N				0				P				Q			
n quantum no.		1	2		3			4				5				6				7			
l quantum no.		0	0	1	0	1	2	0	1	2	3	0	1	2	3	0	1	2	3	0	1	2	3
Atomic No.	Element																						
44	Ru	2	2	6	2	6	10	2	6	7		1											
45	Rh	2	2	6	2	6	10	2	6	8		1											
46	Pd	2	2	6	2	6	10	2	6	10		0											
47	Ag	2	2	6	2	6	10	2	6	10		1											
48	Cd	2	2	6	1	6	10	2	6	10		2											
49	In	2	2	6	2	6	10	2	6	10		2	1										
50	Sn	2	2	6	2	6	10	2	6	10		2	2										
51	Sb	2	2	6	2	6	10	2	6	10		2	3										
52	Te	2	2	6	2	6	10	2	6	10		2	4										
53	I	2	2	6	2	6	10	2	6	10		2	5										
54	Xe	2	2	6	2	6	10	2	6	10		2	6										
55	Cs	2	2	6	2	6	10	2	6	10		2	6			1							
56	Ba	2	2	6	2	6	10	2	6	10		2	6			2							
57	La	2	2	6	2	6	10	2	6	10		2	6	1		2							
58	Ce	2	2	6	2	6	10	2	6	10	2	2	6			2							
59	Pr	2	2	6	2	6	10	2	6	10	3	2	6			2							
60	Nd	2	2	6	2	6	10	2	6	10	4	2	6			2							
61	Pm	2	2	6	2	6	10	2	6	10	5	2	6			2							
62	Sm	2	2	6	2	6	10	2	6	10	6	2	6			2							
63	Eu	2	2	6	2	6	10	2	6	10	7	2	6			2							
64	Gd	2	2	6	2	6	10	2	6	10	7	2	6	1		2							
65	Tb	2	2	6	2	6	10	2	6	10	9	2	6			2							
66	Dy	2	2	6	2	6	10	2	6	10	10	2	6			2							
67	Ho	2	2	6	2	6	10	2	6	10	11	2	6			2							
68	Er	2	2	6	2	6	10	2	6	10	12	2	6			2							
69	Tm	2	2	6	2	6	10	2	6	10	13	2	6			2							
70	Yb	2	2	6	2	6	10	2	6	10	14	2	6			2							
71	Lu	2	2	6	2	6	10	2	6	10	14	2	6	1		2							
72	Hf	2	2	6	2	6	10	2	6	10	14	2	6	2		2							
73	Ta	2	2	6	2	6	10	2	6	10	14	2	6	3		2							
74	W	2	2	6	2	6	10	2	6	10	14	2	6	4		2							
75	Re	2	2	6	2	6	10	2	6	10	14	2	6	5		2							
76	Os	2	2	6	2	6	10	2	6	10	14	2	6	6		2							
77	Ir	2	2	6	2	6	10	2	6	10	14	2	6	7		2							
78	Pt	2	2	6	2	6	10	2	6	10	14	2	6	9		1							
79	Au	2	2	6	2	6	10	2	6	10	14	2	6	10		1							
80	Hg	2	2	6	2	6	10	2	6	10	14	2	6	10		2							
81	Tl	2	2	6	2	6	10	2	6	10	14	2	6	10		2	1						
82	Pb	2	2	6	2	6	10	2	6	10	14	2	6	10		2	2						
83	Bi	2	2	6	2	6	10	2	6	10	14	2	6	10		2	3						
84	Po	2	2	6	2	6	10	2	6	10	14	2	6	10		2	4						
85	At	2	2	6	2	6	10	2	6	10	14	2	6	10		2	5						
86	Rn	2	2	6	2	6	10	2	6	10	14	2	6	10		2	6						
87	Fr	2	2	6	2	6	10	2	6	10	14	2	6	10		2	6			1			
88	Ra	2	2	6	2	6	10	2	6	10	14	2	6	10		2	6			2			

Shell	K	L		M			N				O				P				Q			
n quantum no.	1	2		3			4				5				6				7			
l quantum no.	0	0	1	0	1	2	0	1	2	3	0	1	2	3	0	1	2	3	0	1	2	3
Atomic No. Element																						
89 Ac	2	2	6	2	6	10	2	6	10	14	2	6	10		2	6	1		2			
90 Th	2	2	6	2	6	10	2	6	10	14	2	6	10		2	6	2		2			
91 Pa	2	2	6	2	6	10	2	6	10	14	2	6	10	2	2	6	1		2			
92 U	2	2	6	2	6	10	2	6	10	14	2	6	10	3	2	6	1		2			
93 Np	2	2	6	2	6	10	2	6	10	14	2	6	10	4	2	6	1		2			
94 Pu	2	2	6	2	6	10	2	6	10	14	2	6	10	6	2	6			2			
95 Am	2	2	6	2	6	10	2	6	10	14	2	6	10	7	2	6			2			
96 Cm	2	2	6	2	6	10	2	6	10	14	2	6	10	7	2	6	1		2			
97 Bk	2	2	6	2	6	10	2	6	10	14	2	6	10	9	2	6			2			
98 Cf	2	2	6	2	6	10	2	6	10	14	2	6	10	10	2	6			2			
99 Es	2	2	6	2	6	10	2	6	10	14	2	6	10	11	2	6			2			
100 Fm	2	2	6	2	6	10	2	6	10	14	2	6	10	12	2	6			2			
101 Md	2	2	6	2	6	10	2	6	10	14	2	6	10	13	2	6			2			
102 No	2	2	6	2	6	10	2	6	10	14	2	6	10	14	2	6			2			
103 Lr	2	2	6	2	6	10	2	6	10	14	2	6	10	14	2	6	1		2			
104 —	2	2	6	2	6	10	2	6	10	14	2	6	10	14	2	6	2		2			

From Weast 1986, p. B–4.

English name	Symbol	Foreign name	Language
antimony	Sb	stibium	Latin
copper	Cu	cuprum	Latin
gold	Au	aurum	Latin
iron	Fe	ferrum	Latin
lead	Pb	plumbum	Latin
mercury	Hg	hydrargyrum	Latin
potassium	K	kalium	Latin
silver	Ag	argentum	Latin
sodium	Na	natrun	Arabic
tin	Sn	stannum	Latin
tungsten	W	wolfram	German

Element	Solar system	Solar atmosphere	CI chondrites	Ordinary chondrites	Whole Earth	Crust (average)	Crust (continental)	Pyrolite	Basalt (ave.)	Granite (ave.)	Shale (ave.)
Na	1.74	1.52	1.97	1.67	1.36	5.37	5.94	0.74	3.44	5.02	5.36
Mg	32.88	31.40	32.58	34.04	28.48	3.97	3.22	41.28	7.54	1.07	4.12
Al	2.60	2.61	2.57	2.58	2.21	15.44	16.57	3.29	13.26	15.00	15.28
Si	30.59	35.28	30.54	36.36	29.48	52.65	56.32	37.07	40.88	63.78	52.00
P	0.32	—	0.37	0.26	0.18	0.20	0.16	—	0.26	0.16	0.25
K	0.11	0.11	0.10	0.13	0.10	4.92	3.40	0.19	1.03	6.64	4.89
Ca	1.87	1.76	2.08	1.85	1.53	6.89	6.40	4.14	12.87	2.96	7.32
Ti	0.07	0.09	0.07	0.08	0.06	0.84	0.69	0.75	2.35	0.45	1.18
Cr	0.41	0.39	0.36	0.33	0.26	0.02	0.01	0.52	—	—	—
Mn	0.29	0.21	0.25	0.27	0.22	0.18	0.12	0.19	0.23	0.18	—
Fe	27.53	25.05	27.57	21.35	33.79	9.50	7.26	11.56	18.13	4.83	9.61
Co	0.07	0.06	0.07	0.05	0.12	0.01	—	—	—	—	—
Ni	1.51	1.52	1.45	1.02	2.21	0.01	—	0.28	—	—	—
	99.99	100.00	99.98	99.99	100.00	100.00	100.09	100.01	99.99	100.09	100.01

From Clark 1966, p. 4, Table 1.1; Ringwood 1966; Dodd 1981, p. 19, Table 2.1; Anders and Ebihara 1982; Mason and Moore 1982, p. 14, Table 2.3; p. 44, Table; p. 46–47, Table 3.5; p. 52, Table 3.9; Weaver and Tarney 1984, p. 576, Table 2.

Prefix	Symbol	Number	Name
hexa-	H	10^{18}	quintillion
penta-	P	10^{15}	quadrillion
tera-	T	10^{12}	trillion
giga-	G	10^{9}	billion
mega-	M	10^{6}	million
kilo-	K	10^{3}	thousand
hecto-	h	10^{2}	hundred
deca-	D	10^{1}	ten
mono-	-	10^{0}	one
deci-	d	10^{-1}	tenth
centi-	c	10^{-2}	hundredth
milli-	m	10^{-3}	thousandth
micro-	μ	10^{-6}	millionth
nano-	n	10^{-9}	billionth
pico-	p	10^{-12}	trillionth
femto-	f	10^{-15}	quadrillionth
atto-	a	10^{-18}	quintillionth

Gems
physical properties

Xl System = crystal system; C = cubic; H = hexagonal; M = monoclinic; O = orthorhombic; T = trigonal; Tc = triclinic; Tt = tetragonal; A = amorphous. G = density relative to water at 3.98°C. H = hardness. n = refractive index.

Name	Composition	Xl System	G	H	n	Color	Coloring agent
Amber	resin	A	2–2.5	1.05	1.54	yellow-brown	organics
Beryl	$Be_3Al_2(Si_6O_{18})$	H	2.64–2.8	7½–8	1.57–1.61		
aquamarine	—	—	—	—	—	pale blue	Fe^{2+} Fe^{3+}
emerald	—	—	—	—	—	green	Cr^{3+}
heliodore	—	—	—	—	—	yellow	Fe^{3+}
morganite	—	—	—	—	—	pink	Mn^{2+}
Chrysoberyl	$BeAl_2O_4$	O	3.65–3.8	8½	1.75		
alexandrite	—	—	—	—	—	green-red	Cr^{3+}
cat's eye	—	—	—	—	—	opalescent	—
Corundum	Al_2O_3	T	3.99	9	1.77		
oriental emerald	—	—	—	—	—	green	Fe^{2+} Ti^{4+}
oriental topaz	—	—	—	—	—	yellow	Fe^{2+} Fe^{3+}
ruby	—	—	—	—	—	red	Cr^{3+}
sapphire	—	—	—	—	—	blue	Fe^{2+} Ti^{4+}
Diamond	C	C	3.51	10	2.42		
	—	—	—	—	—	white	—
	—	—	—	—	—	blue	—
	—	—	—	—	—	yellow	—
Garnets							
almandine	$Fe_3Al_2Si_3O_{12}$	C	4.32	7	1.83	red	Fe^{2+}
andradite	$Ca_3Fe_2Si_3O_{12}$	C	3.86	7	1.89	reddish	Fe^{3+}
grossularite	$Ca_3Al_2Si_3O_{12}$	C	3.59	6½	1.73	white to brown	V^{3+}
pyrope	$Mg_3Al_2Si_3O_{12}$	C	3.58	7	1.71	red	—
spessartite	$Mn_3Al_2Si_3O_{12}$	C	4.19	7	10.80	dark red	Mn^{2+}
uvarovite	$Ca_3Cr_2Si_3O_{12}$	C	3.90	7½	1.87	green	—
Jade							
jadeite	$NaAl(SiO_3)_2$	M	3.3	6½–7	—	green	—
nephrite							
actinolite	$Ca_2(Mg,Fe)_5Si_8O_{22}(OH)_2$	M	3.1–3.3	5–6	1.65	green	—
tremolite	$Ca_2Mg_5Si_8O_{22}(OH)_2$	M	3.0–3.2	5–6	1.61	white-gray	—
Lazurite	$(Na,Ca)_8(AlSiO_4)_6 \cdot$	C	2.40–2.45	5–5½	1.5	blue	S_3
(lapis lazuli)	$(SO_4,S,Cl)_2$						
Malachite	$Cu_2CO_3(OH)_2$	M	3.90–4.03	3½–4	1.88	green	—
Mother of pearl	$CaCO_3$ (aragonite)	O	2.95	3½–4	1.68	iridescent	—
Olivine	$(Mg,Fe)_2SiO_4$	O	3.27–4.37	6½–7	1.69		
chrysolite	—	—	—	—	—	yellow-green	Fe^{2+}
Opal	$SiO_2 \cdot nH_2O$	A	2.0–2.25	5–6	1.44	iridescent	—
Pearl	$CaCO_3$ (aragonite)	O	2.95	3½–4	1.68	iridescent	organics
Quartz	SiO_2	T,H	2.65	7	1.54		
eucrystalline							
amethyst	—	—	—	—	—	purple	Fe^3
citrine	—	—	—	—	—	yellow	Fe^{3+}
rock crystal	—	—	—	—	—	colorless	—
rose quartz	—	—	—	—	—	pink	—
smoky	—	—	—	—	—	brown	Al^{3+}
microcrystalline							
agate	—	—	—	—	—	variegated	—
chalcedony	—	—	—	—	—	translucent	—
cornelian	—	—	—	—	—	red	Fe oxides
flint	—	—	—	—	—	gray	—
jasper	—	—	—	—	—	red-green	Fe oxides
onyx	—	—	—	—	—	banded	—

Xl System = crystal system; C = cubic; H = hexagonal; M = monoclinic; O = orthorhombic;
T = trigonal; Tc = triclinic; Tt = tetragonal; A = amorphous. G = density relative to water at 3.98°C.
H = hardness. n = refractive index.

Name	Composition	Xl System	G	H	n	Color	Coloring agent
Spinel	$MgAl_2O_4$	C	3.5–4.1	8	1.72	blue	Co^{2+}
	—	—	—	—	—	green	—
	—	—	—	—	—	red	Co^{2+}
	—	—	—	—	—	yellow	—
Topaz	$Al_2SiO_4(OH,F)_2$	O	3.4–3.6	8	1.61–1.63	blue	—
	—	—	—	—	—	colorless	—
	—	—	—	—	—	green	—
	—	—	—	—	—	red	—
	—	—	—	—	—	yellow	—
Tourmaline	$(Na,Ca)(Li,Mg,Al) \cdot$ $(Al,Fe,Mn)_6(BO_3)_3 \cdot$ $Si_6O_{18}(OH)_4$	T	3.0–3.25	7–7½	1.64–1.68		
achroite	—	—	—	—	—	colorless	—
Brazilian emerald	—	—	—	—	—	green	—
indicolite	—	—	—	—	—	blue	—
peridot of Ceylon	—	—	—	—	—	yellow	—
rubellite	—	—	—	—	—	red	Mn^{3+}
Turquoise	$CuAl_6(PO_4)_4(OH)_8 \cdot 5H_2O$	Tc	2.6–2.8	6	1.62	azure	Cu^{2+}
Zircon	$ZrSiO_4$	Tt	4.68	7½	1.92–1.96	colorless	—
—	—	—	—	—	—	blue	—
hyacinth	—	—	—	—	—	orange	—

Geological time scale

Era	Period		Age (y B.P.)	Major Events
CENOZOIC	QUATERNARY	HOLOCENE	0	
			45	Beginning of Atomic age (December 2, 1942, 21.45 ZULU)
			3,000	Beginning of the Iron Age
			10,000	
		PLEISTOCENE	11,600	Intense deglaciation; giant floods down the Mississippi Valley
			18,000	Maximum of the last ice age
			125,000	Temperature maximum of the last interglacial: appearance of *Homo sapiens sapiens* and *Homo sapiens neanderthalensis*
			400,000	Disappearance of *Homo erectus*
			$1.5 \cdot 10^6$	Disappearance of *Homo habilis;* appearance of *Homo erectus*
			1.6	Appearance of *Hyalinea baltica*
	TERTIARY	Pliocene	2.0	Appearance of *Homo erectus*
			3.0	Appearance of *Australopithecus africanus*
			3.2	Closing of the Central American isthmus; beginning of extensive northern glaciation
			5.1	Gibraltar passage opens
		Miocene		The Mediterranean is isolated; salt deposits on its floor
			14	The Antarctic sheet reaches the ocean; Rhamapithecus
			24.6	Alpine orogenesis apex
		Oligocene	30	Fayum beds: Aegyptopithecus
			38.0	Sudden expansion of the Antarctic ice sheet Separation of Australia from Antarctica Appearance of Artiodactyls, Perissodactyls, and apes
		Eocene		
			54.9	
		Paleocene		Appearance of globorotalids; radiation of placental mammals; first primates; angiosperms spread; Laramide orogeny
			65	Giant asteroidal impact; extinction of Cycadeoidales, globotruncanids, ammonoids, belemnoids, ichthyosaurs, plesiosaurs, dinosaurs
MESOZOIC	Cretaceous			First angiosperm and marsupials; opening of the South Atlantic; the White Cliffs of Dover
			144	
	Jurassic			Opening of the North Atlantic; first coccoliths and planktonic Foraminifera
			213	
	Triassic			Palisades sill, New York State; Carrara marble; appearance of dinosaurs, lizards, turtles; first mammals; first birds
			248	

Era	Period	Age (y B.P.)	Major Events
PALEOZOIC	Permian		Appalachian–Allenghanian/Hercynian–Variscan orogenesis; New Red Sandstone in Europe; glaciation in the southern hemisphere; extinction of tetracorals, cystoids, placoderms
		286	
	Carboniferous		Widespread formation of coal; cyclothems; first reptiles, winged insects
		360	
	Devonian		Old Red Sandstone; Queenston–Juniata red beds; first sharks; first amphibia
		408	Taconic–Caledonian orogeny
	Silurian		Lockport dolostone; first bony fishes; first trees
		438	
	Ordovician		First corals; first vertebrates (jawless fishes)
		505	
	Cambrian		Appearance of trilobites, brachiopods, echinoderms, and shelled mollusks
		590	Appearance of Archaeocyatha
	Ediacaran		
PROTEROZOIC		630	Appearance of metazoa
		$1.7 \cdot 10^9$	Increasing O_2 in atmosphere; appearance of eucaryota
		2.7	Oldest stromatolites
ARCHEAN		3.5	Earliest bacteria (heterotrophs)
		3.8	Oldest terrestrial rocks
HADEAN		4.6	Age of the meteorites and oldest lunar rocks; formation of the solar system
		4.7	Formation of the elements in the region of the solar system
GAMOWIAN			Formation of stars and galaxies Evolution of the elements Evolution of matter–antimatter Evolution of the four natural forces
		16.6	Rapid expansion
PLANCKIAN			The first $5.390 \cdot 10^{-44}$ seconds: evolution of space, time, and energy
		16.6	Beginning of the present universe

Heat production by radioactive elements

	Heat production (cal/y) per gram of parent element
Isotope	
^{238}U series	0.71
^{235}U series	4.3
^{232}Th series	0.20
^{40}K	0.21
^{87}Rb	$1.3 \cdot 10^{-8}$
Element	
U	0.73
Th	0.20
K	$2.7 \cdot 10^{-5}$
Rb	$3.6 \cdot 10^{-7}$

Heat production by radioactive elements in rocks

	Element concentration (10^{-6})			Heat production (10^{-6} cal g^{-1} y^{-1})			
Rock type	U	Th	K	U	Th	K	Total
granite	4.7	2.0	40,000	3.4	4	1.08	8.48
diorite	2.6	9.0	25,000	1.9	1.8	0.67	4.37
gabbro	0.9	2.7	4,600	0.66	0.5	0.12	1.28
peridotite	0.015	0.01	300	0.011	0.002	0.008	0.020
chondrites	0.012	0.04	850	0.0088	0.008	0.0023	0.0191

From Clark 1966, p. 522–534; Wetherill 1966, p. 517, Table 23.7; Garland 1979, p. 325, Table 22.2.

Latin	Greek lowercase	Greek uppercase	Latin	Greek lowercase	Greek uppercase
a	α	A	o	o	O
b	β	B	p	π	Π
c	γ	Γ	r, rh	ρ	P
d	δ	Δ	s	σ, ς	Σ
e	ε, ϵ	E	t	τ	T
z	ζ	Z	y	υ	Υ
e	η	H	f	φ, ϕ	Φ
th	ϑ, θ	Θ	ch	χ	X
i	ι	I	ps	ψ	Ψ
k	$\varkappa, \kappa$	K	o	ω	Ω
l	λ	Λ	ng	$\gamma\gamma$	$\Gamma\Gamma$
m	μ	M	nk	$\gamma\varkappa$	ΓK
n	ν	N	nx	$\gamma\xi$	$\Gamma\Xi$
x	ξ	Ξ	nch	$\gamma\chi$	ΓX

Hadrons
quark structure (flavor only)

Family	Symbol	Quark structure
Mesons	π^+	$\bar{u}d$
	π^0	$(d\bar{d}-u\bar{u})/2^{1/2}$
	π^-	$d\bar{u}$
	K^+	$u\bar{s}$
	K^-	$s\bar{u}$
	K^0	$s\bar{d}, d\bar{s}$
	D^+	$c\bar{d}$
	D^-	$d\bar{c}$
	D^0	$u\bar{c}$
	F^+	$c\bar{s}$
	F^-	$s\bar{c}$
	B^-	$b\bar{u}$
	B^0	$b\bar{d}$
Baryons	p	uud
	n	udd
	Λ	uds
	Σ^+	uus
	Σ^-	dds
	Σ^0	uds
	Ξ^-	dss
	Ξ^0	uss
	Ω^-	sss
	Λ_c^+	udc

From Close 1979, p. 39, Table 3.2; Perkins 1982, p. 182–187; Halzen and Martin 1984, p. 59.

ρ = mean density; V_P = mean velocity of P waves.

Name	Intrusive	Extrusive
granite	70% SiO_2 quartz–feldspar–biotite $\rho = 2.667$ $V_P = 5.6$ km/s	rhyolite
granodiorite	65% SiO_2 feldspar–quartz–biotite $\rho = 2.716$ $V_P = 5.7$ km/s	dacite
diorite	55% SiO_2 feldspar–amphibole–pyroxene $\rho = 2.839$ $V_P = 5.8$ km/s	andesite
gabbro	50% SiO_2 pyroxene–feldspar $\rho = 2.976$ $V_P = 6.6$ km/s	basalt
peridotite	40% SiO_2 olivine–pyroxene $\rho = 3.234$ $V_P = 8.0$ km/s	—

From Clark 1966.

Z	Element	Energy (eV)	Z	Element	Energy (eV)	Z	Element	Energy (eV)
1	H	13.6057	36	Kr	13.999	71	Lu	5.426
2	He	24.587	37	Rb	4.177	72	Hf	7.0
3	Li	5.392	38	Sr	5.695	73	Ta	7.89
4	Be	9.322	39	Y	6.38	74	W	7.98
5	B	8.298	40	Zr	6.84	75	Re	7.88
6	C	11.260	41	Nb	6.88	76	Os	8.7
7	N	14.534	42	Mo	7.099	77	Ir	9.1
8	O	13.618	43	Te	7.28	78	Pt	9.0
9	F	17.422	44	Ru	7.37	79	Au	9.225
10	Ne	21.564	45	Rh	7.46	80	Hg	10.437
11	Na	5.139	46	Pd	8.34	81	Tl	6.108
12	Mg	7.646	47	Ag	7.576	82	Pb	7.416
13	Al	5.986	48	Cd	8.993	83	Bi	7.289
14	Si	8.151	49	In	5.786	84	Po	8.42
15	P	10.486	50	Sn	7.344	85	At	—
16	S	10.360	51	Sb	8.641	86	Rn	10.748
17	Cl	12.967	52	Te	9.099	87	Fr	—
18	Ar	15.759	53	I	10.451	88	Ra	5.279
19	K	4.341	54	Xe	12.130	89	Ac	6.9
20	Ca	6.113	55	Cs	3.894	90	Th	6.08
21	Sc	6.54	56	Ba	5.212	91	Pa	5.89
22	Ti	6.82	57	La	5.577	92	U	6.05
23	V	6.74	58	Ce	5.47	93	Np	6.19
24	Cr	6.766	59	Pr	5.42	94	Pu	6.06
25	Mn	7.435	60	Nd	5.49	95	Am	5.993
26	Fe	7.870	61	Pm	5.55	96	Cm	6.02
27	Co	7.86	62	Sm	5.63	97	Bk	6.23
28	Ni	7.635	63	Eu	5.67	98	Cf	6.30
29	Cu	7.726	64	Gd	6.14	99	Es	6.42
30	Zn	9.394	65	Tb	5.85	100	Fm	6.50
31	Ga	5.999	66	Dy	5.93	101	Md	6.58
32	Ge	7.899	67	Ho	6.02	102	No	6.65
33	As	9.81	68	Er	6.10			
34	Se	9.752	69	Tm	6.18			
35	Br	11.814	70	Yb	6.254			

From Weast 1986, p. E-76-77; other sources.

Isotope chart

The isotopes are plotted on a grid in which the neutron number increases along the abscissa and the proton number along the ordinate. All naturally occurring isotopes and isotopes important in natural processes, including element formation, are entered. If an isotope exists in more than one isomeric state, only the more stable isomer is given.

The following data, from top to bottom in each square, are given for each isotope:
—element symbol and isotopic number
—mass of the neutral atom (u)
—natural abundance (percent) in common terrestrial matter
—decay mode, if any, with its relative frequency in percent if more than one mode is present (α = alpha decay; β^+ = beta-plus decay; β^- = beta-minus decay; ϵ = electron capture; IT = isomeric transition; SF = spontaneous fission).
—half-life (y = years; d = days, h = hours; m = minutes; s = seconds; ms = milliseconds; μs = microseconds).

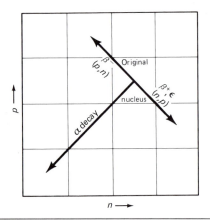

From Tuli 1985; Wapstra and Audi 1985.

286

Isotope chart cells:

- Ne22 — 21.991383 — 9.22 — | |
- Ne21 — 20.993843 — 0.27 — | |
- Ne20 — 19.992436 — 90.51 — | |
- F19 — 18.998403 — 100 — | |
- O18 — 17.999160 — 0.200 — | |
- O17 — 16.999131 — 0.038 — | |
- O16 — 15.9949146 — 99.762 — | |
- N15 — 15.000109 — 0.37 — | |
- C14 — 14.0032420 — β^- — 5730 y
- O15 — 15.003065 — β^+ — 122.2 s
- N14 — 14.0030740 — 99.63 — | |
- C13 — 13.0033548 — 1.10 — | |
- N13 — 13.005739 — β^+ — 9.965 m
- C12 — 12.0000000 — 98.90 — | |
- B11 — 11.009305 — 80.1 — | |
- Be10 — 10.013534 — β^- — $1.6 \cdot 10^6$ y
- B10 — 10.012937 — 19.9 — | |
- Be9 — 9.012182 — 100.0 — | |
- Li8 — 8.022486 — β^- — 0.838 s
- Be8 — 8.005305 — 2α — $\sim 1 \cdot 10^{-16}$ s
- Li7 — 7.016003 — 92.5 — | |
- He6 — 6.018886 — β^- — 0.807 s
- Be7 — 7.016928 — ε — 53.3 d
- Li6 — 6.015121 — 7.5 — | |
- He5 — 5.0122 — n,α — $2 \cdot 10^{-21}$ s
- Li5 — 5.0125 — p,α — $\sim 10^{-21}$ s
- He4 — 4.0026032 — 99.999862 — | |
- H3 — 3.0160493 — β^- — 12.33 y
- He3 — 3.0160293 — 0.000138 — | |
- H2 — 2.01410178 — 0.015 — | |
- n — 1.00866490 — β^- — 10.5 m
- H1 — 1.00782503 — 99.985 — | |

Chart of the Nuclides (segment)

Ca (Calcium)
- Ca40 — 39.962591 — 96.941 — | —
- Ca41 — 40.962278 — ε — 103,000 y
- Ca42 — 41.958618 — 0.647 — | —
- Ca43 — 42.958766 — 0.135 — | —
- Ca44 — 43.958766 — 2.086 — | —
- Ca45 — 44.956185 — β− — 164 d

K (Potassium)
- K39 — 38.963707 — 93.2581 — | —
- K40 — 39.963999 — 0.0117 — β−(89.30), ε(10.70) — 1.277·10⁹ y
- K41 — 40.961825 — 6.7302 — | —
- K42 — 41.962402 — β− — 12.36 h

Ar (Argon)
- Ar36 — 35.9675455 — 0.337 — | —
- Ar37 — 36.966776 — ε — 35.04 d
- Ar38 — 37.962732 — 0.063 — | —
- Ar39 — 38.964314 — β− — 269 y
- Ar40 — 39.962384 — 99.600 — | —

Cl (Chlorine)
- Cl34 — 33.9737629 — β+ — 32.23 m
- Cl35 — 34.9688527 — 75.77 — | —
- Cl36 — 35.9683069 — β−(98.10), ε(1.90) — 301,000 y
- Cl37 — 36.9659026 — 24.23 — | —
- Cl38 — 37.9680105 — β− — 37.2 m
- Cl39 — 38.968005 — β− — 55.6 m

S (Sulfur)
- S32 — 31.972071 — 95.02 — | —
- S33 — 32.971458 — 0.75 — | —
- S34 — 33.9678667 — 4.21 — | —
- S35 — 34.969032 — β− — 87.5 d
- S36 — 35.9670806 — 0.02 — | —
- S37 — 36.9711255 — β− — 5.05 m
- S38 — 37.971162 — β− — 2.84 h

P (Phosphorus)
- P31 — 30.973762 — 100 — | —
- P32 — 31.973907 — β− — 14.26 d
- P33 — 32.971725 — β− — 25.3 d

Si (Silicon)
- Si28 — 27.976927 — 92.23 — | —
- Si29 — 28.976495 — 4.67 — | —
- Si30 — 29.973770 — 3.10 — | —
- Si31 — 30.975362 — β− — 2.62 h
- Si32 — 31.974148 — β− — 105 y

Al (Aluminum)
- Al26 — 25.986892 — β+ — 720,000 y
- Al27 — 26.981539 — 100.0 — | —
- Al28 — 27.981910 — β− — 2.241 m

Mg (Magnesium)
- Mg24 — 23.985042 — 78.99 — | —
- Mg25 — 24.985837 — 10.00 — | —
- Mg26 — 25.982594 — 11.01 — | —
- Mg27 — 26.984341 — β− — 9.462 m
- Mg28 — 27.983877 — β− — 20.90 h

Na (Sodium)
- Na22 — 21.994434 — β+ — 2.602 y
- Na23 — 22.989768 — 100.0 — | —
- Na24 — 23.990961 — β− — 15.02 h

Ne (Neon)
- Ne20 — 19.992436 — 90.51 — | —
- Ne21 — 20.993843 — 0.27 — | —
- Ne22 — 21.991383 — 9.22 — | —

Zn

Zn63 62.933214 β⁺ — 38.1 m	Zn64 63.929145 48.6 — | | |	Zn65 64.929243 — ε — 243.9 d	Zn66 65.926035 27.9 — | | |	Zn67 66.927129 4.1 — | | |
Zn68 67.924846 18.8 — | | |	Zn69 68.926552 — IT (99.97) β⁻ (0.03) 13.76 h			

Cu

- Cu63 62.929599 69.17 — | | |
- Cu64 63.929766 β⁺ (62.90) β⁻ (37.10) 12.701 h
- Cu65 64.927793 30.83 — | | |

Ni

- Ni58 57.935346 68.27 — | | |
- Ni59 58.934349 ε 75,000 y
- Ni60 59.930788 26.10 — | | |
- Ni61 60.931058 1.13 — | | |
- Ni62 61.928346 3.59 — | | |
- Ni63 62.929670 β⁻ 100 y
- Ni64 63.927968 0.91 — | | |

Co

- Co59 58.933198 100 — | | |
- Co60 59.933820 β⁻ 5.271 y

Fe

- Fe54 53.939613 5.8 — | | |
- Fe55 54.938296 ε 2.68 y
- Fe56 55.934939 91.72 — | | |
- Fe57 56.935396 2.2 — | | |
- Fe58 57.93277 0.28 — | | |
- Fe59 58.934877 β⁻ 44.49 d
- Fe60 59.934078 β⁻ $1.49 \cdot 10^6$

Mn

- Mn53 52.941291 ε $3.7 \cdot 10^6$ y
- Mn54 53.940361 ε 312.5 d
- Mn55 54.938047 100 — | | |

Cr

- Cr50 49.946046 4.35 — | | |
- Cr51 50.944768 ε 27.70 d
- Cr52 51.940510 83.79 — | | |
- Cr53 52.940651 9.50 — | | |
- Cr54 53.938882 2.36 — | | |

V

- V50 49.947161 0.250 β⁻ $\sim 1.5 \cdot 10^{17}$ y
- V51 50.943962 99.750 — | | |

Ti

- Ti46 45.952629 8.0 — | | |
- Ti47 46.951764 7.3 — | | |
- Ti48 47.947947 73.8 — | | |
- Ti49 48.947871 5.5 — | | |
- Ti50 49.944792 5.4 — | | |

Sc

- Sc45 44.955910 100 — | | |
- Sc46 45.955170 β⁻ 83.83 d
- Sc47 46.952409 β⁻ 3.345 d

Ca

- Ca44 43.955481 2.086 — | | |
- Ca45 44.956185 β⁻ 163 d
- Ca46 45.953689 0.004 — | | |
- Ca47 46.954543 β⁻ 4.535 d
- Ca48 47.952533 0.187 — | | |

Chart of the nuclides (Z = 30 to 40 region). Each cell lists: nuclide, atomic mass, isotopic abundance or decay mode (%), decay type, and half-life.

Zr (Zirconium)

Nuclide	Mass	Abundance/Decay	Type	Half-life
Zr93	92.906474	—	β^-	$1.5 \cdot 10^6$ y
Zr92	91.905039	17.15	—	—
Zr91	90.905644	11.22	—	—
Zr90	89.904703	51.45	—	—

Y (Yttrium)

Nuclide	Mass	Abundance/Decay	Type	Half-life
Y90	89.907152	—	β^-	64.1 h
Y89	88.905849	100	—	—

Sr (Strontium)

Nuclide	Mass	Abundance/Decay	Type	Half-life
Sr90	89.907738	—	β^-	28.6 y
Sr89	88.907450	—	β^-	50.5 d
Sr88	87.905619	82.58	—	—
Sr87	86.908884	7.00	—	—
Sr86	85.909267	9.86	—	—
Sr85	84.912937	—	ϵ	64.84 d
Sr84	83.913430	0.56	—	—

Rb (Rubidium)

Nuclide	Mass	Abundance/Decay	Type	Half-life
Rb87	86.909187	27.83	β^-	$48 \cdot 10^9$ y
Rb86	85.911172	—	β^-	18.66 d
Rb85	84.911794	72.17	—	—

Kr (Krypton)

Nuclide	Mass	Abundance/Decay	Type	Half-life
Kr86	85.910616	17.3	—	—
Kr85	84.912531	—	β^-	10.72 y
Kr84	83.911507	57.0	—	—
Kr83	82.914135	11.5	—	—
Kr82	81.91348	11.6	—	—
Kr81	80.91659	—	ϵ	210.000 y
Kr80	79.91638	2.25	—	—
Kr79	78.92008	—	β^+ ϵ	35.0 h
Kr78	77.92040	0.35	—	—

Se (Selenium)

Nuclide	Mass	Abundance/Decay	Type	Half-life
Se82	81.916698	9.2	—	$1.4 \cdot 10^{20}$ y
Se81	80.917991	—	IT β^-	57.25 m
Se80	79.916520	49.7	—	—
Se79	78.918498	—	β^-	≤ 65.000 y
Se78	77.917308	23.6	—	—
Se77	76.919912	7.6	—	—
Se76	75.919212	9.0	—	—
Se75	74.922521	—	ϵ	119.77 d
Se74	73.922475	0.9	—	—

Br (Bromine)

Nuclide	Mass	Abundance/Decay	Type	Half-life
Br81	80.916289	49.31	—	—
Br80	79.918528	—	IT β^-	4.42 h
Br79	78.918336	50.69	—	—

As (Arsenic)

Nuclide	Mass	Abundance/Decay	Type	Half-life
As75	74.921594	100	—	—

Ge (Germanium)

Nuclide	Mass	Abundance/Decay	Type	Half-life
Ge76	75.921402	7.8	—	—
Ge75	74.922858	—	β^-	82.8 m
Ge74	73.921177	36.5	—	—
Ge73	72.923463	7.8	—	—
Ge72	71.922079	27.4	—	—
Ge71	70.924954	—	ϵ	11.8 d
Ge70	69.924250	20.5	—	—

Ga (Gallium)

Nuclide	Mass	Abundance/Decay	Type	Half-life
Ga71	70.924701	39.9	—	—
Ga70	69.926028	—	β^- (99.59) ϵ (0.41)	21.15 m
Ga69	68.925580	60.1	—	—

Zn (Zinc)

Nuclide	Mass	Abundance/Decay	Type	Half-life
Zn70	69.925325	0.6	—	—
Zn69	68.926552	—	IT (99.97) β^- (0.03)	13.76 h
Zn68	67.924846	18.8	—	—

Isotope chart data (each cell: nuclide, atomic mass, % abundance, decay mode, half-life):

Cadmium (Cd)

Cd106	Cd107	Cd108	Cd109	Cd110	Cd111	Cd112	Cd113
105.90646	106.90661	107.90418	108.90495	109.903005	110.904182	111.902757	112.904400
1.25	—	0.89	—	12.49	12.80	24.13	12.22
—	ε	—	ε	—	—	—	β^-
—	6.5 h	—	462.9 d	—	—	—	$9.3 \cdot 10^{15}$ y

Silver (Ag)

Ag107	Ag108	Ag109
106.90509	107.90595	108.904756
51.83	—	48.17
—	ε (91.30) IT (8.70)	—
—	130 y	—

Palladium (Pd)

Pd102	Pd103	Pd104	Pd105	Pd106	Pd107	Pd108	Pd109	Pd110
101.90563	102.90611	103.90403	104.90508	105.90348	106.90513	107.903895	108.905954	109.90517
1.02	—	11.14	22.33	27.33	—	26.46	—	11.72
—	ε	—	—	—	β^-	—	β^-	—
—	16.99 d	—	—	—	$6.5 \cdot 10^6$ y	—	13.7 h	—

Rhodium (Rh)

Rh103
102.905500
100
—
—

Ruthenium (Ru)

Ru96	Ru97	Ru98	Ru99	Ru100	Ru101	Ru102	Ru103	Ru104
95.90760	96.90756	97.90529	98.905939	99.904219	100.905582	101.904348	102.906323	103.905424
5.52	—	1.88	12.7	12.6	17.0	31.6	—	18.7
—	ε	—	—	—	—	—	β^-	—
—	2.9 d	—	—	—	—	—	39.26 d	—

Technetium (Tc)

Tc97	Tc98	Tc99
96.906364	97.907215	98.906254
—	—	—
ε	β^-	β^-
$2.6 \cdot 10^6$ y	$4.2 \cdot 10^6$ y	213,000 y

Molybdenum (Mo)

Mo92	Mo93	Mo94	Mo95	Mo96	Mo97	Mo98	Mo99	Mo100
91.906808	92.906813	93.905085	94.905841	95.904678	96.906020	97.905407	98.907711	99.90748
14.84	—	9.25	15.92	16.68	9.55	24.13	—	9.63
—	ε	—	—	—	—	—	β^-	—
—	3500 y	—	—	—	—	—	66.0 h	—

Niobium (Nb)

Nb93	Nb94	Nb95
92.906377	93.907281	94.906835
100	—	—
—	β^-	β^-
—	20,000 y	34.97 d

Zirconium (Zr)

Zr90	Zr91	Zr92	Zr93	Zr94	Zr95	Zr96
89.904703	90.905644	91.905039	92.906474	93.906315	94.908042	95.908275
51.45	11.22	17.15	—	17.38	—	2.80
—	—	—	β^-	—	β^-	—
—	—	—	$1.5 \cdot 10^6$ y	—	64.02 d	—

Chart of the nuclides (Cd–Te region).

	108	109	110	111	112	113	114	115	116	117	118	119	120	121	122	123	124	125	126	127
Te													Te120 119.90405 0.09 — — —	Te121 120.90495 — IT (88.6) ε (11.4) 154 d	Te122 121.903050 2.60 — — —	Te123 122.904271 0.91 ε — 1.3·10^{13} y	Te124 123.902818 4.82 — — —	Te125 124.904428 7.14 — — —	Te126 125.903309 18.95 — — —	Te127 126.905221 — IT (97.6) β⁻ (2.4) 109 d
Sb														Sb121 120.903821 57.3	Sb122 121.905179 — β⁻ (97.62) ε (2.38) 2.70 d	Sb123 122.904216 42.7				
Sn					Sn112 111.904827 0.97 — — —	Sn113 112.905176 — ε — 115.1 d	Sn114 113.902784 0.65 — — —	Sn115 114.903348 0.36 — — —	Sn116 115.901747 14.53 — — —	Sn117 116.902956 7.68 — — —	Sn118 117.901609 24.22 — — —	Sn119 118.903311 8.58 — — —	Sn120 119.902199 32.59 — — —	Sn121 120.904239 — IT (77.6) β⁻ (22.4) 55 y	Sn122 121.903440 4.63 — — —	Sn123 122.905722 — β⁻ — 129.2 d	Sn124 123.905274 5.79 — — —			
In						In113 112.904061 4.3 — — —	In114 113.904916 — IT (95.7) ε (4.3) 49.51 d	In115 114.903882 95.7 β⁻ — 441·10^{12} y											
Cd	Cd108 107.90418 0.89 — — —	Cd109 108.90495 — ε — 462.9 d	Cd110 109.903005 12.49 — — —	Cd111 110.904182 12.80 — — —	Cd112 111.902757 24.13 — — —	Cd113 112.904400 12.22 β⁻ — 9.3·10^{15} y	Cd114 113.903357 28.73 — — —	Cd115 114.905430 — β⁻ — 44.6 d	Cd116 115.904755 7.49 — — —										

	130	131	132	133	134	135	136	137	138
Ba	Ba130 129.90628 0.106 — —	Ba131 130.90690 ε 11.8 d	Ba132 131.90504 0.101 — —	Ba133 132.90599 ε 10.74 y	Ba134 133.90449 2.417 — —	Ba135 134.90566 6.592 — —	Ba136 135.90455 7.854 — —	Ba137 136.90581 11.23 — —	Ba138 137.90523 71.70 — —

	131	132	133	134	135
Cs	Cs131 130.90544 ε 9.69 d	Cs132 131.90643 ε(98) β⁻(2) 6.47 d	Cs133 132.90543 100 — —	Cs134 133.90670 β⁻ 2.062 y	Cs135 134.90588 β⁻ $3 \cdot 10^6$ y

	124	125	126	127	128	129	130	131	132	133	134	135	136
Xe	Xe124 123.905894 0.10 — —	Xe125 124.906397 ε 16.9 h	Xe126 125.90428 0.09 — —	Xe127 126.90518 ε 36.4 d	Xe128 127.903531 1.91 — —	Xe129 128.904780 26.4 — —	Xe130 129.903509 4.1 — —	Xe131 130.905072 21.2 — —	Xe132 131.904144 26.9 — —	Xe133 132.90589 β⁻ 5.24 d	Xe134 133.90540 10.4 — —	Xe135 134.90713 β⁻ 9.09 h	Xe136 135.90721 8.9 — —

	125	126	127	128	129
I	I125 124.904620 ε 60.1 d	I126 125.905624 ε(56.3) β⁻(43.7) 13.0 d	I127 126.904473 100 — —	I128 127.905810 β⁻(93.1) ε(6.9) 25.0 m	I129 128.904986 β⁻ $15.7 \cdot 10^6$ y

	122	123	124	125	126	127	128	129	130
Te	Te122 121.903050 2.60 — —	Te123 122.904271 0.91 ε $1.3 \cdot 10^{13}$ y	Te124 123.902818 4.82 — —	Te125 124.904428 7.14 — —	Te126 125.903309 18.95 — —	Te127 126.905221 β⁻(97.6) IT(2.4) 109 d	Te128 127.904463 31.69 β⁻ $>8 \cdot 10^{24}$ y	Te129 128.906594 IT(64) β⁻(36) 33.4 d	Te130 129.906229 33.80 β⁻ $2.5 \cdot 10^{21}$ y

Chart of the nuclides (Gd, Eu, Sm, Pm, Nd, Pr, Ce, La, Ba)

Gd (Gadolinium)

Gd152	Gd153	Gd154	Gd155	Gd156	Gd157
151.919786	152.921745	153.920861	154.922618	155.922118	156.923956
0.20	—	2.18	14.80	20.47	15.65
α	ε	—	—	—	—
$1.1 \cdot 10^{14}$ y	241.6 d	—	—	—	—

Eu (Europium)

Eu151	Eu152	Eu153
150.919847	151.921742	152.921225
47.8	ε(72.08) β−(27.92)	52.2
—	13.33 y	—
—		—

Sm (Samarium)

Sm144	Sm145	Sm146	Sm147	Sm148	Sm149	Sm150	Sm151	Sm152	Sm153	Sm154
143.911998	144.913409	145.91305	146.914894	147.914819	148.917180	149.917273	150.919929	151.919728	152.922094	153.922205
3.1	—	—	15.0	11.3	13.8	7.4	—	26.7	—	22.7
—	ε	α	α	α	α	—	β−	—	β−	—
—	340 d	$103 \cdot 10^{6}$ y	$106 \cdot 10^{9}$ y	$7 \cdot 10^{15}$ y	$>2 \cdot 10^{15}$ y	—	90 y	—	46.7 h	—

Pm (Promethium)

Pm143	Pm144	Pm145	Pm146	Pm147	Pm148	Pm149
142.910930	143.912588	144.912743	145.91471	146.915135	147.91747	148.918332
—	—	—	—	—	—	—
ε	ε	ε	ε(66.1) β−(33.9)	β−	β−(95.4) IT(4.6)	β−
265 d	363 d	17.7 y	5.53 y	2.6234 y	41.3 d	53.1 h

Nd (Neodymium)

Nd142	Nd143	Nd144	Nd145	Nd146	Nd147	Nd148	Nd149	Nd150
141.907719	142.909810	143.910083	144.912570	145.913113	146.916097	147.916889	148.920145	149.920887
27.13	12.18	23.80	8.30	17.19	—	5.76	—	5.64
—	—	α	—	—	β−	—	β−	—
—	—	$2.1 \cdot 10^{15}$ y	$>6 \cdot 10^{16}$ y	—	10.98 d	—	103.5 m	—

Pr (Praseodymium)

Pr141
140.907647
100
—
—

Ce (Cerium)

Ce136	Ce137	Ce138	Ce139	Ce140	Ce141	Ce142
135.9071	136.9078	137.90598	138.90663	139.905433	140.908271	141.909241
0.19	—	0.25	—	88.48	—	11.08
—	IT(99.22) ε(0.78)	—	ε	—	β−	α
—	34.4 h	—	137.7 d	—	32.501 d	$>5 \cdot 10^{16}$ y

La (Lanthanum)

La137	La138	La139
136.9065	137.907105	138.906347
—	0.09	99.91
ε	ε(66.7) β−(33.3)	—
60.000 y	$128 \cdot 10^{9}$ y	—

Ba (Barium)

Ba134	Ba135	Ba136	Ba137	Ba138
133.90449	134.90566	135.90455	136.90581	137.90523
2.417	6.592	7.854	11.23	71.70
—	—	—	—	—
—	—	—	—	—

Yb

Yb168	Yb169	Yb170	Yb171	Yb172	Yb173	Yb174	Yb175
167.933894	168.935186	169.934759	170.936323	171.936378	172.938208	173.938859	174.941273
0.13	ε	3.05	14.3	21.9	16.12	31.8	β−
—	32.02 d	—	—	—	—	—	4.19 d
—		—	—	—	—	—	

Tm

Tm169 / 168.934212 / 100 / — / —

Er

Er162	Er163	Er164	Er165	Er166	Er167	Er168	Er169	Er170
161.928775	162.930030	163.929198	164.930723	165.930290	166.932046	167.932368	168.934588	169.935461
0.14	ε β+	1.61	ε	33.6	22.95	26.8	β−	14.9
—	75.0 m	—	10.36 h	—	—	—	9.40 d	—
—		—		—	—	—		—

Ho

Ho163	Ho164	Ho165
162.928731	163.930285	164.930319
ε	ε (58) β− (42)	100
>10 y	29 m	—

Dy

Dy156	Dy157	Dy158	Dy159	Dy160	Dy161	Dy162	Dy163	Dy164
155.92428	156.92546	157.924403	158.925735	159.925193	160.926930	161.926795	162.928728	163.929171
0.06	ε	0.10	ε	2.34	18.9	25.5	24.9	28.2
—	8.1 h	—	144.4 d	—	—	—	—	—
—		—		—	—	—	—	—

Tb

Tb157	Tb158	Tb159
156.924023	157.925411	158.925342
ε	ε (82) β− (18)	100
150 y	150 y	—
		—

Gd

Gd154	Gd155	Gd156	Gd157	Gd158	Gd159	Gd160
153.920861	154.922618	155.922118	156.923956	157.924099	158.926384	159.927049
2.18	14.80	20.47	15.65	24.84	β−	21.86
—	—	—	—	—	18.56 h	—
—	—	—	—	—		—

Chart of the nuclides (segment covering Yb–Pt).

Nuclide	Atomic mass	Abundance / %	Decay	Half-life
Pt195	194.964766	33.8	—	—
Pt194	193.962655	32.9	—	—
Pt193	192.962977	—	ε	50 y
Pt192	191.961019	0.79	—	—
Pt191	190.96166	—	ε	2.9 d
Pt190	189.95992	0.01	α	6·10¹¹ y
Ir193	192.962917	62.7	—	—
Ir192	191.962580	—	IT	241 y
Ir191	190.960584	37.3	—	—
Os192	191.961467	41.0	—	—
Os191	190.960920	—	β⁻	15.4 d
Os190	189.958436	26.4	—	—
Os189	188.958137	16.1	—	—
Os188	187.955830	13.3	—	—
Os187	186.955741	1.6	—	—
Os186	185.953830	1.58	α	2.0·10¹⁵ y
Os185	184.954041	—	ε	93.6 d
Os184	183.952488	0.02		>1·10¹⁷ y
Re187	186.955744	62.60	β⁻	50·10⁹ y
Re186	185.954984	—	IT	200.000 y
Re185	184.952951	37.40	—	—
W186	185.954357	28.6	—	—
W185	184.953416	—	β⁻	75.1 d
W184	183.950928	30.67	—	—
W183	182.950220	14.3	—	—
W182	181.948202	26.3	—	—
W181	180.94819	—	ε	121.0 d
W180	179.946701	0.13		>1.1·10¹⁵ y
Ta181	180.947992	99.988	—	—
Ta180	179.947462	0.012	ε	>1.2·10¹⁵ y
Hf180	179.946546	35.100	—	—
Hf179	178.945812	13.629	—	—
Hf178	177.943696	27.297	—	—
Hf177	176.943217	18.606	—	—
Hf176	175.941406	5.206	—	—
Hf175	174.941507	—	ε	70 d
Hf174	173.940044	0.162	α	2.0·10¹⁵ y
Lu176	175.942679	2.59	β⁻	36·10⁹ y
Lu175	174.940770	97.41	—	—
Yb176	175.942564	12.7	—	—
Yb175	174.941273	—	β⁻	4.19 d
Yb174	173.938859	31.8	—	—
Yb173	172.938208	16.12	—	—
Yb172	171.936378	21.9	—	—

Isotope	Mass	Abundance / %	Decay	Half-life
Pb203	202.973365		ϵ	51.88 h
Pb204	203.973020	1.4	α	$\geq 1.4\cdot10^{17}$ y
Pb205	204.974458		ϵ	$15.2\cdot10^{6}$ y
Pb206	205.974440	24.1	—	—
Pb207	206.975872	22.1	—	—
Pb208	207.976627	52.4	—	—
Pb209	208.981065		β^-	3.25 h
Pb210	209.984163		β^-	22.3 y
Pb211	210.988735		β^-	36.1 m
Pb212	211.991871		β^-	10.64 h
Pb213	212.9965		β^-	10.2 m
Tl203	202.972320	29.524	—	—
Tl204	203.973839		β^- (97.45) ϵ (2.55)	3.78 y
Tl205	204.974401	70.476	—	—
Tl206	205.976084		β^-	4.20 m
Tl207	206.977404		β^-	4.77 m
Tl208	207.98199		β^-	3.053 m
Tl209	208.98533		β^-	2.2 m
Tl210	209.99006		β^-	1.30 m
Hg196	195.965807	0.14	—	—
Hg197	196.967187		ϵ	64.1 h
Hg198	197.966743	10.02	—	—
Hg199	198.968254	16.84	—	—
Hg200	199.968300	23.13	—	—
Hg201	200.970277	13.22	—	—
Hg202	201.970617	29.80	—	—
Hg203	202.972848		β^-	46.60 d
Hg204	203.973467	6.85	—	—
Hg205	204.976047		β^-	5.2 m
Hg206	205.977489		β^-	8.15 m
Au197	196.966543	100	—	—
Au198	197.968217		β^-	2.696 d
Pt194	193.962655	32.9	—	—
Pt195	194.964766	33.8	—	—
Pt196	195.964926	25.3	—	—
Pt197	196.967315		β^-	18.3 h
Pt198	197.967787	7.2	—	—

Nuclide chart (each cell lists: nuclide, atomic mass, decay mode(s) with percentages, half-life)

Element	A=205	206	207	208	209	210	211	212	213	214	215	216	217	218	219	220	221
Rn														Rn218 218.005580 α — 35 ms	Rn219 219.009479 α — 3.96 s	Rn220 220.01137 α — 55.6 s	Rn221 221.0155 β⁻ (78) α (22) 25 m
At											At215 214.99864 α — 0.10 ms	At216 216.00239 α — 0.30 ms	At217 217.00469 α — 32.3 ms	At218 218.00868 α (99.9) β⁻ (0.1) ~2 s	At219 219.0113 α (97) β⁻ (3) 0.9 m		
Po				Po208 207.981222 α — 2.898 y	Po209 208.982404 α (99.74) ε (0.26) 102 y	Po210 209.982848 α — 138.376 d	Po211 210.986627 α — 0.516 s	Po212 211.988842 α — 45 s	Po213 212.992833 α — 4.2 μs	Po214 213.995176 α — 164.3 μs	Po215 214.999419 α — 1.780 ms	Po216 216.00189 α — 0.15 s	Po217 217.0063 α — <10 s	Po218 218.008966 α (99.98) β⁻ (0.02) 3.11 m			
Bi		Bi206 205.978478 ε — 6.243 d	Bi207 206.978446 ε — 32.2 y	Bi208 207.979717 ε — 368,000 y	Bi209 208.980374 100 — —	Bi210 209.984095 β⁻ — 5.013 d	Bi211 210.98725 α (99.72) β⁻ (0.28) 2.14 m	Bi212 211.99125 β⁻ (64.06) α (35.94) 60.55 m	Bi213 212.99436 β⁻ (97.84) α (2.16) 45.6 m	Bi214 213.99869 β⁻ (99.98) α (0.02) 19.9 m	Bi215 215.001836 β⁻ — 7.4 m						
Pb	Pb205 204.974458 — — 15.2·10⁶ y	Pb206 205.974440 24.1 — —	Pb207 206.975872 22.1 — —	Pb208 207.976627 52.4 — —	Pb209 208.981065 β⁻ — 3.25 h	Pb210 209.984163 β⁻ — 22.3 y	Pb211 210.988735 β⁻ — 36.1 m	Pb212 211.991871 β⁻ — 10.64 h	Pb213 212.9965 β⁻ — 10.2 m	Pb214 213.999798 β⁻ — 26.8 m							

Isotope	Mass	Abundance / branching	Decay	Half-life
Np237	237.048168		α	2.14·10⁶ y
Np238	238.050941		β−	2.117 d
Np239	239.052933		β−	2.355 d
Np240	240.0560		β−	61.9 m
U232	232.03713		α	69 y
U233	233.039628		α	159.200 y
U234	234.040947	0.0055	α	245.000 y
U235	235.043924	0.7200	α	704·10⁶ y
U236	236.045563		α	23.42·10⁶ y
U237	237.048725		β−	6.75 d
U238	238.050785	99.2745	α	4.468·10⁹ y
U239	239.054290		β−	23.5 m
Pa231	231.035880		α	32.760 y
Pa232	232.03856		β−	1.31 d
Pa233	233.040242		β−	27.0 d
Pa234	234.04330		β−	6.70 h
Th226	226.02488		α	31 m
Th227	227.027703		α	18.718 d
Th228	228.028715		α	1.913 y
Th229	229.031755		α	7340 y
Th230	230.033128		α	75.380 y
Th231	231.036298		β−	25.52 h
Th232	232.038051	100	α	14.05·10⁹ y
Th233	233.041577		β−	22.3 m
Th234	234.04359		β−	24.10 d
Ac225	225.02320		α	10.0 d
Ac226	226.026084	β− (82.79) ε (17.20) α (0.01)		29 h
Ac227	227.027750	β− (98.62) α (1.38)		21.773 y
Ac228	228.031015		β−	6.13 h
Ra222	222.01535		α	38 s
Ra223	223.018501		α	11.434 d
Ra224	224.020186		α	3.66 d
Ra225	225.023604		β−	14.8 d
Ra226	226.025403		α	1600 y
Ra227	227.029171		β−	42.2 m
Ra228	228.031064		β−	5.75 y
Fr221	221.01423		α	4.9 m
Fr222	222.0175		β−	14.4 m
Fr223	223.019733	β− (99.99) α (0.01)		21.8 m
Rn218	218.005580		α	35 ms
Rn219	219.009479		α	3.96 s
Rn220	220.01137		α	55.6 s
Rn221	221.0155	β− (78) α (22)		25 m
Rn222	222.017571		α	3.8235 d
At217	217.00469		α	32.3 ms
At218	218.00868	α (99.90) β− (0.10)		2 s
At219	219.0113	α (97) β− (3)		0.9 m

Md258
258.0986
—
α
55 d

Fm257
257.09510
—
α (99.79)
SF (0.21)
100.5 d

Fm256
256.09177
—
SF (91.9)
α (8.1)
2.63

Es255
255.0903
—
β⁻ (92)
α (8)
40 d

Cf254
254.08732
—
SF (99.69)
α (0.31)
60.5 d

Fm255
255.08995
—
α
20.1 h

Es254
254.08802
—
α
275.5 d

Cf253
253.08513
—
β⁻ (99.69)
α (0.31)
17.8 d

Fm254
254.08685
—
α (99.94)
SF (0.06)
3.240 d

Es253
253.084818
—
α
20.47 d

Cf252
252.081621
—
α (96.91)
SF (3.09)
2.64 y

Cm250
250.07835
—
SF (~65)
α (~28)
β⁻ (~7)
7400 y

Fm253
253.085173
—
ε (88)
α (12)
3.0 d

Cf251
251.079580
—
α
900 y

Bk250
250.07831
—
β⁻
3.22 h

Cm249
249.075948
—
β⁻
64.15 m

Cf250
250.076400
—
α (99.92)
SF (0.08)
13.1 y

Bk249
249.074980
—
β⁻
320 d

Cm248
248.072343
—
α (91.74)
SF (8.26)
340,000 y

Pu246
246.07017
—
β⁻
10.85 d

Cf249
249.074845
—
α
351 y

Bk248
248.07311
—
α (>70)
β⁻ (<30)
>9 y

Cm247
247.070347
—
α
15.6·10⁶ y

Am246
246.06977
—
—
39 m

Pu245
245.06782
—
β⁻
10.5 h

Cf248
248.072183
—
α
334 d

Bk247
247.070300
—
α
1400 y

Cm246
246.067218
—
α (99.97)
SF (0.03)
4700 y

Am245
245.066444
—
β⁻
2.05 h

Pu244
244.064199
—
α (99.88)
SF (0.12)
80.8·10⁶ y

Cf247
247.0710
—
ε (99.96)
α (0.03)
3.11 h

Bk246
246.0687
—
ε
1.80 d

Cm245
245.065483
—
α
8500 y

Am244
244.064279
—
β⁻
10.1 h

Pu243
243.061998
—
β⁻
4.956 h

Cf246
246.068800
—
α
35.7 h

Bk245
245.066357
—
ε (99.88)
α (0.12)
4.94 d

Cm244
244.062747
—
α
18.10 y

Am243
243.061375
—
α
7380 y

Pu242
242.058737
—
α
376,000 y

U240
240.056587
—
β⁻
14.1 h

Cm243
243.061382
—
α (99.76)
ε (0.24)
28.5 y

Am242
242.059542
—
β⁻ (82.7)
ε (17.3)
16.02 h

Pu241
241.056846
—
β⁻
14.35 y

Np240
240.0560
—
β⁻
61.9 m

U239
239.054290
—
β⁻
23.5 m

Cm242
242.058830
—
α
162.8 d

Am241
241.056823
—
α
432 y

Pu240
240.053808
—
α
6570 y

Np239
239.052933
—
β⁻
2.355 d

U238
238.050785
99.2745
α
4.468·10⁹ y

Pu239
239.052158
—
α
24,120 y

Np238
238.050941
—
β⁻
2.117 d

U237
237.048725
—
β⁻
6.75 d

Pu238
238.049555
—
α
87.74 y

Np237
237.048168
—
α
2.14·10⁶ y

U236
236.045563
—
α
23.42·10⁶ y

U235
235.043924
0.7200
α
704·10⁶ y

U234
234.040947
0.0055
α
245,000 y

Length

Egyptian
cubit   52.4 cm

Greek
cubit   52.67 cm
digit $= 1/40$ of $(2 \text{ cubit}^2)^{1/2}$ 1.86 cm
foot $= 3/5$ of cubit   31.6 cm
stadion $= 600$ ft   189.60 m

Roman
digit   1.84 cm
palm $= 4$ digits   7.37 cm
foot $= 4$ palms   29.49 cm
step $= 5$ feet   1.475 m
stadium $= 125$ steps   184.31 m
mile $= 1000$ steps   1475.0 m

Capacity

Egyptian
ro   59.6 cm^3
hen $= 8$ ro   477 cm^3
hennu $= 4$ hen   1.908 liter
apt $= 10$ hennu   19.08 liter
tama $= 4$ apt   76.32 liter
sa $= 25$ tama   1.908 m^3

Greek
kyathos   47.5 cm^3
kotyle $= 6$ kyathoi   285 cm^3
khous $= 12$ kotylai   3.420 liter
metretes $= 12$ khoes   41.04 liter

Roman
quartarius   145.25 cm^3
sextarius $= 4$ quartarii   581 cm^3
conglus $= 6$ sextarii   3.486 liter
urna $= 4$ conglii   13.944 liter
amphora $= 2$ urnae   27.888 liter

Weight

Egyptian
shekel   8.40 g

Greek
khalkous   0.09 g
obelos $= 8$ khalkoi   0.72 g
drachma $= 6$ obeloi   4.34 g
mina $= 100$ drachmai   434 g
talanton $= 60$ minai   26.040 kg

Measures
ancient *(continued)*

Weight *(continued)*

Roman	
siliqua	0.19 g
scripulum = 6 siliquae	1.13 g
sextula = 4 scripula	4.54 g
uncia = 6 sextulae	27.27 g
libra = 12 unciae	327.24 g

From Petrie 1952, v. 15, p. 142–145.

Meteorites
classification and mineralogy

Name	Major minerals
Stony (92.8%)	
chondrites (85.7%)	
carbonaceous (5.7%)	olivine, pyroxene, serpentine, sulfates, organic compounds, 1–9% H_2O
ordinary (67.6%)	pyroxenes, olivine, plagioclase, Fe–Ni
other (12.4%)	pyroxenes, olivine, Fe–Ni
achondrites (7.1%)	
Ca-rich (4.7%)	pyroxene, olivine, Ca-plagioclase
Ca-poor (2.4%)	pyroxene, olivine
Stony-iron (1.5%)	
mesosiderites (0.9%)	pyroxene, plagioclase, Fe–Ni, troilite
pallasites (0.5%)	olivine, Fe–Ni, troilite, schreibersite, chromite
others (0.1)	olivine, pyroxene, Fe–Ni
Iron (5.7%)	
hexahedrites (0.6%)	kamacite
octahedrites (4.3%)	kamacite, taenite
ataxites (0.8%)	kamacite, taenite

From Glass 1982, p. 96, Table 4.2.

Physical properties. Xl system = crystal system (C = cubic; H = hexagonal; M = monoclinic; O = ortho-rhombic; T = trigonal; Tc = triclinic; Tt = tetragonal; A = amorphous). G = density (relative to water at 3.98°C). H = hardness. n = refractive index.

Name	Composition	Xl System	G	H	n	Remarks
Acanthite	Ag_2S	M,C	7.3	2–2½	—	—
Acmite	$NaFe^{3+}(SiO_3)_2$	M	3.5	6–6½	1.82	—
Actinolite	$Ca_2(Mg,Fe)_5Si_8O_{22}(OH)_2$	M	3.1–3.3	5–6	1.65	—
Adularia	$KAlSi_3O_8$	M	—	—	—	Translucent
Aegirine	—	—	—	—	—	Acmite
Agate	—	—	—	—	—	Concentrically layered chalcedony
Alabaster	—	—	—	—	—	Cryptocrystalline gypsum
Albite	$NaAlSi_3O_8$	Tc	2.62	6	1.53	Na end–member of plagioclase series
Alexandrite	—	—	—	—	—	Gem chrysoberyl
Alkali feldspar	—	—	—	—	—	Na or K feldspar
Allanite	$(Ce,Ca)_3(Fe^{2+},Fe^{3+})Al_2O\cdot(SiO_4)(Si_2O_7)(OH)$	M	3.5–4.2	5½–6	1.70–1.81	—
Almandine	$Fe_3Al_2Si_3O_{12}$	C	4.32	7	1.83	A garnet
Almandite	—	—	—	—	—	Almandine
Alunite	$KAl_3(SO_4)_2(OH)_6$	T	2.6–2.8	4	1.57	—
Amalgam	—	—	—	—	—	An alloy of Hg with Ag or Au
Amblygonite	$LiAlFPO_4$	Tc	3.0–3.1	6	1.60	—
Amethyst	SiO_2	—	—	—	—	Purple quartz
Amphiboles	$Q_{2-3}R_5(Si,Al)_8O_{22}(OH)_2$	—	—	—	—	A group of minerals where Q = Mg,Fe²⁺,Ca,Na R = Mg,Fe²⁺,Fe³⁺,Al
Analcime	$NaAlSi_2O_6\cdot H_2O$	C	2.27	5–5½	1.48–1.49	—
Anatase	TiO_2	Tt	3.9	5½–6	2.6	—
Andalusite	Al_2SiO_5	O	3.14–3.20	7½	1.64	—
Andesine	$Ab_{70}An_{30}-Ab_{50}-An_{50}$	Tc	2.69	6	1.55	A plagioclase
Andradite	$Ca_3Fe_2Si_3O_{12}$	C	3.86	7	1.89	A garnet
Anglesite	$PbSO_4$	O	6.2–6.4	3	1.88	—
Anhydrite	$CaSO_4$	O	2.89–2.98	3–3½	1.58	—
Ankerite	$CaFe(CO_3)_2$	T	2.95–3	3½	1.70–1.75	—
Anorthite	$CaAl_2Si_2O_8$	Tc	2.76	6	1.58	Ca end–member of the plagioclase series
Anorthoclase	$(K,Na)AlSi_3O_8-NaAlSi_3O_8$	Tc	2.58	6	1.53	An alkali feldspar
Anthophyllite	$(Mg,Fe)_7Si_8O_{22}(OH)_2$	O	2.85–3.2	5½–6	1.61–1.71	An amphibole
Antigorite	$Mg_3Si_2O_5(OH)_4$	M	2.5–2.6	4	1.55	Platy serpentine
Antimony	Sb	R	6.7	3	—	—
Antlerite	$Cu_3SO_4(OH)_4$	O	3.9	3½–4	1.74	—
Apatite	$Ca_5(PO_4)_3(F,Cl,OH)$	H	3.15–3.20	5	1.63	—

Minerals *(continued)*

Physical properties. Xl system = crystal system (C = cubic; H = hexagonal; M = monoclinic; O = orthorhombic; T = trigonal; Tc = triclinic; Tt = tetragonal; A = amorphous). G = density (relative to water at 3.98°C). H = hardness. n = refractive index.

Name	Composition	Xl System	G	H	n	Remarks
Apophyllite	$KCa_4(Si_4O_{10})_2F \cdot 8H_2O$	Tt	2.3–2.4	4½–5	1.54	—
Aquamarine	—	—	—	—	—	Gem beryl
Aragonite	$CaCO_3$	O	2.93	3½–4	1.68	—
Argentite	—	—	—	—	—	Acanthite
Arsenic	As	T	5.7	3½	—	—
Arsenopyrite	FeAsS	M	6.07	5½–6	—	—
Asbestos	—	—	—	—	—	Commercial name for a group of fibrous silicates (see chrysotile, crocidolite)
Atacamite	$Cu_2Cl(OH)_3$	O	3.75–3.77	3–3½	1.86	—
Attapulgite	—	—	—	—	—	Palygorskite
Augite	$(Ca,Na)(Mg,Fe,Al)[(Si,Al)O_3]_2$		M3.2–3.5	5–6	1.67–1.73	A pyroxene
Autunite	$Ca(UO_2)_2(PO_4)_2 \cdot 10–12H_2O$	Tt	3.1–3.2	2–2½	1.58	—
Azurite	$Cu_3(CO_3)_2(OH)_2$	M	3.77	3½–4	1.76	—
Baddeleyite	ZrO_2	M	5.4–6.0	6.5	2.13–2.20	—
Barite	$BaSO_4$	O	4.5	3–3½	1.64	—
Bauxite	—	—	—	—	—	A mixture of Al oxides and hydroxides
Beidellite	$(Ca,Na)_{0.3}Al_2(OH)_2(Al,Si)_4O_{10} \cdot (H_2O)_4$	M	2–3	1–2	—	Member of the montmorillonite group
Bentonite	—	—	—	—	—	Montmorillonite and colloidal silica produced by devitrification of volcanic ash
Beryl	$Be_3Al_2(Si_6O_{18})$	H	2.64–2.8	7½–8	1.57–1.61	—
Biotite	$K(Mg,Fe)_3(AlSi_3O_{10})(OH)_2$	M	2.8–3.2	2½–3	1.61–1.70	The black mica
Bismuth	Bi	T	9.8	2–2½	—	—
Bismuthinite	Bi_2S_3	O	6.78	2	—	—
Bloodstone	—	—	—	—	—	Heliotrope
Boehmite	$\gamma AlO(OH)$	O	3.1–3.6	3½–4	1.65	γ-phase of diaspore
Bog iron ore	—	—	—	—	—	Iron hydroxides, mainly limonite
Boracite	$Mg_3ClB_7O_{13}$	O	2.9–3.0	7	1.66	—
Borax	$Na_2B_4O_5(OH)_4 \cdot H_2O$	M	1.7	2–2½	1.47	—
Bornite	Cu_5FeS_4	Tt,C	5.06–5.08	3	—	—
Brochantite	$Cu_4SO_4(OH)_6$	M	3.9	3½–4	1.78	—
Bronzite	$(Mg,Fe)SiO_3$	O	3.3	5½	1.68	An orthopyroxene
Brucite	$Mg(OH)_2$	T	2.39	2½	1.57	—
Bytownite	$Ab_{30}An_{70}–Ab_{10}An_{90}$	Tc	2.74	6	1.57	A plagioclase

Physical properties. Xl system = crystal system (C = cubic; H = hexagonal; M = monoclinic; O = ortho-rhombic; T = trigonal; Tc = triclinic; Tt = tetragonal; A = amorphous). G = density (relative to water at 3.98°C). H = hardness. n = refractive index.

Name	Composition	Xl System	G	H	n	Remarks
Calcite	$CaCO_3$	T	2.71	3	1.66	—
Cancrinite	$Na_6Ca(CO_3)(AlSiO_4)_6 \cdot 2H_2O$	H	2.45	5–6	1.52	A feldspathoid
Carnallite	$KMgCl_3 \cdot 6H_2O$	O	1.6	1	1.48	—
Carnelian	—	—	—	—	—	Improper spelling for Cornelian
Carnotite	$K_2(UO_2)_2(VO_4)_2 \cdot 3H_2O$	M	4.7–5	—	1.93	Powdery incrustations
Cassiterite	SnO_2	Tt	6.8–7.1	6–7	2.00	—
Cat's eye	—	—	—	—	—	Gem variety of chrysoberyl
Celestite	$SrSO_4$	O	3.95–3.97	3–3½	1.62	—
Cerargyrite	$AgCl$	C	5.5	2–3	2.07	—
Cerussite	$PbCO_3$	O	6.55	3–3½	2.08	—
Chabazite	$Ca_2Al_2Si_4O_{12} \cdot 6H_2O$	T	2.05–2.15	4–5	1.48	—
Chalcanthite	$CuSO_4 \cdot 5H_2O$	Tc	2.12–2.30	2½	1.54	—
Chalcedony	—	—	—	—	—	Semitranslucent microcrystalline quartz
Chalcocite	Cu_2S	O,H	5.5–5.8	2½–3	—	—
Chalcopyrite	$CuFeS_2$	Tt	4.1–4.3	3½–4	—	—
Chalcosiderite	$CuFe_6(PO_4)_4(OH)_8 \cdot 4H_2O$	Tc	3.22	4½	1.84	—
Chalk	—	—	—	—	—	An aggregate of small calcitic particles
Chert	SiO_2	—	2.65	7	1.54	Opaque, compact, microcrystalline quartz
Chiastolite	—	—	—	—	—	A variety of andalusite with carbonaceous impurities
Chlorapatite	$Ca_5(PO_4)_3Cl$	—	—	—	—	A Cl-rich apatite
Chlorargyrite	$AgCl$	C	5.5	2–3	2.07	—
Chlorite	$(Mg,Fe)_3(Si,Al)_4O_{10} \cdot (OH)_2(Mg,Fe)_3(OH)_6$	M,Tc	2.6–3.3	2–2½	1.57–1.67	—
Chloritoid	$(Fe,Mg)Al_4O_2(SiO_4)_2(OH)_4$	M,Tc	3.5–3.8	6½	1.72–1.73	—
Chondrodite	$Mg_5(SiO_4)_2(F,OH)_2$	M	3.1–3.2	6–6½	1.60–1.63	—
Chromite	$FeCr_2O_4$	C	4.6	5½	2.16	—
Chrysoberyl	$BeAl_2O_4$	O	3.65–3.8	8½	1.75	—
Chrysocolla	$Cu_2H_2(Si_2O_5)(OH)_4$	—	2.0–2.4	2–4	1.4	Cryptocrystalline or amorphous
Chrysolite	—	—	—	—	—	A variety of olivine with 10–30 mol percent of Fe_2SiO_4
Chrysotile	$Mg_3Si_2O_5(OH)_4$	M	2.5–2.6	4	1.55	Fibrous serpentine (an asbestos)

Minerals (continued)

Physical properties. Xl system = crystal system (C = cubic; H = hexagonal; M = monoclinic; O = ortho-rhombic; T = trigonal; Tc = triclinic; Tt = tetragonal; A = amorphous). G = density (relative to water at 3.98°C). H = hardness. n = refractive index.

Name	Composition	Xl System	G	H	n	Remarks
Cinnabar	H_2S	T	8.10	2½	2.81	—
Citrine	SiO_2	—	—	—	—	Yellow quartz
Clinoenstatite	$MgSiO_3$	M	3.19	6	1.66	A clinopyroxene
Clinohypersthene	$(Mg,Fe)SiO_3$	M	3.4–3.5	5–6	1.68–1.72	A clinopyroxene
Clinozoisite	$Ca_2Al_3O(SiO_4)Si_2O_7(OH)$	M	3.25–3.37	6–6½	1.67–1.72	—
Cobaltite	$(Co,Fe)AsS$	O	6.33	5½	—	—
Coesite	SiO_2	M	2.915	7	1.49	High-pressure phase of silica
Colemanite	$CaB_3O_4(OH)_3 \cdot H_2O$	M	2.42	4–4½	1.59	—
Collophane	—	—	—	—	—	Cryptocrystalline apatite
Columbite	$(Fe,Mn)Nb_2O_6$	O	5.2–7.3	6	—	—
Copper	Cu	C	8.9	2½–3	—	—
Cordierite	$(Mg,Fe)_2Al_4Si_5O_{18} \cdot nH_2O$	O	2.60–2.66	7–7½	1.53–1.57	—
Cornelian	—	—	—	—	—	Red chalcedony
Corundum	Al_2O_3	T	3.99	9	1.77	—
Coulsonite	FeV_2O_4	C	—	5½–6½	—	A spinel
Covellite	CuS	H	4.60–4.76	1½–2	—	—
Cristobalite	SiO_2	Tt,C	2.32	6½	1.48	High-temperature phase of silica
Crocidolite	$Na(Mg,Fe^{2+})_3Fe_2^{3+}Si_8O_{22}(OH)_2$	M	3.2–3.3	4	1.70	Blue amphibole asbestos
Crocoite	$PbCrO_4$	M	5.9–6.1	2½–3	2.36	—
Cryolite	Na_3AlF_6	M	2.95–3.0	2½	1.34	—
Cummingtonite	$(Mg,Fe)_7Si_8O_{22}(OH)_2$	M	3.1–3.3	5½–6	1.66–1.68	—
Cuprite	Cu_2O	C	6.1	3½–4	—	—
Dahllite	$Ca_5(PO_4,CO_3)_3(OH)$	H	3.2–3.3	5	—	Carbonate-hydroxylapatite; cf. francolite
Datolite	$CaB(SiO_4)(OH)$	M	2.8–3.0	5–5½	1.65	—
Diallage	—	—	—	—	—	A lamellar variety of augite or diopside
Diamod	C	C	3.51	10	2.42	High-pressure phase of graphite
Diaspore	$\alpha AlO(OH)$	O	3.35–3.45	6½–7	1.72	Cf. Boehmite
Diatomaceous earth	—	—	—	—	—	Diatomite
Diatomite	—	—	—	—	—	An aggregated of diatom frustules
Diopside	$CaMg(SiO_3)_2$	M	3.3	5–6	1.67	A clinopyroxene
Dioptase	$Cu_6(Si_6O_{18}) \cdot 6H_2O$	H	3.3	5	1.65	—
Dolomite	$CaMg(CO_3)_2$	T	2.87	3½–4	1.68	—
Electron	—	—	—	—	—	See Electrum

Physical properties. Xl system = crystal system (C = cubic; H = hexagonal; M = monoclinic; O = orthorhombic; T = trigonal; Tc = triclinic; Tt = tetragonal; A = amorphous). G = density (relative to water at 3.98°C). H = hardness. n = refractive index.

Name	Composition	Xl System	G	H	n	Remarks
Electrum	—	—	—	—	—	A natural alloy of Au (80%) and Ag (20%)
Emerald	—	—	—	—	—	Gem beryl
Emery	—	—	—	—	—	Corundum with Fe oxides
Enargite	Cu_3AsS_4	O	4.45	3	—	—
Enstatite	$MgSiO_3$	O	3.2	5½	1.65	—
Epidote	$Ca_2(Al,Fe)Al_2O(SiO_4)\cdot(Si_2O_7)(OH)$	M	3.25–3.50	6–7	1.72–1.78	—
Epsomite	$MgSO_4\cdot7H_2O$	O	1.75	2–2½	1.46	—
Epsom salt	—	—	—	—	—	See Epsomite
Euclase	$BeAl(SiO_4)(OH)$	M	3.1	7½	1.66	—
Fayalite	Fe_2SiO_4	O	4.39	6½	1.86	Fe end-member of the olivine group
Feldspars	$QAl(Al,Si)Si_2O_8$	—	—	—	—	A group of minerals where Q = K,Na,Ca,Ba
Feldspathoids	—	—	—	—	—	A group of low-silica Na,K,Ca Al-silicates, the most common being leucite, nepheline, cancrinite, sodalite
Fergusonite	$(REE,Fe)NbO_4$	Tt	5.8	5½–6	2.07	—
Flint	SiO_2	—	—	—	—	Homogeneous microcrystalline quartz
Fluorapatite	$Ca_5(PO_4)_3F$	—	—	—	—	A F-rich apatite
Fluorite	CaF_2	C	3.18	4	1.43	—
Forsterite	Mg_2SiO_4	O	3.2	6½	1.63	Mg end-member of the olivine group
Francolite	$Ca_5(PO_4,CO_3)_3F$	H	3.1–3.2	5	—	Carbonate-fluorapatite; cf. dahllite
Franklinite	$(Zn,Fe,Mn)(Fe,Mn)_2O_4$	C	5.15	6	—	—
Gadolinite	$YFeBe_2(SiO_4)_2O_2$	M	4–4.5	6½–7	1.79	—
Gahnite	$ZnAl_2O_4$	C	4.55	7½–8	1.80	—
Galena	PbS	C	7.6	2½		—
Garnet	$Q_3R_2(SiO_4)_3$	C	3.5–4.3	6½–7½	1.71–1.88	A group of minerals where Q = Ca, Mg, Fe^{2+}, Mn^{2+} R = Al, Fe^{3+}, Mn^{3+}, V^{3+}, Cr, Ti, Zr
Garnierite	$(Ni,Mg)_3Si_2O_5(OH)_4$	M	2.2–2.8	2–3	1.59	—
Gaylussite	$Na_2Ca(CO_3)_2\cdot5H_2O$	M	1.99	2–3	1.52	—
Geyserite	—	—	—	—	—	Opaline silica incrustations in hot springs
Gibbsite	$Al(OH)_3$	M	2.3–2.4	2½–3½	1.57	—

Minerals *(continued)*

Physical properties. Xl system = crystal system (C = cubic; H = hexagonal; M = monoclinic; O = orthorhombic; T = trigonal; Tc = triclinic; Tt = tetragonal; A = amorphous). G = density (relative to water at 3.98°C). H = hardness. n = refractive index.

Name	Composition	Xl System	G	H	n	Remarks
Glauconite	$(K,Na)(Al,Fe,Mg)_2(Al,Si)_4 \cdot O_{10}(OH)_2$	M	2.4	2	1.62	—
Glaucophane	$Na_2(Mg,Fe)_3Al_2Si_8O_{22}(OH)_2$	M	3.1–3.3	6–6½	1.62–1.67	An amphibole
Goethite	$\alpha FeO(OH)$	O	4.37	5–5½	2.39	—
Gold	Au	C	19.32	2½–3	—	—
Goldmanite	$Ca_3V_2Si_3O_{12}$	C	—	—	—	A garnet
Graphite	C	H	2.27	1–2	—	—
Greenalite	$(Fe,Mg)_3Si_2O_5(OH)_4$	M	3.2	—	1.67	—
Greenockite	CdS	H	4.9	3–3½	—	—
Grossularite	$Ca_3Al_2Si_3O_{12}$	C	3.59	6½	1.73	A garnet
Grünerite	$Fe_7Si_8O_{22}(OH)_2$	M	3.6	6	1.71	—
Gypsum	$CaSO_4 \cdot 2H_2O$	M	2.32	2	1.52	—
Halite	NaCl	C	2.16	2½	1.54	—
Halloysite	$Al_2Si_2O_5(OH)_4$ and $Al_2Si_2O_5(OH)_4 \cdot 2H_2O$	M	2.0–2.2	1.2	1.54	A clay mineral
Harmotome	$Ba(Al_2Si_6O_{16}) \cdot 6H_2O$	M	2.45	4½	1.51	A zeolite
Hausmannite	Mn_3O_4	Tt	4.84	5½–6	2.15–2.46	—
Haüynite	$(Na,Ca)_{4-8}(AlSiO_4)_6(SO_4)_{1-2}$	C	2.4–2.5	5½–6	1.5	A feldspathoid
Hedenbergite	$CaFe(SiO_3)_2$	M	3.55	5–6	1.73	Ca end-member of the clinopyroxene group
Heliotrope	—	—	—	—	—	Red and green chalcedony.
Hematite	αFe_2O_3	R	5.27	5½–6½	—	—
Hemimorphite	$Zn_4(Si_2O_7)(OH)_2 \cdot H_2O$	O	3.4–3.5	4½–5	1.62	—
Hercynite	$FeAl_2O_4$	C	4.39	7½–8	1.80	Fe spinel
Hessite	Ag_2Te	C,M	8.4	2½–3	—	—
Heulandite	$(Na,Ca)_{2-3}Al_3(Al,Si)_2 \cdot Si_{13}O_{36} \cdot 12H_2O$	M	2.18–2.20	3½–4	1.48	—
Hornblende	$(Ca,Na)_{2-3}(Mg,Fe,Al)_5(Si,Al)_8 \cdot O_{22}(OH)_2$	M	3.0	5–6	1.62–1.72	The commonest amphibole
Hyacinth	—	—	—	—	—	Gem zircon
Hyalite	—	—	—	—	—	Colorless opal
Hydroxylapatite	$Ca_5(PO_4)_3(OH)$	—	—	—	—	—
Hypersthene	$(Mg,Fe)SiO_3$	O	3.4–3.5	5–6	1.68–1.73	An orthopyroxene
Iceland spar	—	—	—	—	—	Pure and transparent calcite
Idocrase	—	—	—	—	—	See vesuvianite
Illite	$(K,H_3O)(Al,Mg,Fe)_2 \cdot (Si,Al)_4O_{10}[(OH)_2,H_2O]$	M	2.6–2.9	1–2	1.54–1.63	Predominant clay mineral in mid-latitudes
Ilmenite	$FeTiO_3$	T	4.8	5½–6	—	—
Indicolite	—	—	—	—	—	A blue, gem variety of tourmaline
Iridium	Ir	C	22.65	6–7	—	A platinum-group metal
Iron	Fe	C	7.3–7.9	4½	—	—

Physical properties. Xl system = crystal system (C = cubic; H = hexagonal; M = monoclinic; O = ortho-rhombic; T = trigonal; Tc = triclinic; Tt = tetragonal; A = amorphous). G = density (relative to water at 3.98°C). H = hardness. n = refractive index.

Name	Composition	Xl System	G	H	n	Remarks
Jacinth	—	—	—	—	—	Improper spelling of Hyacinth
Jade	—	—	—	—	—	Microcrystalline jadeite or nephrite
Jadeite	$NaAl(SiO_3)_2$	M	3.3	6½–7	—	Microcrystalline clinopyroxene
Jasper	—	—	—	—	—	Microcrystalline red quartz
Kainite	$MgSO_4 \cdot KCl \cdot 3H_2O$	M	2.1	3	1.51	—
Kamacite	$\alpha Fe,Ni$	C	7.3–7.9	4	—	Fe(92–95%)–Ni(5–7%) alloy occurring in meteorites
Kaolin	—	—	—	—	—	A mixture of kaolinite and other clay minerals
Kaolinite	$Al_4Si_4O_{10}(OH)_8$	Tc	2.6	2	1.55–1.57	A clay mineral dominant in the tropics
Kermesite	Sb_2S_2O	M	4.5–4.6	1–1.5	—	—
Kernite	$Na_2B_4O_6(OH)_2 \cdot 3H_2O$	M	1.95	3	1.47	—
K-feldspars	—	—	—	—	—	Orthoclase, microcline, or sanidine
Kieserite	$MgSO_4 \cdot H_2O$	M	2.57	3½	1.53	—
Kyanite	Al_2SiO_5	Tc	3.67	5–7	1.72	—
Labradorite	$Ab_{50}An_{50} - Ab_{30}An_{70}$	Tc	2.71	6	1.56	A plagioclase
Lapis lazuli	—	—	—	—	—	A blue rock consisting mainly of lazurite
Laumontite	$Ca(Al_2Si_4O_{12}) \cdot 4H_2O$	M	2.28	4	1.52	A zeolite
Lawsonite	$CaAl_2(Si_2O_7)(OH)_2 \cdot H_2O$	O	3.09	8	1.67	—
Lazulite	$(Mg,Fe)Al_2(PO_4)_2(OH)_2$	M	3.0–3.1	5–5½	1.64	—
Lazurite	$(Na,Ca)_8(AlSiO_4)_6(SO_4,S,Cl)_2$	C	2.40–2.45	5–5½	1.5	—
Lechatelierite	SiO_2	A	2.2	6–7	1.46	Fused, glassy silica
Lepidocrocite	$\gamma FeO(OH)$	O	4.09	5	2.2	—
Lepidolite	$(K,Rb)(Li,Al)_3(Si,Al)_4O_{10}(OH,F)_2$	M	2.8–2.9	2½–4	1.55–1.59	A mica
Leucite	$KAlSi_2O_6$	Tt,C	2.47	5½–6	1.51	A feldspathoid
Limonite	$FeO(OH) \cdot nH_2O$	A	3.6–4.0	5–5½	—	—
Litharge	PbO	Tt	9.14	2	2.66	—
Lithiophilite	$Li(Mn,Fe)PO_4$	O	3.5	5	1.67	—
Lodestone	—	—	—	—	—	Naturally magnetized magnetite
Maghemite	γFe_2O_3	C	4.88	5	2.52–2.74	—
Magnesite	$MgCO_3$	T	3.0–3.2	3½–5	1.70	—
Magnetite	Fe_3O_4	C	5.20	6	—	—

Minerals *(continued)*

Physical properties. Xl system = crystal system (C = cubic; H = hexagonal; M = monoclinic; O = ortho-rhombic; T = trigonal; Tc = triclinic; Tt = tetragonal; A = amorphous). G = density (relative to water at 3.98°C). H = hardness. n = refractive index.

Name	Composition	Xl System	G	H	n	Remarks
Malachite	$Cu_2CO_3(OH)_2$	M	3.90–4.03	3½–4	1.88	—
Manganite	$MnO(OH)$	M	4.3	4	—	—
Marcasite	FeS_2	O	4.89	6–6½	—	—
Meerschaum	—	—	—	—	—	Massive sepiolite
Melilite	$Ca_2(Mg,Al)(Al,Si)_2O_7$	Tt	2.9–3.0	5–6	1.65	—
Mercury	Hg	—	13.6	0	—	—
Micas	$(K,Na,Ca)(Mg,Fe,Li,Al)_{2-3} \cdot (AlSi)_4O_{10}(OH,F)_2$	M	—	—	—	A group of phyllosilicates
Microcline	$KAlSi_3O_8$	Tc	2.54–2.57	6	1.53	Low-temperature K-feldspar
Microlite	$Ca_2Ta_2O_6(O,OH,F)$	C	5.48–5.56	5½	1.92–1.99	—
Microperthite	—	—	—	—	—	A perthite with thin (5–100 μm) lamellae
Minium	Pb_3O_4	—	8.9–9.2	2½	2.42	—
Molybdenite	MoS_2	H	4.62–4.73	1–1½	—	—
Monazite	$(REE,Th)PO_4$	M	4.6–5.4	5–5½	1.79	—
Monticellite	$CaMgSiO_4$	O	3.2	5	1.65	—
Montmorillonite	$(Na,Ca)(Al,Mg)_6(Si_4O_{10})_3(OH)_6 \cdot nH_2O$	M	2.5	1–1½	1.50–1.64	A clay mineral
Moonstone	—	—	—	—	—	Translucent adularia
Mother of pearl	$CaCO_3$ (aragonite)	O	2.95	3½–4	1.68	Microcrystalline, lamellar aragonite
Mullite	$Al_6Si_2O_{13}$	O	3.23	6–7	1.67	—
Muscovite	$KAl_2(AlSi_3O_{10})(OH)_2$	M	2.76–2.88	2–2½	1.60	A mica
Nacrite	$Al_2Si_2O_5(OH)_4$	M	2.6	2–2½	1.56	—
Natrolite	$Na_2Al_2Si_3O_{10} \cdot 2H_2O$	O	2.25	5–5½	1.48	—
Natron	$Na_2CO_3 \cdot 10H_2O$	M	—	—	—	—
Nepheline	$(Na,K)AlSiO_4$	H	2.60–2.65	5½–6	1.54	—
Nephrite	—	—	—	—	—	Microcrystalline tremolite or actinolite
Niccolite	$NiAs$	H	7.78	5–5½	—	—
Nickeline	—	—	—	—	—	Niccolite
Niter	KNO_3	O	2.09–2.14	2	1.50	—
Nontronite	$Na_{0.33}Fe_2^{3+}(Al_{0.33}Si_{3.67})O_{10}(OH)_2 \cdot nH_2O$	M	2.5	1–1½	1.60	A clay mineral
Oligoclase	$Ab_{90}An_{10}-Ab_{70}An_{30}$	Tc	2.65	6	1.54	A plagioclase
Olivine	$(Mg,Fe)_2SiO_4$	O	3.27–4.37	6½–7	1.69	—
Omphacite	$(Ca,Na)(Mg,Fe,Al)(SiO_3)_2$	M	3.2–3.4	5–6	1.67–1.70	A clinopyroxene
Onyx	—	—	—	—	—	Banded chalcedony
Opal	$SiO_2 \cdot nH_2O$	A	2.0–2.25	5–6	1.44	—

310

Physical properties. Xl system = crystal system (C = cubic; H = hexagonal; M = monoclinic; O = ortho-rhombic; T = trigonal; Tc = triclinic; Tt = tetragonal; A = amorphous). G = density (relative to water at 3.98°C). H = hardness. n = refractive index.

Name	Composition	Xl System	G	H	n	Remarks
Orpiment	As_2S_3	M	3.49	1½–2	2.8	—
Orthite	—	—	—	—	—	Allanite in slender crystals
Orthoclase	$KAlSi_3O_8$	M	2.55	6	1.52	An alkali feldspar
Orthoferrosilite	$FeSiO_3$	O	3.9	6	1.79	Fe end-member of orthopyroxenes
Palladium	Pd	C	11.9	4½–5	—	A platinum-group metal
Palygorskite	$Mg_2(Al,Fe)_2(Si_2O_5)_4(OH)_2 \cdot 4H_2O$	—	—	—	—	A fibrous clay mineral
Paragonite	$NaAl_2(AlSi_3O_{10})(OH)_2$	M	2.85	2	1.60	A mica
Patronite	VS_4	M	—	—	—	—
Pearl	$CaCO_3$ (aragonite)	—	—	—	—	Microcrystalline, lamellar aragonite
Pectolite	$NaCa_2Si_3O_8(OH)$	Tc	2.8	5	1.60	—
Pentlandite	$(Fe,Ni)_9S_8$	C	4.6–5.0	3½–4	—	—
Periclase	MgO	C	3.58	5½	1.73	—
Peridot	—	—	—	—	—	Gem olivine
Perovskite	$CaTiO_3$	O	4.03	5½	2.38	—
Perthite	—	—	—	—	—	Lamellar interspacing of microcline and albite
Petalite	$Li(AlSi_4O_{10})$	M	2.4	6–6½	1.51	—
Phenacite	Be_2SiO_4	R	2.97–3.0	7½–8	1.65	—
Phillipsite	$(K_2Na_2Ca)Al_2Si_4O_{12} \cdot 4$–$5H_2O$	M	2.2	4½–5	1.50	A zeolite
Phlogopite	$KMg_3(AlSi_3O_{10})(OH)_2$	M	2.86	2½–3	1.56–1.64	—
Phosphorite	—	—	—	—	—	A sedimentary rock consisting mainly of phosphatic and calcitic minerals and bioclasts
Pigeonite	$(Ca,Mg,Fe^{2+})(Mg,Fe^{2+})(SiO_3)_2$	M	3.30–3.46	6	1.64–1.72	A clinopyroxene
Pitchblende	—	—	—	—	—	Massive uraninite
Plagioclase	$Ab_{100}An_0 - Ab_0An_{100}$	Tc	2.62–2.76	6	1.53–1.59	Complete solid solution from albite ($NaAlSi_3O_8$) to anorthite ($CaAl_2Si_2O_8$)
Platinum	Pt	C	21.45	4–4½	—	—
Pollucite	$CsAlSi_2O_6 \cdot H_2O$	C	2.9	6½	1.52	—
Prehnite	$Ca_2Al_2Si_3O_{10}(OH)_2$	O	2.8–2.95	6–6½	1.63	—
Proustite	Ag_3AsS_3	T	5.57	2–2½	3.09	—
Pseudowollastonite	$CaSiO_3$	Tc	—	—	—	High-temperature phase of wollastonite
Psilomelane	$BaMn^{2+}Mn_8^{4+}O_{16}(OH)_4$	O	3.7–4.7	5–6	—	—
Pyrargyrite	Ag_3SbS_2	T	5.85	2–2½	3.08	—
Pyrite	FeS_2	C	5.02	6–6½	—	—

Minerals (*continued*)

Physical properties. Xl system = crystal system (C = cubic; H = hexagonal; M = monoclinic; O = ortho-rhombic; T = trigonal; Tc = triclinic; Tt = tetragonal; A = amorphous). G = density (relative to water at 3.98°C). H = hardness. n = refractive index.

Name	Composition	Xl System	G	H	n	Remarks
Pyrochlore	$(Ca,Na)_2(Nb,Ta)_2O_6(O,OH,F)$	C	4.3	5	—	—
Pyrolusite	MnO_2	Tt	4.75	1–2	—	—
Pyromorphite	$Pb_5(PO_4)_3Cl$	H	7.04	3½–4	2.06	—
Pyrope	$Mg_3Al_2Si_3O_{12}$	C	3.58	7	1.71	A garnet
Pyrophyllite	$AlSi_4O_{10}(OH)_2$	M	2.8	1–2	1.59	—
Pyroxenes	$QRSi_2O_6$ $(Q = Ca,Na,Mg,Fe^{2+})$ $(R = Mg,Fe^{2+}Fe^{3+},Fe,Cr,Mn,Al)$	—	—	—	—	A group of Ca,Na,Mg,Fe-silicates
Pyrrhotite	$F_{0.8-1}S$	M,H	4.58–4.65	4	—	—
Quartz	SiO_2	T,H	2.65	7	1.54	—
Realgar	AsS	M	3.48	1½–2	2.60	—
Rhodochrosite	$MnCO_3$	T	3.5–3.7	3½–4	1.82	—
Rhodonite	$MnSiO_3$	Tc	3.4–3.7	5½–6	1.73–1.75	—
Riebeckite	$Na_2(Mg,Fe^{2+})_3Fe_2^{3+}Si_8O_{22}(OH)_2$	M	3.4	5	1.66–1.71	—
Rock crystal	—	—	—	—	—	Megacrystalline quartz
Rock salt	—	—	—	—	—	Halite
Rubellite	—	—	—	—	—	Red to pink tourmaline
Ruby	—	—	—	—	—	Red gem corundum
Ruby spinel	—	—	—	—	—	Red gem spinel
Rutile	TiO_2	Tt	4.25	6–6½	2.61	—
Saltpeter	—	—	—	—	—	Niter
Sanidine	$KAlSi_3O_8$	M	2.56–2.62	6	1.53	High temperature K-feldspar
Saponite	$(0.5Ca,Na)_{0.33}(Mg,Fe)_3(Si_{3.67}Al_{0.33})O_{10}(OH)_2 \cdot 4(H_2O)$	M	2.5	1–1½	1.52	A montmorillonitic clay
Sapphire	—	—	—	—	—	Blue gem corundum
Scapolite	$3NaAlSi_3O_8 \cdot NaCl$ to $3CaAl_2Si_2O_8 \cdot CaCO_3$	Tt	2.55–2.74	5–6	1.55–1.60	—
Scheelite	$CaWO_4$	Tt	5.9–6.1	4½–5	1.92	—
Scolecite	$CaAl_2Si_3O_{10} \cdot 3H_2O$	M	2.2	5–5½	1.52	A zeolite
Selenite	—	—	—	—	—	Megacrystalline gypsum
Sepiolite	$Mg_4(Si_2O_5)_3(OH) \cdot 4H_2O$	O	2.0	2–2½	1.52	A fibrous clay mineral
Sericite	—	—	—	—	—	Fine-grained mica
Serpentine	$(Mg,Fe)_3Si_2O_5(OH)_4$	M,O	2.5–2.6	3–5	1.55	—
Siderite	$FeCO_3$	T	3.94	3½–4	1.88	—
Sillimanite	Al_2SiO_5	O	3.25	6–7	1.66	—
Silver	Ag	C	10.5	2½–3	—	—
Smithsonite	$ZnCO_3$	T	4.30–4.45	4–4½	1.85	—
Smoky quartz	—	—	—	—	—	Brown gem quartz
Soapstone	—	—	—	—	—	Steatite
Sodalite	$Na_8(AlSiO_4)_6Cl_2$	C	2.15–2.30	5½–6	1.48	—

Physical properties. Xl system = crystal system (C = cubic; H = hexagonal; M = monoclinic; O = ortho-rhombic; T = trigonal; Tc = triclinic; Tt = tetragonal; A = amorphous). G = density (relative to water at 3.98°C). H = hardness. n = refractive index.

Name	Composition	Xl System	G	H	n	Remarks
Sperrylite	$PtAs_2$	C	10.50	6–7	—	—
Spessartine	$Mn_3Al_2Si_3O_{12}$	C	4.19	7	1.80	A garnet
Sphalerite	ZnS	C	3.9–4.1	3½–4	2.37	—
Sphene	$CaTiO(SiO_4)$	M	3.40–3.55	5–5½	1.91	—
Spinel	$MgAl_2O_4$	C	3.5–4.1	8	1.72	—
Spinel group	QR_2O_4	C	—	—	—	A group of mineral where Q = Mg, Fe^{2+},Fe^{3+},Mn,Zn R = Al,Fe^{2+},Fe^{3+}, Ti^{4+},Cr,V
Spodumene	$LiAl(SiO_3)_2$	M	3.15–3.20	6½–7	1.67	—
Stannite	Cu_2FeSnS_4	Tt	4.3–4.5	4	—	—
Staurolite	$(Mg,Fe)_2Al_9Si_4O_{22}(O,OH)_2$	M	3.65–3.75	7–7½	1.75	—
Steatite	—	—	—	—	—	Massive, compact talc
Stibnite	Sb_2S_3	O	4.52–4.62	2.0	—	—
Stilbite	$CaAl_2Si_7O_{18}\cdot 7H_2O$	M	2.1–2.2	3½–4	1.50	—
Stishovite	SiO_2	Tt	4.28	7	1.80	High-pressure phase of silica
Strontianite	$SrCO_3$	O	3.7	3½–4	1.67	—
Sulfur	S	O	2.07	1½–2½	2.04	—
Sylvanite	$(Au,Ag)Te_2$	M	8.0–8.2	1½–2	—	—
Sylvite	KCl	C	1.99	2	1.49	—
Taenite	γFe,Ni	C	7.8–8.2	5	—	Fe(35–70%)–Ni(30–65%) alloy occurring in meteorites
Talc	$Mg_3Si_4O_{10}(OH)_2$	M	2.7–2.8	1	1.59	—
Tantalite	$(Fe,Mn)(Ta,Nb)_2O_6$	O	6.5	6	—	—
Tanzanite	—	—	—	—	—	Blue gem zoisite
Tennantite	$Cu_{12}As_4S_{13}$	C	4.6–5.1	3–4½	—	—
Tenorite	CuO	Tc	6.5	3–4	—	—
Tetrahedrite	$Cu_{12}Sb_4S_{13}$	C	4.6–5.1	3–4½	—	—
Thorianite	ThO_2	C	9.7	6½	—	—
Thorite	$ThSiO_4$	Tt	5.3	5	1.8	—
Tiger eye	—	—	—	—	—	A gem variety of quartz
Tin	Sn	Tt	7.3	2	—	—
Titanaugite	$Ca(Mg,Fe,Ti)(Si,Al)_2O_6$	M	3.3	6	1.7	Cf. augite
Titanite	—	—	—	—	—	Sphene
Topaz	$Al_2SiO_4(OH,F)_2$	O	3.4–3.6	8	1.61–1.63	—

Minerals (*continued*)

Physical properties. Xl system = crystal system (C = cubic; H = hexagonal; M = monoclinic; O = ortho-rhombic; T = trigonal; Tc = triclinic; Tt = tetragonal; A = amorphous). G = density (relative to water at 3.98°C). H = hardness. n = refractive index.

Name	Composition	Xl System	G	H	n	Remarks
Tourmaline	$(Na,Ca)(Li,Mg,Al) \cdot (Al,Fe,Mn)_6(BO_3)_3Si_6O_{18}(OH)_4$	T	3.0–3.25	7–7½	1.64–1.68	—
Travertine	—	—	—	—	—	Hardened freshwater limestone
Tremolite	$Ca_2Mg_5Si_8O_{22}(OH)_2$	M	3.0–3.2	5–6	1.61	—
Tridymite	SiO_2	M,O	2.26	7	1.47	High-temperature polymorph of quartz
Troilite	FeS	H	4.83	4	—	—
Trona	$Na_2Co_3 \cdot NaHCO_3 \cdot 2H_2O$	M	2.13	3	1.49	—
Tufa	—	—	—	—	—	Soft fresh-water limestone
Turquoise	$CuAl_6(PO_4)_4(OH)_8 \cdot 4H_2O$	Tc	2.6–2.8	6	1.62	—
Ulexite	$NaCaB_5O_6(OH)_6 \cdot 5H_2O$	Tc	1.96	1–2½	1.50	—
Ulvöspinel	Fe_2TiO_4	C	4.78	7½–8	—	—
Uraninite	UO_2	C	7.5–9.7	5½	—	—
Uvarovite	$Ca_3Cr_2Si_3O_{12}$	C	3.90	7½	1.87	A garnet
Vanadinite	$Pb_5(VO_4)_3Cl$	H	6.9	3	2.25–2.42	—
Vermiculite	$(Mg,Fe,Al)_3(Al,Si)_4O_{10}(OH)_2 \cdot 4H_2O$	M	2.4	1½	1.55–1.58	—
Vesuvianite	$Ca_{10}Mg_2Al_4(SiO_4)_5(Si_2O_7)_2(OH)_4$	Tt	3.35–3.45	6½	1.70–1.75	—
Wavellite	$Al_3(PO_4)_2(OH)_3 \cdot 5H_2O$	O	2.36	3½–4	1.54	—
Wernerite	—	—	—	—	—	Scapolite
Willemite	Zn_2SiO_4	T	3.9–4.2	5½	1.69	—
Witherite	$BaCO_3$	O	4.3	3½	1.68	—
Wolframite	$(Fe,Mn)WO_4$	M	7.0–7.5	4–4½	—	—
Wollastonite	$CaSiO_3$	Tc	2.8–2.9	5–5½	1.63	—
Wulfenite	$PbMoO_4$	Tt	6.8	3	2.40	—
Wurtzite	ZnS	H	3.98	4	2.35	—
Wüstite	FeO	—	—	—	—	—
Xanthophyllite	$Ca(Mg,Al)_3(Al_2Si_2O_{10})(OH)_2$	M	3–3.1	3½	1.65	—
Xenotime	YPO_4	Tt	4.3–4.7	4–5	1.72–1.83	—
Zeolites	—	—	—	—	—	Hydrous Na,K,Ca aluminosilicates
Zincite	ZnO	H	5.68	4	2.01	—
Zircon	$ZrSiO_4$	Tt	4.68	7½	1.92–1.96	—
Zoisite	$Ca_2Al_3Si_3O_{12}(OH)$	O	3.35	6	1.69	—

From Hurlbut and Klein 1977; Bates and Jackson 1980; Berry, Mason, and Dietrich 1983; Lof 1983; other sources.

Mineral	Chemical composition
Quartz	SiO_2
Feldspars	
Orthoclase	$KAlSi_3O_8$
Albite	$NaAlSi_3O_8$
Anorthite	$CaAl_2Si_2O_8$
Feldspathoids	
Leucite	$KAlSi_2O_6$
Nepheline	$(Na,K)AlSiO_4$
Pyroxenes	
Enstatite	$MgSiO_3$
Hypersthene	$(Mg,Fe)SiO_3$
Diopside	$CaMg(SiO_3)_2$
Hedenbergite	$CaFe(SiO_3)_2$
Augite	$(Ca,Na)(Mg,Fe,Al)[(Si,Al)O_3]_2$
Amphibole	
Hornblende	$(Ca,Na)_{2-3}(Mg,Fe,Al)_5(Si,Al)_8O_{22}(OH)_2$
Olivine	$(Mg,Fe)_2SiO_4$
Micas	
Muscovite	$KAl_2(AlSi_3O_{10})(OH)_2$
Biotite	$K(Mg,Fe)_3(AlSi_3O_{10})(OH)_2$
Clays	
Chlorite	$(Mg,Fe)_3(Si,Al)_4O_{10}(OH)_2(Mg,Fe)_3(OH)_6$
Illite	$(K,H_3O)(Al,Mg,Fe)_2(Si,Al)_4O_{10}[(OH)_2,H_2O]$
Montmorillonite	$(Na,Ca)(Al,Mg)_6(Si_4O_{10})_3(OH)_6 \cdot nH_2O$
Kaolinite	$Al_4Si_4O_{10}(OH)_8$
Carbonates	
Calcite	$CaCO_3$
Aragonite	$CaCO_3$
Dolomite	$CaMg(CO_3)_2$

H_2	H_2O	CH_2CO
H_3^+	H_2S	CH_2NH
	HCN	CH_2NCN
CH^+	HNC	CH_2CHCN
C_2	HCO^+	
CN	HN_2	CH_3CH
CO	HNO	CH_3CN
CS	HCS	CH_3OH
	OCS	CH_3CHO
NO		CH_3COOH
NS	H_2CS	CH_3NH_2
OH	H_2CO	CH_3C_2H
	HNCO	
SiO	HNCS	CH_3C_3N
SiS	HCOOH	CH_3SH
	NH_2CH	
SO	NH_2CN	CH_3CH_2CN
	NH_2CHO	CH_3CH_2OH
	NH_3	CH_3OCH_3
	N_2H^+	
		CH_4
		C_2H
		C_2H_2
		C_2H_3CN
		C_2H_5OH
		C_2H_5CN
		C_3N
		C_4H
		H_2CNH
		HC_3N
		HC_5N
		HC_7N
		HC_9N

Note. Numerous ionic and isotopic species of the above molecules have also been identified.

From Hernbst 1978, p. 89, Table 1; Linke et al. 1979, p. L140, Table 1, p. L141, Table 2; Mann and Williams 1980, p. 722–724, Table 1; Dalgarno 1985, p. 8, Table 1; Bally 1986.

H = 1.

10^{-2}–10^{-5}	10^{-6}	10^{-7}	10^{-8}	10^{-9}	10^{-10}
H_2	HCN	OH	CH	HNCO	H_2CS
CO	HNC	CS	CN	NH_2CN	HCOOH
H_2O	NH_3	SO	OCS	CH_3CH_2	CH_2NH
		HCO^+	H_2S		$HCONH_2$
		N_2^+	HCO		CH_3NH_2
		C_2H	H_2CO		C_2H_3CN
		SO_2	HC_3N		CH_3CHO
		CH_3OH			HC_5N
					CH_3COOH
					CH_3C_3N
					CH_3OCH_3
					C_2H_5OH

From Hernbst 1978, p. 90, Table 2; Guelin 1985, p. 38, Table 2.

Crater name	Age (10^9 y)	Rhegolith thickness (m)
Tycho	0.27	—
Copernicus	0.9	—
Oceanus Procellarum	3.16–3.36	3
Palus Putredinis	3.3	5
Mare Fecunditatis	3.45	—
Mare Tranquillitatis	3.6–3.7	5
Taurus–Littrow	3.8	—
Fra Mauro	3.85–3.96	8
Apennine Front	4.0	—
Highlands	4.4	10–15

From Taylor 1975, p. 87, Table 3-8; Glass 1982, p. 225, p. 7.18.

**Moon
major rock types**

	Mineral composition (%)					
Rock type	Olivine	Orthopyroxene	Clinopyroxene	Ca-plagioclase	Tridymite/ cristobalite	Opaques
Maria						
olivine basalt	6–20	—	35–63	15–27	0–2	4–15
quartz basalt	—	—	45–68	24–35	2–7	2–10
high-Al basalt	—	—	50	40	0–2	3–7
high-K basalt	0–5	—	45–55	20–40	10–15	1–5
low-K basalt	0–5	—	40–50	30–40	10–15	1–5
high-Ti basalt	<5	—	45–55	25–30	15–25	—
Terrae						
anorthositic gabbro/ highland basalt	9	20	—	70	—	1
Fra Mauro basalt	10	30	7	53	—	—
anorthosite	—	—	—	90–98	—	—

From Taylor 1975, p. 128, Table 4.1, p. 234–237.

**Moon
major rock types
chemical composition (atom percent)**

Element	Anorthositic gabbro	High-titanium basalt	Low-titanium basalt
Na	0.41	0.40	0.58
Mg	6.07	9.00	7.04
Al	25.45	7.88	11.42
Si	37.79	30.56	37.49
P	—	—	—
K	0.04	0.07	<0.07
Ca	22.17	13.92	17.83
Ti	0.29	12.46	1.03
Cr	0.13	0.56	0.18
Mn	0.11	0.35	0.33
Fe	7.52	24.79	24.15
Co	—	—	—
Ni	—	—	—
	99.98	99.99	100.12

From Glass 1982, p. 210–211, Tables 7.7 and 7.9.

a = lunar radius toward Earth; b = lunar radius along lunar orbit; c = lunar polar radius.

age	$= 4.6 \cdot 10^6$ y
age of lunar rocks	
maria	$= 3.1$ to $3.8 \cdot 10^9$ y
terrae	$= 3.7$ to $4.6 \cdot 10^9$ y
albedo	$= 0.068$
atmosphere	
composition	
Ne	$= 40\%$
Ar	$= 40\%$
He	$= 20\%$
pressure	$= 2 \cdot 10^{-14}$ bar
	$= 1.5 \cdot 10^{-11}$ torr
density (mean)	$= 3.343$ g cm^{-3}
diameter (mean)	$= 3576.4$ km
	$= 0.011596$ light sec
distance from Earth (mean)	$= 384{,}401$ km
	$= 1.2822237$ light sec
distance from Earth (range)	$= 356{,}400$ to $406{,}700$ km
eccentricity of orbit	$= 0.0549$
escape velocity	$= 2.38$ km s^{-1}
evection	
displacement of selenocentric longitude	$= 1°16'20.4''$
period	$= 31.807$ d
gravity	
anomalies	$= \pm 200$ mgals
wavelength of anomalies	$= \sim 900$ km
surface (mean)	$= 1.62$ m s^{-2}
heat flow (Apollo 15 site)	$= 3 \cdot 10^{-2}$ W m^{-2}
	$= 0.7 \cdot 10^{-6}$ cal cm^{-2} s^{-1}
inclination of equator to orbit	$= 6.68°$
inclination of orbit	
to Earth's equator	$= 18°17'$ to $28°35'$
to ecliptic (mean)	$= 5°8'43'$
range	$= \pm 9'$
period	$= 173$ d
lunar day (mean transit interval)	$= $ 24h 50m 28s
lunar months	
anomalistic (perigee to perigee)	$= 27.55455$ d
draconic	$= $ nodical
nodical (node to node)	$= 27.21222$ d
sidereal	$= 27.32167$ d$_E$
synodical (new moon to new moon)	$= 29.5305883$ d$_E$
lunar year	$= 12$ synodical months
	$= 354.3672504$ d$_E$
magnetic field intensity	$= 0.2 \cdot 10^{-5}$ gauss
mass	$= 7.349 \cdot 10^{22}$ kg
	$= 1/81.286$ of Earth's mass
nutation period	$= 18.61$ tropical years
physical libration	
displacement in selenocentric latitude	$= \pm 0.02$
period	$= 1$ y

a = lunar radius toward Earth; b = lunar radius along lunar orbit; c = lunar polar radius.

physical libration *(continued)*	
displacement in selenocentric longitude	$= \pm 0.04$
period	$= 6$ y
radii	
mean $(b + c)/2$	$= 1738.2$ km
	$= 0.27283$ of the Earth's mean equatorial radius
$a - c$	$= 1.09$ km
$a - b$	$= 0.31$ km
$b - c$	$= 0.78$ km
rhegolith	
bearing strength (a few cm below surface)	$= 1$ kg cm^{-2}
particle size (range of means)	$= 0.004$–0.8 mm
sidereal motion	$= 13°10'34.89''$ d$_E^{-1}$
sidereal period	$= 27.32166140 + 0.00000016 T$ d$_E$
soil (see *rhegolith*)	
temperatures	
equator	
surface, noon	$= 400$ K
surface, night minimum	$= 115$ K
at 1 m of depth in rhegolith	$= 230$ K
transit interval (mean)	$= 24$h 50m 28.2s
volume	$= 2.1998 \cdot 10^{19}$ m^3
year (see *lunar year*)	

From Allen 1976, p. 147–148; Glass 1982, p. 224–225; other sources.

Modern	Greek	Roman	Arabic	Modern	Greek	Roman	Arabic
1	A	I	١	30	Λ	XXX	٣٠
2	B	II	٢	40	M	XL	٤٠
3	Γ	III	٣	50	N	L	٥٠
4	Δ	IV	٤	60	Ξ	LX	٦٠
5	E	V	٥	70	O	LXX	٧٠
6	F	VI	٦	80	Π	LXXX	٨٠
7	Z	VII	٧	90	Q	XC	٩٠
8	H	VIII	٨	100	P	C	١٠٠
9	Θ	IX	٩	200	Σ	CC	٢٠٠
10	I	X	١٠	300	T	CCC	٣٠٠
11	IA	XI	١١	400	Υ	CD, CCCC	٤٠٠
12	IB	XII	١٢	500	Φ	D	٥٠٠
13	IΓ	XIII	١٣	600	X	DC	٦٠٠
14	IΔ	XIV	١٤	700	Ψ	DCC	٧٠٠
15	IE	XV	١٥	800	Ω	DCCC	٨٠٠
16	IF	XVI	١٦	900	Ψ	CM, DCCCC	٩٠٠
17	IZ	XVII	١٧	1,000	/A	(I), M	١٠٠٠
18	IH	XVIII	١٨	5,000	/E	$\overline{V}$	٥٠٠٠
19	IΘ	XIX	١٩	10,000	/I	((I)), $\overline{X}$	١٠٠٠٠
20	K	XX	٢٠	100,000	/P	(((I))), $\overline{C}$	١٠٠٠٠٠
				1,000,000	—	((((I)))), $\overline{M}$	١٠٠٠٠٠٠

From Smith 1952, v. 16, p. 610–614.

Current name	Transport (10^6 m³/s)
Antarctic Circumpolar	125
Antilles	12
Benguela	16
Brazil	10
California	10
Canaries	16
Caribbean	26
Florida	26
Gulf Stream	55
Kuroshio	65
Labrador	6
North Atlantic	10
North Equatorial Countercurrent (Pacific)	45
North Equatorial Current (Pacific)	45

From Sverdrup et al. 1942, p. 617, 629, 684, 727.

Group IA	IIA	IIIA IIIB	IVA IVB	VA VB	VIA VIB	VIIA VIIB	VIII			IB	IIB	IIIB IIIA	IVB IVA	VB VA	VIB VIA	VIIB VIIA	VIIIA	Orbit

KEY TO CHART

Atomic Number → 50 +2
Symbol → Sn +4
1983 Atomic Weight → 118.71
Electron Configuration → 18 18 4
Oxidation States

New notation — IUPAC form
Previous IUPAC form
CAS version

1 Group IA +1 −1	2 IIA											13 IIIB IIIA	14 IVB IVA	15 VB VA	16 VIB VIA	17 VIIB VIIA	18 VIIIA 0	Orbit
1 H 1.00794 1																	2 He 4.00260 2	K
3 Li +1 6.941 2-1	4 Be +2 9.01218 2-2											5 B +3 10.81 2-3	6 C +3 12.011 2-4	7 N −3 14.0067 2-5	8 O −2 15.9994 2-6	9 F −1 18.9984 2-7	10 Ne 0 20.179 2-8	K-L
11 Na +1 22.9898 2-8-1	12 Mg +2 24.305 2-8-2	3 IIIA IIIB	4 IVA IVB	5 VA VB	6 VIA VIB	7 VIIA VIIB	8 VIII	9 VIII	10 VIII	11 IB	12 IIB	13 Al +3 26.9815 2-8-3	14 Si 28.0855 2-8-4	15 P 30.9738 2-8-5	16 S 32.06 2-8-6	17 Cl 35.453 2-8-7	18 Ar 0 39.948 2-8-8	K-L-M
19 K +1 39.0983 -8-8-1	20 Ca +2 40.08 -8-8-2	21 Sc +3 44.9559 -8-9-2	22 Ti +3 +4 47.88 -8-10-2	23 V +2 +3 +4 +5 50.9415 -8-11-2	24 Cr +2 +3 +6 51.996 -8-13-1	25 Mn +2 +3 +4 +6 +7 54.9380 -8-13-2	26 Fe +2 +3 55.847 -8-14-2	27 Co +2 +3 58.9332 -8-15-2	28 Ni +2 +3 58.69 -8-16-2	29 Cu +1 +2 63.546 -8-18-1	30 Zn +2 65.39 -8-18-2	31 Ga +3 69.72 -8-18-3	32 Ge +2 +4 72.59 -8-18-4	33 As +3 +5 −3 74.9216 -8-18-5	34 Se +4 +6 −2 78.96 -8-18-6	35 Br +1 +5 −1 79.904 -8-18-7	36 Kr 0 83.80 -8-18-8	-L-M-N
37 Rb +1 85.4678 -18-8-1	38 Sr +2 87.62 -18-8-2	39 Y +3 88.9059 -18-9-2	40 Zr +4 91.224 -18-10-2	41 Nb +3 +5 92.9064 -18-12-1	42 Mo +6 95.94 -18-13-1	43 Tc +7 (98) -18-13-1	44 Ru +3 101.07 -18-15-1	45 Rh +3 102.906 -18-16-1	46 Pd +2 +4 106.42 -18-18-0	47 Ag +1 107.868 -18-18-1	48 Cd +2 112.41 -18-18-2	49 In +3 114.82 -18-18-3	50 Sn +2 +4 118.71 -18-18-4	51 Sb +3 +5 −3 121.75 -18-18-5	52 Te +4 +6 −2 127.60 -18-18-6	53 I +1 +5 +7 −1 126.905 -18-18-7	54 Xe 0 131.29 -18-18-8	-M-N-O
55 Cs +1 132.905 -18-8-1	56 Ba +2 137.33 -18-8-2	57* La +3 138.906 -18-9-2	72 Hf +4 178.49 -32-10-2	73 Ta +5 180.948 -32-11-2	74 W +6 183.85 -32-12-2	75 Re +4 +6 +7 186.207 -32-13-2	76 Os +3 +4 +6 190.2 -32-14-2	77 Ir +3 +4 192.22 -32-15-2	78 Pt +2 +4 195.08 -32-16-2	79 Au +1 +3 196.967 -32-18-1	80 Hg +1 +2 200.59 -32-18-2	81 Tl +1 +3 204.383 -32-18-3	82 Pb +2 +4 207.2 -32-18-4	83 Bi +3 +5 208.980 -32-18-5	84 Po +2 +4 (209) -32-18-6	85 At (210) -32-18-7	86 Rn 0 (222) -32-18-8	-N-O-P
87 Fr +1 (223) -18-8-1	88 Ra +2 226.025 -18-8-2	89** Ac +3 227.028 -18-9-2	104 Unq +4 (261) -32-10-2	105 Unp (262) -32-11-2	106 Unh (263) -32-12-2	107 Uns (262) -32-13-2												O P Q

•Lanthanides

58 Ce +3 +4 140.12 -20-8-2	59 Pr +3 140.908 -21-8-2	60 Nd +3 144.24 -22-8-2	61 Pm +3 (145) -23-8-2	62 Sm +2 +3 150.36 -24-8-2	63 Eu +2 +3 151.96 -25-8-2	64 Gd +3 157.25 -25-9-2	65 Tb +3 158.925 -27-8-2	66 Dy +3 162.50 -28-8-2	67 Ho +3 164.930 -29-8-2	68 Er +3 167.26 -30-8-2	69 Tm +3 168.934 -31-8-2	70 Yb +2 +3 173.04 -32-8-2	71 Lu +3 174.967 -32-9-2	N O P

••Actinides

90 Th +4 232.038 -18-10-2	91 Pa +5 +4 231.036 -20-9-2	92 U +3 +4 +5 +6 238.029 -21-9-2	93 Np +3 +4 +5 +6 237.048 -22-9-2	94 Pu +3 +4 +5 +6 (244) -24-8-2	95 Am +3 +4 +5 +6 (243) -25-8-2	96 Cm +3 (247) -25-9-2	97 Bk +3 +4 (247) -27-8-2	98 Cf +3 (251) -28-8-2	99 Es +3 (252) -29-8-2	100 Fm +3 (257) -30-8-2	101 Md +2 +3 (258) -31-8-2	102 No +2 +3 (259) -32-8-2	103 Lr +3 (260) -32-9-2	O P Q

Numbers in parentheses are mass numbers of most stable isotope of that element. (Weast 1986, inner cover).

Name	Composition	Boiling point (°C)
Rhigolene	C_4H_{10}, C_5H_{12}	18–21
Petroleum ether	C_5H_{12}, C_6H_{14}	40–60
Gasoline	C_6H_{14} to $C_{10}H_{22}$	60–200
Naphtha	C_6H_{14}, C_7H_{16}	70–90
Ligroin	C_7H_{16}, C_8H_{18}	90–120
Benzine	C_8H_{18}, C_9H_{20}	120–150
Kerosene	C_9H_{20} to $C_{16}H_{34}$	150–300
Lubricating oils:		
light	$C_{12}H_{26}$ to $C_{20}H_{42}$	300+
medium	$C_{16}H_{34}$ to $C_{22}H_{46}$	,,
heavy	$C_{18}H_{38}$ to $C_{20}H_{42}$	,,
Petrolatum	$C_{20}H_{42}$ to $C_{30}H_{62}$	,,
Asphalt	$C_{30}H_{62}$ to $>C_{90}H_{182}$	,,

pH scale

pH $= -\log_{10}$ concentration of H^+ ions in water, in moles/liter (25°C).

	pH	Concentration of H^+ ions (mol/l)	Concentration of OH^- ions (mol/l)	Number of ions per liter	
				H^+	OH^-
Acid	0	10^0	10^{-14}	6×10^{23}	6×10^9
	1	10^{-1}	10^{-13}	6×10^{22}	6×10^{10}
	2	10^{-2}	10^{-12}	6×10^{21}	6×10^{11}
	3	10^{-3}	10^{-11}	6×10^{20}	6×10^{12}
	4	10^{-4}	10^{-10}	6×10^{19}	6×10^{13}
	5	10^{-5}	10^{-9}	6×10^{18}	6×10^{14}
	6	10^{-6}	10^{-8}	6×10^{17}	6×10^{15}
Neutral	7	10^{-7}	10^{-7}	6×10^{16}	6×10^{16}
Basic	8	10^{-8}	10^{-6}	6×10^{15}	6×10^{17}
	9	10^{-9}	10^{-5}	6×10^{14}	6×10^{18}
	10	10^{-10}	10^{-4}	6×10^{13}	6×10^{19}
	11	10^{-11}	10^{-3}	6×10^{12}	6×10^{20}
	12	10^{-12}	10^{-2}	6×10^{11}	6×10^{21}
	13	10^{-13}	10^{-1}	6×10^{10}	6×10^{22}
	14	10^{-14}	10^0	6×10^9	6×10^{23}

Pressure = 10^{-3} atm.

Mineral	Temperature (°C)
corundum (Al_2O_3)	1470
Fe–Si (metallic)	1185
diopside ($CaMgSi_2O_6$)	1165
forsterite (Mg_2SiO_4)	1160
enstatite ($MgSiO_3$)	1078
andalusite (Al_2SiO_5)	795
jadeite ($NaAlSi_2O_6$)	507
titanite ($CaTiSiO_5$)	501
troilite (FeS)	430
magnetite (Fe_3O_4)	130
carbonaceous compounds	100–200
hydrated Mg-silicates	0–100
ice	<0

From Ringwood 1975, p. 556, Table 16.3; Lattimer and Grossman 1978, p. 172, Table 1.

ρ = mean density of body in g/cm^3; g = surface gravity in m/s^2; P = surface pressure in bars; T = surface temperature in K; gases in volume percent.

Body	ρ	g	P	T	gas	%
Mercury	5.48	3.95	$2\cdot10^{-15}$	440	He	98
					H	2
Venus	5.243	8.88	90	730	CO_2	96.0
					N_2	3.5
					SO_2	0.01
					Ar	0.01
					H_2O	var.
Earth	5.515	9.81	1	288	N_2	78.084
					O_2	20.946
					H_2O	var.
					Ar	0.934
Moon	3.343	1.62	$2\cdot10^{-14}$	257	Ne	40
					Ar	40
					He	20
Mars	3.970	3.73	0.007	218	CO_2	95.7
					N_2	2.7
					Ar	1.6
Jupiter	1.33	23.20	$\gg100$	165	H_2	90
					He	10
Saturn	0.67	8.77	$\gg100$	140	H_2	90
					He	10
Uranus	1.31	9.46	$\gg100$	57	H_2	90
					He	10
Neptune	1.65	1.37	$\gg100$	57	H_2	90
					He	10
Pluto	0.9 (?)	—	—	42	CH_4	100

From Pollack and Yung 1980; Cruikshank and Silvaggio 1980.

Name	Mean distance from Sun (AU)	Eccentricity	Perihelion distance (10^6 km)	Aphelion distance (10^6 km)	Sidereal period tropical years	days
Mercury	0.387099	0.2056	46.0	69.9	0.24085	87.969
Venus	0.723332	0.0068	107.4	108.8	0.61521	224.701
Earth	1.000000	0.0167	147.1	152.1	1.00004	365.256
Mars	1.523688	0.0934	206.6	249.1	1.88089	686.980
Jupiter	5.202561	0.0485	740.5	815.8	11.8623	4,322.71
Saturn	9.554747	0.0556	1349	1504	29.4577	10,759.5
Uranus	19.21814	0.0472	2738	3002	84.0139	30,685
Neptune	30.10957	0.0086	4463	4537	164.79	60,190
Pluto	39.44	0.248	4443	7375	248.5	90,800

Name	Inclination of orbit over ecliptic (degrees)	Sidereal rotation period (d = days) (h = hours)	Inclination of equator to orbit (degrees)	Equatorial radius km	Earth = 1	Oblateness	Mean orbital velocity (km/s)	Mass (10^{24} kg)
Mercury	7.00	58.65 d	0.0	2,439	0.382	0.000	47.89	0.3302
Venus	3.39	243.01 d[a]	3.4	6,051.4	0.949	0.000	35.03	4.871
Earth	0.00	23.9345 h	23.44	6,378.164	1.000	0.00334894	29.79	5.9737
Mars	1.85	24.6229 h	23.98	3,398	0.533	0.0059	24.13	0.6421
Jupiter	1.30	9.841 h[b]	3.08	71,492	11.21	0.0637	13.06	1899.728
Saturn	2.49	10.233 h[b]	29.00	60,268	9.45	0.102	9.64	568.8
Uranus	0.77	17.24 h[a,b]	97.92	25,400	3.98	0.024	6.81	86.9
Neptune	1.77	15.8 h[b]	28.8	24,750	3.81	0.0266	5.43	103.0
Pluto	17.17	6.3874 d	50 (?)	1,145	0.18	?	4.74	0.0115

Name	Volume (10^{24} m^3)	Mean density (g/cm^3)	Equatorial surface gravity (m/s^2)[c]	Equatorial escape velocity (km/s)	Albedo	Number of satellites
Mercury	0.0603	5.48	3.78	4.3	0.06	0
Venus	0.929	5.243	8.60	10.3	0.72	0
Earth	1.0834	5.515	9.78	11.2	0.39	1
Mars	0.1617	3.970	3.72	5.0	0.16	2
Jupiter	1460.841	1.33	23.12	59.5	0.70	16
Saturn	851.832	0.67	9.05	35.6	0.75	17
Uranus	66,238	1.31	7.77	21.2	0.90	15
Neptune	62,263	1.65	11.00	23.6	0.82	2
Pluto	0.0000063	1.84	0.59	1.33	0.61	1

[a]Retrograde.

[b]At equator.

[c]Including centrifugal term.

From Beatty, O'Leary, and Chaikin 1982; Lindal et al. 1985; other sources.

a = inner edge; b = outer edge.

Jupiter (equatorial radius = 71,492 km)	ring radius (10^3 km)	*Saturn* (equatorial radius = 60,268 km)	ring radius (10^3 km)	*Uranus* (equatorial radius = 25,400 km)	ring radius (10^3 km)
Secondary Ring I	a = (71.492) b = 122.8	D ring	a = 67.0	Ring 1986U2R	= 37-39.5
		Guerin division (width)	= (1.2)	6	= 41.85
Primary Ring	a = 122.8 b = 129.2			5	= 42.24
		C ring	a = 73.2 b = 91.7	4	= 42.58
Secondary Ring 2	a = 251.7 b = (359)	B ring	a = 91.7 b = 117.5	α	= 44.73
				β	= 45.67
		Cassini division (width)	= 3.5	η	= 47.18
		A ring	a = 121.0	γ	= 47.63
		Encke division	= 133.5	δ	= 48.31
		A ring	b = 136.2	1986U1R	= 50.04
		F ring (average)	= 140.6	ϵ	= 51.16
		G ring (average)	= 170.0		
		E ring	a = 181.0 b = (480)		
thickness (km)	<30		<0.2		?
particle size	~ μm		~ cm to m		~ cm to m?
total mass (kg)	~ 10^7-10^{17}		~ 10^{17}-10^{19}		~ 10^{14}-10^{16}

From Beatty, O'Leary, and Chaikin 1982, p. 219; Collins et al. 1985; Elliott and Nicholson 1985, p. 37, Table 3.

Primordial radionuclides
extant

Radionuclide	Half-life (y)	% of element	Radionuclide	Half-life (y)	% of element
^{40}K	$1.277 \cdot 10^9$	0.0117	^{149}Sm	$>2 \cdot 10^{15}$	13.8
^{50}V	$\sim 150 \cdot 10^{15}$	0.250	^{152}Gd	$110 \cdot 10^{12}$	0.20
^{82}Se	$140 \cdot 10^{18}$	9.2	^{174}Hf	$2.0 \cdot 10^{15}$	0.162
^{87}Rb	$48 \cdot 10^9$	27.83	^{176}Lu	$36 \cdot 10^9$	2.59
^{113}Cd	$9.3 \cdot 10^{15}$	12.22	^{180}Ta	$>1.2 \cdot 10^{15}$	0.012
^{115}In	$441 \cdot 10^{12}$	95.7	^{180}W	$>1.1 \cdot 10^{15}$	0.13
^{123}Te	$13 \cdot 10^{12}$	0.91	^{184}Os	$>100 \cdot 10^{15}$	0.02
^{128}Te	$>8 \cdot 10^{24}$	31.69	^{186}Os	$2.0 \cdot 10^{15}$	1.58
^{130}Te	$2.5 \cdot 10^{21}$	33.80	^{187}Re	$50 \cdot 10^9$	62.60
^{138}La	$128 \cdot 10^9$	0.09	^{190}Pt	$600 \cdot 10^9$	0.01
^{142}Ce	$>50 \cdot 10^{15}$	11.08	^{204}Pb	$\geq 140 \cdot 10^{15}$	1.4
^{144}Nd	$2.1 \cdot 10^{15}$	23.80	^{232}Th	$14.05 \cdot 10^9$	100.00
^{145}Nd	$>60 \cdot 10^{15}$	8.30	^{235}U	$704 \cdot 10^6$	0.7200
^{147}Sm	$106 \cdot 10^9$	15.0	^{238}U	$4.468 \cdot 10^9$	99.2745
^{148}Sm	$7 \cdot 10^{15}$	11.3			

Quarks

Each quark comes in three "colors" (red, green, and blue). Anti-quarks have opposite baryon number (B), charge (electron charge = 1), strangeness (S), charm (C), and beauty (B^*).

Flavor	B	Spin $(h/2\pi)$	Mass (u)	Charge	S	C	B^*
up (u)	1/3	1/2	0.4	$+2/3$	0	0	0
down (d)	1/3	1/2	0.4	$-1/3$	0	0	0
strange (s)	1/3	1/2	0.5	$-1/3$	-1	0	0
charm (c)	1/3	1/2	1.8	$+2/3$	0	1	0
bottom (b)	1/3	1/2	5	$-1/3$	0	0	-1
top (t)	1/3	1/2	~ 50	$+2/3$	0	0	0

From Perkins 1982; Halzen and Martin 1984.

Planet	Name	Mean radius or dimensions (km)	Semimajor axis of orbit (1000 km)	Orbital inclination (degrees)	Sidereal period (days)	Mass (10^{21} kg)	Mean density (g/cm^3)
Earth	Moon	1738.2	384.401	5.15	27.32166	73.49	3.343
Mars	Phobos	19 × 21 × 27	9.37853	2.5	0.31891	0.0000096	1.8 ± 0.5
	Deimos	11 × 12 × 15	23.45981	2.0	1.26244	0.0000020	2.0 ± 0.6
Jupiter	Metis, J3	20?	128.2	0?	0.2948	—	—
	Adrastea, J1	15?	128.5	0?	0.29826	—	—
	Amalthea, J5	120	181.3	0.455	0.49818	—	—
	Thebe, J2	40?	223.0	1–2	0.67455	—	—
	Io, J1	1816	412.6	0.027	1.75032	89.169	3.55
	Europa, J2	1563	670.9	0.468	3.511181	48.730	3.04
	Ganymede, JIII	2638	1070	0.183	7.154533	149.000	1.93
	Callisto, JIV	2410	1883	0.253	16.689019	106.400	1.81
	Leda, JXIII	1–7	11110	26.7	238.7	—	—
	Himalia, JVI	85	11470	27.6	250.6	—	—
	Lysithea, JX	3–16	11710	29.0	260	—	—
	Elara, JVII	40	11740	24.8	260.1	—	—
	Ananke, JXII	3–14	20700	147	617	—	—
	Carme, JXI	4–20	22350	163	692	—	—
	Pasiphae, JVIII	4–23	23300	147	735	—	—
	Sinope, JIX	3–18	23700	156	758	—	—
Saturn	1980 S28	10 × 20	137.670	0.3	0.60192	—	—
	1980 S27	40 × 50 × 70	139.350	0.0	0.61300	—	—
	1980 S26	35 × 45 × 55	141.700	0.05	0.62854	—	—
	1980 S3	50 × 60 × 70	151.422	0.34	0.69433	—	—
	1980 S1	80 × 100 × 110	151.472	0.14	0.69466	—	—
	Mimas, S1	196	185.540	1.517	0.94242	0.0045?	1.4?
	Enceladus, S2	250	238.040	0.023	1.37021	0.084	1.2?
	Tethys, S3	530	294.670	1.093	1.88779	0.755	1.21
	1980 S13	13 × 14 × 17	294.670	1.1	1.88779	—	—
	1980 S25	11 × 11 × 17	294.670	1.1	1.88779	—	—
	Dione, S4	560	377.420	0.023	2.73692	1.050	1.43

Planet	Name	Mean radius or dimensions (km)	Semimajor axis of orbit (1000 km)	Orbital inclination (degrees)	Sidereal period (days)	Mass (10^{21} kg)	Mean density (g/cm^3)
Saturn *(continued)*							
	1980 S6	15 × 16 × 18	378.060	0?	2.73692	—	—
	Rhea, S5	765	527.100	0.35	4.5175	2.490	1.33
	Titan, S6	2575	1221.860	0.33	15.9454	134.570	1.88
	Hyperion, S7	110 × 130 × 205	1481.0	0.4	21.2767	—	—
	Iapetus, S8	730	3560.8	14.7	79.3308	1.880	1.16
	Phoebe, S9	110	12954.0	150.1	550.45	—	—
Uranus	1986U7	7.5	49.3	—	0.33	—	—
	1986U8	10	53.8	—	0.37	—	—
	1986U9	28	59.2	—	0.43	—	—
	1986U3	35	61.8	—	0.46	—	—
	1986U6	25	62.7	—	0.47	—	—
	1986U2	35	64.6	—	0.49	—	—
	1986U1	45	66.1	—	0.51	—	—
	1986U4	25	69.9	—	0.56	—	—
	1985U5	25	75.3	—	0.62	—	—
	1985U1	85	86.0	—	0.76	—	—
	Miranda, U5	242	129.9	4.22	1.41349	0.077	1.3?
	Ariel, U1	580	190.9	0.31	2.52038	1.307	1.6
	Umbriel, U2	595	266.0	0.36	4.14418	1.235	1.4
	Titania, U3	805	436.3	0.14	8.70587	3.496	1.6
	Oberon, U4	775	583.4	0.10	13.46325	2.925	1.5
	1986 U6-U15	—	—	—	—	—	—
Neptune	Triton	1600	355	159.945	5.876844	34?	2?
	Nereid	150?	5562	27.64	365.21	0.028??	2?
Pluto	Charon	642	19.7	94	6.3867	1.6?	1.84

From Burnham 1981; Morrison 1981; Beatty, O'Leary, and Chaikin 1982; Dermott and Nicholson 1986; Smith et al. 1986; JPL, unpublished.

Element	Abundances (mg/kg)	Principal species	Residence time (years)
H	110,000	H_2O	—
He	0.0000072	He	—
Li	0.17	Li^+	$20 \cdot 10^6$
Be	0.0000006	—	150
B	4.45	$B(OH)_3$; $B(OH)_2^-$	—
C	28	HCO_3^-; H_2CO_3; CO_3^{2-}	—
	2	organic compounds in solution	
N	0.67	NO_3^-; NO_2^-; NH_4^+	—
	15.5	N_2 in solution	
O	857,000	H_2O; O_2; SO_4^{2-}; other anions	
	6	O_2 in solution	
F	1.2	F^-	—
Ne	0.0000120	Ne	—
Na	10,800	Na^+	$260 \cdot 10^6$
Mg	1,290	Mg^{2+}; $MgSO_4$	$45 \cdot 10^6$
Al	0.01	—	100
Si	2.9	$Si(OH)_4$; $Si(OH)_3O^-$	8000
P	0.088	HPO_4^{2-}; $H_2PO_4^-$; H_3PO_4; PO_4^{3-}	—
S	904	SO_4^{2-}	—
Cl	19,400	Cl^-	—
Ar	0.45	Ar	—
K	392	K^-	$11 \cdot 10^6$
Ca	411	Ca^{2+}; $CaSO_4$	$8 \cdot 10^6$
Sc	<0.000004	—	5600
Ti	0.001	—	160
V	0.0019	$VO_2(OH)_3^{2-}$	10,000
Cr	0.0002	—	350
Mn	0.0019	Mn^{2+}; $MnSO_4$	1400
Fe	0.0034	$Fe(OH)_3$	140
Co	0.00039	Co^{2+}; $CoSO_4$	18,000
Ni	0.0066	Ni^{2+}; $NiSO_4$	18,000
Cu	0.023	Cu^{2+}; $CuSO_4$	50,000
Zn	0.011	Zn^{2+}; $ZnSO_4$	180,000
Ga	0.000030	—	1400
Ge	0.000060	$Ge(OH)_4$; $Ge(OH)_3^-$	7000
As	0.0026	$HAsO_4^{2-}$; $H_2AsO_4^-$; H_3AsO_4; H_3AsO_3	—
Se	0.000090	SeO_4^{2-}	—
Br	67.3	Br^-	—
Kr	0.00021	Kr	—
Rb	0.120	Rb^+	270,000
Sr	8.1	Sr^{2+}, $SrSO_4$	$19 \cdot 10^6$
Y	0.000003	—	7500
Zr	0.000026	—	—
Nb	0.000015	—	300
Mo	0.010	MoO_4^{2-}	500,000
Ru	0.0000007	—	—
Rh	—	—	—
Pd	—	—	—
Ag	0.00028	$AgCl_2^-$; $AgCl_3^{2-}$	$2.1 \cdot 10^6$
Cd	0.00011	Cd^{2+}; $CdSO_4$	500,000
In	0.000004	—	—

Element	Abundances (mg/kg)	Principal species	Residence time (years)
Sn	0.00081	—	100,000
Sb	0.00033	—	350,000
Te	—	—	—
I	0.064	IO_3^-; I^-	—
Xe	0.000047	Xe	—
Cs	0.0003	Cs^+	40,000
Ba	0.021	Ba^{2+}; $BaSO_4$	84,000
La	0.0000029	—	440
Ce	0.0000013	—	80
Pr	0.00000064	—	320
Nd	0.0000023	—	270
Pm	—	—	—
Sm	0.00000042	—	180
Eu	0.000000114	—	300
Gd	0.0000006	—	260
Tb	0.0000009	—	—
Dy	0.00000073	—	460
Ho	0.00000022	—	530
Er	0.00000061	—	690
Tm	0.00000013	—	1800
Yb	0.00000052	—	530
Lu	0.00000012	—	450
Hf	<0.000008	—	—
Ta	<0.0000025	—	—
W	<0.000001	WO_4^{2-}	1000
Re	0.0000084	—	—
Os	—	—	—
Ir	—	—	—
Pt	—	—	—
Au	0.000011	$AuCl_4^-$	560,000
Hg	0.00015	$HgCl_3^-$; $HgCl_4^{2-}$	42,000
Tl	<0.00001	Tl^+	—
Pb	0.00003	Pb^{2+}; $PbSO_4$	2000
Bi	0.00002	—	450,000
Po	—	—	—
At	—	—	—
Rn	$6 \cdot 10^{-16}$	Rn	—
Fr	—	—	—
Ra	$1.0 \cdot 10^{-10}$	Ra^{2+}; $RaSO_4$	—
Ac	—	—	—
Th	0.0000015	—	350
Pa	$2.0 \cdot 10^{-13}$	—	—
U	0.0033	$UO_2(CO_3)_3^{4-}$	500,000

From Turekian 1968; Horne 1969, p. 153–155, Table 5.3; Hood and Pytkowicz 1974.

Concentration (g/kg) of major ions in sea water of 35‰ salinity.

Ion	Average value	Ion	Average value
chloride	19.353	bicarbonate	0.145
sodium	10.77	bromide	0.0673
sulfate	2.712	strontium	0.0080
magnesium	1.295	boron	0.0046
calcium	0.412	fluoride	0.0013
potassium	0.399		

From Hood and Pytkowicz 1974, p. 4, Table 1.2-1.

**Sedimentary rocks
average mineral composition**

Mineral (% by mass)	Sandstone	Graywacke	Shale	Limestone
quartz	82	36	20	0
feldspars	5	26	15	0
muscovite	8	6	10	0
biotite	0	2	0	0
pyroxenes and amphiboles	0	5	0	0
clay minerals	2	22	52	14
calcite and dolomite	3	3	3	86

From Wedepohl 1969, p. 262, Table 8.5.

**Sediments
and derived sedimentary and metamorphic rocks**

Sediment	Sedimentary rock	Metamorphic rock
boulders ⎫ ⎬ pebbles ⎭	conglomerate	conglomerite
sand ⎧ sandstone ⎨ ⎩ graywacke		quartzite gneiss
silt ⎫ siltsone ⎬ clay ⎭ shale		schist
calcareous sand ⎫ ⎬ calcareous mud ⎭	limestone	marble

P and S wave velocities (km/s) in sediments and rocks at standard pressure (except as indicated).

Sediments	v_P	v_S
globigerina ooze	1.5	
red clay	1.5	—
loess	0.3–0.6	—
soil	0.3–0.9	—
clay	1.1–2.5	—
sand	1.0–2.0	0.5
sandstone	1.4–4.3	—
shale	2.1–3.5	—
chalk	2.1–4.2	—
marl	0.8–1.8	—
limestone:		
soft	1.7–4.2	—
hard	2.8–6.4	—
Metamorphic rocks		
slate	4.3	2.9
schist	4.9	3.3
quartzite	6.1	—
gneiss	3.5–7.5	—
dolomite	3.5–6.9	—
marble	3.8–6.9	2.0–3.9
Igneous rocks		
granite	5.0–6.0	2.9–3.2
granodiorite	4.6–4.9	3.1–3.2
diorite	5.8	3.6
andesite (35 bars)	5.2	2.7
gabbro (200 bars)	6.5–6.7	3.4–3.5
diabase	5.8–6.6	—
basalt	5.0–6.4	2.7–3.2
eclogite	8.0	4.3
dunite (200 bars)	8.6	4.4

From Press 1966, p. 197–204, Tables 9.1–9.5.

$v_P = P$ wave velocity; $v_S = S$ wave velocity.

Rock	Density (g/cm³)	Pressure (bars)						
		10	500	1000	2000	4000	6000	10,000
granite	2.630							
v_P		4.50	5.76	6.03	6.18	6.29	6.35	6.43
v_S		2.76	3.37	3.48	3.57	3.64	3.67	3.71
gabbro	2.953							
v_P		6.30	6.89	7.00	7.06	7.12	7.17	7.22
v_S		3.59	3.70	3.73	3.76	3.79	3.82	3.84
eclogite	3.465							
v_P		6.56	—	7.50	7.61	7.71	7.77	7.86
v_S		3.61	4.37	4.43	4.69	4.54	4.57	4.61
dunite	3.290							
v_P		7.31	7.73	7.84	7.92	8.00	8.06	8.14
v_S		3.92	4.05	4.07	4.10	4.14	4.17	4.22

From Press 1966, p. 198–201, Tables 9.2 and 9.3.

Sound velocities

Velocity of sound (*P* waves) in different substances at 25°C (unless otherwise indicated).

Substance	Velocity (km/s)	Substance	Velocity (km/s)
air, −40°C	0.30622	lead	2.16
air, −20°C	0.31909	limestone, soft	1.7–4.2
air, 0°C	0.33129	limestone, hard	2.8–6.4
air, 20°C	0.34337	loam	0.8–1.8
air, 40°C	0.35489	marble	3.8–6.9
aluminum	6.420	nickel	6.040
argon, 20°C	0.319	nitrogen (N_2), 0°C	0.334
basalt	5.0–6.4	oxygen (O_2), 0°C	0.316
beryllium	12.890	peridotite	7.8–8.4
carbon (diamond)	18.1	quartz	5.75
carbon dioxide, 0°C	0.259	quartzite	6.1
chalk	2.1–4.2	red clay	1.5
clay	1.1–2.5	sand	1.0–2.0
copper	4.760	sandstone	1.4–4.3
diorite	5.8	schist	4.9
dolomite	3.5–6.9	shale	2.1–3.5
eclogite	8.0	slate	4.3
gabbro, 200 bars	6.5–6.7	soil	0.1–0.2
globigerina-ooze	1.5	steel, 1% C	5.940
gneiss	3.5–7.5	water, fresh, 0°C	1.4035
granite	5.0–5.9	water, fresh, 25°C	1.509
helium, 0°C	0.965	water, sea, 0°C, 35‰ salinity, 1 atm	1.4491
hydrogen, 0°C	1.284	water, sea, 0°C, 35‰ salinity, 500 atm	1.5349
ice	3.43	water, sea, 25°C, 35‰ salinity, 0 atm	1.5346
iron	5.950	water vapor (134°C)	0.494

From Press 1966, p. 197–198, Table 9.1; Moses 1978, p. 528–529, Table 1k5; Weast 1986.

*Identifies excited state.

Reaction	Energy produced (MeV)
Proton–proton Chain	
$^1H + {}^1H \rightarrow H^2 + e^+ + \nu_e$	1.179
$^2H + {}^1H \rightarrow {}^3He + \gamma$	5.493
$^3He + {}^3He \rightarrow {}^4He + 2\,{}^1H$	12.859 or
$^3He + {}^4He \rightarrow {}^7Be + \gamma$	1.586
$^7Be + e^- \rightarrow {}^7Li + \bar{\nu}_e$	0.061
$^7Li + {}^1H \rightarrow 2\,{}^4He$	17.347 or
$^3He + {}^4He \rightarrow {}^7Be + \gamma$	1.586
$^7Be + {}^1H \rightarrow {}^8B + \gamma$	0.135
$^8B \rightarrow {}^8Be + e^+ + \nu_e$	10.78
$^8Be \rightarrow 2\,{}^4He$	0.095
Carbon–Nitrogen–Oxygen Cycle	
$^{12}C + {}^1H \rightarrow {}^{13}N + \gamma$	1.944
$^{13}N \rightarrow {}^{13}C + e^+ + \nu_e$	1.511
$^{13}C + {}^1H \rightarrow {}^{14}N + \gamma$	7.550
$^{14}N + {}^1H \rightarrow {}^{15}O + \gamma$	7.293
$^{15}O \rightarrow {}^{15}N + e^+ + \nu_e$	1.761
$^{15}N + {}^1H \rightarrow {}^{12}C + {}^4He$	4.965
$^{15}N + {}^1H \rightarrow {}^{16}O + \gamma$	12.126
$^{16}O + {}^1H \rightarrow {}^{17}F + \gamma$	0.061
$^{17}F \rightarrow {}^{17}O + e^+ + \nu_e$	1.822
$^{17}O + {}^1H \rightarrow {}^{14}N + {}^4He$	1.193
Triple-alpha Process	
$^4He + {}^4He \rightarrow {}^8Be$	-0.0921
$^8Be + {}^4He \rightarrow {}^{12}C^*$	-0.286
$^{12}C^* \rightarrow {}^{12}C + \gamma$	7.656

From Lang 1980, p. 421–423.

m_V = apparent visual magnitude measured photoelectrically to match the yellow color-sensitivity of the eye; M_V = absolute visual magnitude; v = variable (m_V and M_V are maximum values).

Name		Coordinates				m_V	M_V	Spectral type	Distance (l.y.)
		R.A.		Decl.					
		h	m	deg.	min				
Sun	—	—	—	—	—	−26.73	+4.84	G2	0.0000158
Sirius	α Canis Maioris A	06	44.2	−16	42	−1.45	+1.45	A1	8.7
Canopus	α Carinae	06	23.5	−52	41	−0.72	−3.1	F0	98
Arcturus	α Bootis	14	14.8	+19	17	−0.06	−0.3	K2	36
Vega	α Lyrae	18	36.2	+38	46	0.04	+0.5	A0	26.5
Rigil ⎰	α Centauri A	14	38.4	−60	46	0.01	+4.39	G2	4.3
Kentaurus ⎱	α Centauri B	14	38.4	−60	46	1.40	+5.8	K4	4.3
Capella	α Aurigae	05	15.2	+45	59	0.05	−0.6	G8	45
Rigel	β Orionis A	05	13.6	−08	13	0.14v	−7.1	B8	900
Procyon	α Canis Minoris A	07	38.2	+05	17	0.37	+2.7	F5	11.3
Achernar	α Eridani	01	37.0	−57	20	0.51	−2.3	B3	118
Hadar	β Centauri AB	14	02.4	−60	16	0.63v	−5.2	B1	490
Betelgeuse	α Orionis	05	54.0	+07	24	0.41v	−5.6	M2	520
Altair	α Aquilae	19	49.8	+08	49	0.77	+2.2	A7	16.5
Aldebaran	α Tauri A	04	34.8	+16	28	0.86v	−0.7	K5	68
Spica	α Virginis	13	24.1	−11	03	0.91v	−3.3	B1	220
Antares	α Scorpii A	16	28.2	−26	23	0.92v	−5.1	M1.5	520
Fomalhaut	α Piscium A	22	56.5	−29	44	1.15	+2.0	A3	22.6
Pollux	β Geminorum	07	44.1	+28	05	1.16	+1.0	K0	35
Deneb	α Cygni	20	40.7	+45	12	1.26	−7.1	A2	1600
Beta Crucis	β Crucis	12	46.6	−59	35	1.28v	−4.6	B0.5	490
Regulus	α Leonis A	10	07.3	+12	04	1.36	−0.7	B7	84
Alpha Crucis ⎰	α Crucis A	12	25.4	−62	59	1.39	−3.9	B0.5	370
⎱	α Crucis B	12	25.4	−62	59	1.86	−3.4	B1	370
Adhara	ε Canis Maioris A	06	57.8	−28	57	1.48	−5.1	B2	680

From Bishop et al. 1986.

α	activity		b	barn
	alpha particle			semiminor axis of an elliptical orbit
	angular acceleration		B	boron
	attenuation coefficient			magnetic flux density, magnetic induction
	fine structure constant			susceptance
	isotopic fractionation factor		**B**	magnetic flux density, magnetic induction
	right ascension		b^I	galactic latitude in the old IAU system
a	absorbance			
	acceleration		b^{II}	galactic latitude in the new IAU system
	activity			
A	*annus* (Latin for "year")		Ba	barium
	optical depth		B.C.	before Christ, referring to the time preceding 0h 0m 0s of January 1, A.D. 1
	semimajor axis of an elliptical orbit			
	ampere			
	area		Be	beryllium
	atomic mass number		BeV	billion electron volt ($= 10^9$ eV)
	avogadro		Bi	bismuth
	azimuth		BIF	banded iron formation
	Helmholtz energy ($= U - TS$)		Bk	berkelium
Å	angstrom		bp	boiling point
Ab	albite		B.P.	before the present, indicating the time before A.D. 1950
ac	alternating current			
Ac	actinium		Br	bromine
A.C.	Ante Christum, Latin for "before Christ"		BTU	British thermal unit
ACS	American Chemical Society		c	curie
A.D.	*Anno Domini*, Latin for "year of the Lord," referring to any year after the birth of Christ taken as having occurred at the beginning of A.D. 1			specific heat capacity
				speed of light *in vacuo*
				speed of sound
			C	capacitance
ADP	adenosine diphosphate			carbon
ae	aeon ($= 10^9$ y)			Celsius, centigrade
Ag	*argentum*, Latin for "silver"			coulomb
AIP	American Institute of Physics			heat capacity
Al	aluminum		Ca	calcium
Am	americium		cal	calorie
AM	amplitude modulation		cal_{IT}	International Table calorie
amp	ampere		cc	cubic centimeter
AMP	adenosine monophosphate		CCD	carbonate compensation depth
amu	atomic unit (¹⁄₁₆ of the mass of ^{16}O)		cd	candela
An	anorthite		Cd	cadmium
a_0	Bohr radius		CD	compensation depth
Ar	argon		Ce	cerium
	aryl		CERN	Conseil Européen Recherches Nucleaires
As	arsenic			
ASTM	American Society for Testing and Materials		cf.	confer, Latin for "compare with"
			Cf	californium
At	astatine		CGS	centimeter–gram–second
atm	atmosphere		CIPW	Cross-Iddings-Pirrson-Washington, who devised the norm system of expressing the mineral composition of rocks
ATP	adenosine triphosphate			
Au	*aurum*, Latin for "gold"			
av.	average			
β	beta particle		Cl	chlorine
β^+	beta plus particle ($=$ positron)		cm	centimeter
β^-	beta minus particle ($=$ electron)			

Cm	curium		E	modulus of elasticity (Young's modulus)
Co	cobalt		**E**	electric field strength
COP	coefficient of performance		EDTA	ethylenediaminetetraacetic acid
cos	cosine		e.g.	*exempli gratia,* Latin for "for example"
cosec	cosecant			
cot	cotangent		E_k	kinetic energy
covers	coversine		emf	electromotive force
cP	centipoise		emu	electromagnetic unit
C_p	heat capacity at constant pressure		E_p	potential energy
cps	cycles per second		EPM	Electron Paramagnetic Resonance
Cr	chromium		Er	erbium
CRT	cathode-ray tube		ERTS	Earth Resource Technology Satellite
Cs	cesium		Es	einsteinium
ct	carat		ESCA	Electron Spectroscopy for Chemical Analysis
csc	cosecant			
Cu	copper		ESR	Electron Spin Resonance
C_v	heat capacity at constant volume		esu	electrostatic unit
			ET	ephemeris time
δ	declination		Eu	europium
∂	partial derivative		eV	electron volt
d	day		exp	exponential
	density, relative (usually referred to that of water at 3.98 or 4.0°C and 1 atm pressure)		η	viscosity
			θ, ϑ	phase angle volume strain (bulk strain) $(= V/V_0)$
	dextrorotatory			
	diameter		ϕ	angular displacement
	diffuse, identifying the orbital of an atomic electron with orbital angular momentum $= 2$			latitude
				phase angle
			Φ	magnetic flux
D	deuterium		f	focal length
	dextral chirality			force
	diffusion coefficient			frequency
D	electric displacement			fundamental, identifying the orbital of an atomic electron with an orbital angular momentum number $= 3$
da-	deca-, deka-			
dam	decameter			
dB	decibel		F	Fahrenheit
dc	direct current			farad (unit of capacitance in the MKS system)
d_E	ephemeris day			
dl	dextro-levo, indicating a racemic mixture			faraday $(=$ Faraday constant)
				Faraday constant $(=$ charge of 1 mol of electrons $= 96,484.56$ coulombs)
dm	decimeter			
DNA	deoxyribonucleic acid			fluorine
dpm	disintegrations per minute			force
dps	disintegrations per second			formality
DSDP	Deep-Sea Drilling Project		Fe	iron
Dy	dysprosium		FET	field-effect transistor
dyn	dyne		FI	felsic index
			Fm	fermium
ϵ	permittivity $(= \mathbf{D/E})$		FM	frequency modulation
ϵ_0	permittivity constant		fps	foot-pound-second
e	eccentricity		Fr	francium
	electron charge		ft	foot
	linear strain (relative elongation) $(= 1/1_0)$		γ	activity coefficient
	$2.7182818284590\ldots$			gamma $(= 10^{-5}$ oersted$)$

γ	gyromagnetic ratio		IC	integrated circuit
	photon		i.e.	*id est,* Latin for "that is"
	surface tension		in.	inch
Γ	gamma ($= 10^{-5}$ oersted)		In	indium
	width of resonant state		Ir	iridium
g	gram		IR	infrared
	gravitational acceleration of the Earth			insoluble residue
G	conductance		IUGG	International Union of Geology and Geophysics
	Gauss		IUGS	International Union of Geological Sciences
	Gibbs free energy		IUPAC	International Union of Pure and Applied Chemistry
	gravitational constant			
Ga	gallium			
g-cal	gram-calorie			
Gd	gadolinium		j	electric current density
Ge	germanium			total angular momentum quantum number
GeV	gigaelectronvolt ($= 10^9$ eV)			$(-1)^{1/2}$
GMAT	Greenwich Mean Astronomical Time		**j**	electric current density
GMT	Greenwich Mean Time		J	electric current density
g_0	standard gravitational acceleration of the Earth			joule
				total angular momentum quantum number
GSA	Geological Society of America		JD	Julian date
GST	Greenwich Sidereal Time			Julian day
Gy	gigayear ($= 10^9$ y)			
			κ	electric conductivity ($= \mathbf{J}/\mathbf{E}$)
h	celestial altitude		k	Boltzmann constant
	hecto-			kilo-
	hour		K	equilibrium constant
	Planck's constant			kelvin
$\hbar$	h bar ($= h/2\pi$)			kinetic energy
H	enthalpy ($= U + pV$)			potassium
	Hamiltonian		kb	kilobar
	henry		k-cal	kilocalorie
	hydrogen		keV	kiloelectronvolt ($= 10^3$ eV)
H	magnetic field strength		kg	kilogram
HI	neutral hydrogen		km	kilometer
HII	ionized hydrogen		Kr	krypton
He	helium		kt	karat
Hf	hafnium		kV	kilovolt
HFU	heat flow unit		kWh	kilowatthour
Hg	mercury			
H_0	Hubble constant		λ	decay constant
Ho	holmium			wavelength
hp	horsepower		λ_c	electron Compton wavelength
hr	hour		$\lambda_{C,n}$	neutron Compton wavelength
Hz	hertz		$\lambda_{C,p}$	proton Compton wavelength
			l	length
i	electric current density			levorotatory
	inclination			liter
	$(-1)^{1/2}$			orbital angular momentum quantum number
I	electric current		L	angular momentum
	iodine			inductance
	ionic strength			Lagrangian
	luminous intensity			lambert
	moment of inertia			
IAT	International Atomic Time			
IAU	International Astronomical Union			

L	left-handed chirality
	luminance
	luminosity
	self-inductance
	sinistral chirality
l^I	galactic longitude in the old IAU system
l^{II}	galactic longitude in the new IAU system
L_{12}	mutual induction
La	lanthanum
lb	pound
lbf	pound force
LED	light-emitting diode
Li	lithium
LIL	large-ion lithophyle element
lm	lumen
ln	logarithm (natural)
log	logarithm (common)
Lr	lawrencium
LST	local sidereal time
LT	local time
Lu	lutetium
lx	lux
ly	langley
l.y.	light year
μ	ionic strength
	magnetic moment
	magneton
	micron
	permeability ($= B/H$)
	proper motion
	viscosity (dynamic)
μ_μ	muon magnetic moment
μ_B	Bohr magneton
μ_e	electron magnetic moment
μF	microfarad
$\mu\mu F$	micromicrofarad
μm	micrometer
μ_N	nuclear magneton
μ_0	permeability constant
μ_p	proton magnetic moment
μ_r	relative permeability ($= \mu/\mu_0$)
m	apparent magnitude
	mass
	meter
	minute
	molal concentration
m	electromagnetic moment ($= -E_p/B$)
m-	meta-
M	absolute magnitude
	magnetization
	mega ($= 10^6$)

M	Messier number
	molar concentration
	mutual induction
M	magnetization [$= (B/\mu_0) - $ **H**]
Ma	mega-annus ($= 10^6$ y)
m_{bol}	bolometric magnitude
md	millidarcy
Md	mendelevium
m_e	electron rest mass
MeV	million electronvolts
Mg	magnesium
mgal	milligal
MHW	mean high water
mi	mile
MKS	meter-kilogram-second
MKSA	meter-kilogram-second-ampere
MKSΩ	meter-kilogram-second-ohm
m_l	magnetic orbital quantum number
ml	milliliter
MLW	mean low water
m_μ	muon rest mass
mmf	magnetomotive force
mmHg	millimeters of mercury
m_n	neutron rest mass
Mn	manganese
m_0	rest mass
$M_\odot$	solar mass
Mo	molybdenum
mol	mole
MORB	mid-oceanic ridge basalt
MOS	metal-oxide semiconductor
MOSFET	metal-oxide semiconductor field-effect transistor
m_π	pion rest mass
m_p	proton rest mass
Mpc	megaparsec ($= 10^6$ parsecs)
m_{pe}	apparent photoelectric magnitude
m_{pg}	apparent photographic magnitude
m_{pv}	apparent photovisual magnitude
mRNA	messenger RNA
m_s	magnetic spin angular momentum quantum number
MSL	mean sea level
Mt.	mount
	mountain
MTL	mean tide level
mV	millivolt
m_{vis}	apparent visual magnitude
M_{vis}	absolute visual magnitude
mW	milliwatt
MW	megawatt
MWL	mean water level
Mx	maxwell
m.y.	million years

ν	frequency
	neutrino
	viscosity, kinematic ($= \eta/\rho$)
n	index of refraction
	neutron
	principal quantum number
N	newton
	nitrogen
	normal concentration
	north-seeking pole of magnetic dipole
N_A	Avogadro number
Na	sodium
NAD	nicotinamide adenine dinucleotide
NADH	reduced NAD
NADP	nicotinamide adenine dinucleotide phosphate
NADPH	reduced NADP
NAP	nonarboreal pollen
Nb	niobium
Nd	neodymium
Ne	neon
ng	nanogram
NGC	New General Catalogue of Nebulae and Stars
Ni	nickel
nm	nanometer
NMR	nuclear magnetic resonance
n.n.	*nomen nudum,* Latin for "naked name"
N_{Nu}	Nusselt number
No	nobelium
Np	neper
	neptunium
N_{Pr}	Prandtl number (convection)
N_{Re}	Reynolds number
NRM	natural remanent magnetization
nt	nit
o-	ortho-
O	oxygen
OD	ordnance datum
Oe	oersted
Os	osmium
π	3.141592653589 ...
II	osmotic pressure
p	momentum
	pressure
	principal, identifying the orbital of an atomic electron with orbital angular momentum $= 1$
	proton
P	parity
	permeance
	phosphorus

P	poise
	primary or pressure wave
P	dielectric polarization ($= \mathbf{D} - \epsilon_0\mathbf{E}$)
p-	para-
Pa	pascal
	protactinium
Pb	lead
pc	parsec
Pd	palladium
PDB	Peedee Belemnite, a standard to which O and C isotope abundances are referred
pdl	poundal
PDR	Precision Depth Recorder
pH	p(otential of) H(ydrogen)
pK	$-\log_{10}$ of ionization constant K
Pm	promethium
Po	polonium
*p*OH	$-\log_{10}$ of OH$^-$ ion concentration in aqueous solution
ppm	parts per million
ppt	parts per thousand
Pr	praseodymium
	total Prandtl number
Pr_M	Prandtl number (mass diffusion)
PSI	pounds per square inch
P_{syn}	synodic period
Pt	platinum
Pu	plutonium
q	perihelion distance
Q	aphelion
	electric charge
	heat
	quality factor
QCD	quantum chromodynamics
QED	quantum electrodynamics
Q.E.D.	*quod erat demonstrandum,* Latin for "which was to be demonstrated"
QSO	quasi-stellar object
q.v.	*quod vide,* Latin for "which you should see"
ρ	density (mass/volume)
	electric charge density (volumetric)
	resistivity ($= \mathbf{E/j}$)
r	radius
R	cosmic scale factor
	gas constant
	radical
	resistance
$\Re$	reluctance
R_∞	Rydberg constant
Ra	radium
RA	right ascension

Symbols and abbreviations *(continued)*

RAM	random-access memory
Rb	rubidium
Re	rhenium
REE	rare earth elements
rem	roentgen-equivalent-man
Rh	rhodium
rms	root-mean-square
Rn	radon
RNA	ribonucleic acid
ROM	read-only memory
Ru	ruthenium
σ	electric conductivity
	electric charge density (surface)
	neutron capture cross section
	Poisson ratio
	standard deviation
	Stefan–Boltzmann constant
Σ	summation
s	second
	sharp, identifying the orbital of an atomic electron with orbital angular momentum = 0
S	entropy
	secondary or shear wave
	siemens
	south-seeking pole of magnetic dipole
	strangeness number
	sulfur
s_A	atomic second
sb	stilb
Sb	antimony
Sc	scandium
SCR	silicon-controlled rectifier
s_E	ephemeris second
Se	selenium
sec	secant
SEM	scanning electron microscope
Si	silicon
SI	Système International (d'Unités)
sin	sine
s.l.	*sensu lato,* Latin for "in the broader sense"
Sm	samarium
SMOW	Standard Mean Ocean Water
Sn	tin
SNU	solar neutrino unit
S°	standard absolute entropy
sp.	species (sing.)
spp.	species (pl.)
sr	steradian
Sr	strontium
s.s.	*sensu stricto,* Latin for "in the strict sense"
St	stoke

statC	statcoulomb
STP	standard temperature and pressure
τ	hour angle
	mean life
	shear stress
t	temperature (Celsius)
	time
T	absolute temperature
	period $(= 1/\nu)$
	tesla
	tritium
$t_{1/2}$	half-life
t_a	atomic time
Ta	tantalum
TAI	Temps Atomique International
tan	tangent
Tb	terbium
Tc	technetium
t_E	ephemeris time
T_{eff}	effective temperature
Te	tellurium
Th	thorium
Ti	titanium
Tl	thallium
TL	thermoluminescence
Tm	thulium
torr	torricelli
TRM	thermoremanent magnetization
tRNA	transfer RNA
T_U	Universal Time
u	atomic mass unit ($\frac{1}{12}$ of mass of ^{12}C)
	velocity
U	internal energy
	uranium
UBV	ultraviolet-blue-visual
USGS	United States Geological Survey
UT	Universal Time
UV	ultraviolet
v	velocity
V	electric potential
	potential energy
	vanadium
	volt
	voltage
	volume
vers	versine
v_P	velocity of the P waves
VRM	viscous remanent magnetization
v_S	velocity of the S waves
W	energy
	tungsten
	watt
	work

Wb	weber		ζ	zenith distance
Whr	watt-hour		z	redshift parameter
			Z	ac impedance ($= R + iX$)
χ	magnetic susceptibility ($= \mu_r - 1$)			atomic number
χ_e	electric susceptibility ($= \epsilon_r - 1$)			valence
X	reactance			ZULU
X_C	capacitative reactance		Zn	zinc
X_{CL}	capacitative-inductive reactance		Zr	zirconium
Xe	xenon		ZULU	Greenwich Mean Time
X_L	inductive reactance			
XPS	X-ray Photoelectron Spectroscopy		ω	angular frequency ($= 2\pi f$)
				angular velocity ($= d\phi/dt$)
y	year			circular frequency ($= 2\pi\nu$)
Y	admittance ($= 1/Z$)		Ω	ohm
	Young's modulus			solid angle
	yttrium			
Yb	ytterbium			
yd	yard			

The classification of living and fossil groups presented here is based on Margulis 1981, p. 353–363. Two superkingdoms (Procaryota and Eucaryota) and five kingdoms (Monera, Protoctista, Fungi, Animalia, Plantae) are recognized. Exclusively fossil taxa are included. No stratigraphic range is given for taxa that left unverified fossil record.

Superkingdom Procaryota (Archean-Recent)
  Kingdom Monera (Archean-Recent)
    Phylum   1. Aphragmabacteria (unable to form cell walls; *Mycoplasma*)
            2. Chemoautotrophs
                Class 1. Sulfur-oxidizing bacteria *(Thiobacillus)*
                Class 2. Ammonia-oxidizing bacteria *(Nitrobacter)*
                Class 3. Iron-oxidizing bacteria *(Ferrobacillus)*
            3. Thiopneutes (anaerobic reducers of sulfate or sulfur to H_2S; *Desulfovibrio*)
            4. Metanocreatrices (methane-synthesizers; *Methanobacterium*)
            5. Fermenting bacteria (unable to synthesize porphirins; *Clostridium*)
            6. Spirochaetae (faculative or obligate anaerobes; *Spirochaeta, Treponema*)
            7. Thiorhodaceae (green and purple anaerobic protosynthesizers using bacteriochlorophyll and chlorobium chlorophyll; *Chlorobium, Chromatium*)
            8. Athiorhodaceae (nonsulfur anaerobic or facultative anaerobic photosinthesizers using bacteriochlorophyll and chlorobium chlorophyll, Archean-Recent; *Rhodospirillum*)
            9. Cyanobacteria (blue-green algae, aerobic or facultative anaerobes; use chlorophyll a and other pigments, Archean-Recent; *Nostoc, Oscillatoria*)
          10. Prochlorophyta (procaryotic green algae; *Prochloron*)
          11. Nitrogen-fixing aerobic bacteria *(Azobacter)*
          12. Pseudomonads (aerobic heterotrophs; *Pseudomonas*)
          13. Aeroendospora (aerobic endospore bacteria; *Bacillus*)
          14. Micrococci (aerobes with Krebs; *Paracoccus, Sarcina*)
          15. Omnibacteria (aerobic heterotrophs; *Acetobacter, Caulobacter, Escherichia, Leptothrix, Neisseria, Salmonella, Spirillum*)
          16. Actinobacteria *(Actinomyces, Streptomyces)*
          17. Myxobacteria (heterotrophic aerobic gliding bacteria; *Beggiatoa, Saprospira*)

Superkingdom Eucaryota (Proterozoic-Recent)
  Kingdom Proctoctista (single-celled microorganisms and their immediate multicellular descendants; Proterozoic-Recent)
    Phylum   1. Caryoblastea (amitotic amoebae; *Pelomyxa*)
            2. Dinoflagellata (dinoflagellates, Triassic-Recent; *Gymnodinium, Noctiluca*)
            3. Sarcodina and Rhizopoda *(Amoeba, Difflugia)*
            4. Chrysophyta (golden-yellow algae; *Dinobryon, Synura*)
            5. Euglenophyta *(Euglena)*
            6. Cryptophyta (cryptomonads; *Cryptomonas*)
            7. Zoomastigina (animal flagellates; *Diplomonas, Trichomonas*)
            8. Eumastigophyta *(Vischeria)*
            9. Bacillariophyta (diatoms, Cretaceous-Recent; *Coscinodiscus, Nitschia*)
          10. Haptophyta (Coccolithophoridae, Triassic-Recent; *Coccolithus, Emiliania*)
          11. Actinopoda (heliozoans and radiolaria, Cambrian-Recent; *Lampocyrtis, Pterocanium*)
          12. Foraminifera (Cambrian-Recent; *Globigerina, Globorotalia, Cibicides, Nodosaria*)
          13. Gamophyta and desmids *(Spirogyra)*
          14. Ciliophora (ciliates; *Paramecium*)
          15. Cnidosporidia (parasites; *Myxobolus, Nosema*)
          16. Apicomplexa (parasites; *Plasmodium*)
          17. Xanthophyta (yellow-green algae; *Botrydium, Vaucheria*)
          18. Rhodophyta (red algae, Cambrian-Recent; *Archaeolithothamnium, Lithothamnium, Lithophyllum, Solenopora*)

19. Chlorophyta (green algae, Archean-Recent; *Botryococcus, Chara, Halimeda, Oedogonium, Penicillus, Sargassum, Spyrogyra, Ulotrix, Valonia*)
20. Phaeophyta (brown algae, Silurian-Recent; *Fucus, Laminaria, Protaxites*)
21. Labyrinthulamycota (slime nets; *Labyrinthula*)
22. Acrasiomycota (cellular slime molds; *Acrasia*)
23. Myxomycota (plasmodial noncellular slime molds; *Dictyostelium, Polyspondylium*)
24. Plasmodiophoromycota *(Polymyxa)*
25. Hyphochytridiomycota *(Rhyzidiomyces)*
26. Chytridiomycota *(Olpidium)*
27. Oomycota *(Saprolegnia)*

Kingdom Fungi

Phylum 1. Zygomycota (zygomycetes; *Phycomyces*)
2. Ascomycota (sac fungi, including yeasts, molds, truffles, etc.; *Ascobolus, Aspergillus, Neurospora, Penicillium, Tuber* and other truffles)
3. Basidiomycota (club fungi, including rusts, smuts, mushrooms, etc.; *Agaricus, Amanita* and other mushrooms, *Puccinia*)
4. Deuteromycota (fungi imperfecti; *Candida, Monilia*)
5. Mycophycophyta [lichens, consisting of a fungus + a blue-green (often *Nostoc*) or a green alga (often *Trebouxia* or *Pseudotrebouxia*)]

Kingdom Animalia
Subkingdom Parazoa

Phylum 1. Placozoa (no polarity or bilateral symmetry; *Trichoplax*)
2. Porifera (sponges; Cambrian-Recent)
  Class 1. Heteractinida (calcareous spicules; Cambrian-Permian)
  2. Calcarea (calcareous spicules, Carboniferous-Recent; *Clathrina, Eudea, Leucosolenia*)
  3. Demospongiae (spongin with or without siliceous spicules, Cambrian-Recent; *Cliona,* boring sponge; *Euspongia,* horny sponge; *Hippospongia,* bath sponge)
  4. Sclerospongiae (aragonitic skeleton; Ordovician-Recent)
    Order 1. Stromatoporida (Ordovician-Devonian; *Stromatopora*)
    2. Ceratoporellida (Caribbean; *Asterosclera, Ceratoporella*)
    3. Tabulofungida (Pacific; *Stromatospongia*)
  5. Hexactinellida (triaxial siliceous spicules, Cambrian-Recent; *Hexactinella, Hydnoceras, Ventriculites*)
3. Archaeocyatha (Early-Middle Cambrian)
  Class 1. Monocyatha (single-walled skeleton, Early-Middle Cambrian; *Monocyathus*)
  2. Archaeocyatha (double-walled skeleton, Early-Middle Cambrian; *Archaeocyathellus*)

Subkingdom Eumetazoa
Branch Radiata

4. Cnidaria (coelenterates; Ediacaran-Recent)
  Class 1. Hydrozoa (Ediacaran-Recent)
    Order 1. Hydroida (Ediacaran-Recent; *Hydra, Obelia*)
    2. Hydrocorallina (Triassic-Recent; *Millepora, Stylaster*)
    3. Trachylina (Ediacaran-Recent; *Cyclomedusa, Gonionemus, Kirklandia, Olindias*)
    4. Siphonophora (Cambrian-Recent; *Physalia, Velella*)
  2. Scyphozoa (jellyfish, Ediacaran-Recent; *Aurelia, Conomedusites*)
  3. Conulata (Ediacaran-Triassic; *Conomedusites, Conularia*)

        4. Anthozoa (corals and sea anemones, Ediacaran-Recent)
           Subclass 1. Tabulata (tabulate corals, Ordovician-Permian; *Favosites,*
                  Halysites, Syringopora, Tubipora)
               2. Rugosa (tetracorals, Ordovician-Permian; *Lithostrotion,*
                  Zaphrentis)
               3. Schizocorallia (Ordovician-Jurassic; *Tetradium)*
               4. Zoantharia (hexacorals; Triassic-Recent)
                  Order 1. Actinaria (sea anemones, Cambrian-Recent; *Actinia,*
                      Edwardsia, Mackenzia)
                     2. Scleractinia (stony corals; Triassic-Recent; *Acropora,*
                      Fungia, Montastrea, Porites)
                     3. Zoanthidea (some resemblance to extinct Rugosa in
                      the arrangement of septa; *Zoanthus)*
                     4. Antipatharia (black corals; *Antipathes)*
                     5. Ceriantharia *(Cerianthus)*
           5. Alcyonaria (octocorals; Ediacaran-Recent)
                  Order 1. Stolonifera *(Tubipora)*
                     2. Telestacea *(Telesto)*
                     3. Alcyonacea (the soft corals, Cretaceous-Recent;
                      Alcyonium, Xenia)
                     4. Coenothecalia (Cretaceous-Recent; *Heliopora)*
                     5. Gorgonacea (the horny corals, Cretaceous-Recent;
                      Corallium, the red coral; *Gorgonia,* the sea fan;
                      Plexaura, the sea whip)
                     6. Pennatulacea (sea pens, Ediacaran-Recent;
                      Pennatula)
     5. Ctenophora (comb jellies; *Cestum, Folia)*
Branch Bilateria
   Grade Acoelomata (lack coelom)
      6. Mesozoa (small, parasitic, worm-like organisms; *Rhopalura)*
      7. Platyhelminthes (flatworms)
         Class 1. Turbellaria (planarians; *Dugesia)*
              2. Trematoda (flukes; *Fasciola)*
              3. Cestoda (tape worms; *Taenia)*
      8. Nemertina (ribbon worms, Recent; *Lineus)*
      9. Gnathostomulida [small (<3.5 mm) marine worms; *Gnathostomula*]
   Grade Pseudocoelomata
     10. Gastrotricha (microscopic pseudocoelomates; *Chaetonotus)*
     11. Kinorhyncha (tiny marine animals with segmented cuticle; *Echinoderes)*
     12. Loricifera (tiny marine animals with Loricate abdomen; *Pliciloricus)*
     13. Acanthocephala (spiny-headed worms; *Echinorhynchus)*
     14. Nematoda (round worms; *Ascaris, Trichina)*
     15. Nematomorpha (Gordiacea) (horsehair worms; *Gordius)*
     16. Entoprocta (endoproct bryozoids; *Urnatella)*
   Grade Coelomata (with mesodermal coelom)
     17. Ectoprocta (Bryozoa; Late Cambrian-Recent; *Bugula)*
     18. Phoronida (tubular mud-dweller, marine; Cambrian ?—Recent; *Phoronis)*
     19. Brachiopoda (Cambrian-Recent)
        Class 1. Inarticulata (Cambrian-Recent; *Lingula)*
              2. Articulata (Cambrian-Recent; *Productus, Spirifer, Terebratula)*
     20. Mollusca (Cambrian-Recent)
        Class 1. Aplacophora (without shell; *Neomenia)*
              2. Monoplacophora (Cambrian-Recent; *Neopilina)*
              3. Polyplacophora (chitons, Cambrian-Recent; *Chiton)*
              4. Scaphopoda (Ordovician-Recent; *Dentalium)*

     5. Hyolitha (Cambrian-Permian; *Ceratotheca, Hyolithes*)

     6. Gastropoda (Cambrian-Recent; *Batillaria, Cerithium, Haliotis, Helix, Littorina, Murex, Natica, Oliva, Patella, Purpura, Strombus*)

     7. Bivalvia (pelecypods, Ordovician-Recent; *Arca, Anomia, Astarte, Codakia, Cyprina, Lucina, Macoma, Mactra, Mercenaria, Mya, Mytilus, Natica, Pecten, Solen, Tellina*)

     8. Cephalopoda (Cambrian-Recent)

         Order 1. Nautiloidea (nautiloids, Cambrian-Recent; *Nautilus*)

             2. Ammonoidea (ammonoids, Ordovician-Cretaceous; *Ceratites, Turrilites*)

             3. Belemnoidea (belemnoids, Mississippian-Cretaceous; *Belemnites*)

             4. Sepiodiea (cuttlefishes, Jurassic-Recent; *Belosepia, Sepia, Spirula, Spirulirostra*)

             5. Teuthoidea (squids, Jurassic-Recent; *Loligo*)

             6. Octopoda (octopi, Cretaceous-Recent; *Octopus*)

21. Priapulida (worm-like marine animals; *Priapulus*)

22. Siphunculida (peanut worms; *Dendrostoma*)

23. Echiurida (sea cucumbers; *Echiurus*)

24. Annelida (worms and worm-like animals; Ediacaran-Recent)

     Class 1. Oligochaeta (oligochaete worms, Silurian ?—Recent; *Lumbricus*)

           2. Polychaeta (polychaete worms, Ediacaran-Recent; *Dickinsonia, Nereis, Serpula, Spirorbis, Spriggina*)

           3. Clitellata (Hirudinea) (leeches, *Hirudo*)

25. Pentastomida (worm-like parasites; *Linguatula*)

26. Tardigrada (microscopic, bilaterally symmetrical animals; *Macrobiotus*)

27. Onychophora (with features of both Annelida and Arthropoda, Cambrian-Recent; *Peripatus*)

28. Arthropoda (arthropods)

     Subphylum 1. Chelicerata (Cambrian-Recent)

         Class 1. Trilobita (trilobites; Cambrian-Permian)

             Order 1. Eodiscida (Lower-Middle Cambrian; *Eodiscus*)

                 2. Agnostida (Lower Cambrian-Ordovician; *Agnostus*)

                 3. Olenellida (Lower Cambrian; *Holmia, Olenellus*)

                 4. Proparia (Lower Ordovician-Devonian; *Calymene, Phacops*)

                 5. Opisthoparia (Lower Cambrian-Permian; *Bumastus, Scutellum*)

           2. Merostomata

             Order 1. Aglaspida (Cambrian-Ordovician; *Aglaspella, Strabops*)

                 2. Eurypterida (Ordovician-Permian; *Eurypterus*)

                 3. Xiphosura (Ordovician-Recent; *Limulus*)

           3. Aracnida (spiders, scorpions, ticks; Silurian-Recent)

           4. Pycnogonida (sea spiders, Devonian-Recent; *Palaeopantopus*)

     Subphylum 2. Mandibulata (Cambrian-Recent)

         Class 1. Crustacea (Cambrian-Recent)

             Order 1. Branchiopoda (small, mostly fresh-water; Devonian-Recent; *Artemia, Daphnia*)

                 2. Ostracoda (ostracods, Cambrian-Recent; *Cypris*)

                 3. Cirripedia (barnacles, Silurian-Recent; *Balanus*)

                 4. Malacostraca (crabs, crayfishes, lobsters, shrimps; Carboniferous-Recent)

                 5. Copepoda (copepods; *Calanus*)

            2. Myriapoda (centipedes, millipedes)

3. Insecta (insects; Devonian-Recent)
29. Pogonophora (body in three segments, each with separate coelom; no mouth, digestive canal, anus; marine; *Lamellisabella*)
30. Vestimentifera (tubular body with anterior sheath; *Lamellibrachia*)
31. Echinodermata (Cambrian-Recent)
    Subphylum Haplozoa
        Class 1. Cyamoidea (Middle Cambrian; *Peridionites*)
             2. Cycloidea (Middle Cambrian; *Cymbionites*)
    Subphylum Pelmatozoa
        Class 1. Cystoidea (Ordovician-Permian; *Aristocystites*)
             2. Blastoidea (Ordovician-Permian; *Pentremites*)
             3. Eocrinoidea (Cambrian-Ordovician; *Cryptocrinus, Macrocystella*)
             4. Paracrinoidea (Ordovician; *Amygdalocystites, Canadocystis, Comarocystites*)
             5. Crinoidea (sea lilies, Ordovician-Recent; *Antedon, Pentacrinus*)
             6. Edrioasteroidea (Cambrian-Pennsylvanian; *Edrioaster, Stromatocystites*)
             7. Carpoidea (Cambrian-Devonian; *Dendrocystis*)
             8. Machaeridia (Ordovician-Devonian; *Lepidocoleus, Turrilepas*)
             9. Somasteroidea (Earliest Ordovician, *Villebrunaster*)
           10. Asteroidea (starfishes, Ordovician-Recent; *Asterias*)
           11. Auluroidea (Ordovician-Mississippian; *Lysophiura, Streptophiura*)
           12. Ophiuroidea (brittle stars, Mississippian-Recent; *Amphipholis*)
           13. Echinoidea (sea urchins, sand dollars; Ordovician-Recent; *Cidaris, Echinus, Strongylocentrotus*)
32. Chaetognatha (arrowworms, marine; *Sagitta*)
33. Conodonta (eel-like animals known mainly from their teeth, Ordovician-Permian; *Belodus, Falcodus, Paltodus*)
34. Hemichordata
    Class 1. Enteropneusta *(Balanoglossus)*
          2. Pterobranchia
             Order 1. Rhabdopleurida (Cretaceous-Recent; *Rhabdopleura*)
                   2. Cephalodiscida (Ordovician ?—Recent; *Cephalodiscus*)
          3. Graptolithina (Cambrian-Mississippian; *Dendrograptus, Dictyonema, Monographus, Tetragraptus*)
35. Chordata
    Subphylum 1. Urochordata (tunicates, ascidians; *Ciona*)
              2. Cephalochordata (*Branchiostoma,* formerly *Amphioxus*)
              3. Craniata (Vertebrata) (Ordovician-Recent)
                  Class 1. Agnatha [jawless fishes, Ordovician-Recent; *Cephalaspis, Myxine* (hagfish), *Pteromyzon* (lamprey)]
                      2. Acanthodii (spiny fishes with jaws, Silurian-Permian; *Acanthodes*)
                      3. Placodermi (jawed, often armored fishes, Silurian-Mississippian; *Coccosteus*)
                      4. Chodrichthyes (cartilaginous fish, Devonian-Recent; *Squalus, Raja*)
                      5. Osteichthyes (bony fishes, Devonian-Recent)
                          Subclass 1. Actinopterygii [ray-finned fishes, Devonian-Recent; *Acipenser* (sturgeon), *Perca, Salmo*]
                                 2. Sarcopterygii (air-breathing, lobe-finned fishes)
                                      Order 1. Crossopterygii (ancestors to amphibians, Devonian-Recent; *Latimeria*)
                                          2. Dipnoi (lungfishes, Devonian-Recent; *Protopterus*)

6. Amphibia (amphibians, Mississippian-Recent; *Bufo, Rana, Seymouria*)
7. Reptilia (reptiles, Pennsylvanian-Recent; *Brontosaurus, Diplodocus, Ichthyosaurus, Pteranodon, Pterodactylus, Rhamphorhynchus, Sphenodon, Stegosaurus, Testudo, Tyrannosaurus*)
8. Aves (birds, Jurassic-Recent; *Archaeopteryx, Dinornis, Hesperornis, Ichthyornis, Aquila, Columba, Gallus*)
9. Mammalia (mammals; Triassic-Recent)

    Subclass 1. Protheria

            Order 1. Monotremata (egg-laying mammals, *Ornithorhynchus, Tachyglossus*)

      2. Allotheria

            Order 1. Multituberculata (Jurassic-Paleocene; *Bolodon, Psalodon*)

                2. Pantotheria (Jurassic; *Docodon, Melanodon*)

      3. Metatheria

            Order 1. Marsupialia [marsupials, Cretaceous-Recent; *Didelphis* (opossum), *Macropus* (kangaroo)]

      4. Eutheria [placental mammals, Cretaceous-Recent; *Bos* (ox), *Canis* (dog), *Cebus* (New World monkeys), *Cercopithecus* (Old World monkeys), *Cervus* (stag), *Dasypus* (armadillo), *Delphinus* (dolphin), *Elephas* (Indian elephant), *Equus* (horse), *Felis* (cat, lynx, ocelot, puma), *Gorilla, Homo, Lemur, Loxodonta* (African elephant), *Mus* (mouse), *Myotis* (bat), *Orcinus* (killer whale), *Panthera* (leopard, lion, tiger), *Phoca* (seal), *Phocaena* (porpoise), *Rattus* (rat), *Sorex* (shrew), *Talpa* (mole), *Tarsius, Tupaia* (tree shrew), *Tursipos* (bottlenosed dolphin)]

Kingdom Plantae

  Grade Bryophyta (no true roots, stem, leaves)

    Phylum   1. Bryophyta (Carboniferous-Recent)

              Class 1. Anthocerotae (hornworts)

                  2. Hepaticae (liverworts)

                  3. Musci (mosses)

            2. Psylophyta (Devonian; *Psylophyton*)

  Grade Tracheophyta (vascular plants)

            3. Lycopodiophyta (Devonian-Recent; *Lepidodendron, Sigillaria, Lycopodium, Selaginella*)

            4. Sphenophyta (Equisetophyta) (Devonian-Recent; horsetails; *Calamites, Equisetum*)

            5. Polypodiophyta (ferns; Devonian-Recent)

            6. Cycadophyta

              Class 1. Lyginopteridales (seed ferns, Carboniferous-Recent; *Glossopteris*)

                  2. Cycadales (cycads, Permian-Recent; *Cycas*)

                  3. Bennettitales (Permian-Oligocene; *Zamites*)

            7. Gingkophyta (maidenhair tree, Triassic-Recent; *Gingko*)

            8. Coniferophyta (Devonian-Recent)

Order 1. Cordaitales (Devonian-Jurassic; *Cordaites*)
   2. Pinales (conifers, Jurassic-Recent; *Abies, Cedrus, Cupressus, Larix, Picea, Pinus, Sequoia, Tsuga*)
   3. Taxales (yews, Jurassic-Recent; *Taxus*)
   4. Gneticae (some climbing shrubs and small tropical trees)
   5. Caytonicae (ancestral angiosperms; Early Jurassic-Early Cretaceous)
9. Angiospermophyta (angiosperms, the flowering plants; Cretaceous-Recent)
 Class 1. Liliatae (Monocotyledonae) (grasses, sedges, lilies, palms, orchids; Cretaceous-Recent)
    2. Magnoliatae (Dicotyledonae) (oaks, elms, sycamores, poplars, birch trees, roses, legumes, cactuses; Cretaceous-Recent)

Asterisks identify fundamental SI units. Symbols and abbreviations are explained in the next table.

Quantity	System	Name	Symbol	Definition
absorbed dose	SI	gray	Gy	$J\ kg^{-1}$
acceleration	SI	—	—	$m\ s^{-2}$
	CGS	—	—	$cm\ s^{-2}$
amount	SI, CGS	mole*	mol	one Avogadro number ($= 6.0221367 \cdot 10^{23}$) of items
angle				
plane	SI, CGS	radian*	rad	angle, with vertex at center of circle, which subtends a segment on the circumference equal to the radius
solid	SI, CGS	steradian*	sr	solid angle, with vertex at center of sphere, which subtends an area on the surface equal to square of radius
angular acceleration	SI, CGS	—	—	$rad\ s^{-2}$
angular momentum	SI	—	—	$kg\ m^2\ s^{-1}$
	CGS	—	—	$g\ cm^2\ s^{-1}$
angular velocity	SI, CGS	—	—	$rad\ s^{-1}$
area	SI	—	—	m^2
	CGS	—	—	cm^2
	SI	barn	b	$10^{-28}\ m^2$
	CGS	barn	b	$10^{-24}\ cm^2$
capacitance	SI	farad	F	$C\ V^{-1}$
	CGS_{esu}	statfarad	statF	$statC\ statV^{-1}$
charge	SI	coulomb	C	$A\ s$
	CGS_{esu}	statcoulomb	statC	charge that exerts the force of 1 dyne on an equal charge at a distance of 1 cm $= 10\ C\ c^{-1}$
conductance	SI	siemens	S	$\Omega^{-1} = A\ V^{-1}$
conductivity				
electric	SI	—	s	$A\ V^{-1}\ m^{-1}$
thermal	SI	—	k	$J\ m^{-1}\ s^{-1}\ K^{-1}$
current	SI	ampere*	A	current that, if maintained along two straight, parallel conductors of negligible cross section and infinite length placed 1 m apart *in vacuo*, would cause each to produce on the other a force of 10^{-7} newtons per meter of length
	CGS_{emu}	abampere	aA	current that, if maintained along two straight, parallel conductors of negligible cross section and infinite length placed 1 cm apart *in vacuo*, would cause each to produce on the other a force of 1 dyne per cm of length; $1aA = 10\ A$
	CGS_{esu}	statampere	statA	$aA\ c^{-1} = 10\ A\ c^{-1}$
current density	SI	—	j	$A\ m^{-2}$
density	SI	—	—	$kg\ m^{-3}$
	CGS	—	—	$g\ cm^3$
electric charge: see *charge*				
electric field strength	SI	—	E	$V\ m^{-1},\ N\ C^{-1}$
	CGS_{emu}	—	—	$aV\ cm^{-1} = 10^{-8}\ V\ cm^{-1}$
	CGS_{esu}	—	—	$statV\ cm^{-1} = 10^{-8}\ c\ V\ cm^{-1}$

Units *(continued)*

Asterisks identify fundamental SI units. Symbols and abbreviations are explained in the next table.

Quantity	System	Name	Symbol	Definition
electric potential: see *potential*				
electromotive force	SI	volt	V	$J\,C^{-1}$
	CGS_{emu}	abvolt	aV	$erg\ aC^{-1} = 10^{-8}\ V$
	CGS_{esu}	statvolt	statV	$erg\ statC^{-1} = 10^{-8}\ c\ V$
energy	SI	joule	J	N m
	CGS	erg	erg	dyn cm
	—	electron volt	eV	$1.6021892 \cdot 10^{-19}$ J
	—	million electron volt	MeV	$1.6021892 \cdot 10^{-13}$ J
	—	calorie internat.	cal_{IT}	4.1868 J (exactly) $2.6131745 \cdot 10^{13}$ MeV
		atomic mass unit	u	$931.49432\ MeV = 1.49244 \cdot 10^{-10}$ J
	—	gram	g	$5.609544 \cdot 10^{26}$ MeV
	—	megaton	Mt	10^{15} cal (exactly) $= 4.1868 \cdot 10^{15}$ J (exactly)
enthalpy	—	—	H	J
entropy	—	—	S	$J\,K^{-1}$
force	SI	newton	N	$kg\ m\ s^{-2}$
	CGS	dyne	dyn	$g\ cm\ s^{-2}$
frequency	SI, CGS	hertz	Hz	s^{-1}
gravitational accel.	SI	—	—	$m\ s^{-2}$
	CGS	galileo	gal	$cm\ s^{-2}$
illuminance	SI	lux	lx	$lm\ m^{-2}$
	CGS	lambert	L	$lm\ cm^{-2}$
impulse	SI	—	—	$kg\ m\ s^{-1}$
	CGS	—	—	$g\ cm\ s^{-1}$
inductance	SI	henry	H	$V\ s\ A^{-1}$
	CGS_{esu}	stathenry	statH	$statV\ s\ statA^{-1} = 10^{-9}\ c^2\ H$
	CGS_{emu}	abhenry	aH	$aV\ s\ aA^{-1} = 10^{-9}\ H$
irradiance	SI	—	—	$W\ m^{-2}$
length	SI	meter*	m	distance traveled by light in 1/299,792,458 s
	—	decimeter	dm	10^{-1} m
	CGS	centimeter	cm	10^{-2} m
	—	millimeter	mm	10^{-3} m
	—	micrometer	μm	10^{-6} m
	—	angstrom	Å	10^{-10} m
	—	nanometer	nm	10^{-9} m
	—	femtometer	fm	10^{-15} m
	—	decameter	da	10 m
	—	kilometer	km	10^{3} m
		astronomical unit	AU	mean distance of the Earth from the Sun $= 149,597,870.7$ km
	—	light year	l. y.	distance traveled by light in 1 tropical year $= 9.4605284 \cdot 10^{12}$ km
	—	parsec	pc	distance at which 1 AU subtends 1 second of arc $= 3.261633$ l.y.
		megaparsec	Mpc	10^6 parsecs $= 3.261633 \cdot 10^6$ l.y.

Asterisks identify fundamental SI units. Symbols and abbreviations are explained in the next table.

Quantity	System	Name	Symbol	Definition
luminance	SI	nit	nt	$cd\ m^{-2}$
	CGS	stilb	sb	$cd\ cm^{-2}$
luminosity: see *luminous intensity*				
luminous flux	SI	lumen	lm	cd sr
luminous flux density: see *illuminance*				
luminous intensity	SI	candela*	cd	luminous intensity of $1/683\ W\ sr^{-1}$ emitted by a monochromatic source radiating at the frequency of $540 \cdot 10^{12}$ Hz
magnetic field strength	SI	—	H	$A\ m^{-1}$
	CGS_{emu}	oersted	Oe	$\frac{1}{4}\pi\ aA\ cm^{-1} = 10^3/4\pi\ A\ m^{-1}$
magnetic flux	SI	weber	Wb	V s
	CGS_{emu}	maxwell	Mx	$aV\ s = 10^{-8}\ Wb$
magnetic flux density	SI	tesla	T	$Wb\ m^{-2}$
	CGS_{emu}	gauss	G	$Mx\ cm^{-2} = 10^{-4}\ T$
magnetic induction: see *magnetic flux density*				
magnetomotive force	SI	ampere-turn	A	A
	CGS_{emu}	gilbert	—	$\frac{1}{4}\pi\ aA$, Oe cm
mass	SI	kilogram*	kg	mass of the Pt-Ir International Prototype Kilogram kept at Sèvres, S.-et-O., France
	CGS	gram	g	$10^{-3}\ kg$
	—	milligram	mg	$10^{-3}\ g$
	—	microgram	μg	$10^{-6}\ g$
	—	nanogram	ng	$10^{-9}\ g$
	—	picogram	pg	$10^{-12}\ g$
	—	quintal	q	$10^2\ kg$
	—	ton	t	$10^3\ kg$
moment of inertia: see *rotational inertia*				
momentum	SI	—	—	$kg\ m\ s^{-1}$
	CGS	—	—	$g\ cm\ s^{-1}$
permeability	SI	—	μ	$H\ m^{-1}$
permittivity	SI	—	ϵ	$F\ m^{-1}$
potential	SI	volt	V	$J\ C^{-1}$
	CGS_{emu}	abvolt	aV	$erg\ aC^{-1} = 10^{-8}\ V$
	CGS_{esu}	statvolt	statV	$erg\ statC^{-1} = 10^{-8}\ c\ V$
power	SI	watt	W	$J\ s^{-1}$
pressure	SI	pascal	Pa	$N\ m^{-2}$
	CGS	microbar	μb	$dyn\ cm^{-2}$
	—	bar	b	$10^6\ dyn\ cm^{-2}$
radiant energy	SI	joule	J	N m
radiant flux	SI	watt	W	$J\ s^{-1}$
radiant flux density: see *irradiance*				
radiant intensity	SI	—		$W\ sr^{-1}$
radiant power: see *radiant flux*				
radiation	—	radiation	rad	radiation energy absorption of 100 erg/g
	—	roetgen equivalent man	rem	amount of radiation as damaging to human body as 1 roentgen of hard x-rays

Units *(continued)*

Asterisks identify fundamental SI units. Symbols and abbreviations are explained in the next table.

Quantity	System	Name	Symbol	Definition
radiation *(cont.)*	—	roentgen	R	amount of radiation that produces, in 1 cm^3 of dry air at 0°C and 760 mmHg, ions carrying 1 statcoulomb of electricity of either sign
radioactivity	—	curie	Ci	quantity of a radioactive substance that produces $3.7 \cdot 10^{10}$ dps
resistance	SI	ohm	Ω	V A^{-1}
	CGS$_{emu}$	abohm	aΩ	aV aA^{-1} = 10^{-9} Ω
	CGS$_{esu}$	statohm	statΩ	statV statA^{-1} = 10^{-8} c^2 Ω
resistivity	SI	—	ρ	Ω m
surface tension	SI	—	γ	N m^{-1}
	CGS	—	—	dyn cm^{-1}
temperature	SI, CGS	kelvin*	K	temperature interval equal to 1/273.16 of the absolute temperature of the triple point of pure water
time	SI, CGS	second*	s	time interval equal to the duration of 9,192,631,770 periods of the radiation associated with the transition between the two hyperfine levels of the ground state of ^{133}Cs
	—	ephemeris second	s$_E$	1/31,556,925.9747 of tropical year 1900
	—	ephemeris day	d$_E$	86,400 s$_E$
	—	tropical year	—	time interval between successive vernal equinoxes = 31,556,925.9747 − 0.530T s$_E$ = 365.242199 − 0.000006T d$_E$ = 365.242193 (A.D. 1986 to A.D. 2002) (T = tropical centuries since 1900.00)
	—	sidereal year	—	time required for the longitude of a distant star to increase by 360° = 31,558,149.984 + 0.010T s$_E$ = 365.256365 d$_E$ (T = tropical centuries since 1900.00)
velocity	SI	—	—	m s^{-1}
	CGS	—	—	cm s^{-1}
viscosity				
dynamic	SI	poiseuille	Pl	kg m^{-1} s^{-1}
	CGS	poise	P	g cm^{-1} s^{-1}
kinematic	SI	—	—	m^2 s^{-1}
	CGS	stoke	St	cm^2 s^{-1}
voltage	SI	volt	V	J C^{-1}
	CGS$_{emu}$	abvolt	aV	erg aC^{-1} = 10^{-8} V
	CGS$_{esu}$	statvolt	statV	erg statC^{-1} = 10^{-8} c V
volume	SI	—	—	m^3
	CGS	—	—	cm^3
volumetric flow rate	SI	—	—	m^3 s^{-1}
volumetric heat release	SI	—	—	W m^{-3}
work: see *energy*				

l = length; m = mass; t = time; i = electric current; T = temperature; cd = candela; sr = steradian.

Symbol	Name	Dimensions	Symbol	Name	Dimensions
A	ampere	i	m	meter	l
Å	angstrom	l	μb	microbar	$ml^{-1}t^{-2}$
aA	abampere	i	MeV	million electron volts	ml^2t^{-2}
aH	abhenry	$ml^2t^{-2}i^{-2}$	μg	microgram	m
AU	astronomical unit	l	mg	milligram	m
aV	abvolt	$ml^2t^{-3}i^{-1}$	μm	micrometer	l
aΩ	abohm	$ml^2t^{-3}i^{-2}$	mm	millimeter	l
b	bar	$ml^{-1}t^{-2}$	mol	mole	0
b	barn	l^2	Mpc	megaparsec	l
c	speed of light	lt^{-1}	Mt	megaton	ml^2t^{-2}
C	coulomb	ti	Mx	maxwell	$ml^2t^{-2}i^{-1}$
cal	calorie	ml^2t^{-2}	N	newton	mlt^{-2}
cal$_{IT}$	international calorie	ml^2t^{-2}	nt	nit	$cd\,l^{-2}$
cd	candela	cd	ng	nanogram	m
Ci	curie	t^{-1}	Oe	oersted	$l^{-1}i$
cm	centimeter	l	P	poise	$ml^{-1}t^{-1}$
cSt	centistoke	l^2t^{-1}	Pa	pascal	$ml^{-1}t^{-2}$
d$_E$	ephemeris day	t	pc	parsec	l
da	decameter	l	pg	picogram	m
dm	decimeter	l	Pl	poiseuille	$ml^{-1}t^{-1}$
dps	decays per second	t^{-1}	q	quintal	m
dyn	dyne	mlt^{-2}	ρ	density	ml^{-3}
ϵ	permittivity	$m^{-1}l^{-3}t^4i^2$	ρ	resistivity	$ml^3t^{-3}i^{-2}$
eV	electron volt	ml^2t^{-2}	R	roentgen	l^2t^{-2}
F	farad	$m^{-1}l^{-2}t^4i^2$	rad	radian	0
fm	femtometer	l	rad	radiation	l^2t^{-2}
g	gram	m	rem	roentgen-equivalent-man	l^2t^{-2}
G	gauss	$mt^{-2}i^{-1}$	σ	electric conductivity	$m^{-1}l^{-3}t^3i^2$
gal	galileo	lt^{-2}	s	second	t
H	henry	$ml^2t^{-2}i^{-2}$	S	entropy	$ml^2t^{-2}T^{-1}$
H	enthalpy	ml^2t^{-2}	S	siemens	$m^{-1}l^{-2}t^3i^2$
H	magnetic field strength	$l^{-1}i$	s$_E$	ephemeris second	t
Hz	hertz	t^{-1}	sb	stilb	$cd\,l^{-2}$
j	current density	$l^{-2}i$	sr	steradian	0
J	joule	ml^2t^{-2}	St	stoke	l^2t^{-1}
k	thermal conductivity	$mlt^{-3}T^{-1}$	t	ton	m
K	kelvin	T	T	centuries from A.D. 1900.0	t
kg	kilogram	m	T	temperature	T
km	kilometer	l	T	tesla	$mt^{-2}i^{-1}$
L	lambert	$cd\,l^{-2}$	u	atomic mass unit	m
lm	lumen	$cd\,sr^{-1}$	V	volt	$ml^2t^{-3}i^{-1}$
lx	lux	$cd\,sr^{-1}l^{-2}$	W	watt	ml^2t^{-3}
l.y.	light year	l	Wb	weber	$ml^2t^{-2}i^{-1}$
μ	permeability	$mlt^{-2}i^{-2}$	Ω	ohm	$ml^2t^{-3}i^{-2}$

Water density

Density of pure air-free water as a function of temperature (at atmospheric pressure from $-10°$ to $+100°C$) (t = temperature in °C; ρ = density in g/cm^3).

t	ρ	t	ρ	t	ρ
−10	0.997907	15	0.9991006	80	0.97180
−5	0.999148	20	0.9982057	90	0.96531
0	0.9998396	25	0.9970472	100	0.95835
1	0.9998985	30	0.9956504	150	0.91718
2	0.9999398	35	0.9940357	200	0.86490
3	0.9999642	40	0.9922195	250	0.79879
3.98	1.000000	45	0.99021	300	0.71264
4	0.9999720	50	0.98804	350	0.57495
5	0.9999639	60	0.98321	374.15	0.30675
10	0.9997001	70	0.97778		

From McKinney and Lindsay 1972, p. 2.152–2.153, Table 21.2; Ražnjević 1976, Table 44.1-7.

Boiling point
    0.5 atm   = 80.9
    1 atm   = 100.00°C
                                      = 373.15 K
    2 atm   = 119.6°C
    5 atm   = 151.1
    10 atm   = 170.0
    25 atm   = 222.9

Compressibility coefficient $(\partial \ln \rho/\partial p)_T$ $(10^{-6}$ $bar^{-1})$
    0°C   = 50.8850
    45°C (minimum)   = 44.1536
    100°C   = 49.019

Critical density (1/critical volume)   = 0.30675 g cm^{-3}
Critical pressure   = 221.297 bar
                                       = 218.40 atm
Critical temperature   = 374.15°C
                                       = 647.30 K
Critical volume (1/critical density)   = 3.25998 cm^3 g^{-1}

ΔH_{fusion} at 0°C   = 333.6 J g^{-1}
                                       = 79.68 cal g^{-1}
                                       = 1.43543 kcal mol^{-1}
ΔH_{vap} at 0°C   = 2500.00 J g^{-1}
                                       = 597.11 cal g^{-1}
                                       = 10.757 kcal mol^{-1}
ΔH_{vap} at 100°C   = 2256.37 J g^{-1}
                                       = 538.92 cal g^{-1}
                                       = 9.709 kcal mol^{-1}
ΔH_{vap} at 374.15°C   = 0.00
Density (g cm^{-3})
    1 atmosphere
        0°C   = 0.9998396
        3.98°C (highest)   = 1.0000000
        25°C   = 0.9770472
        50°C   = 0.98804
        100°C   = 0.95838
        374.15°C   = 0.30675
    25°C
        1 kbar (liquid)   = 0.9632
        5 kbar (liquid)   = 0.970
        25 kbar (solid)   = 0.707
        50 kbar (solid)   = 0.618

Dielectric constant
    0°C   = 87.740
    25°C   = 77.738

Dissociation constant $[H^+][OH^-]/[H_2O]$ at 25°C,   = $1.0 \cdot 10^{-14}$
    1 atm
Dissociation energy (25°C)
    H−OH   = 119 cal mol^{-1}
                                       = 498 kJ mol^{-1}
    H−O   = 102.2 kcal mol^{-1}
                                       = 427.9 kJ mol^{-1}

Heat capacity (*see* Thermal capacity)
Heat of fusion (*see* ΔH_{fusion})
Heat of vaporization (*see* ΔH_{vap})

Ice
  density
    0°C   = 0.91647 g cm^{-3}
    −175°C   = 0.94 g cm^{-3}
  crystallographic parameters (0°C)
    a   = 4.5239 Å
    c   = 7.3690 Å
  length of hydrogen bond   = 2.765 Å
  heat conductivity   = 0.00535 cal °C^{-1} cm^{-1} s^{-1}
  thermal capacity (−2.2°C) = 0.5018 cal g^{-1} °C^{-1}
  water vapor pressure over ice
    0°C   = 0.006107 bar
                            = 4.581 mmHg
    −10°C   = 0.02607 bar
                            = 1.946 mmHg
    −25°C   = 0.0006370 bar
                            = 0.4778 mmHg
    −50°C   = 0.00003947 bar
                            = 0.02961 mmHg

Ionic concentration (25°C, 1 atm)
  [H$^+$] = [OH$^-$]   = 1.004 · 10^{-7} mol liter^{-1}
  [H$_2$O]   = 55.34 mol liter^{-1}

Ionization potentials (eV)
  1st   = 12.62
  2nd   = 14.5
  3rd   = 16.2
  4th   = 18.0

Ion product [H$^+$][OH$^-$]
  (25°C, 1 atm)   = 1.0^{-14}

Melting point (1 atm)   = 0.00°C
                            = 273.15 K

Molecular properties
  bond length   = 0.95718 Å
  bond angle   = 104.523°
  bond strength (25°C)
    H−OH   = 5.2 eV
                            = 119 kcal mol^{-1}
    H−O   = 4.4 eV
                            = 102.2 kcal mol^{-1}
  dipole moment   = 1.84 · 10^{-18} esu
                            = 0.383 e Å
  molecular mass (u)
    H$_2$O   = 18.0153
    ^{1}H^{2}HO   = 19.0213
    H$_2$^{18}O   = 20.0150
    D$_2$O   = 20.0276
  zero-point vibrational energy = 0.575 eV

Refractive index (sodium D line, $\lambda = 5892.6$ Å)
   0°C   = 1.33395
   25°C   = 1.33251

Self-diffusion (10^{-7} m^2 s^{-1})
   5°C   = 1.4
   25°C   = 2.6

Solubility of atmospheric gases in water (ml/l)
   0°C
      air   = 29.18
      air oxygen   = 10.19
      CO_2 = 1713
      N_2 = 23.54
      O_2 = 48.89
      Ar   = 52.8
   25°C
      air   = 17.08
      air oxygen   = 5.78
      CO_2 = 759
      N_2 = 14.34
      O_2 = 28.31
      Ar   = 30.5

Surface tension vs. air (10^{-2} N/m)
   0°C   = 7.583
   25°C   = 7.214
   50°C   = 6.845
   100°C   = 6.180

Thermal capacity at 1 atm (cal g^{-1} °C^{-1})
   0°C   = 1.00738
   35°C (minimum)   = 0.99795
   65°C   = 1.00000
   100°C   = 1.00697

Thermal expansion coefficient $(-\partial \ln \rho/\partial T)_p$ (10^{-6}/°C)
   0°C   = -68.05
   25°C   = 257.21
   100°C   = 750.14

Triple point
   temperature   = 0.01°C
                                     = 273.16 K
   vapor pressure   = 4.57 mm Hg

Vapor pressure
   0°C   = 0.006107 bar
                                     = 4.581 mmHg
   25°   = 0.031676 bar
                                     = 23.759 mmHg
   50°C   = 0.12338 bar
                                     = 92.545 mmHg
   75°C   = 0.38553 bar
                                     = 289.17 mmHg
   100°C   = 1.01325 bar
                                   = 760.00 mmHg

Water
physical properties *(continued)*

Vapor pressure *(cont.)*

374.15°C	= 221.23 bar
	= 165,936 mmHg

Viscosity (cP)

0°C	= 1.7916
25°C	= 0.8903
50°C	= 0.5471
75°C	= 0.3775
100°C	= 0.2820

From Dorsey 1968; Nemethy and Scheraga 1964; Eisenberg and Kauzman 1969; Korson et al. 1969; Kennedy and Keeler 1972; Stull 1972; Ražnjević 1976; Dean 1985; Weast 1986.

Water
world reservoirs

Reservoir	Total water Volume (10^3 km³)	Total water Percent	Fresh water Volume (10^3 km³)	Fresh water Percent
atmosphere	13.3	0.00095	13.3	0.1540
streams	1.2	0.000087	1.2	0.0139
freshwater lakes	125.0	0.00906	125.0	1.4475
saline lakes and inland seas	104	0.0075	—	—
ice sheets and glaciers	29	0.0021	29	0.3358
soil and vadose water	67	0.00486	67	0.7759
groundwater				
<800 m deep	4,200	0.3045	4,200	48.6364
>800 m deep	4,200	0.3045	4,200	48.6364
ocean	1,370,323	99.3663	—	—
Total	1,379,062.5	99.9999	8,635.5	99.9999

From Robinove 1972, p. 533, Table 1.

References for Tables and Illustrations

Aguilar-Benitez M. et al. 1984. Tables of particle properties. *Review of Modern Physics,* v. 56, p. S10–S31.

Allen C.W. 1976. *Astrophysical Quantities.* The Athlone Press, University of London, 206 p.

Anders E. and Ebihara M. 1982. Solar-system abundances of the elements. *Geochimica et Cosmochimica Acta,* v. 46, p. 2363–2380.

Bally J. 1986. Interstellar molecular clouds. *Science,* v. 232, p. 185–193.

Barbon R., Ciatti F., and Rosino L. 1974. On the light curve of Type I supernovae. In: Batalli Cosmovici C., Ed., *Supernovae and Supernovae Remnants,* D. Reidel Publ. Co., Dordrecht, Holland, p. 99–102.

Barbon R., Ciatti, F., and Rosino L. 1974a. Recent observations of supernovae at Asiago. In: Batalli Cosmovici C., Ed., *Supernovae and Supernovae Remnants,* D. Reidel Publ. Co., Dordrecht, Holland, p. 115–118.

Bates R.L. and Jackson J.A., Eds., 1980. *Glossary of Geology.* American Geological Institute, Falls Church, Virginia, 749 p.

Beatty J.K., O'Leary B., and Chaikin A. 1982. *The New Solar System.* Sky Publishing Corporation and Cambridge University Press, 240 p.

Bender D.F. 1979. Osculating orbital elements of the asteroids. In: Gerhels T., Ed., *Asteroids.* University of Arizona Press, Tucson, Arizona, p. 1014–1039.

Berger, A.L. 1978. Long-term variations of caloric insolation resulting from the Earth's orbital elements. *Quaternary Research,* v. 9, p. 139–167.

Berry L.G., Mason B., and Dietrich R.V. 1983. *Mineralogy.* W.H. Freeman and Co., San Francisco, 561 p.

Bishop R.L., Ed., 1986. *Observer's Handbook.* Royal Astronomical Society of Canada, Toronto, Ontario, 184 p.

Bowell E., Gehrels T., and Zellner B. 1979. Magnitudes, colors, types, and adopted diameters of the asteroids. In: Gehrels T., Ed., *Asteroids,* University of Arizona Press, Tucson, Arizona, p. 1108–1129.

Bowers R.L. and Deeming T. 1984. *Astrophysics.* Jones and Bartlett Publ., Boston, Massachusetts, vol. 1, p. 1–344, vol. 2, p. 345–619.

Bragg W.L. 1955. *Atomic Structure of Minerals.* Cornell University Press, Ithaca, New York, 3 vols.

Brownlow A.H. 1979. *Geochemistry.* Prentice-Hall, Englewood Cliffs, New Jersey, 498 p.

Calvin M. 1969. *Chemical Evolution.* Oxford University Press, New York and Oxford, 278 p.

Clark S.P. Jr., Ed., 1966. *Handbook of Physical Constants.* Geological Society of America, Memoir 97, 587 p.

Clark S.P. Jr. 1966. Composition of rocks. In: Clark S.P. Jr., Ed., *Handbook of Physical Constants,* Geological Society of America, Memoir 97, p. 1–5.

Close F.E. 1979. *An Introduction to Quarks and Partons.* Academic Press, New York, 481 p.

Cohen E.R. and Taylor B.N. 1986. The 1986 adjustment of the fundamental physical constants. CODATA Bulletin 63, International Council of Scientific Unions, Paris, France, 32 p.

Collins S.A., Diner J., Garneau G.W., Lane A.L., Miner E.D., Synnott S.P., Terrile R.J., Holberg J.B., Smith B.A., and Tyler G.L. 1985. Atlas of Saturn's rings. In: Greenber R. and Branic A., Eds., *Planetary Rings,* University of Arizona Press, Tucson, Arizona, p. 737–743.

Cruikshank D.P. and Silvaggio P.M. 1980. The surface and atmosphere of Pluto. *Icarus,* v. 41, p. 96–102.

Curtis H. 1983. *Biology.* Worth Publishers, Inc., New York, 1159 p.

Dalgarno A. 1985. Molecular astrophysics. In: Diercksen G.F.H., Huebner W.F., and Langhoff P.W., Eds., *Molecular Astrophysics,* D. Reidel Publishing Co., Dordrecht, Holland, p. 3–22.

Dean J.A., Ed., 1985. *Lange's Handbook of Chemistry.* McGraw-Hill Book Co., New York, 1861 p.

Dermott S.F. and Nicholson P.D. 1986. Masses of the satellites of Uranus. *Nature,* v. 319, p. 115–120.

Dietrich G., Kalle K., Krauss W., and Siedler G. 1975. *Allgemeine Meereskunde.* Gebrüder Borntraeger, Berlin, 593 p.

Dodd R.T. 1981. *Meteorites—A Petrologic-Chemical Synthesis.* Cambridge University Press, Cambridge, England, 368 p.

Dorsey N.E. 1968. *Properties of Ordinary Water-Substance.* Hafner Publishing Co., New York, 673 p.

Ehlers E.G. and Blatt H. 1982. *Petrology—Igneous, Sedimentary, and Metamorphic.* W.H. Freeman and Company, San Francisco, 732 p.

Eisenberg D. and Kauzman W. 1969. *The Structure and Properties of Water.* Oxford University Press, Oxford, 296 p.

Elliot J.L. and Nicholson P.D. 1985. The rings of Uranus. In: Greenberg R. and Bahic A., *Planetary Rings,* University of Arizona Press, Tucson, Arizona, p. 25–72.

Eisner L. 1985. Radiation, thermal. In: Besancon R.M., Ed., *The Encyclopedia of Physics,* Van Nostrand Reinhold Co., New York, p. 1016–1023.

Faure G. 1986. *Principles of Isotope Geology.* John Wiley & Sons, New York, 589 p.

Forsythe W.E. 1964. *Smithsonian Physical Tables.* Smithsonian Institution, Washington, D.C., 827 p.

Garland G.D. 1979. *Introduction to Geophysics.* W.B. Saunders Company, Philadelphia, 494 p.

Glass B.P. 1982. *Introduction to Planetary Geology.* Cambridge University Press, Cambridge, 469 p.

Guelin M. 1985. Chemical composition and molecular abundances of molecular clouds. In: Diercksen J.F.H., Huebner W.F., and Langhoff P.W., *Molecular Astrophysics,* D. Reidel Publishing Co., Dordrecht, Holland, p. 23–43.

Halzen F. and Martin A.D. 1984. *Quarks and Leptons.* John Wiley & Sons, New York, 396 p.

Harland W.B., Cox A.V., Llewellyn P.G., Pickton C.A.G., Smith A.G., and Walters R. 1982. *A Geologic Time Scale.* Cambridge University Press, Cambridge, 131 p.

Hernbst E. 1978. The current state of interstellar chemistry of dense clouds. In: Gehrels T., Ed., *Protostars and Planets,* University of Arizona Press, Tucson, Arizona, p. 88–99.

Hewitt A. and Burbidge G. 1980. A revised optical catalog of Quasi-Stellar Objects. *Astrophysical Journal.* Supplement Series v. 43, p. 57–158.

Holland H.D. 1978. *The Chemistry of the Atmosphere and Oceans.* John Wiley & Sons, New York, 352 p.

Hood D. and Pytkowicz R.M. 1974. Chemical Oceanography. In: Smith F.G.W., Ed., *Handbook of Marine Science.* CRC Press, Boca Raton, Florida, v. 1, p. 1–70.

Horne R.A. 1969. *Marine Chemistry.* Wiley-Interscience, New York, 568 p.

Hurlbut C.S. Jr. and Klein C. 1977. *Manual of Mineralogy.* John Wiley & Sons, New York, 532 p.

Hutchison R. 1983. *The Search for our Beginning.* Oxford University Press, Oxford, England, 164 p.

Irvine W.M. and Hjalmarson A. 1983. Comets, interstellar molecules, and the origin of life. In: Ponnamperuma C., Ed., *Cosmochemistry and the Origin of Life.* D. Reidel Publ. Co., Dordrecht, Holland, p. 113–142.

Jayaraman A. and Cohen L.H. 1970. Phase diagrams in high-pressure research. In: Alper A.M., Ed., *Phase Diagrams.* Academic Press, New York, vol. 1, p. 245–293.

Jones H.S. 1961. *General Astronomy.* Edward Arnold & Co., London, 456 p.

Kathren R.L. 1984. *Radioactivity in the Environment: Sources, Distribution, and Surveillance.* Harwood Academic Publishers, Chur, Switzerland, 397 p.

Kennedy G.C. and Keeler R.N. 1972. Compressibility. In: Gray D.E., Ed., *American Institute of Physics Handbook,* McGraw-Hill Book Co., New York, p. 4.38–4.104.

King R.C. and Stansfield W.D. 1985. *A Dictionary of Genetics.* Oxford University Press, New York and Oxford, 480 p.

Korson L., Drost-Hansen W., and Millero F.J. 1969. Viscosity of water at various temperatures. *Journal of Physical Chemistry,* v. 73, p. 34–38.

Krauskopf K.B. 1979. *Introduction to Geochemistry.* McGraw-Hill Book Co., New York, 617 p.

Lal D. 1974. Radionuclides: cosmic-ray produced. In: Fairbridge R.W., Ed., *Encyclopedia of Geochemistry and Environmental Sciences.* Van Nostrand Reinhold Company, New York, p. 996–1066.

Lang K.R. 1980. *Astrophysical Formulae.* Springer-Verlag, Berlin, 783 p.

Lattimer J.M. and Grossman L. 1978. Chemical condensation sequence in supernova ejecta. *The Moon and the Planets,* vol. 19, p. 169–184.

Lindal G.F., Sweetnam D.N., and Eshleman V.R. 1985. The atmosphere of Saturn: an analysis of the voyager radio occultation measurements. *Astronomical Journal,* v. 90, p. 1136–1146.

Linke R.A., Frerking M.A., and Thaddeus P. 1979. Interstellar methyl mercaptan. *Astrophysical Journal,* v. 234, p. L139–L142.

List R.J. 1964. *Smithsonian Meteorological Tables.* Smithsonian Institution Press, Washington, D.C., 527 p.

Lof P. 1983. *Minerals of the World.* Elsevier, Amsterdam.

Loveland W.D. 1972. Nuclear fission. In: Gray D.E., Ed., *American Institute of Physics Handbook.* McGraw-Hill Book Co., New York, p. 8.253–8.276.

Mann A.P.C. and Williams D.A. 1980. A list of interstellar molecules. *Nature,* v. 283, p. 721–725.

Margulis L. 1981. *Symbiosis in Cell Evolution.* W.H. Freeman and Company, San Francisco, 419 p.

Mason B. and Moore C.B. 1982. *Principles of Geochemistry.* John Wiley and Sons, New York, 344 p.

McKinney J.E. and Lindsay R. 1972. Density and compressibility of liquids. In: Gray D.E., Ed., *American Institute of Physics Handbook,* McGraw-Hill Book Co., New York, p. 2.148–2.187.

McQuillin R. and Ardus D.A. 1977. *Exploring the Geology of the Shelf Seas.* Graham and Trotman, London, 234 p.

Moore W.J. 1950. *Physical Chemistry.* Prentice-Hall, New York, 592 p.

Moses A.J. 1978. *The Practising Scientist's Handbook.* Van Nostrand Reinhold Co., New York, 1292 p.

National Oceanic and Atmospheric Administration 1966. *U.S. Standard Atmosphere Supplements, 1966.* U.S. Government Printing Office, Washington, D.C.

National Oceanic and Atmospheric Administration, 1976. *U.S. Standard Atmosphere, 1976.* U.S. Government Printing Office, Washington, D.C., 227 p.

Nemethy G. and Scheraga H.A. 1964. Structure of water and hydrophobic bonding in proteins. IV. The thermodynamic properties of liquid deuterium oxide. *Journal of Chemical Physics,* v. 41, p. 680–689.

Payne-Gaposchkin C. 1954. *Introduction to Astronomy.* Prentice-Hall, Englewood, New Jersey, 508 p.

Perkins D.H. 1982. *Introduction to High Energy Physics.* Addison-Wesley Publishing Company, Reading, Massachusetts, 437 p.

Petrie W.M.F. 1952. Measures and weights, ancient. *Encyclopedia Britannica,* v. 15, p. 142–145.

Pollack J.B. and Yung Y.L. 1980. Origin and evolution of planetary atmospheres. *Annual Reviews of Earth and Planetary Sciences,* v. 8, p. 425–487.

Press F. 1966. Seismic velocities. In: Clark S.P. Jr., Ed., *Handbook of Physical Constants,* Geological Society of America, Memoir 97, p. 195–218.

Raznjevic K. 1976. *Handbook of Thermodynamic Tables and Charts.* McGraw-Hill Book Company, New York, 392 p.

Rigutti M. 1984. *A Hundred Billion Stars.* MIT Press, Cambridge, Massachusetts, 285 p.

Ringwood A.E. 1966. The chemical composition and origin of the Earth. In: Hurley P., Ed., *Advances in Earth Sciences.* MIT Press, Cambridge, Massachusetts, p. 287–356.

Ringwood A.E. 1975. *Composition and Petrology of the Earth's Mantle.* McGraw-Hill Book Co., New York, 618 p.

Ringwood A.E. 1979. *Origin of the Earth and the Moon.* Springer-Verlag, New York, 295 p.

Robie R.A., Bethke P.M., Toulmin M.S., and Edwards J.L. 1966. X-ray crystallographic data, densities, and molar volumes of minerals. In: Clark S.P. Jr., Ed., *Handbook of Physical Constants,* Geological Society of America, Memoir 97, p. 27–73.

Robinove C.J. 1972. Hydrology. In: Fairbridge R.W., Ed., *Encyclopedia of Geochemistry and Environmental Sciences,* Van Nostrand Reinhold, New York, p. 531–535.

Schmid H.S. and Koch K.R. 1972. Geodetic data. In: Gray D.E., Ed., *American Institute of Physics Handbook,* McGraw-Hill Book Co., New York, p. 2.92–2.102.

Schule J.J. Jr. 1966. Sea state. In: Fairbridge R.W., Ed., *Encyclopedia of Oceanography,* Dowden, Hutchinson, and Ross, Stroudsburg, Pennsylvania, p. 786–792.

Sellers W.D. 1965. *Physical Climatology.* University of Chicago Press, Chicago, 272 p.

Siever R. 1983. The dynamic Earth. *Scientific American,* v. 249, No. 3, p. 46–55.

Simpson J.A. 1983. Elemental and isotopic composition of galactic cosmic rays. *Annual Reviews of Nuclear and Particle Science,* v. 33, p. 323–381.

Smith B.A., Soderblom L.A., Beebe R., Bliss D., Boyce J.M., Brahic A., Briggs G.A., Brown R.H., Collins S.A., Cook II A.F., Croft S.K., Cuzzi J.N., Danielson G.E., Davies M.E., Dowling T.E., Godfrey D., Hansen C.J., Harris C., Hunt G.E., Ingersoll A.P., Johnson T.V., Krauss R.J., Masursky H., Morrison D., Owen T., Plescia J., Pollack J.B., Porco C.C., Rages K., Sagan C., Shoemaker E.M., Sromovsky L.A., Stoker C., Strom R.G., Suomi V.E., Synnott S.P., Terrile R.J., Thomas P., Thompson V.R., and Veverka J. 1986. Voyager 2 in the Uranian system: imaging science results. *Science,* v. 233, p. 43–64.

Smith D.E. 1952. Numerals. *Encyclopedia Britannica,* v. 16, p. 610–614.

Stacey F.D. 1977. *Physics of the Earth.* John Wiley & Sons, New York, 414 p.

Strahler A.N. 1971. *The Earth Sciences.* Harper and Row, New York, 824 p.

Strauss H.J. and Kaufman M. 1976. *Handbook for Chemical Technicians.* McGraw-Hill Book Co., New York, 454 p.

Stull D.R. 1972. Vapor pressure. In: Gray D.E., Ed., *American Institute of Physics Handbook,* McGraw-Hill Book Co., New York, p. 4.261–4.315.

Sverdrup H.U., Johnson M.W., and Fleming R.H. 1942. *The Oceans.* Prentice-Hall, Englewood Cliffs, New Jersey, 1087 p.

Taylor S.R. 1975. *Lunar Science: a Post-Apollo View.* Pergamon Press, New York, 372 p.

Trent H.M., Stone D.E., and Lindsay R.B. 1972. Density of solids. In: Gray D.E., Ed., *American Institute of Physics Handbook,* McGraw-Hill Book Co., New York, p. 2.19–2.37.

Tuli J.K. 1985. *Nuclear Wallet Cards.* National Nuclear Data Center, Brookhaven National Laboratory, Upton, New York, 62 p.

Turcotte D.L. and Schubert G. 1982. *Geodynamics.* John Wiley & Sons, New York, 450 p.

Turekian K.K. 1968. *Oceans.* Prentice-Hall, Englewood Cliffs, New Jersey.

U.S. Standard Atmosphere Supplements 1966. See *National Oceanic and Atmospheric Administration 1966.*

U.S. Standard Atmosphere 1976. See *National Oceanic and Atmospheric Administration 1976.*

Walker F.W., Miller D.G., and Feiner F. 1984. *Chart of the Nuclides.* General Electric Co., San Jose, California, 59 p.

Wapstra A.H. and Audi G. 1985. Atomic mass table. *Nuclear Physics,* v. A-432, p. 1–54.

Weast R.C., Ed., 1986. *CRC Handbook of Chemistry and Physics.* CRC Press, Boca Raton, Florida, 2408 p.

Weaver B.L. and Tarney J. 1984. Empirical approach to estimating the composition of the continental crust. *Nature,* v. 310, p. 575–577.

Weaver H.A., Mumma M.J., Larson H.P., and Davis D.S. 1986. Post-perihelion observations of water in Comet Halley. *Nature,* v. 324, p. 441–444.

Wedepohl K.H., Ed., 1969. *Handbook of Geochemistry.* Springer-Verlag, Berlin.

Wetherill G.W. 1966. Radioactive decay constants and energies. In: Clark S.P. Jr., Ed., *Handbook of Physical Constants,* Geological Society of America, Memoir 97, p. 513–519.

Whipple F.L. 1985. Present status of the icy conglomerate model. In: Klinger J., Benest D., Dollfus A., and Smoluchowski R., *Ices in the Solar System.* D. Reidel Publishing Co., Dordrecht, Holland, p. 343–366.